OCEANOGRAPHY and MARINE BIOLOGY

AN ANNUAL REVIEW

Volume 52

OCEANOGRAPHY and MARINE BIOLOGY

AN ANNUAL REVIEW

Volume 52

Editors

R. N. Hughes

Bangor University
Bangor, Gwynedd, United Kingdom
r.n.hughes@bangor.ac.uk

D. J. Hughes

Scottish Association for Marine Science
Oban, Argyll, United Kingdom
david.hughes@sams.ac.uk

I. P. Smith

School of Life Sciences
College of Medical, Veterinary and
Life Sciences, University of Glasgow
United Kingdom
philip.smith@millport.gla.ac.uk

CRC Press
Taylor & Francis Group
Boca Raton London New York

CRC Press is an imprint of the
Taylor & Francis Group, an **informa** business

International Standard Serial Number: 0078-3218

CRC Press
Taylor & Francis Group
6000 Broken Sound Parkway NW, Suite 300
Boca Raton, FL 33487-2742

Printed on acid-free paper
Version Date: 20140206

International Standard Book Number-13: 978-1-4822-2059-9 (Hardback)

Visit the Taylor & Francis Web site at
http://www.taylorandfrancis.com

and the CRC Press Web site at
http://www.crcpress.com

Contents

Preface

The 52nd volume of OMBAR contains six reviews ranging in subject matter from aspects of coral reef ecology and autecology of ghost crabs, through the possible role of biogenic sulphur in climate control, to the potential of 'citizen scientists' in large-scale monitoring programmes.

Together, the reviews reflect an editorial policy of presenting diverse topics to a broadly based readership. OMBAR welcomes suggestions from potential authors for topics that could form the basis of appropriate reviews. Contributions from physical, chemical, and biological oceanographers that seek to inform both oceanographers and marine biologists are especially welcome. Because the annual publication schedule constrains the timetable for submission, evaluation, and acceptance of manuscripts, potential contributors are advised to contact the editors at an early stage of manuscript preparation. Contact details are listed on the title page of this volume.

The editors gratefully acknowledge the willingness and speed with which authors complied with the editors' suggestions and requests and the efficiency of CRC Press, especially Marsha Hecht, Jill Jurgensen, and John Sulzycki, in ensuring the timely appearance of this volume.

Oceanography and Marine Biology: An Annual Review, 2014, **52**, 1-80

Taylor & Francis

PAIR-FORMATION IN CORAL REEF FISHES: AN ECOLOGICAL PERSPECTIVE

SIMON J. BRANDL[1,2] & DAVID R. BELLWOOD[1,2]

[1]*Australian Research Council Centre of Excellence for Coral Reef Studies, James Cook University, Townsville, Queensland 4811, Australia*
[2]*School of Marine and Tropical Biology, James Cook University, Townsville, Queensland 4811, Australia*
E-mail: simon.brandl@my.jcu.edu.au (corresponding author), david.bellwood@jcu.edu.au

Pair-formation is a common social system among animals. However, the use of the term 'pair-formation' is often ambiguous, and the assumed reproductive importance of pairing often supersedes consideration of aspects that are more social or ecological. This review provides a new social-ecological definition of pair-formation, examines the prevalence of pair-formation among coral reef fishes, and assesses the ecological and reproductive characteristics of pair-forming reef fishes. Of 1981 Indo-Pacific reef fish species examined in this review, 341 (17.2%) are reported to live in pairs. Pair forming has been reported in 29 families, with 5 families (Malacanthidae, Chaetodontidae, Siganidae, Syngnathidae, Ptereleotridae) having more than half of their species reported to form pairs. Two traits appear to favour the formation of social, cooperative pairs: (1) foraging on small, benthic, and relatively immobile prey; and (2) living in burrows. In contrast, there are limited similarities among pair-forming species with regard to their mating system or spawning mode. It appears that the basis of pair-formation in reef fishes is complex and may involve a range of ecological factors related to food procurement and predation risk.

Introduction

Of all the ecosystems on the planet, coral reefs are among the most diverse. Providing structural complexity and heterogeneous habitats for countless organisms, coral reefs harbour approximately 5000 species of fish (Bellwood et al. 2012), 1400 species of corals (Baird et al. 2009), and at least 165,000 species of reef-associated invertebrates (Stella et al. 2011). This exceptional species richness is matched by a comparable variety of behavioural traits, and almost every known social interaction can be observed in reef-dwelling organisms.

Although social and reproductive behaviours have elicited intensive theoretical and empirical examination, they are inherently difficult to study. Unlike morphological characteristics, which are usually conspicuous in a species' external appearance and often relatively consistent among individuals, social and reproductive traits are more variable and responsive to local environmental factors (Yahner 1978) and therefore have the potential for extreme plasticity within narrow taxonomic units such as species or families (e.g., Colin & Bell 1991, Morgan & Kramer 2004, Wong et al. 2005). Nevertheless, social and reproductive behaviours are important determinants of an organism's ecology, and a sound understanding of these traits, including their underlying environmental drivers and functional consequences, is crucial for organismic and community ecology (Orians 1961). Meaningful evaluations of social and reproductive behaviours, however, are often stymied by the close relationship between social organizations and reproduction and the resulting ambiguity of the terms 'social system' and 'mating system' (Emlen & Oring 1977, Robertson & Hoffmann 1977, Neudecker & Lobel 1982, Hourigan 1989, Reynolds 1996). Throughout the animal kingdom, social groupings commonly pave the way for reproductive activity (e.g., Fricke 1980, Getz & Hofmann 1986). Nevertheless, there are well-known examples in which social associations appear to be unrelated to reproductive behaviour and instead appear to be linked to ecological benefits, such as increased safety or foraging success (e.g., Robertson et al. 1976, Parrish & Edelstein-Keshet 1999, Morgan & Kramer 2004). Although invariably interrelated, this review distinguishes between ecological factors (i.e., factors related to survival and energy intake) and reproductive factors (i.e., factors enhancing reproductive output or fertilization rate or ensuring mate availability).

The distinction between social and mating systems is particularly unclear if the social unit is a pair, that is, two individuals of the same species. In animals, sexual reproduction usually requires two individual parents, and in many species, adults form transient breeding pairs whose sole purpose is reproduction (e.g., Liske & Davis 1984). It may be for this reason that, in the current literature, pair-formation (i.e., the prolonged association with only one other conspecific) is, with few exceptions (e.g., Gwinner et al. 1994, Pratchett et al. 2006, Young et al. 2008, Brandl & Bellwood 2013a), predominantly linked to a monogamous mating system and circumstances that led to the evolution of monogamy (Emlen & Oring 1977, Wittenberger & Tilson 1980, Barlow 1984, Fricke 1986, Reavis & Barlow 1998, Whiteman & Côté 2003, 2004, Reavis & Copus 2011).

Pair-formation is described in many taxa, ranging from unicellular organisms to higher vertebrates. However, understanding of the significance of pairing varies markedly among taxa (Table 1). Pairing has been extensively studied in birds, for which the manifest link between pairing and monogamous mating is well established (Orians 1969, Wittenberger & Tilson 1980, Black 2001, but see Westneat & Stewart 2003). The significance of prolonged pairing beyond the act of reproduction in birds is easily explained by biparental care (e.g., Wittenberger & Tilson 1980, Reynolds 1996, Adkins-Regan 2002). In contrast, most marine fishes display virtually no characteristics commonly associated with monogamous mating (Wittenberger & Tilson 1980, Reynolds 1996), and evidence for 'genetic monogamy' (i.e., exclusive reproduction with only one mate) is rare. A comprehensive review of monogamy in marine fishes (Whiteman & Côté 2004) found evidence for genetic monogamy in only 14.6% of the 164 species described as monogamous in the literature; the majority (64%) display 'social monogamy', described as "a social coalition with no implications for exclusive mating" (Whiteman & Côté 2004, p. 352). Although this definition suggests that these pairs are a 'social coalition' rather than a mating pair, environmental factors leading to the formation of social pairs and the possible ecological consequences beyond reproductive advantages or constraints are largely unexplored. This is particularly intriguing, as pair-formation appears to be a common social system in marine fishes, especially on tropical coral reefs (Randall et al. 1997, Allen et al. 2003, Froese & Pauly 2013).

The formation of long-term pairs in coral reef fishes has been a widely studied phenomenon. The majority of studies examining reef fish pairs are case studies that seek to explain why species lacking biparental care and exhibiting potential for polygamous mating apparently restrict themselves to a single reproductive partner (e.g., Fricke & Fricke 1977, Gore 1983, Fricke 1986, Barlow 1987, Donaldson 1989, Herold & Clark 1993, Kuwamura et al. 1993, Reavis 1997a,b, Reavis & Barlow 1998, Takegaki 2000, Harding et al. 2003, Whiteman & Côté 2003, Pratchett et al. 2006, Sogabe et al. 2007, Reavis & Copus 2011). The most commonly cited reasons for the evolution of monogamy are (1) environmental factors that prevent the sequestration of multiple mates, (2) paternal egg tending leading to mutual mate guarding, and (3) sparse populations. In contrast, few studies have examined possible ecological correlates of pairing as a social system (Robertson et al. 1979, Hourigan 1989, Pratchett et al. 2006, Brandl & Bellwood 2013a,b). These studies suggest increased feeding efficiency, improved territorial defence, and increased vigilance for predators as possible ecological benefits of pair-formation. However, there has, to date, been no holistic approach

Table 1 Pair-formation in animals

Kingdom	Phylum	Class	Order	Example	Source
Protista	Ciliophora	Ciliatea	Peniculida	*Paramecium primaurelia*	Delmonte Corrado et al. 1997
		Oligohymenophorea	Hymenostomatida	*Tetrahymena pyriformis*	Bruns & Brussard 1974
Animalia	Acanthocephala	Eoacanthocephala	Neoechinorhynchida	*Neoechinorhynchus pseudomydis*	Cable & Hopp 1954
	Platyhelminthes	Trematoda	Strigeidida	*Schistosoma* spp.	Beltran & Boissier 2008
	Annelida	Polychaeta	Phyllodocida	*Harmothoe imbricata*	Daly 1973
	Arthropoda	Merostomata	Xiphosura	*Limulus polyphemus*	Botton & Loveland 1992
		Crustacea	Malacostraca	*Homarus americanus*	Atema et al. 1979
		Insecta	Mecoptera	*Panorpa* spp.	Thornhill 1979
	Mollusca	Gastropoda	—	*Ovula ovum*	Kei 2010
	Chordata	Chondrichthyes	Carcharhiniformes	*Carcharhinus melanopterus*	Johnson & Nelson 1978
		Actinopterygii	Perciformes	*Amatitiania nigrofasciata*	Mackereth & Keenleyside 1993
		Amphibia	Anura	*Rana sylvatica*	Howard & Kluge 1985
		Reptilia	Squamata	*Tiliqua rugosa*	Leu et al. 2011
		Aves	Passeriforma	*Taeniopygia guttata*	Adkins-Regan 2002
		Mammalia	Rodentia	*Galea monasteriensis*	Adrian et al. 2008

Note: Pair-formation is a common behaviour from protists to mammals.

to examining common ecological traits among pairing species. Yet, such an approach may enable inferences to be made about environmental determinants of pair-formation and will provide more-detailed insights into the functional ecology of pairing species. This review, therefore, aims to provide an overview of pairing in coral reef fishes, following a social-ecological definition of the term 'pair-formation'. To do so, (1) an ecologically orientated definition of the term 'pair-formation' is offered, and (2) a meta-analysis of pair-formation, ecological traits, and reproductive modes in coral reef fishes of the Indo-Pacific is presented to identify common characteristics of pairing species. These characteristics are then discussed from an ecological perspective.

The definition of 'pair-formation'

Initially, the definition of the term 'pair-formation' appears to be straightforward, describing an association between two individuals. However, this degree of simplicity is only appropriate for molecular pair bonds. When applied to animals, pair-formation requires more careful definition. To develop a clear definition to facilitate investigation of pair-formation in coral reef fishes, we have made an initial assessment of the use of the term 'pair-formation' in relation to animals. To this end, a basic meta-analysis was performed, using the ISI (currently Thomson Reuters) Web of

Knowledge bibliographic database to search for studies with the term 'pair-formation' in the title. The search was specified by including the word 'animals' in the 'topic' field to exclude a large body of literature about pair-formation in molecules. A total of 168 studies was found (from the years 1969 to 2012), with 67 off topic or duplicates. The remaining 101 studies were divided into four different categories: (1) articles describing copulation or spawning and associated courtship or premating behaviours; (2) articles describing pair-formation beyond copulation or courtship, including post-mating behaviour and the function of long-term pairs; (3) articles about the conjugation of unicellular organisms, including all studies on unicellular eukaryotes and their form of 'mating'; and (4) articles describing pairs without any reproductive association.

Of the 101 studies, half (50.5%) were restricted to the process of copulation or spawning, courtship, or precopulatory mate guarding. Another 33.7% described reproductive pairs but examined post-copulatory behaviour or long-term pair-formation (when the pair bond is maintained over more than one reproductive season). A further 11.9% of articles described conjugation of unicellular organisms, and just 4% of all 101 studies related to cooperative 'pairs' without a clear reproductive purpose (Figure 1). Because more than half of the studies on pair-formation described only copulation or precopulatory behaviour, referring to a 'mating system' rather than a 'social system', a more precise definition of pair-formation is needed that incorporates both ecological and reproductive components. Pair-formation must be distinguished from pure mating and premating behaviour such as courtship or precopulatory mate guarding (Kortlandt 1995), as the latter only pave the way for copulation (e.g., Hartnoll & Smith 1978, Johnson & Nelson 1978, Burpee & Sakaluk 1993). As an example, in spawning aggregations of numerous families of reef fishes (e.g., Mullidae, Serranidae, Siganidae), actual spawning occurs between members of a pair, temporarily separating from the aggregation (Johannes 1981, Samoilys & Squire 1994, Domeier & Colin 1997). Likewise, many other species (e.g., Acanthuridae) spawn in pairs but remain solitary throughout their lives within

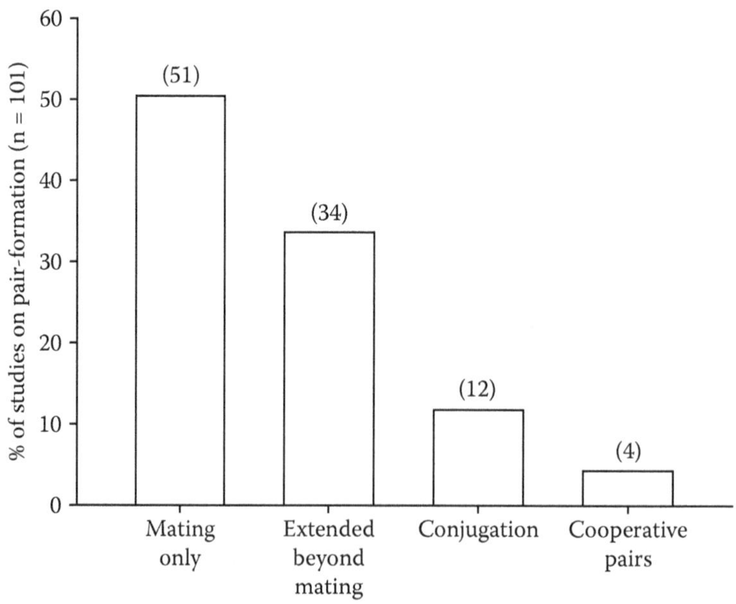

Figure 1 Proportions of studies investigating 'pair-formation' in animals. Papers are assigned to different categories, reflecting the use of the term 'pair-formation'. Numbers are based on the ISI Web of Knowledge database in November 2011.

their own individual feeding territories (Robertson et al. 1979, Robertson 1983). These 'pairs' are therefore restricted to brief encounters that only serve to fertilize eggs and have no implications for daily survival.

With few exceptions (see Ahlgren et al. 2011), there are no clear ecological implications of these transient pairing associations. Hence, if the formation of a pair is restricted to copulation only, the term 'pair-formation' may be misleading, as it does not describe a social system but simply the process of reproduction, that is, a mating system. Therefore, we suggest that the term 'pair-formation' should be limited to associations that are maintained beyond the process of reproduction, such as (1) cooperation in rearing offspring (e.g., Robertson 1973, Cox et al. 1993, Adrian et al. 2008, Young et al. 2008); (2) mutual maintenance of a dwelling or territory (e.g., Linsenmair & Linsenmair 1971, Fricke 1986, Mathews 2002); (3) increased predator avoidance through shared vigilance (Gwinner et al. 1994, Pratchett et al. 2006); or (4) improved feeding efficiency (e.g., Robertson et al. 1979, Hourigan 1989, Gwinner et al. 1994). In contrast, pairing for copulation alone is perhaps more appropriately termed a 'mating pair'.

An apparent issue arises for pairs in which mate guarding occurs after mating, as it initially appears solely to benefit the guarding individual in terms of its reproductive success. However, pair-formation in these species can be ecologically meaningful. For example, female lobsters (*Homarus americanus*) experience lower predation rates after moulting for reproduction (Atema et al. 1979), and female water striders (*Gerris remigis*) exhibit higher foraging rates when guarded by males after copulation (Wilcox 1984). Both of these benefits are seen in the Australian sleepy lizard, *Tiliqua rugosa*, with the female able to dedicate more time to foraging efforts, and predation risk is reduced through increased vigilance of the male (Bull & Pamula 1998, Leu et al. 2011). Thus, to address ecological implications of associating with a single individual, we suggest that pair-formation be defined as an association between two conspecific individuals, maintained beyond (before and/or after) the time required for reproductive activity leading to fertilization. With this definition of pair-formation as a social system, one can begin to explore the role of this behaviour in coral reef fishes.

Pair-formation in coral reef fishes

To explore pair-formation in reef fishes in greater detail, the Indo-Pacific coral reef fish fauna was examined. The primary goal was to understand the basis of pair-formation in fishes. As a first step, we sought to identify ecological traits that are shared by pair-forming species. Often, ecological traits are correlated, and if evolutionarily successful, they occur repeatedly among distantly related taxonomic groups (Westneat et al. 2005). For example, foraging in large aggregations appears to be a beneficial trait for species that feed on pelagic zooplankton. As such, aggregations frequently occur in distantly related taxa, such as the Caesionidae, Labridae, Pomacentridae, and Serranidae (e.g., Hamner et al. 1988, Hobson 1991). Similarly, cryptobenthic fishes are almost exclusively known to spawn adhesive, benthic eggs that are deposited in a cave, burrow, or nesting site (e.g., Fishelson 1976, Thresher 1984, Hernaman & Munday 2007). We therefore reviewed a comprehensive set of traits among Indo-Pacific reef fishes (extracted from Randall et al. 1997, Allen et al. 2003, Randall 2005). The traits encompassed social systems, trophic affiliation, strategies to avoid predators, and reproductive characteristics (Appendix). Specifically, each species was classified as either pairing or non-pairing, assigned to a trophic category, and classified with regard to other ecological characteristics (maximum size, maintenance of burrows, nocturnal or diurnal activity). In addition, reproductive characteristics such as the spawning mode (pair or group spawning); the nature of gamete release (broadcast, demersal, pouch brooding, mouth brooding, egg scattering, gelatinous egg-mass spawning); and the mating system were recorded. While other variables such as longevity, age of maturation, hermaphroditism, and seasonal reproduction may be important, data on these factors were too sparse to allow for detailed comparisons and were therefore not included in this review.

The prevalence of pairing behaviour in reef fishes

Of 1981 species of coral reef fish in 79 families, 341 species are reported to form pairs; 1561 species have no records of pair-formation. No information was available for 79 species, which are predominantly small, cryptobenthic species. An average of 18.7% of species within a family are reported to form pairs, ranging from 100% in the Malacanthidae, Microdesmidae, Monocentridae, Pegasidae, Solenostomidae, and Zanclidae to 0% in 42 families (Figure 2). To account for extreme proportions in families with low numbers of species, families with fewer than five species are not considered further here.

Five families of reef fishes contain more than 50% of the total pair-forming species. Belonging to two different orders (Perciformes and Syngnathiformes), these include tilefishes (Malacanthidae, 100% of the examined species in pairs); butterflyfishes (Chaetodontidae, 83.3%); rabbitfishes (Siganidae, 60.0%); sea horses and pipefishes (Syngnathidae, 56.4%); and dartfishes (Ptereleotridae, 55.5%) (Figure 3). The two following families with relatively high proportions of pair-forming species, lizardfishes (Synodontidae, 50.0%) and pufferfishes (Tetraodontidae, 40.7%), belong to different orders: the Aulopiformes and Tetraodontiformes, respectively. Given that the seven families with the highest proportions of pairing species belong to four different orders, the tendency to arrange in pairs appears to have evolved independently across several phylogenetically distinct lineages. Furthermore, there are major differences between families within the same order. For

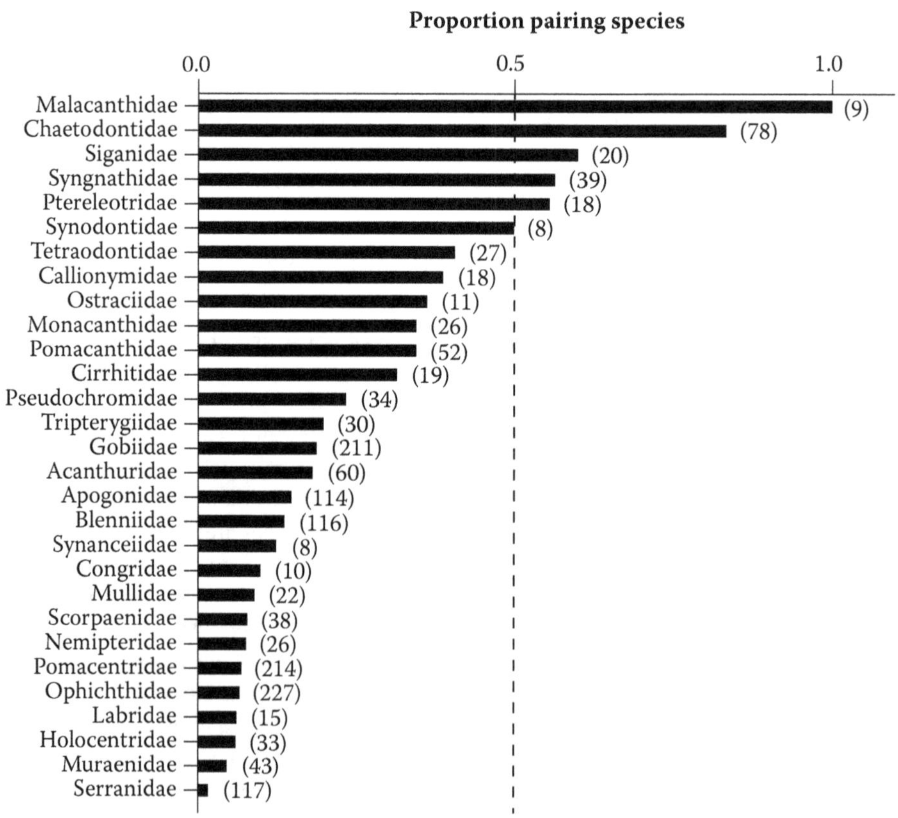

Figure 2 The prevalence of pair-formation in coral reef fish families. Only families with more than five species are considered. The dashed line marks 50% prevalence. The number of species in each family is given in parentheses.

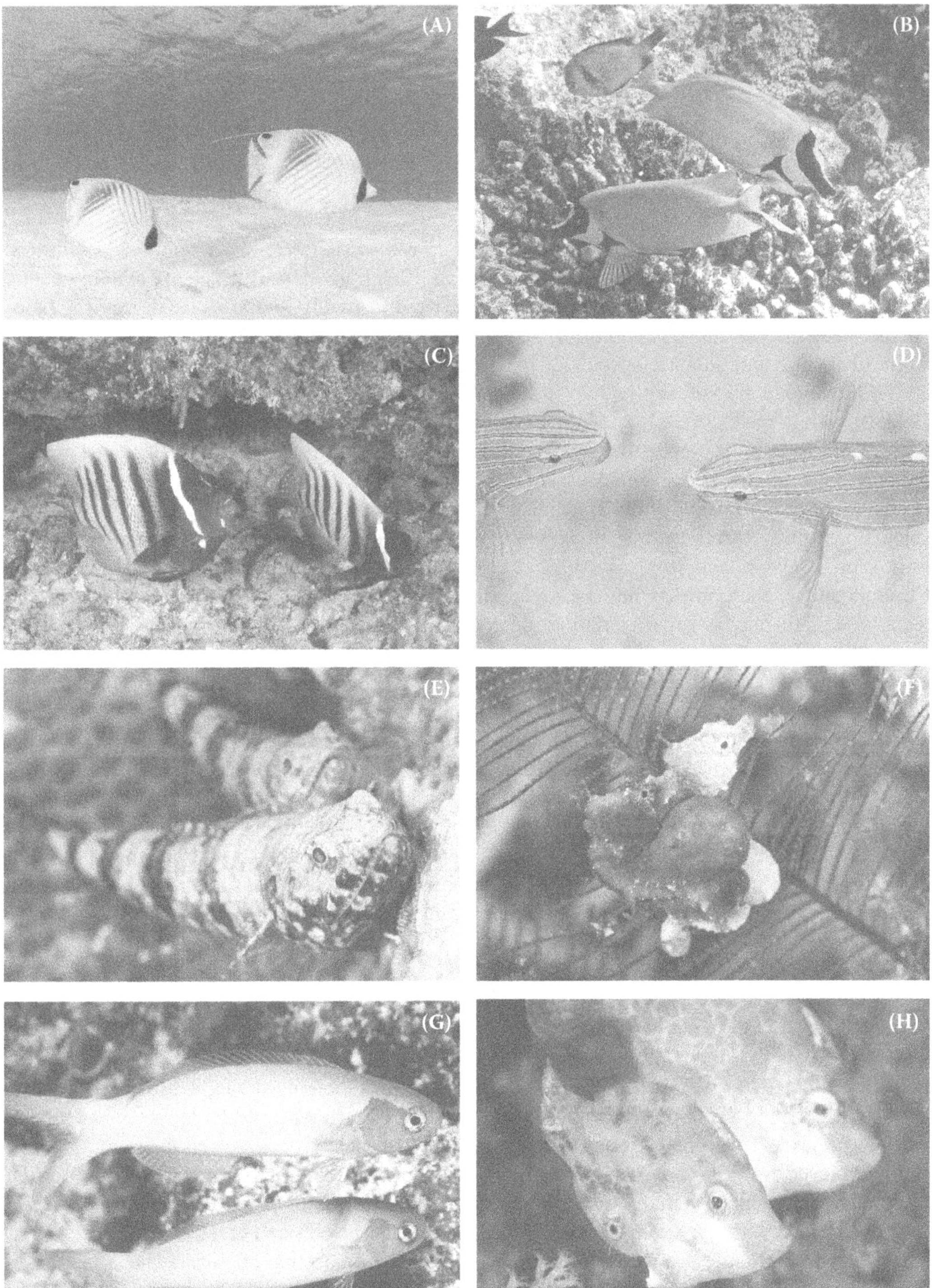

Figure 3 (See also colour figure in the insert) Examples of pair-forming species in the Chaetodontidae (A), Siganidae (B), Pomacanthidae (C), Gobiidae (D), Synodontidae (E), Syngnathidae (F), Malacanthidae (G), and Monacanthidae (H). (Photographs by J.P. Krajewski, B. Halstead, and S.J. Brandl.)

example, triggerfishes (Balistidae) and porcupinefishes (Diodontidae) do not comprise any pairing species, whereas pair-formation is relatively common in pufferfishes (Tetraodontidae, 40.7%), box- and cowfishes (Ostraciidae, 36.3%), and filefishes (Monacanthidae, 34.6%). All these families belong to a single order, the Tetraodontiformes. A similar situation occurs in the Syngnathiformes, for which sea horses and pipefishes are in marked contrast to trumpetfishes (Aulostomidae) and cornetfishes (Fistulariidae), both of which have no pairing species. Overall, based on the most recently assembled phylogeny of fishes (Near et al. 2012), pairing appears to have arisen at least 13 times among reef fishes.

Narrowing the taxonomic scale reveals that there is also significant variation within families. In some cases, entire lineages are pair forming, and there are clear distinctions among genera. In the Malacanthidae, for example, pairing is restricted to highly reef-associated, tropical Indo-Pacific species in the genera *Hoplolatilus* and *Malacanthus* (Clark & Pohle 1992, Clark et al. 1998), while temperate and Atlantic species of the genera *Branchiostegus*, *Caulolatilus*, *Lopholatilus*, and *Malacanthus* are solitary or live in colonies (Able et al. 1982, Ross 1982, Baird & Baird 1992, Mitamura et al. 2005). In the Tetraodontidae, pairing is restricted to the genus *Canthigaster* (Kobayashi 1986, Sikkel 1990), the genus with the smallest species in the family. In other families, the prevalence of pairing behaviour varies within genera. For example, there is a clear dichotomy in *Siganus*, the single genus of the Siganidae, with pair-formation reported only for reef-associated, colourful tropical species, while drab, mangrove-associated, estuarine and subtropical species form schools (Woodland 1990, Borsa et al. 2007, Brandl & Bellwood 2013b). Likewise, in the Chaetodontidae, there are pairing and non-pairing species in the dominant genus *Chaetodon*, regardless of their phylogenetic relationships (Hourigan 1989, Roberts & Ormond 1992, Kelley et al. 2013). In the Syngnathidae, both pipefishes (genera *Corythoichthys*, *Dunckerocampus*, *Doryrhamphus*) and sea horses (genus *Hippocampus*) have several species that form pairs, while others live solitarily or in groups (Allen et al. 2003, Foster & Vincent 2004, Sogabe & Yanagisawa 2008). Given the occurrence of pairing behaviour in a diverse array of orders and families and the high variation within closely related taxa, it appears that pair-formation has arisen repeatedly over many millions of years. With such diverse groups involved, there is ample opportunity for a critical evaluation of the ecological and reproductive role of pair-formation in coral reef fishes.

In ecology, the three main drivers of social behaviour are associated with feeding, predation avoidance, and reproduction. To identify ecological and reproductive traits that are associated with pair-formation in reef fishes, we explore each of these components separately.

Trophic ecology of pair-forming fishes

The acquisition of food is a crucial process in the life history of animals and their prey. Consequently, species are often classified within certain trophic groups (e.g., Williams & Hatcher 1983, Green & Bellwood 2009, Cheal et al. 2012), where species with similar foraging strategies are grouped based on their major prey items. Interestingly, there is a clear pattern with regard to the prevalence of pairing species within trophic groups (Figure 4). In three trophic groups (spongivores, corallivores, and microinvertevores), more than half of the species form pairs. This is a remarkable proportion considering the relatively small number of pairing species (341) compared to non-pairing species (1640; i.e., 17.2% pairing species). Two additional groups, omnivores and herbivores, also have relatively high proportions of pairing species (>25%), while the proportion of pairing species is low in planktivores, macroinvertevores, piscivores, carnivores, and detritivores.

The major pattern that emerges from these results is the link between small or sedentary prey items and pairing behaviour (Figure 4), as all trophic groups that forage on such prey have high proportions of pairing species. Their prey includes coral (polyps, mucus); microinvertebrates (e.g., harpacticoid copepods, amphipods, small polychaetes, molluscs); filamentous or fleshy algae; sponges; or all of those mentioned (omnivores). Features common to all these prey types are that

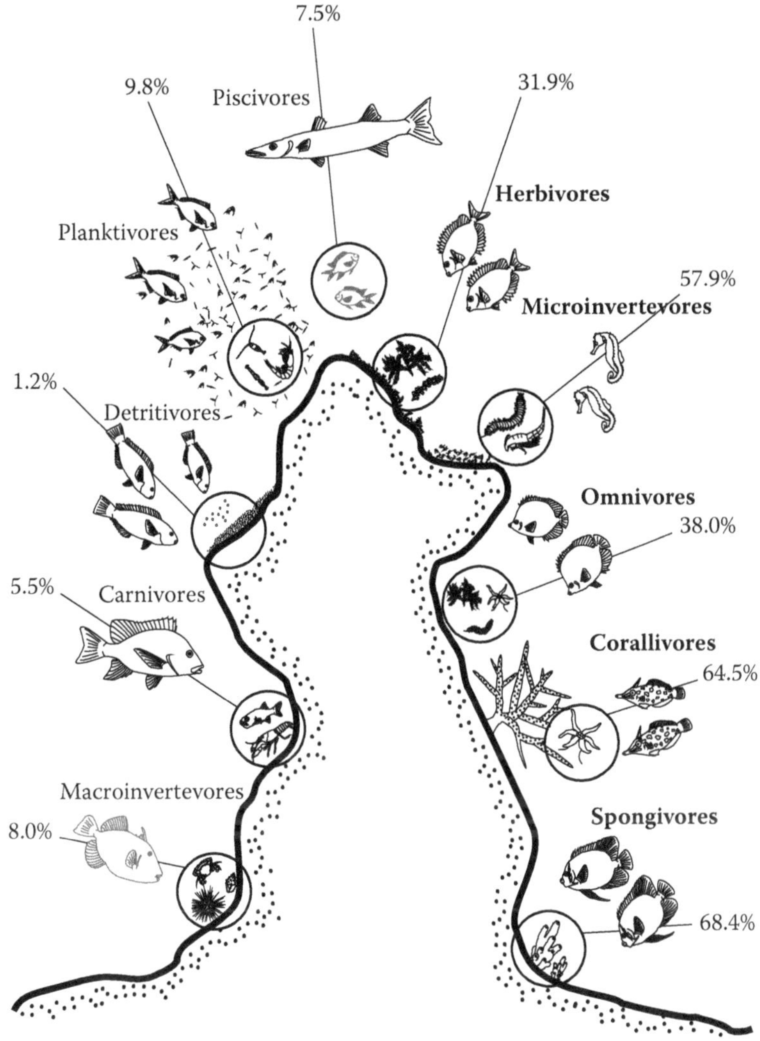

Figure 4 The prevalence of pair-formation in major trophic groups of coral reef fishes. Percentages indicate the proportion of pair-forming species within each trophic group. Groups with more than 25% pairing species are marked in bold. Black circles indicate prey items.

their predators require relatively high visual acuity, and the occupation of topographically complex microhabitats on coral reefs. Consequently, the majority of pairing species are reported to be diurnal (96.4%) and are, on average, smaller than non-pairing species (mean maximum total length ± SE: 162.6 ± 6.4 mm vs. 278.0 ± 8.6 mm, respectively). Likewise, in 19 of 29 families that include both pairing and non-pairing species, pairing species were on average smaller than their non-pairing counterparts, suggesting that size differences are not solely due to phylogenetic effects.

The association of pair-formation with particular trophic groups may be because of the relatively immobile nature and distribution of prey. Although some benthic microinvertebrates are motile, their movements are restricted to small spatial scales, making these organisms effectively a near-stationary food source on a whole-reef scale (Klumpp et al. 1988, Kramer et al. 2012a). Widely distributed prey has been associated with the evolution of monogamous mating systems in animals, as the range of movement required to feed on such prey prevents the monopolization of multiple mates (Emlen & Oring 1977, Whiteman & Côté 2004, Reavis & Copus 2011). Such an association has been proposed for butterflyfishes, distinguishing between mobile, planktivorous, schooling species and demersal, corallivorous species that occur predominantly in pairs (Reese 1975, Fricke 1986), leading to the suggestion that environmental factors—food availability in particular—can be used to predict the social system in this family (Hourigan 1989). Several authors have argued that evenly distributed, stationary, and predictable food sources promote pairing behaviour in butterflyfishes, as males are not able to sequester more than one female (e.g., Hourigan 1989, Reavis & Copus 2011). This is in accordance with theoretical expectations based on the evolution of monogamy (e.g., Emlen & Oring 1977) and is consistent with the high proportion of corallivores that occur in pairs (Figure 4).

However, there are other factors that may also favour the formation of pairs in these trophic groups. One key aspect might be the size and benthic nature of prey items. As noted, with the exception of some large species of algae and sponges, the majority of organisms preyed on by pair-rich trophic groups are relatively small (e.g., coral polyps, filamentous algae, harpacticoid copepods) and are located in a microtopographically highly complex and heterogeneous environment (Mundy 2000). Therefore, species foraging on these items may require highly dexterous movements and fine-scale interactions with the substratum to obtain their prey. This has been documented in several pair-forming groups, such as butterflyfishes and rabbitfishes (Motta 1988, Ferry-Graham et al. 2001, Fox & Bellwood 2013, Brandl & Bellwood in press), and other studies have highlighted the particular ability of some species to exploit complex microhabitats (Hobson 1975, Robertson & Gaines 1986, Motta 1988, Montgomery et al. 1989, Ferry-Graham et al. 2001, Fox & Bellwood 2013, Brandl & Bellwood in press). Interestingly, many species in the latter group are known to form pairs. This may be because of a rapid reduction in feeding efficiency with increasing group size in these niches (White & Warner 2007a). A loss of foraging efficiency may be exacerbated by reduced accessibility of prey or increased handling time of small benthic prey (Pratchett et al. 2006, Brandl & Bellwood 2013a,b, Fox & Bellwood 2013). The majority of small coral reef fishes devote only a limited time to reproduction, and they probably need to continue feeding during reproductive periods. In such circumstances, one would anticipate selection against large groups and more frequent pair-formation (Ford & Swearer 2013a).

Another specialized trophic mode that appears to be associated with pair-formation is the maintenance of cleaning stations. Several species of reef fish 'clean clients' (remove ectoparasites and dead skin from other fish) and perform cleaning tasks in cooperative pairs (Bshary et al. 2008). Besides the most well-known species in the labrid genus *Labroides*, which commonly live in harems or pairs (Allen et al. 2003), there are species within the Gobiidae (e.g., *Elacatinus evelynae*) and Syngnathidae (e.g., *Dunckerocampus pessuliferus*) that are commonly found in pairs and maintain cleaning stations (Allen et al. 2003, Harding et al. 2003, Whiteman & Côté 2003). In most of these species, it appears that the cleaning service provided by a pair is superior to cleaning by solitary individuals (Whiteman & Côté 2003, Bshary et al. 2008). This benefits both individuals in a simple way, as clients are more likely to visit high-quality cleaning stations (Whiteman & Côté 2003). While this has well-known ramifications for the mating system and reproductive success (Bshary et al. 2008), it also provides a good example of cooperative social pairing that affords ecological benefits to both members of the pair.

Overall, although there may be benefits in pairing, it is not clear why all these species occur as pairs rather than in small groups. Perhaps there are other factors operating that limit social group sizes, such as predation risk. It has been suggested that for certain species larger groups and shoaling behaviour may actually increase the overall predation risk when exposed to more than one guild of predators (Ford & Swearer 2013a,b).

Predation avoidance in pair-forming fishes

Mortality caused by predation is an important agent shaping species' behaviour and is tightly linked to an animal's ecology (Holbrook & Schmitt 2004, Almany & Webster 2006, Holmes &

McCormick 2011). Predation risk can influence behaviour in different ways, such as restricting the movement of prey fishes to small areas around shelter sites (Reavis 1997b, Madin et al. 2012, Welsh & Bellwood 2012a) or away from unstable, complex habitats (Hoey & Bellwood 2011); or driving prey to form large aggregations to avoid predation on the individual (Pitcher & Parrish 1993, White & Warner 2007b). Thus, the threat of predation, in synergy with the need to forage and reproduce, can either decrease (e.g., restriction of movements; Welsh & Bellwood 2012a) or increase (e.g., formation of large schools; Welsh & Bellwood 2012b) the size and movement of social groups. This raises the question: Are there any circumstances where a pair may be the preferred group size for avoiding predation?

Pair-formation in reef fishes has rarely been linked to anti-predation strategies. However, several lines of evidence suggest that it may play a role. For mobile species, the individual risk of predation is likely to increase with decreasing group size (but see Ford & Swearer 2013a,b). In large groups, vigilance is shared between numerous individuals, providing security for each group member (Pitcher & Parrish 1993, White & Warner 2007b). Thus, in theory, in terms of overall predation risk, individuals in pairs should be more vulnerable than those in schools but less vulnerable than solitary individuals (Pratchett et al. 2006). One way that pairing fishes may compensate for the higher risk is with morphological or behavioural adaptations. It is striking that virtually all families of mobile (non-burrowing) reef fishes with high proportions of pairing species exhibit conspicuous physical adaptations to avoid predation. This includes large venomous spines (Siganidae, Pomacanthidae); exceptionally deep bodies and bright colouration (Chaetodontidae, Pomacanthidae); and body inflation, extremely tough skin or bony plates encasing the body, and toxicity (Tetraodontidae, Ostraciidae, Monacanthidae) (Hixon 1991). Although some of these features are also found in solitary or schooling species (e.g., Acanthuridae, Balistidae), some families with few pairing species appear to have fewer morphological adaptations to avoid predation (e.g., Mullidae, Nemipteridae, Labridae, Caesionidae). In these predominantly schooling species, speed appears to be the major determinant of individual survival, as an individual often only needs to be faster than a single adjacent individual to avoid predation. Within families, there appears to be a similar trend. For example, within the Chaetodontidae, Acanthuridae, and Siganidae, pairing species possess deeper bodies (Brandl & Bellwood 2013b) and may exhibit brighter colouration and more anti-predator features, such as eye stripes, than non-pairing species (Kelley et al. 2013).

Behavioural adaptations offer an additional means of reducing vulnerability to predation in pairs. Most mobile pairing species are relatively slow swimmers (Fulton 2007), which makes predator avoidance through high-speed escape rather unlikely, although a deep body (Brandl & Bellwood 2013b) may facilitate rapid changes of direction as a possible means of predator avoidance. The most commonly cited predation-avoiding benefit of pair-formation in mobile reef fishes is to spawn with the respective partner. The permanent availability of a reproductive partner releases these species from the need to undertake dangerous, predation-prone ventures to spawning sites in search of a mate (Robertson et al. 1979, Herold & Clark 1993). Nevertheless, many pairing species still migrate to spawning sites, such as *Siganus punctatus* (Johannes 1981) and *Chaetodon lunulatus* (= *trifasciatus*) (Yabuta 1997).

It has been suggested that the close association between pairing butterflyfishes and rabbitfishes and the reef matrix—in other words, swimming within the complex interstices of the reef—reduces vulnerability to predation (Hourigan 1989, Borsa et al. 2007). Although the mechanistic basis for this has not been explored, recent observations provide an indication of the possible link between pairing, reef complexity, and avoiding predation. A recent study of rabbitfishes raised the possibility of shared vigilance between pair members, with one member of a pair often observed 'hanging' in the water column in a tail-down vertical orientation, apparently scanning the surroundings, while the other feeds (Brandl & Bellwood 2013a, Fox & Bellwood 2013) (Figure 5A). Interestingly, exactly the same posture and behaviour have been described for pairing tilefishes, with one member of a pair hanging tail-down in the water column while the other one engages in other activities and

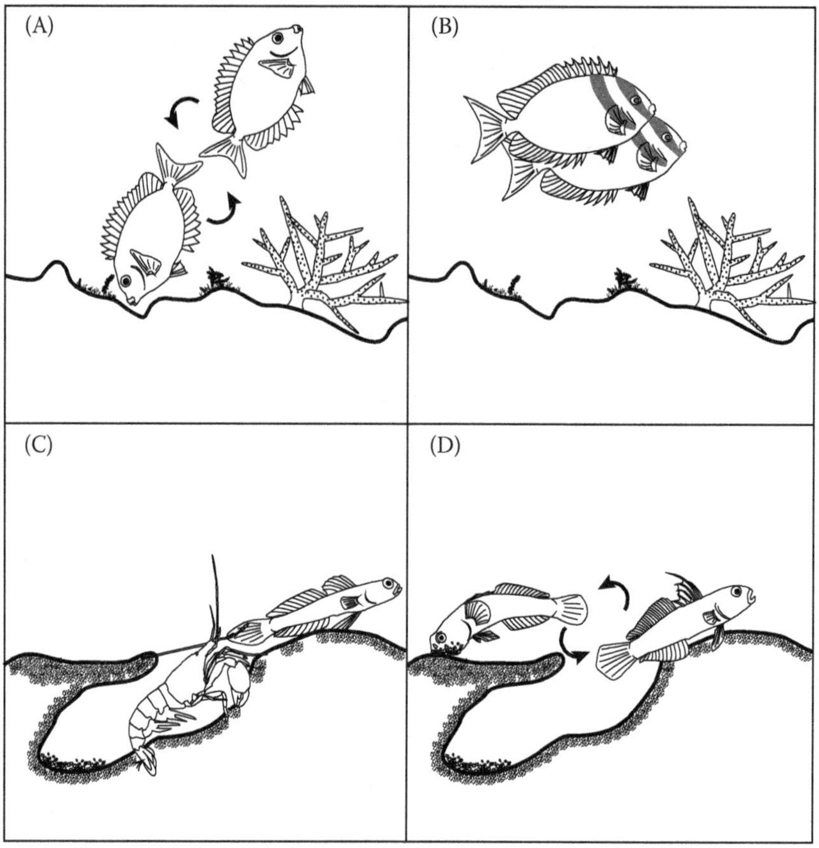

Figure 5 Observed behavioural adaptations of pair-forming species in mobile (A and B) and burrowing (C and D) species: (A) Shared vigilance in rabbitfishes. One individual is assuming a 'tail-down' position, apparently scanning the surroundings, while the other individual is feeding. (B) Synchronized swimming in rabbitfishes. Individuals arrange in a manner that creates the illusion of a single larger fish. (C) Shared labour in shrimp-associated, burrowing gobies. The fish is vigilant while the shrimp is maintaining the burrow and using its antennae to maintain contact with the goby. (D) Shared labour in burrowing gobies. One individual is remaining close to the burrow entrance, looking out for predators, while the other individual is feeding further away from the burrow.

often escapes immediately prior to the arrival of a predator (Clark et al. 1998). The close resemblance of the shared vigilance behaviour observed in rabbitfishes and tilefishes suggests that it may be a common predator avoidance behaviour in pairing species. A similar type of shared vigilance (although lacking the tail-down posture) has been described in the gobies *Valenciennea helsdingenii* and *V. longipinnis*, in which pair members seem to alternate between vigilance and foraging (Takegaki & Nakazono 1999, Clark et al. 2000). This has also been suggested for the pairing butterflyfish *Chaetodon melannotus* (Pratchett et al. 2006). In addition, several species of pair-forming rabbitfishes (*Siganus puellus*, *S. doliatus*, *S. punctatus*) have been observed swimming in a synchronized fashion, aligning their bodies in a manner that creates the illusion of one, significantly larger fish (also *S. stellatus*; J.H. Choat, personal communication 2013) (Figure 5B). This is likely to decrease the risk of predation by smaller predators, but this possibility needs to be explored in further detail. Thus, it appears that pairing reef fishes may avoid predation through a range of behavioural responses that are tailored to their particular social system.

Further support for a non-reproductive role for pair-formation is provided by the presence of homosexual and immature pairs in pairing reef fishes. Homosexual pairs have been found in the butterflyfishes *Chaetodon capistratus* (Gore 1983), *C. madagaskariensis* (= *chrysurus*) (Fricke 1986), *C. lunulatus*, and *C. melannotus* (Pratchett et al. 2006); the surgeonfish *Acanthurus triostegus* (Robertson et al. 1979); the gobies *Valenciennea muralis* and *V. strigata* (Pratchett et al. 2006); and the rabbitfish *Siganus doliatus*, in which 25% of pairs were homosexual (Brandl & Bellwood 2013a). This is also consistent with reports of non-reproductive, mixed-species pairs. Mixed-species pairs have been reported in the blennies *Petroscirtes fallax* and *Meiacanthus lineatus* (Allen et al. 2003) and the tilefishes *Hoplolatilus cuniculus*, *H. chlupatyi*, *H. marcosi*, and *H. purpureus* (Clark et al. 1998). All these examples question a purely reproductive function of pair-formation in the respective species and suggest a crucial role of pairing for daily survival. Mortality rates of paired versus non-paired individuals have yet to be reported for any of these species, but predation risk may be the most important driver of this behaviour (Reavis & Barlow 1998).

While anti-predator behaviour in mobile pairing species is poorly understood, the benefits of pairing are much clearer in bottom-dwelling or burrowing species (Hixon 1991, Depczynski & Bellwood 2004, Forrester & Steele 2004, Hernaman & Munday 2005). Of the 102 species that inhabit and maintain permanent burrows, 50 commonly occur in pairs, with many of them strongly paired or monogamous (Reavis 1997a,b, Clark et al. 1998, Reavis & Barlow 1998, Takegaki & Nakazono 1999, Clark et al. 2000, Pratchett et al. 2006, Hernaman & Munday 2007). While almost half of the burrowing species form pairs, it is noteworthy that 47 of the remaining 52 species live in close association with shrimps of the family Alpheidae. In these latter, shrimp-associated species, there is a clear division of labour between the shrimp and the goby; the shrimp is responsible for digging and maintaining the burrow (Figure 5C), and the goby acts as a sentinel, dedicating extensive time to vigilance and warning the shrimp through rapid tail flicks if a predator is approaching (Karplus 1987, Thompson 2004, 2005). This relationship is mutualistic and obligate. Solitary gobies experience rapid mortality through predation, and solitary shrimps grow more slowly because of reduced foraging time (Karplus et al. 1972, Thompson 2004, 2005). While this relationship should be considered a symbiotic relationship rather than pair-formation, it demonstrates a key aspect of the ecology of burrowing species: The maintenance of a permanent burrow is greatly facilitated by the presence of a second individual. Overall, 95.1% of all fish species that maintain permanent burrows do so with a partner, be it fish or shrimp. Division of labour between pair members of burrowing fishes is well documented (Clark et al. 1998, Reavis & Barlow 1998, Takegaki & Nakazono 1999, Clark et al. 2000) and mainly involves the partitioning of diurnal tasks (burrow maintenance, vigilance, foraging) (Figure 5D). Thus, for these species, the pair bond may be the essential feature for daily survival, suggesting a strong ecological basis for the establishment of pairs. This is further reinforced by the presence of non-reproductive pairs in burrowing gobies (Pratchett et al. 2006) and tilefishes (Clark et al. 1998). It is also seen in other organisms that maintain burrows as pairs, including shrimps of the genus *Lysiosquilla* (Christy & Salmon 1991).

Thus, in summary, predation appears to be a significant factor that may influence the formation of pairs in numerous species, and there appears to be a clear link between pairing and predation in species that maintain a permanent burrow. Although several behavioural traits also suggest anti-predation benefits of pairing in non-burrowing mobile species, the role of pairing behaviour for predator avoidance in these species is poorly understood.

Reproductive characteristics of pair-forming fishes

Coral reef fishes exhibit virtually every form of mating system known in animals, from monogamous mating to mass spawning aggregations, where up to 100,000 individuals of a single species spawn simultaneously (Robertson 1983, Colin & Bell 1991, Sadovy de Mitcheson et al. 2008). Spawning modes include broadcast spawning, benthic-clutch spawning, egg scattering, pouch brooding, mouth brooding, the release of gelatinous floating egg rafts, and even live bearing (Thresher 1984). If the function of pairing behaviour were solely reproductive, one would expect to find several unifying reproductive traits among pairing species, resulting in the following three hypotheses: First, prolonged association with a single partner should result in a mating system restricted to pair members (i.e., pair spawning), which, second, should lead to high proportions of monogamy in pairing species. Third, given the close link between parental care and pairing (Barlow 1981) and the advantages of aggregations for broadcast spawning (Thresher 1984, Sadovy de Mitcheson & Colin 2012), pairing species should predominantly spawn benthic eggs and have greater parental investment.

Initially, it appears that these hypotheses are supported. Among pairing reef fishes, 284 of 341 (83.3%) are reported to have a mating system based on pairs—that is, courtship and fertilization occurs between two individuals—while the remaining 57 species spawn in groups or aggregations. However, the majority of non-pairing species (69.1%) also mate in pairs. Pairing species that are reported to spawn in aggregations rather than pairs are predominantly rabbitfishes, some surgeonfishes, and a few butterflyfishes. In surgeonfishes, this may be the result of relatively high plasticity in social and mating systems, observed throughout the family (Robertson et al. 1979, Robertson 1983). In butterflyfishes, pair spawning has been observed in *Chaetodon nippon* (Suzuki et al. 1980), *C. multicinctus* (Lobel 1989), *C. madagaskariensis* (= *chrysurus*) (Fricke 1986), *C. citrinellus*, *C. unimaculatus*, *C. ornatissimus* (Sancho et al. 2000), *C. lunulatus* (= *trifasciatus*) (Yabuta 1997), and *C. rainfordi* (Thresher 1984), providing relatively good support for the hypothesis that permanently paired species should have a mating system based on pairs (Emlen & Oring 1977, Whiteman & Côté 2004). In contrast, anecdotal reports of spawning aggregations in several strongly paired species, such as *C. ephippium*, *C. lunulatus*, and *C. melannotus*, appear to be incongruous (Claydon 2004, Yabuta 2007). However, although these species may aggregate, they may ultimately spawn in pairs with numerous other pairs, utilizing a common location with favourable currents for egg dispersal (Bell & Colin 1986, Hixon 1991). In addition, the total fertilization rate of such pairs might be even higher in an aggregation (Petersen et al. 2001), particularly as interference in pair matings appears to be common in butterflyfishes (Suzuki et al. 1980, Neudecker & Lobel 1982, Lobel 1989).

Reliable records for rabbitfish spawning are not yet available (Woodland 1990). However, the few reports of spawning in rabbitfishes suggest that large spawning aggregations may be the common way of mating in this family (Johannes 1981, Hara et al. 1986, Domeier & Colin 1997, Hoque et al. 1999, Harahap et al. 2001, Sadovy de Mitcheson & Colin 2012). This is particularly surprising because rabbitfishes are one of the most commonly and strongly paired families (Woodland 1990, Borsa et al. 2007, Brandl & Bellwood 2013a,b). As in butterflyfishes, it is possible that siganid pairs also spawn in aggregations (Johannes 1981, Woodland 1990). However, there are no benefits in terms of egg dispersal because rabbitfishes produce negatively buoyant, adhesive eggs (Woodland 1990). Thus, given the reports of homosexual pairs (Brandl & Bellwood 2013a) and the suggested anti-predator and feeding strategies of pairing rabbitfishes, reproductive factors may only be partially responsible for pairing in this family.

Overall, most pairing species also reproduce in pairs. However, any links between reproduction and pairing may need to be interpreted with caution, given the few noteworthy exceptions as well as the general tendency of reef fishes to reproduce in pairs (69.1% of non-pairing species). Reproduction is probably an important factor of pairing in many species, but it appears to be only one of a range of potential drivers.

The second hypothesis suggests that pairing leads to monogamy. However, only 25.2% of all species that are known to form pairs are monogamous. True genetic monogamy in reef fishes has rarely been reported (Barlow 1981, Whiteman & Côté 2004), although many studies infer exclusive mating with a single partner from field observations, replacement experiments, or aquarium studies (e.g., Barlow 1987, Herold & Clark 1993, Hess 1993, Reavis & Barlow 1998, Whiteman

& Côté 2003, Reavis & Copus 2011). Interestingly, some species, such as jawfishes of the genus *Opisthognathus* (Hess 1993) or the anemonefishes *Amphiprion clarkii*, *A. frenatus*, and *A. perideraion* (Hirose 1995), appear to be monogamous but live either solitarily (jawfishes) or in groups (anemonefishes). These species are not pair forming *per se* and represent 15.2% of all monogamous species. Therefore, although monogamy appears to be a relatively good indicator of pair-formation, it is not obligatory. Conversely, pairing species are not necessarily monogamous: The majority (74.8%) of pair-forming species are not known to be monogamous.

The third hypothesis predicts that pairing species should predominantly spawn benthic eggs, while spawning in aggregations should be the predominant spawning mode in non-pairing fishes. Guarding or brooding eggs requires high levels of investment by the parents and is usually rewarded by greater control of parentage and higher survival rates in offspring (Jones & Avise 1997). This high investment is regarded as a characteristic of pair mating, leading to the evolution of monogamy (Wittenberger & Tilson 1980). Thus, if pair-formation were driven by reproduction alone, one would expect all species with parental care to form pairs. Benthic-clutch spawners have a high potential for parental care, as eggs are usually attached to the substratum within a restricted territory (e.g., Balistidae; Kawase 2002), cave (e.g., Blenniidae; Fishelson 1976), or burrow (e.g., Gobiidae; Reavis 1997a,b) and require intensive parental care (e.g., Hernaman & Munday 2007). Likewise, the specialized systems of mouth brooding and pouch brooding require high parental investment (e.g., Barlow 1981), and such species could therefore be assumed to have a high potential for pair-formation, while egg-scattering and broadcast-spawning species appear to have a low capacity for pair-formation (Johannes 1981, Thresher 1984).

Surprisingly, the prevalence of pairing in the different spawning modes did not support the third hypothesis. Mouth brooding had the lowest proportion of pairing species, followed by broadcast spawning and benthic-clutch spawning. Pairing prevalence was highest in egg-scattering and pouch-brooding species (Figure 6). Mouth brooding is largely restricted to cardinalfishes (Apogonidae) and a few other lineages, such as jawfishes (Opisthognathidae; Hess 1993). Most apogonids live in aggregations, only forming 'transient' breeding pairs (Kuwamura 1985). In aggregating species (e.g., *Apogon notatus*; Allen et al. 2003), the females frequently desert the males after spawning, resulting in a solely courting and copulating pair, and hence a mating pair (Kuwamura 1985). While in some pair-forming species, the female may engage in the defence of the eggs after spawning (e.g., *A. doederleini*; Kuwamura 1985), thus providing protection for the male, most pairs in cardinalfishes are probably mating pairs. Similarly, in jawfishes, individuals are solitary and only pair for courtship and spawning, after which the solitary male orally incubates the eggs (Hess 1993). This is contrary to the suggestion that high parental investment will favour pair-formation. Similarly, the almost-equal proportions of pairing and non-pairing species in broadcast and demersal spawning species, and the high prevalence of pairing in egg-scattering species, suggest that pairing is not significantly linked with parental investment. The two families with the highest proportion of pair-forming species, tilefishes and butterflyfishes, both are broadcast spawners producing pelagic eggs (Thresher 1984, Clark et al. 1988), while egg scattering is the most widespread spawning mode in rabbitfishes, the family with the third-highest prevalence of pairing (Thresher 1984, Woodland 1990).

The only spawning mode that appears to support the hypothesis that high parental investment should lead to pairing and, subsequently, monogamy is pouch brooding, a specialized form of breeding in the Syngnathidae, by which the male incubates egg clutches in a pouch located on its ventral surface (Vincent & Sadler 1995). Many syngnathids form pairs (Allen et al. 2003), and there is substantial evidence that numerous species are monogamous, maintaining their pair bond beyond the reproductive season (Vincent & Sadler 1995, Jones et al. 1998, Kvarnemo et al. 2000, Sogabe & Yanagisawa 2008). However, overall, with the exception of the Syngnathidae, there is limited evidence in support of a direct correlation between pair-formation and the spawning mode of reef fishes.

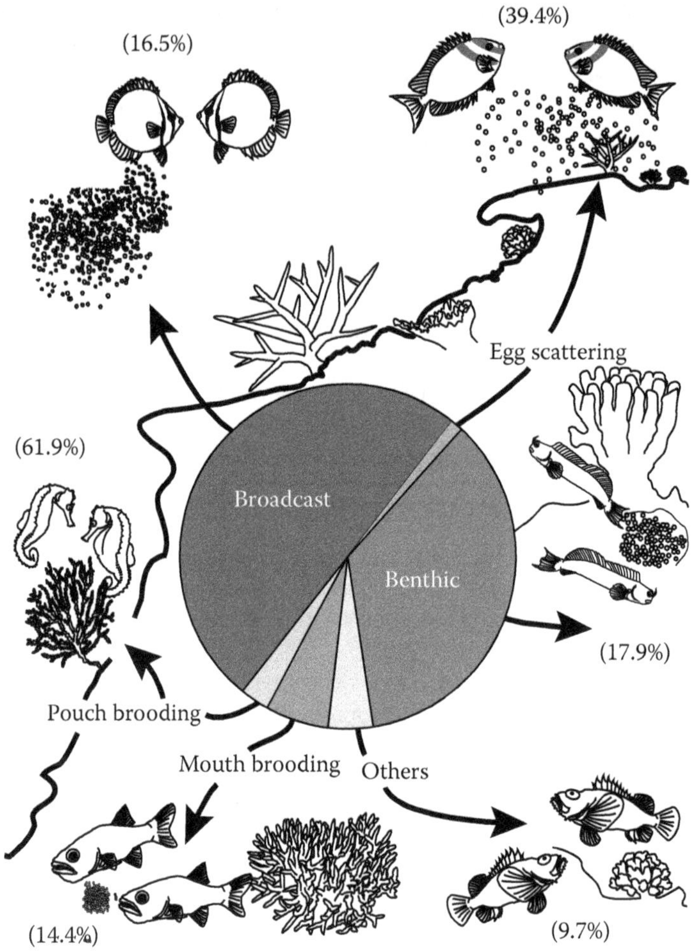

Figure 6 The prevalence of pair-formation in different spawning modes. Proportions indicate the percentage of pair-forming species within each spawning mode. The pie chart indicates the contribution of each spawning mode to the overall species pool.

Conclusion and future directions

The formation of pairs is a common social system for animals. Yet, the definition of pair-formation as a social system is ambiguous and often confused with reproductive interactions between only two individuals, which may be more appropriately termed a mating pair. Following a more ecologically orientated definition, pair-formation is identified as a common trait among coral reef fishes, occurring across a wide range of phylogenetic lineages. However, the processes that have led to the evolution of this social system are poorly understood. This is particularly true for ecological factors that may be correlated with pairing behaviour. Most research to date has focused on the evolution of monogamous mating (i.e., a specialized mating system) in pairing species rather than the implications of pair-formation for daily survival, including food acquisition or reducing predation risk. Furthermore, the majority of studies on pair-formation in reef fishes have focused on three families—the Chaetodontidae, Gobiidae, and Syngnathidae—while other families with a high proportion of pairing species, such as the Malacanthidae, Siganidae, Ptereleotridae, and Synodontidae, have received comparatively little attention.

This review has revealed several similarities in the ecology of pairing reef fishes, identifying two major traits that appear to be common among pair-forming species, regardless of phylogenetic relationships. First, pair-formation appears to be beneficial for species that forage on small, benthic, and relatively immobile prey, including coral polyps, sponges, filamentous algae, or microinvertebrates. This is consistent with the theoretical framework for the evolution of monogamous mating, which explains monogamy as a result of widely distributed resources and the inability of males to sequester multiple females. On reefs, the size and benthic nature of these resources may also influence the feeding efficiency of species that forage on such prey. As a consequence, associations with only one individual may be the preferred group size in these species. Second, the maintenance of permanent burrows appears to be almost exclusively restricted to species that live in association with a partner. Although not all of these species can be considered to be pairing *per se*, these findings highlight the apparent necessity of a cooperative partner for the maintenance of permanent burrows. In both cases, the allocation of tasks between pair members appears to be important to avoid predation and maximize daily energy intake.

In contrast, the review has identified few aspects of the reproductive biology of reef fishes that help to explain pairing. Three hypotheses, based on the assumptions that pairing species should exhibit similarities in their reproductive behaviour, were not supported. While most pairing species also reproduce in pairs, there are some notable exceptions to this pattern, and pair mating appears to be a common mating system among coral reef fishes in general. Thus, there is no clear link between pair-formation and reproduction in pairs. Monogamous mating, while a good indicator for pair-formation, is reported in only a quarter of all pair-forming fishes. Finally, although parental investment appears to favour pairing in pouch-brooding species, there was no clear association between these two traits across a broad range of families.

Based on the finding of common ecological traits among pairing species and the unexpectedly weak correlations between pair-formation and reproductive traits, we suggest that pairing behaviour in reef fishes may be strongly linked to ecological factors, benefitting daily survival and food acquisition. However, as opposed to well-defined theories and numerous empirical studies investigating the reproductive biology of pairing species, the environmental circumstances and ecological benefits of pairing species remain poorly understood. Given the potential importance of these ecological aspects, this promises to be an interesting and exciting avenue for future research.

Acknowledgements

This study was supported by the Australian Research Council (D.R.B.). We thank O. Bellwood, B.J. Bergseth, J.M. Casey, R. Cates, S.A. Rohlfing, and J.Q. Welsh for assistance and J.P. Krajewski and B. Halstead for providing photographic material.

References

Able, K.W., Grimes, C.B., Cooper, R.A. & Uzmann, J.R. 1982. Burrow construction and behavior of tilefish, *Lopholatilus chamaeleonticeps*, in Hudson Submarine Canyon. *Environmental Biology of Fishes* **7**, 199–205.

Adkins-Regan, E. 2002. Development of sexual partner preference in the zebra finch: a socially monogamous, pair-bonding animal. *Archives of Sexual Behavior* **31**, 27–33.

Adrian, O., Kaiser, S., Sachser, N., Jandewerth, P., Löttker, P., Epplen, J.T. & Hennessy, M.B. 2008. Female influences on pair-formation, reproduction and male stress responses in a monogamous cavy (*Galea monasteriensis*). *Hormones and Behavior* **53**, 403–412.

Ahlgren, J., Åbjörnsson, K. & Brönmark, C. 2011. The influence of predator regime on the behaviour and mortality of a freshwater amphipod, *Gammarus pulex*. *Hydrobiologia* **671**, 39–49.

Allen, G.R. 1987 Descriptions of three new pseudochromid fishes of the genus *Pseudoplesiops* from Australia and surrounding regions. *Records of the Western Australian Museum* **13**, 249–261.

Allen, G.R. 1991. *Damselfishes of the World*. Melle, Germany: Mergus.
Allen, G.R. & Munday, P.L. 1995. Description of four new gobies (Gobiidae) from the western Pacific Ocean. *Revue Française d'Aquariologie* **22**, 99–104.
Allen, G.R. & Rajasuriya, A. 1995. *Chrysiptera kuiteri*, a new species of damselfish (Pomacentridae) from Indonesia and Sri Lanka. *Records of the Western Australian Museum* **17**, 283–286.
Allen, G.R., Steene, R.C. & Allen, M. 1998. *Guide to Angelfishes and Butterflyfishes*. Perth, Australia: Tropical Reef Research.
Allen, G.R., Steene, R.C., Humann, P. & DeLoach, N. 2003. *Reef Fish Identification: Tropical Pacific*. Jacksonville, Florida: New World Publications.
Almany, G.R. & Webster, M.S. 2006. The predation gauntlet: early post-settlement mortality in reef fishes. *Coral Reefs* **25**, 19–22.
Ang, T.Z. & Manica, A. 2010. Benefits and costs of dominance in the angelfish *Centropyge bicolor*. *Ethology* **116**, 855–865.
Atema, J., Jacobson, S., Karnofsky, E., Oleszko-Szuts, S. & Stein, L. 1979. Pair-formation in the lobster *Homarus americanus*: behavioral development, pheromones and mating. *Marine Behaviour and Physiology* **6**, 277–296.
Aydin, M. 2011. Growth, reproduction and diet of pufferfish (*Lagocephalus sceleratus*, Gmelin, 1789) from Turkey's Mediterranean Sea coast. *Turkish Journal of Fisheries and Aquatic Sciences* **11**, 589–596.
Baird, A.H., Guest, J.R. & Willis, B.L. 2009. Systematic and biogeographical patterns in the reproductive biology of scleractinian corals. *Annual Review of Ecology, Evolution, and Systematics* **40**, 551–571.
Baird, T.A. & Baird, T.D. 1992. Colony formation and some possible benefits and costs of gregarious living in the territorial sand tilefish, *Malacanthus plumieri*. *Bulletin of Marine Science* **50**, 56–65.
Barlow, G.W. 1974. Extraspecific imposition of social grouping among surgeonfishes (Acanthuridae, Pisces). *Journal of Zoology* **174**, 333–340.
Barlow, G.W. 1981. Patterns of parental investment, dispersal and size among coral-reef fishes. *Environmental Biology of Fishes* **6**, 65–85.
Barlow, G.W. 1984. Patterns of monogamy among teleost fishes. *Archiv für Fischereiwissenschaft* **35**, 75–123.
Barlow, G.W. 1987. Spawning, eggs and larvae of the long-nose filefish *Oxymonacanthus longirostris*, a monogamous corallivore. *Environmental Biology of Fishes* **20**, 183–194.
Barnett, A., Bellwood, D.R. & Hoey, A.S. 2006. Trophic ecomorphology of cardinalfish. *Marine Ecology Progress Series* **322**, 249–257.
Barreiros, J.P., Morato, T., Santos, R.S. & Borba, A.E. 2003. Interannual changes in the diet of the almaco jack *Seriola rivoliana* (Perciformes: Carangidae) from the Azores. *Cybium* **27**, 37–40.
Barros, B., Sakai, Y., Hashimoto, H. & Gushima, K. 2008. Feeding behavior of leaf-like juveniles of the round batfish *Platax orbicularis* (Ephippidae) on reefs of Kuchierabu-jima Island, southern Japan. *Journal of Ethology* **26**, 287–293.
Bell, L.J. & Colin, P.L. 1986. Mass spawning of *Caesio teres* (Pisces: Caesionidae) at Enewetak Atoll, Marshall Islands. *Environmental Biology of Fishes* **15**, 69–74.
Bellwood, D.R., Hoey, A.S., Ackerman, J.L. & Depczynski, M. 2006a. Coral bleaching, reef fish community phase shifts and the resilience of coral reefs. *Global Change Biology* **12**, 1587–1594.
Bellwood, D.R., Hoey, A.S. & Choat, J.H. 2003. Limited functional redundancy in high diversity systems: resilience and ecosystem function on coral reefs. *Ecology Letters* **6**, 281–285.
Bellwood, D.R., Hughes, T.P. & Hoey, A.S. 2006b. Sleeping functional group drives coral-reef recovery. *Current Biology* **16**, 2434–2439.
Bellwood, D.R., Klanten, S., Cowman, P.F., Pratchett, M.S., Konow, N. & van Herwerden, L. 2010. Evolutionary history of the butterflyfishes (f: Chaetodontidae) and the rise of coral feeding fishes. *Journal of Evolutionary Biology* **23**, 335–349.
Bellwood, D.R., Renema, W. & Rosen, B.R. 2012. Biodiversity hotspots, evolution and coral reef biogeography: a review. In *Biotic Evolution and Environmental Change in Southeast Asia*, D.J. Gower et al. (eds). Cambridge, UK: Cambridge University Press, 216–245.
Beltran, S. & Boissier, J. 2008. Schistosome monogamy: who, how, and why? *Trends in Parasitology* **24**, 386–391.
Beumer, J.P. 1978. Feeding ecology of four fishes from a mangrove creek in north Queensland, Australia. *Journal of Fish Biology* **12**, 475–490.

Blaber, S.J.M. 1977. The feeding ecology and relative abundance of mullet (Mugilidae) in Natal and Pondoland estuaries. *Biological Journal of the Linnean Society* **9**, 259–275.
Blaber, S.J.M. & Cyrus, D.P. 1983. The biology of Carangidae (Teleostei) in Natal estuaries. *Journal of Fish Biology* **22**, 173–188.
Blaber, S.J.M., Milton, D.A., Rawlinson, N.J.F., Tiroba, G. & Nichols, P.V. 1990. Diets of lagoon fishes of the Solomon Islands: predators of tuna baitfish and trophic effects of baitfishing on the subsistence fishery. *Fisheries Research* **8**, 263–286.
Black, J.M. 2001. Fitness consequences of long-term pair bonds in barnacle geese: monogamy in the extreme. *Behavioral Ecology* **12**, 640–645.
Boaden, A.E. & Kingsford, M.J. 2012. Diel behaviour and trophic ecology of *Scolopsis bilineatus* (Nemipteridae). *Coral Reefs* **31**, 871–883.
Bonin, M.C., Almany, G.R. & Jones, G.P. 2011. Contrasting effects of habitat loss and fragmentation on coral-associated reef fishes. *Ecology* **92**, 1503–1512.
Borsa, P., Lemer, S. & Aurell, D. 2007. Pattern of lineage diversification in rabbitfishes. *Molecular Phylogenetics and Evolution* **44**, 427–435.
Botton, M.L. & Loveland, R.E. 1992. Body size, morphological constraints, and mated pair-formation in four populations of horseshoe crabs (*Limulus polyphemus*) along a geographic cline. *Marine Biology* **112**, 409–415.
Brandl, S.J. & Bellwood, D.R. 2013a. Pair formation in the herbivorous rabbitfish *Siganus doliatus*. *Journal of Fish Biology* **82**, 2031–2044.
Brandl, S.J. & Bellwood, D.R. 2013b. Morphology, sociality, and ecology: can morphology predict pairing behaviour in coral reef fishes? *Coral Reefs* **32**, 835–846.
Brandl, S.J. & Bellwood, D.R. in press. Individual-based analyses reveal limited functional overlap in a coral reef fish community. *Journal of Animal Ecology* doi: 10.1111/1365-2656.12171
Brewer, D.T., Blaber, S.J.M. & Salini, J.P. 1989. Feeding biology of *Caranx bucculentus* Alleyne and Macleay (Teleostei: Carangidae) in Albatross Bay, Gulf of Carpentaria, with special reference to predation on penaeid prawns. *Australian Journal of Marine and Freshwater Research* **40**, 657–668.
Brooker, R.M., Munday, P.L. & Ainsworth, T.D. 2010. Diets of coral-dwelling fishes of the genus *Gobiodon* with evidence of corallivory. *Journal of Fish Biology* **76**, 2578–2583.
Bruns, P.J. & Brussard, T.B. 1974. Pair-formation in *Tetrahymena pyriformis*—an inducible developmental system. *Journal of Experimental Zoology* **188**, 337–344.
Bshary, R., Grutter, A.S., Willener, A.S.T. & Leimar, O. 2008. Pairs of cooperating cleaner fish provide better service quality than singletons. *Nature* **455**, 964–966.
Bull, C.M. & Pamula, Y. 1998. Enhanced vigilance in monogamous pairs of the lizard *Tiliqua rugosa*. *Behavioral Ecology* **9**, 452–455.
Burpee, D.M. & Sakaluk, S.K. 1993. The effect of pair-formation on diel calling patterns in two cricket species, *Gryllus veletis* and *Gryllodes sigillatus* (Orthoptera, Gryllidae). *Journal of Insect Behavior* **6**, 431–440.
Cable, R.M. & Hopp, R.M. 1954. Acanthocephalan parasites of the genus *Neoechinorhynchus* in North American turtles with the description of two new species. *Journal of Parasitology* **40**, 674–680.
Carlson, B.A. 2012. Feeding activity by the blenny *Exallias brevis* causes multifocal bleaching in corals: comment on Zvuloni et al. (2011). *Marine Ecology Progress Series* **463**, 297–299.
Ceccarelli, D.M. 2007. Modification of benthic communities by territorial damselfish: a multi-species comparison. *Coral Reefs* **26**, 853–866.
Cheal, A., Emslie, M., Miller, I. & Sweatman, H. 2012. The distribution of herbivorous fishes on the Great Barrier Reef. *Marine Biology* **159**, 1143–1154.
Chen, T.C., Ormond, R.F.G. & Mok, H.K. 2001. Feeding and territorial behaviour in juveniles of three co-existing triggerfishes. *Journal of Fish Biology* **59**, 524–532.
Choat, J.H., Robbins, W.D. & Clements, K.D. 2004. The trophic status of herbivorous fishes on coral reefs. *Marine Biology* **145**, 445–454.
Christy, J.H. & Salmon, M. 1991. Comparative studies of reproductive behavior in mantis shrimps and fiddler crabs. *American Zoologist* **31**, 329–337.
Clark, E. & Pohle, J.F. 1992. Monogamy in tilefish. *National Geographic Research and Exploration* **8**, 276–295.
Clark, E. & Pohle, M. 1996. *Trichonotus halstead*, a new sand-diving fish from Papua New Guinea. *Environmental Biology of Fishes* **45**, 1–11.

Clark, E., Pohle, J.F. & Halstead, B. 1998. Ecology and behaviour of tilefishes, *Hoplolatilus starki, H. fronticinctus* and related species (Malacanthidae): non-mound and mound builders. *Environmental Biology of Fishes* **52**, 395–417.

Clark, E., Rabin, J.S. & Holderman, S. 1988. Reproductive behavior and social organization in the sand tilefish, *Malacanthus plumieri. Environmental Biology of Fishes* **22**, 273–286.

Clark, E., Stoll, M.J., Alburn, T.K. & Petzold, R. 2000. Mound-building and feeding behavior of the two-stripe goby *Valenciennea helsdingenii*, in the south Red Sea. *Environmental Biology of Fishes* **57**, 131–141.

Claydon, J. 2004. Spawning aggregations of coral reef fishes: characteristics, hypotheses, threats and management. *Oceanography and Marine Biology: An Annual Review* **42**, 265–302.

Clemente, S., Hernández, J.C., Rodríguez, A. & Brito, A. 2010. Identifying keystone predators and the importance of preserving functional diversity in sublittoral rocky-bottom areas. *Marine Ecology Progress Series* **413**, 55–67.

Clements, K.D. & Choat, J.H. 1997. Comparison of herbivory in the closely-related marine fish genera *Girella* and *Kyphosus. Marine Biology* **127**, 579–586.

Cole, A.J., Pratchett, M.S. & Jones, G.P. 2008. Diversity and functional importance of coral-feeding fishes on tropical coral reefs. *Fish and Fisheries* **9**, 286–307.

Colin, P.L. & Bell, L.J. 1991. Aspects of the spawning of labrid and scarid fishes (Pisces: Labroidei) at Enewetak Atoll, Marshall Islands (North Pacific Ocean) with notes on other families. *Environmental Biology of Fishes* **31**, 229–260.

Connell, S.D. 1998. Patterns of pisciviory by resident predatory reef fish at One Tree Reef, Great Barrier Reef. *Marine and Freshwater Research* **49**, 25–30.

Cox, C.R., Goldschmidt, V.I. & Engelhardt, H.R. 1993. Pair-formation in California condors. *American Zoologist* **33**, 126–138.

Craig, M.T. 2007. Preliminary observations on the life history of the white-streaked grouper, *Epinephelus ongus*, from Okinawa, Japan. *Ichthyological Research* **54**, 81–84.

Cvitanovic, C. & Bellwood, D.R. 2009. Local variation in herbivore feeding activity on an inshore reef of the Great Barrier Reef. *Coral Reefs* **28**, 127–133.

Daly, J.M. 1973. Some relationships between the process of pair-formation and gamete maturation in *Harmothoe imbricata* (L.) (Annelida: Polychaeta). *Marine Behaviour and Physiology* **1**, 277–284.

Daroonchoo, L. 1991. Study on some biology of Mahidol smiling goby (*Mahidolia mystacina* Cuvier and Valenciennes) in Chanthaburi River. In *Proceedings of the Seminar on Fisheries 16–18 September 1991, Bangkok, Thailand.* Bangkok, Thailand: Department of Fisheries, 214–221.

Debenay, J.P., Sigura, A. & Justine, J.L. 2011. Foraminifera in the diet of coral reef fish from the lagoon of New Caledonia: predation, digestion, dispersion. *Revue de Micropaléontologie* **54**, 87–103.

Delmonte Corrado, M.U., Locatelli, D., Paleari, L. & Bottiroli, G. 1997. Lectin-binding sites involved in *Paramecium primaurelia* mating pair-formation. *Journal of Eukaryotic Microbiology* **44**, 603–608.

Depczynski, M. & Bellwood, D.R. 2003. The role of cryptobenthic reef fishes in coral reef trophodynamics. *Marine Ecology Progress Series* **256**, 183–191.

Depczynski, M. & Bellwood, D.R. 2004. Microhabitat utilization patterns in cryptobenthic reef fish communities. *Marine Biology* **145**, 455–463.

DiSalvo, L.H., Randall, J.E. & Cea, A. 2007. Stomach contents and feeding observations of some Easter Island fishes. *Atoll Research Bulletin* **548**, 234–245.

Domeier, M.L. & Colin, P.L. 1997. Tropical reef spawning aggregations: defined and reviewed. *Bulletin of Marine Science* **60**, 698–726.

Donaldson, T.J. 1989. Facultative monogamy in obligate coral-dwelling hawkfishes (Cirrhitidae). *Environmental Biology of Fishes* **26**, 295–302.

Donaldson, T.J. 1990. Reproductive behavior and social organization of some Pacific hawkfishes (Cirrhitidae). *Japanese Journal of Ichthyology* **36**, 439–458.

Elliott, J.P. & Bellwood, D.R. 2003. Alimentary tract morphology and diet in three coral reef fish families. *Journal of Fish Biology* **63**, 1598–1609.

Emlen, S.T. & Oring, L.W. 1977. Ecology, sexual selection, and the evolution of mating systems. *Science* **197**, 215–223.

Farmer, B.M. & Wilson, S.K. 2011. Diet of finfish targeted by fishers in North West Australia and the implications for trophic cascades. *Environmental Biology of Fishes* **91**, 71–85.

Fautin, D.G. & Allen, G.R. 1992. *Field Guide to Anemonefishes and Their Host Sea Anemones*. Perth, Australia: Western Australian Museum.
Feeney, W.E., Lönnstedt, O.M., Bosiger, Y., Martin, J., Jones, G.P., Rowe, R.J. & McCormick, M.I. 2012. High rate of prey consumption in a small predatory fish on coral reefs. *Coral Reefs* **31**, 909–918.
Ferry-Graham, L.A., Bolnick, D.I. & Wainwright, P.C. 2002. Using functional morphology to examine the ecology and evolution of specialization. *Integrative and Comparative Biology* **42**, 265–277.
Ferry-Graham, L.A., Wainwright, P.C., Hulsey, C.D. & Bellwood, D.R. 2001. Evolution and mechanics of long jaws in butterflyfishes (family Chaetodontidae). *Journal of Morphology* **248**, 120–143.
Fishelson, L. 1976. Spawning and larval development of the blenniid fish *Meiacanthus nigrolineatus* from the Red Sea. *Copeia* **4**, 798–800.
Ford, J.R. & Swearer, S.E. 2013a. Two's company, three's a crowd: food and shelter limitation outweigh the benefits of group living in a shoaling fish. *Ecology* **94**, 1069–1077.
Ford, J.R. & Swearer, S.E. 2013b. Shoaling behavior enhances risk of predation from multiple predator guilds in a marine fish. *Oecologia* **172**, 387–397.
Forrester, G.E. & Steele, M.A. 2004. Predators, prey refuges, and the spatial scaling of density-dependent prey mortality. *Ecology* **85**, 1332–1342.
Foster, S. & Vincent, A.C.J. 2004. The life history and ecology of seahorses: implications for conservation and management. *Journal of Fish Biology* **65**, 1–61.
Fox, R.J. & Bellwood, D.R. 2008. Remote video bioassays reveal the potential feeding impact of the rabbitfish *Siganus canaliculatus* (f: Siganidae) on an inner-shelf reef of the Great Barrier Reef. *Coral Reefs* **27**, 605–615.
Fox, R.J. & Bellwood, D.R. 2013. Niche partitioning of feeding microhabitats produces a unique function for herbivorous rabbitfishes (Perciformes, Siganidae) on coral reefs. *Coral Reefs* **32**, 13–23.
Fox, R.J., Sunderland, T.L., Hoey, A.S. & Bellwood, D.R. 2009. Estimating ecosystem function: contrasting roles of closely related herbivorous rabbitfishes (Siganidae) on coral reefs. *Marine Ecology Progress Series* **385**, 261–269.
Frédérich, B., Fabri, G., Lepoint, G., Vandewalle, P. & Parmentier, E. 2009. Trophic niches of thirteen damselfishes (Pomacentridae) at the Grand Récif of Toliara, Madagascar. *Ichthyological Research* **56**, 10–17.
Fricke, H.W. 1980. Control of different mating systems in a coral reef fish by one environmental factor. *Animal Behavior* **28**, 561–569.
Fricke, H.W. 1986. Pair swimming and mutual partner guarding in monogamous butterflyfish (Pisces, Chaetodontidae): a joint advertisement for territory. *Ethology* **73**, 307–333.
Fricke, H.W. & Fricke, S. 1977. Monogamy and sex change by aggressive dominance in coral reef fish. *Nature* **266**, 830–832.
Fricke, R. & Zaiser, M.J. 1982. Redescription of *Diplogrammus xenicus* (Teleostei: Callionymidae) from Miyake-jima, Japan, with ecological notes. *Japanese Journal of Ichthyology* **29**, 253–259.
Froese, R. & Pauly, D. (eds) 2013. *FishBase*. Online. http://www.fishbase.org (accessed April 2013).
Fukumori, K., Okuda, N., Hamaoka, H., Fukumoto, T. Takahashi, D. & Omori, K. 2008. Stable isotopes reveal life history polymorphism in the coastal fish *Apogon notatus*. *Marine Ecology Progress Series* **362**, 279–289.
Fulton, C.J. 2007. Swimming speed performance in coral reef fishes: field validations reveal distinct functional groups. *Coral Reefs* **26**, 217–228.
Getz, L.L. & Hofmann, J.E. 1986. Social organization in free-living prairie voles, *Microtus ochrogaster*: field and laboratory evidence for pair-bonding. *Behavioral Ecology and Sociobiology* **8**, 189–194.
Gluckmann, I. & Vandewalle, P. 1998. Morphofunctional analysis of the feeding apparatus in four Pomacentridae species: *Dascyllus aruanus*, *Chromis retrofasciata*, *Chrysiptera biocellata* and *C. unimaculata*. *Italian Journal of Zoology* **65**, 421–424.
Golani, D. & Galil, B. 1991. Trophic relationships of colonizing and indigenous goatfishes (Mullidae) in the eastern Mediterranean with special emphasis on decapod crustaceans. *Hydrobiologia* **218**, 27–33.
Gore, M.A. 1983. The effect of a flexible spacing system on the social organization of a coral reef fish, *Chaetodon capistratus*. *Behaviour* **85**, 118–145.
Grandcourt, E.M., Al Abdessalaam, T.Z., Francis, F. & Al Shamsi, A. 2004. Population biology and assessment of representatives of the family Carangidae: *Carangoides bajad* and *Gnathanodon speciosus* (Forsskaal, 1775), in the Southern Arabian Gulf. *Fisheries Research* **69**, 331–341.

Green, A.L. & Bellwood, D.R. 2009. *Monitoring Functional Groups of Herbivorous Fishes as Indicators of Coral Reef Resilience—A Practical Guide for Coral Reef Managers in the Asia Pacific Region.* IUCN Working Group on Climate Change and Coral Reefs. Gland, Switzerland: IUCN.

Gronell, A.M. 1984. Courtship, spawning and social organization of the pipefish, *Corythoichthys intestinalis* (Pisces: Syngnathidae) with notes on two congeneric species. *Zeitschrift für Tierpsychologie* **65**, 1–24.

Grubich, J.R., Rice, A.N. & Westneat, M.W. 2008. Functional morphology of bite mechanics in the great barracuda (*Sphyraena barracuda*). *Zoology* **111**, 16–29.

Gwinner, E., Rödl, T. & Schwabl, H. 1994. Pair territoriality of wintering stonechats: behaviour, function and hormones. *Behavioral Ecology and Sociobiology* **34**, 321–327.

Hadley Hansen, P.E. 1986. Revision of the tripterygiid fish genus *Helcogramma*, including descriptions of four new species. *Bulletin of Marine Science* **38**, 313–354.

Hagiwara, K. & Winterbottom, R. 2007. Two new species of *Trimma* (Gobiidae) from the western Pacific. *Bulletin of National Museum of Nature and Science, Series A* **Supplement 1**, 163–174.

Hajisamae, S. 2009. Trophic ecology of bottom fishes assemblage along coastal areas of Thailand. *Estuarine, Coastal and Shelf Science* **82**, 503–514.

Hajisamae, S., Chou, L.M. & Ibrahim, S. 2003. Feeding habits and trophic organization of the fish community in shallow waters of an impacted tropical habitat. *Estuarine, Coastal and Shelf Science* **58**, 89–98.

Hamilton, I.M. & Dill, L.M. 2003. The use of territorial gardening versus kleptoparasitism by a subtropical reef fish (*Kyphosus cornelii*) is influenced by territory defendability. *Behavioral Ecology* **14**, 561–568.

Hamner, W.M., Jones, M.S., Carleton, J.H., Hauri, I.R. & Williams, D.M. 1988. Zooplankton, planktivorous fish and water currents on a windward reef face: Great Barrier Reef, Australia. *Bulletin of Marine Science* **42**, 459–479.

Hara, S., Duray, M.N., Parazo, M. & Taki, Y. 1986. Year-round spawning and seed production of the rabbitfish, *Siganus guttatus*. *Aquaculture* **59**, 259–272.

Harahap, A.P., Takemura, A., Nakamura, S., Rahman, M.S. & Takano, K. 2001. Histological evidence of lunar-synchronized ovarian development and spawning in the spiny rabbitfish *Siganus spinus* (Linnaeus) around the Ryukyus. *Fisheries Science* **67**, 888–893.

Harding, J.A., Almany, G.R., Houck, L.D. & Hixon, M.A. 2003. Experimental analysis of monogamy in the Caribbean cleaner goby, *Gobiosoma evelynae*. *Animal Behaviour* **65**, 865–874.

Harmelin-Vivien, M.L. 1979. *Ichtyofaune des récifs coralliens de Tuléar (Madagascar): écologie et Relations trophiques.* Doctoral thesis, University of Aix-Marseille II, France.

Harmelin-Vivien, M.L. & Bouchon, C. 1976. Feeding behavior of some carnivorous fishes (Serranidae and Scorpaenidae) from Tulear (Madagascar). *Marine Biology* **37**, 329–340.

Hartnoll, R.G. & Smith, S.M. 1978. Pair-formation and the reproductive cycle in *Gammarus duebeni*. *Journal of Natural History* **12**, 501–511.

Hensley, D.A. & Allen, G.R. 1977. A new species of *Abudefduf* (Pisces: Pomacentridae) from the Indo-Australian Archipelago. *Records of the Western Australian Museum* **6**, 107–118.

Herler, J., Munday, P.L. & Hernaman, V. 2011. *Gobies on Coral Reefs. The Biology of Gobies.* Saint Helier, UK: Science Publishers, 493–529.

Hernaman, V. & Munday, P.L. 2005. Life-history characteristics of coral reef gobies. II. Mortality rate, mating system and timing of maturation. *Marine Ecology Progress Series* **290**, 223–237.

Hernaman, V. & Munday, P.L. 2007. Evolution of mating systems in coral reef gobies and constraints on mating system plasticity. *Coral Reefs* **26**, 585–595.

Hernaman, V., Probert, P.K. & Robbins, W.D. 2009. Trophic ecology of coral reef gobies: interspecific, ontogenetic, and seasonal comparison of diet and feeding intensity. *Marine Biology* **156**, 317–330.

Herold, D. & Clark, E. 1993. Monogamy, spawning and skin-shedding of the sea moth, *Eurypegasus draconis* (Pisces: Pegasidae). *Environmental Biology of Fishes* **37**, 219–236.

Hess, H.C. 1993. Male mouthbrooding in jawfishes (Opistognathidae): constraints on polygyny. *Bulletin of Marine Science* **52**, 806–818.

Hiatt, R.W. & Strasburg, D.W. 1960. Ecological relationships of the fish fauna on coral reefs of the Marshall Islands. *Ecological Monographs* **30**, 65–127.

Hirose, Y. 1995. Patterns of pair formation in protandrous anemonefishes, *Amphiprion clarkii*, *A. frenatus* and *A. perideraion*, on coral reefs of Okinawa, Japan. *Environmental Biology of Fishes* **43**, 153–161.

Hixon, M.A. 1991. Predation as a process structuring coral reef fish communities. In *The Ecology of Fishes on Coral Reefs*, P.F. Sale (ed.). San Diego, California: Academic Press, 475–508.

Ho, C.T., Kao, S.J., Dai, C.F., Hsieh, H.L., Shiah, F.K. & Jan, R.Q. 2007. Dietary separation between two blennies and the Pacific gregory in northern Taiwan: evidence from stomach content and stable isotope analyses. *Marine Biology* **151**, 729–736.

Hobson, E.S. 1975. Feeding patterns among tropical reef fishes. *American Scientist* **63**, 382–392.

Hobson, E.S. 1991. Trophic relationships of fishes specialised to feed on zooplankters above coral reefs. In *The Ecology of Fishes on Coral Reefs*, P.F. Sale (ed.). San Diego, California: Academic Press, 69–95.

Hoey, A.S. & Bellwood, D.R. 2011. Suppression of herbivory by macroalgal density: a critical feedback on coral reefs? *Ecology Letters* **14**, 267–273.

Hoey, A.S., Brandl, S.J. & Bellwood, D.R. 2013. Diet and cross-shelf distribution of rabbitfishes (f. Siganidae) on the northern Great Barrier Reef: implications for ecosystem function. *Coral Reefs* doi:10.1007/s00338-013-1043-z

Holbrook, S.J. & Schmitt, R.J. 2004. Population dynamics of a damselfish: effects of a competitor that also is an indirect mutualist. *Ecology* **85**, 979–985.

Holmes, T.H. & McCormick, M.I. 2011. Response across a gradient: behavioural reactions of newly settled fish to predation cues. *Animal Behaviour* **81**, 543–550.

Hoque, M.M., Takemura, A., Matsuyama, M., Matsuura, S. & Takano, K. 1999. Lunar spawning in *Siganus canaliculatus*. *Journal of Fish Biology* **55**, 1213–1222.

Horinouchi, M., Tongnunui, P., Furumitsu, K., Nakamura, Y., Kanou, K., Yamaguchi, A., Okamoto, K. & Sano, M. 2012. Food habits of small fishes in seagrass habitats in Trang, southern Thailand. *Fisheries Science* **78**, 577–587.

Horn, M.H. 1989. Biology of marine herbivorous fishes. *Oceanography and Marine Biology: An Annual Review* **27**, 167–272.

Hourigan, T.F. 1989. Environmental determinants of butterflyfish social systems. *Environmental Biology of Fishes* **25**, 61–78.

Howard, R.D. & Kluge, A.H. 1985. Proximate mechanisms of sexual selection in wood frogs. *Evolution* **39**, 260–277.

Iyer, N. 2012. The effects of background adaptation and food availability on habitat preference of *Corythoichthys flavofasciatus*. Online. http://escholarship.org/uc/item/5182x106 (accessed May 2013).

Job, S.D. & Shand, J. 2001. Spectral sensitivity of larval and juvenile coral reef fishes: implications for feeding in a variable light environment. *Marine Ecology Progress Series* **214**, 267–277.

Johannes, R.E. 1981. *Words of the Lagoon: Fishing and Marine Lore in the Palau District of Micronesia*. Berkeley, California: University of California Press.

Johnson, R.H. & Nelson, D.R. 1978. Copulation and possible olfaction-mediated pair-formation in two species of carcharhinid sharks. *Copeia* **3**, 76–84.

Jones, A.G. & Avise, J.C. 1997. Microsatellite analysis of maternity and the mating system in the Gulf pipefish *Syngnathus scovelli*, a species with male pregnancy and sex-role reversal. *Molecular Ecology* **6**, 203–213.

Jones, A.G., Kvarnemo, C., Moore, G.I., Simmons, L.W. & Avise, J.C. 1998. Microsatellite evidence for monogamy and sex-biased recombination in the Western Australian seahorse *Hippocampus angustus*. *Molecular Ecology* **7**, 1497–1505.

Kadota, T., Osato, J., Hashimoto, H. & Sakai, Y. 2011. Harem structure and female territoriality in the dwarf hawkfish *Cirrhitichthys falco* (Cirrhitidae). *Environmental Biology of Fishes* **92**, 79–88.

Kane, C.N., Brooks, A.J., Holbrook, S.J. & Schmitt, R.J. 2009. The role of microhabitat preference and social organization in determining the spatial distribution of a coral reef fish. *Environmental Biology of Fishes* **84**, 1–10.

Karplus, I. 1979. The tactile communication between *Cryptocentrus steinitzi* (Pisces, Gobiidae) and *Alpheus purpurilenticularis* (Crustacea, Alpheidae). *Zeitschrift für Tierpsychologie* **49**, 173–196.

Karplus, I. 1987. The association between gobiid fishes and burrowing alpheid shrimps. *Oceanography and Marine Biology: An Annual Review* **25**, 507–562.

Karplus, I., Szlep, R. & Tsurnamal, M. 1972. Associative behavior of the fish *Cryptocentrus cryptocentrus* (Gobiidae) and the pistol shrimp *Alpheus djiboutensis* (Alpheidae) in artificial burrows. *Marine Biology* **15**, 95–104.

Kawase, H. 2002. Simplicity and diversity in the reproductive ecology of triggerfish (Balistidae) and filefish (Monacanthidae). In *Proceedings of the International Commemorative Symposium, 70th Anniversary of the Japanese Society of Fisheries Science, 1–5 October 2001, Yokohama. Fisheries Science* **68** Suppl. 1, 119–122.

Kei, K. 2010. Pair-formation and reproductive behavior in the egg cowry *Ovula ovum* (Gastropoda: Ovulidae) in Southern Kyushu, Japan. *Venus* **69**, 49–58.

Kelley, J.L., Fitzpatrick, J.L. & Merilaita, S. 2013. Spots and stripes: ecology and colour pattern evolution in butterflyfishes. *Proceedings of the Royal Society B* **280**, 20122730. doi:10.1098/rspb.2012.2730

Kiso, K. & Mahyam, M.I. 2003. Distribution and feeding habits of juvenile and young John's snapper *Lutjanus johnii* in the Matang mangrove estuary, west coast of Peninsular Malaysia. *Fisheries Science* **69**, 563–568.

Kizhakudan, S.J. & Gomathy, S. 2007. Unusual landings of the bluntnose lizardfish *Trachinocephalus myops* (Forster, 1801) at Chennai, with a note on some aspects of biology. *Journal of the Marine Biological Association of India* **49**, 250–253.

Klumpp, D.W., McKinnon, A.D. & Mundy, C.N. 1988. Motile cryptofauna of a coral reef: abundance, distribution and trophic potential. *Marine Ecology Progress Series* **45**, 95–108.

Kobayashi, D.R. 1986. Social organisation of the spotted sharpnose puffer, *Canthigaster punctatissima* (Tetradontidae). *Environmental Biology of Fishes* **15**, 141–145.

Kohda, M. 1988. Diurnal periodicity of spawning activity of permanently territorial damselfishes (Teleostei: Pomacentridae). *Environmental Biology of Fishes* **21**, 91–100.

Kokita, T. & Mizota, T. 2002. Male secondary sexual traits are hydrodynamic devices for enhancing swimming performance in a monogamous filefish *Paramonacanthus japonicus*. *Journal of Ethology* **20**, 35–42.

Kortlandt, A. 1995. Patterns of pair-formation and nest-building in the European cormorant *Phalacrocorax carbo sinensis*. *Ardea* **83**, 11–25.

Kramer, M.J., Bellwood, D.R. & Bellwood, O. 2012a. Cryptofauna of the epilithic algal matrix on an inshore coral reef, Great Barrier Reef. *Coral Reefs* **31**, 1007–1015.

Kramer, M.J., Bellwood, O. & Bellwood, D.R. 2012b. The trophic importance of algal turfs for coral reef fishes: the crustacean link. *Coral Reefs* **32**, 575–583.

Kuwamura, T. 1985. Social and reproductive behaviour of three mouthbrooding cardinalfishes, *Apogon doederleini*, *A. niger* and *A. notatus*. *Environmental Biology of Fishes* **13**, 17–24.

Kuwamura, T. 1991. Habitat segregation, coexistence or interspecific territoriality between two triggerfishes, *Rhinecanthus aculeatus* and *Sufflamen chrysopterus*, with notes on distribution of other balistids at Sesoko Island, Okinawa. *Galaxea* **10**, 65–78.

Kuwamura, T., Yugo, Y. & Nakashima, Y. 1993. Size-assortative monogamy and paternal egg care in a coral goby *Paragobiodon echinocephalus*. *Ethology* **95**, 65–75.

Kvarnemo, C., Moore, G.I., Jones, A.G., Nelson, W.S. & Avise, J.C. 2000. Monogamous pair bonds and mate switching in the Western Australian seahorse *Hippocampus subelongatus*. *Journal of Evolutionary Biology* **13**, 882–888.

Kwak, S.N., Huh, S.H. & Klumpp, D.W. 2004. Partitioning of food resources among *Sillago japonica*, *Ditremma temmincki*, *Tridentiger trigonocephalus*, *Hippocampus japonicus* and *Petroscirtes breviceps* in an eelgrass, *Zostera marina*, bed. *Environmental Biology of Fishes* **71**, 353–364.

Lek, E., Fairclough, D.V., Platell, M.E., Clarke, K.R., Tweedley, J.R. & Potter, I.C. 2011. To what extent are the dietary compositions of three abundant, co-occurring labrid species different and related to latitude, habitat, body size and season? *Journal of Fish Biology* **78**, 1913–1943.

Letourneur, Y. 1996. Dynamics of fish communities on Reunion fringing reefs, Indian Ocean. I. Patterns of spatial distribution. *Journal of Experimental Marine Biology and Ecology* **195**, 1–30.

Leu, S.T., Kappeler, P.M. & Bull, C.M. 2011. Pair-living in the absence of obligate biparental care in a lizard: trading-off sex and food? *Ethology* **117**, 758–768.

Lewis, A.R. 1998. Effects of experimental coral disturbance on the population dynamics of fishes on large patch reefs. *Journal of Experimental Marine Biology and Ecology* **230**, 91–110.

Leysen, H., Christiaens, J., De Kegel, B., Boone, M.N., Van Hoorebeke, L. & Adriaens, D. 2011. Musculoskeletal structure of the feeding system and implications of snout elongation in *Hippocampus reidi* and *Dunckerocampus dactyliophorus*. *Journal of Fish Biology* **78**, 1799–1823.

Linsenmair, K.E. & Linsenmair, C. 1971. Paarbildung und Paarzusammenhalt bei der monogamen Wüstenassel *Hemilepistus reaumuri* (Crustacea, Isopoda, Oniscoidea). *Zeitschrift für Tierpsychologie* **29**, 134–155.

Liske, E. & Davis, W.J. 1984. Sexual behaviour of the Chinese praying mantis. *Animal Behaviour* **32**, 916–917.

Lobel, P.S. 1989. Spawning behavior of *Chaetodon multicinctus* (Chaetodontidae): pairs and intruders. *Environmental Biology of Fishes* **25**, 125–130.

Longenecker, K. 2007. Devil in the details: high-resolution dietary analysis contradicts a basic assumption of reef-fish diversity models. *Copeia* **3**, 543–555.

Longenecker, K. & Langston, R. 2006. Life history characteristics of a small cardinalfish, *Ostorhinchus rubrimacula* (Percoidei: Apogonidae), from Koro, Fiji. *Pacific Science* **60**, 225–233.
López-Peralta, R.H. & Arcila, C.A.T. 2002. Diet composition of fish species from the southern continental shelf of Colombia. *Naga WorldFish Center Quarterly* **25**, 23–29.
Lourie, S.A. & Randall, J.E. 2003. A new pygmy seahorse, *Hippocampus denise* (Teleostei: Syngnathidae) from the Indo-Pacific. *Zoological Studies* **42**, 284–291.
Mackereth, R.W. & Keenleyside, M.H.A. 1993. Breeding territoriality and pair-formation in the convict cichlid (*Cichlasoma nigrofasciatum*, Pisces, Cichlidae). *Canadian Journal of Zoology* **71**, 960–967.
Madin, E.M.P., Gaines, S.D., Madin, J.S., Link, A.K., Lubchenco, P.J., Selden, R.L. & Warner, R.R. 2012. Do behavioral foraging responses of prey to predators function similarly in restored and pristine foodwebs? *PLoS ONE* **7**, doi:10.1371/journal.pone.0032390
Mantyka, C.S. & Bellwood, D.R. 2007. Macroalgal grazing selectivity among herbivorous coral reef fishes. *Marine Ecology Progress Series* **352**, 177–185.
Marguillier, S., van der Velde, G., Dehairs, F., Hemminga, M.A. & Rajagopal, S. 1997. Trophic relationships in an interlinked mangrove-seagrass ecosystem as traced by $\delta^{13}C$ and $\delta^{15}N$. *Marine Ecology Progress Series* **151**, 115–121.
Marnane, M.J. & Bellwood, D.R. 2002. Diet and nocturnal foraging in cardinalfishes (Apogonidae) at one tree reef, Great Barrier Reef, Australia. *Marine Ecology Progress Series* **231**, 261–268.
Mathews, L.M. 2002. Territorial cooperation and social monogamy: factors affecting intersexual interactions in pair-living snapping shrimp. *Animal Behaviour* **63**, 767–777.
McCormick, M.I. 1995. Fish feeding on mobile benthic invertebrates: influence of spatial variability in habitat associations. *Marine Biology* **121**, 627–637.
Mehta, R.S. 2009. Ecomorphology of the moray bite: relationship between dietary extremes and morphological diversity. *Physiological and Biochemical Zoology* **82**, 90–103.
Mequila, A.T. & Campos, W. 2007. Feeding relationships of dominant fish species in the Visayan Sea. *Science Diliman* **19**, 35–46.
Metian, M., Warnau, M., Chouvelon, T., Pedraza, F., Rodriguez y Baena, A. & Bustamante, P. 2013. Trace element bioaccumulation in reef fish from New Caledonia: influence of trophic groups and risk assessment for consumers. *Marine Environmental Research* doi:10.1016/j.marenvres.2013.03.001
Meyer, C.G., Holland, K.N., Wetherbee, B.M. & Lowe, C.G. 2001. Diet, resource partitioning and gear vulnerability of Hawaiian jacks captured in fishing tournaments. *Fisheries Research* **53**, 105–113.
Meyer-Rochow, V.B. 1976. Some observations on spawning and fecundity in the luminescent fish *Photoblepharon palpebratus*. *Marine Biology* **37**, 325–328.
Mitamura, H., Arai, N., Mitsunaga, Y., Yokota, T., Takeuchi, H., Tsuzaki, T. & Itani, M. 2005. Directed movements and diel burrow fidelity patterns of red tilefish *Branchiostegus japonicus* determined using ultrasonic telemetry. *Fisheries Science* **71**, 491–498.
Montgomery, W.L., Myrberg, A.A. & Fishelson, L. 1989. Feeding ecology of surgeonfishes (Acanthuridae) in the northern Red Sea, with particular reference to *Acanthurus nigrofuscus* (Forsskål). *Journal of Experimental Marine Biology and Ecology* **132**, 179–207.
Mooi, R.D. & Gill, A.C. 2004. Notograptidae, sister to *Acanthoplesiops* Regan (Teleostei: Plesiopidae: Acanthoclininae), with comments on biogeography, diet and morphological convergence with Congrogadinae (Teleostei: Pseudochromidae). *Zoological Journal of the Linnean Society* **141**, 179–205.
Morgan, I.E. & Kramer, D.L. 2004. The social organization of adult blue tangs, *Acanthurus coeruleus*, on a fringing reef, Barbados, West Indies. *Environmental Biology of Fishes* **71**, 261–273.
Motta, P.J. 1988. Functional morphology of the feeding apparatus of ten species of Pacific butterflyfishes (Perciformes, Chaetodontidae): an ecomorphological approach. *Environmental Biology of Fishes* **22**, 39–67.
Moyer, J.T. & Nakazono, A. 1978. Population structure, reproductive behavior and protogynous hermaphroditism in the angelfish *Centropyge interruptus* at Miyake-jima, Japan. *Japanese Journal of Ichthyology* **25**, 25–39.
Moyer, J.T. & Sano, M. 1987. Feeding habits of two sympatric ostraciid fishes at Miyake-jima, Japan. *Ichthyological Research* **34**, 108–112.
Munday, P.L., Pierce, S.J., Jones, G.P. & Larson, H.K. 2002. Habitat use, social organization and reproductive biology of the seawhip goby, *Bryaninops yongei*. *Marine and Freshwater Research* **53**, 769–775.
Mundy, C.N. 2000. An appraisal of methods used in coral recruitment studies. *Coral Reefs* **19**, 124–131.

Muñoz, G., Grutter, A.S. & Cribb, T.H. 2006. Endoparasite communities of five fish species (Labridae: Cheilininae) from Lizard Island: how important is the ecology and phylogeny of the hosts? *Parasitology* **132**, 363–374.

Muñoz, R.C., Currin, C.A. & Whitfield, P.E. 2011. Diet of invasive lionfish on hard bottom reefs of the southeast USA: insights from stomach contents and stable isotopes. *Marine Ecology Progress Series* **432**, 181–193.

Nakamura, Y., Horinouchi, M., Nakai, T. & Sano, M. 2003. Food habits of fishes in a seagrass bed on a fringing coral reef at Iriomote Island, southern Japan. *Ichthyological Research* **50**, 15–22.

Nakane, Y., Yusuke, S. & Mitsuhiko, S. 2011. Food habits of fishes on an exposed sandy beach at Fukiagehama, South-West Kyushu Island, Japan. *Helgoland Marine Research* **65**, 123–131.

Nanami, A. & Yamada, H. 2008. Size and spatial arrangement of home range of checkered snapper *Lutjanus decussatus* (Lutjanidae) in an Okinawan coral reef determined using a portable GPS receiver. *Marine Biology* **153**, 1103–1111.

Nanjo, K., Kohno, H. & Sano, M. 2008. Food habits of fishes in the mangrove estuary of Urauchi River, Iriomote Island, southern Japan. *Fisheries Science* **74**, 1024–1033.

Near, T.J., Eytan, R.I., Dornburg, A., Kuhn, K.L., Moore, J.A., Davis, M.P., Wainwright, P.C., Friedman, M. & Smith, W.L. 2012. Resolution of ray-finned fish phylogeny and timing of diversification. *Proceedings of the National Academy of Sciences of the United States of America* **109**, 13698–13703.

Neudecker, S. & Lobel, P.S. 1982. Mating systems of chaetodontid and pomacanthid fishes at St. Croix. *Zeitschrift für Tierpsychologie* **59**, 299–318.

Öhman, M.C., Munday, P.L., Jones, G.P. & Caley, M.J. 1998. Settlement strategies and distribution patterns of coral-reef fishes. *Journal of Experimental Marine Biology and Ecology* **225**, 219–238.

Orians, G.H. 1961. The ecology of blackbird (*Agelaius*) social systems. *Ecological Monographs* **31**, 285–312.

Orians, G.H. 1969. On the evolution of mating systems in birds and mammals. *American Naturalist* **103**, 589–603.

Parrish, J.K. & Edelstein-Keshet, L. 1999. Complexity, pattern and evolutionary trade-offs in animal aggregation. *Science* **284**, 99–101.

Petersen, C.W., Warner, R.R., Shapiro, D.Y. & Marconato, A. 2001. Components of fertilization success in the bluehead wrasse, *Thalassoma bifasciatum*. *Behavioral Ecology* **12**, 237–245.

Pietsch, T.W. & Grobecker, D.B. 1987. *Frogfishes of the World: Systematics, Zoogeography, and Behavioral Ecology.* Stanford, California: Stanford University Press.

Pitcher, T.J. & Parrish, J.K. 1993. Functions of shoaling behaviour in teleosts. In *Behaviour of Teleost Fishes*, T.J. Pitcher (ed.). London: Chapman and Hall, 363–439.

Pratchett, M.S. 2005. Dietary overlap among coral-feeding butterflyfishes (Chaetodontidae) at Lizard Island, northern Great Barrier Reef. *Marine Biology* **148**, 373–382.

Pratchett, M.S., Gust, N., Goby, G. & Klanten, S.O. 2001. Consumption of coral propagules represents a significant trophic link between corals and reef fish. *Coral Reefs* **20**, 13–17.

Pratchett, M.S., Pradjakusuma, O.A. & Jones, G.P. 2006. Is there a reproductive basis to solitary living versus pair-formation in coral reef fishes? *Coral Reefs* **25**, 85–92.

Pyle, R.L. 2003. *A systematic treatment of the reef-fish family Pomacanthidae (Pisces: Perciformes).* PhD thesis, University of Hawaii, United States of America.

Randall, H.A. & Allen, G.R. 1977. A revision of the damselfish genus *Dascyllus* (Pomacentridae) with the description of a new species. *Records of the Australian Museum* **31**, 349–385.

Randall, J.E. 1980. A survey of ciguatera at Eniwetak and Bikini, Marshall Islands, with notes on the systematics and food habits of ciguatoxic fishes. *Fishery Bulletin* **78**, 201–249.

Randall, J.E. 1998. Revision of the Indo-Pacific squirrelfishes (Beryciformes: Holocentridae: Holocentrinae) of the genus *Sargocentron*, with descriptions of four new species. *Indo-Pacific Fishes* **27**, 1–105.

Randall, J.E., 2004. Revision of the goatfish genus *Parupeneus* (Perciformes: Mullidae), with descriptions of two new species. *Indo-Pacific Fishes* **36**, 64.

Randall, J.E. 2005. *Reef and Shore Fishes of the South Pacific: New Caledonia to Tahiti and the Pitcairn Islands.* Honolulu, Hawaii: University of Hawaii Press.

Randall, J.E., Allen, G.R. & Steene, R.C. 1997. *Fishes of the Great Barrier Reef and Coral Sea.* Bathurst, Australia: Crawford House.

Randall, J.E. & Goren, M. 1993. A review of the gobioid fishes of the Maldives. *Ichthyological Bulletin of the J.L.B. Smith Institute of Ichthyology* **58**, 1–37.

Randall, J.E. & Myers, R.F. 2002. *Parupeneus insularis*, a new central Pacific species of goatfish (Perciformes: Mullidae) of the *P. trifasciatus* complex. *Zoological Studies Taipei* **41**, 431–440.

Randall, J.E. & Struhsaker, P. 1981. *Naso maculatus*, a new species of acanthurid fish from the Hawaiian Islands and Japan. *Copeia* **3**, 553–558.

Reavis, R.H. 1997a. The natural history of a monogamous coral-reef fish, *Valenciennea strigata* (Gobiidae): 1. Abundance, growth, survival and predation. *Environmental Biology of Fishes* **49**, 239–246.

Reavis, R.H. 1997b. The natural history of a monogamous coral reef fish, *Valenciennea strigata* (Gobiidae): 2. Behaviour, mate fidelity and reproductive success. *Environmental Biology of Fishes* **49**, 247–257.

Reavis, R.H. & Barlow, G.W. 1998. Why is the coral reef fish *Valenciennea strigata* (Gobiidae) monogamous? *Behavioral Ecology and Sociobiology* **43**, 229–237.

Reavis, R.H. & Copus, J.M. 2011. Monogamy in a feeding generalist, *Chaetodon trichrous*, the endemic Tahitian butterflyfish. *Environmental Biology of Fishes* **92**, 167–179.

Reece, J.S., Bowen, B.W., Smith, D.G. & Larson, A.F. 2010. Molecular phylogenetics of moray eels (Muraenidae) demonstrates multiple origins of a shell-crushing jaw (*Gymnomuraena*, *Echidna*) and multiple colonizations of the Atlantic Ocean. *Molecular Phylogenetics and Evolution* **57**, 829–835.

Reese, E.S. 1975. A comparative field study of the social behaviour and related ecology of reef fishes of the family Chaetodontidae. *Zeitschrift für Tierpyschologie* **37**, 37–61.

Reynolds, J.D. 1996. Animal breeding systems. *Trends in Ecology and Evolution* **11**, 68–72.

Rimmer, D.W. & Wiebe, W.J. 1987. Fermentative microbial digestion in herbivorous fishes. *Journal of Fish Biology* **31**, 229–236.

Roberts, C.M. 1987. Experimental analysis of resource sharing between herbivorous damselfish and blennies on the Great Barrier Reef. *Journal of Experimental Marine Biology and Ecology* **111**, 61–75.

Roberts, C.M. & Ormond, R.F.G. 1992. Butterflyfish social behaviour, with special reference to the incidence of territoriality: a review. *Environmental Biology of Fishes* **34**, 79–93.

Robertson, D.R. 1973. Field observation on the reproductive behaviour of a pomacentrid fish, *Acanthochromis polyacanthus*. *Zeitschrift für Tierpsychologie* **32**, 319–324.

Robertson, D.R. 1983. On the spawning behavior and spawning cycles of eight surgeonfishes (Acanthuridae) from the Indo-Pacific. *Environmental Biology of Fishes* **9**, 193–223.

Robertson, D.R. & Gaines, S.D. 1986. Interference competition structures habitat use in a local assemblage of coral reef surgeonfishes. *Ecology* **67**, 1372–1383.

Robertson, D.R. & Hoffman, S.G. 1977. The roles of female mate choice and predation in the mating systems of some tropical labroid fishes. *Zeitschrift für Tierpsychologie* **45**, 298–320.

Robertson, D.R., Polunin, N.V.C. & Leighton, K. 1979. The behavioural ecology of three Indian Ocean surgeonfishes (*Acanthurus lineatus*, *A. leucosternon* and *Zebrasoma scopas*): their feeding strategies, and social and mating systems. *Environmental Biology of Fishes* **4**, 125–170.

Robertson, D.R., Sweatman, H.P.A., Fletcher, E.A. & Cleland, M.G. 1976. Schooling as a mechanism for circumventing the territoriality of competitors. *Ecology* **57**, 1208–1220.

Ross, J.L. 1982. Feeding habits of the gray tilefish, *Caulolatilus microps* (Goode and Bean, 1878) from North Carolina and South Carolina waters. *Bulletin of Marine Science* **32**, 448–454.

Russ, G.R. 1984. Distribution and abundance of herbivorous grazing fishes in the central Great Barrier Reef. I. Levels of variability across the entire continental shelf. *Marine Ecology Progress Series* **20**, 23–34.

Russ, G.R. 1987. Is rate of removal of algae by grazers reduced inside territories of tropical damselfishes? *Journal of Experimental Marine Biology and Ecology* **110**, 1–17.

Russell, B.C. 1990. *Nemipterid Fishes of the World (Threadfin Breams, Whiptail Breams, Monocle Breams, Dwarf Monocle Breams, and Coral Breams): Family Nemipteridae: An Annotated and Illustrated Catalogue of Nemipterid Species Known to Date*. FAO Fisheries Synopsis No. 125, Volume 12. Rome: Food and Agriculture Organization of the United Nations.

Sabrah, M.M. & El-Ganainy, A. 2009. Observation on biological traits of striped goatfish (*Upeneus vittatus*) and freckled goatfish (*Upeneus tragula*) from the Gulf of Suez, Egypt. *World Journal of Fish and Marine Sciences* **1**, 121–128.

Sabrah, M.M., El-Ganainy, A.A. & Zaky, M.A. 2006. Biology and toxicity of the pufferfish *Lagocephalus sceleratus* (Gmelin, 1789) from the Gulf of Suez. *Egyptian Journal of Aquatic Research* **32**, 283–297.

Sackley, P.G. & Kaufman, L.S. 1996. Effect of predation on foraging height in a planktivorous coral reef fish, *Chromis nitida*. *Copeia* **3**, 726–729.

Sadovy de Mitcheson, Y. & Colin, P.L. (eds) 2012. *Reef Fish Spawning Aggregations: Biology, Research and Management*. Fish and Fisheries Series 35. Dordrecht, the Netherlands: Springer.

Sadovy de Mitcheson, Y., Cornish, A., Domeier, M., Colin, P.L., Russel, M. & Lindeman, K.C. 2008. A global baseline for spawning aggregations of reef fishes. *Conservation Biology* **22**, 1233–1244.

Sakashita, H. 1992. Sexual dimorphism and food habits of the clingfish, *Diademichthys lineatus*, and its dependence on host sea urchin. *Environmental Biology of Fishes* **34**, 95–101.

Salini, J.P., Blaber, S.J.M. & Brewer, D.T. 1990. Diets of piscivorous fishes in a tropical Australian estuary, with special reference to predation on penaeid prawns. *Marine Biology* **105**, 363–374.

Salini, J.P., Blaber, S.J.M. & Brewer, D.T. 1994. Diets of trawled predatory fish of the Gulf of Carpentaria, Australia, with particular reference to predation on prawns. *Marine and Freshwater Research* **45**, 397–411.

Samoilys, M.A. & Squire, L.C. 1994. Preliminary observations on the spawning behaviour of coral trout, *Plectropomus leopardus* (Pisces: Serranidae), on the Great Barrier Reef. *Bulletin of Marine Science* **54**, 332–342.

Sancho, G., Petersen, C.W. & Lobel, P.S. 2000. Predator-prey relations at a spawning aggregation site of coral reef fishes. *Marine Ecology Progress Series* **203**, 275–288.

Sandin, S.A. & Williams, I. 2010. Trophic classifications of reef fishes from the tropical U.S. Pacific (Version 1.0). San Diego, California: Scripps Institution of Oceanography. Online. http://escholarship.org/uc/item/5394f7m3 (accessed May 2013).

Sano, M. 1989. Feeding habits of Japanese butterflyfishes (Chaetodontidae). *Environmental Biology of Fishes* **25**, 195–203.

Sano, M., Shimizu, M. & Nose, Y. 1984. Food habits of teleostean reef fishes in Okinawa Island, southern Japan. *The University of Tokyo Bulletin* **25**, 1–128.

Schmitz, L. & Wainwright, P.C. 2011. Nocturnality constrains morphological and functional diversity in the eyes of reef fishes. *BMC Evolutionary Biology* **11**, 338.

Shallenberger, R.J. & Madden, W.D. 1973. Luring behavior in the scorpionfish, *Iracundus signifer*. *Behaviour* **47**, 1–2.

Sikkel, P.C. 1990. Social organisation and spawning in the Atlantic sharpnose puffer, *Canthigaster rostrata* (Tetraodontidae). *Environmental Biology of Fishes* **27**, 243–254.

Sikkel, P.C. & Sikkel, N.M. 2012. First report of spawning and social organization in Hawai'ian Ambon Toby, *Canthigaster amboinensis*. *Ichthyological Research* **59**, 394–395.

Silberschneider, V. & Booth, D.J. 2001. Resource use by *Enneapterygius rufopileus* and other rockpool fishes. *Environmental Biology of Fishes* **61**, 195–204.

Sivakumar, R. & Ramaiyan, V. 1987. The morphology of the alimentary tract in relation to food of platycephalids of Porto Novo Coast (Order: Scorpaeniformes). Kochi, India: *CMFRI Bulletin National Symposium on Research and Development in Marine Fisheries Sessions I & II* **44**, 262–266.

Smith, G.C. & Parrish, J.D. 2002. Estuaries as nurseries for the jacks *Caranx ignobilis* and *Caranx melampygus* (Carangidae) in Hawaii. *Estuarine, Coastal and Shelf Science* **55**, 347–359.

Sogabe, A., Matsumoto, K. & Yanagisawa, Y. 2007. Mate change reduces the reproductive rate of males in a monogamous pipefish *Corythoichthys haematopterus*. *Ethology* **113**, 764–771.

Sogabe, A. & Yanagisawa, Y. 2008. Maintenance of pair bond during the non-reproductive season in a monogamous pipefish *Corythoichthys haematopterus*. *Journal of Ethology* **26**, 195–199.

Sreenivasan, P.V. 1978. Observations on the fishery and biology of *Megalaspis cordyla* (Linnaeus) at Vizhinjam. *Indian Journal of Fisheries* **25**, 122–140.

Stella, J.S., Pratchett, M.S., Hutchings, P.A. & Jones, G.P. 2011. Coral-associated invertebrates: diversity, ecological importance and vulnerability to disturbance. *Oceanography and Marine Biology: An Annual Review* **49**, 43–104.

Stewart, B.D. & Jones, G.P. 2001. Associations between the abundance of piscivorous fishes and their prey on coral reefs: implications for prey-fish mortality. *Marine Biology* **138**, 383–397.

Suzuki, K., Tanaka, Y. & Hioki, S. 1980. Spawning behaviour, eggs, and larvae of the butterflyfish, *Chaetodon nippon*, in an aquarium. *Japanese Journal of Ichthyology* **26**, 334–341.

Sweatman, H.P.A. 1993. Tropical snapper (Lutjanidae) that is piscivorous at settlement. *Copeia* **4**, 1137–1139.

Takegaki, T. 2000. Monogamous mating system and spawning cycle in the gobiid fish, *Amblygobius phalaena* (Gobiidae). *Environmental Biology of Fishes* **59**, 61–67.

Takegaki, T. & Nakazono, A. 1999. Division of labour in the monogamous goby, *Valenciennea longipinnis*, in relation to burrowing behaviour. *Ichthyological Research* **46**, 125–129.

Tamaki, A., Miyamoto, S., Yamazaki, T. & Nojima, S. 1992. Abundance pattern of the ghost shrimp *Callianassa japonica* Ortmann (Thalassinidea) and the snake eel *Pisodonophis cancrivorus* Richardson (Pisces, Ophichthidae) and their possible interaction on an intertidal sand flat. *Plankton and Benthos Research* **43**, 11–22.

Tandon, K.K. 1960. The food and feeding habits of *Selaroides leptolepis* (Cuvier and Valenciennes). *Current Science* **29**, 62–63.

Taquet, M., Sancho, G., Dagorn, L., Gaertner, J.C., Itano, D., Aumeeruddy, R. & Peignon, C. 2007. Characterizing fish communities associated with drifting fish aggregating devices (FADs) in the Western Indian Ocean using underwater visual surveys. *Aquatic Living Resources* **20**, 331–341.

Thompson, A.R. 2004. Habitat and mutualism affect the distribution and abundance of a shrimp-associated goby. *Marine and Freshwater Research* **55**, 105–113.

Thompson, A.R. 2005. Dynamics of demographically open mutualists: immigration, intraspecific competition, and predation impact goby populations. *Oecologia* **143**, 61–69.

Thornhill, R. 1979. Male pair-formation pheromones in *Panorpa* scorpionflies (Mecoptera: Panorpidae). *Environmental Entomology* **8**, 886–888.

Thresher, R.E. 1984. *Reproduction in Reef Fishes*. Neptune City, New Jersey: T.F.H. Publications.

Townsend, K.A. & Tibbetts, I.R. 2000. Biomass and distribution of herbivorous blennies in the southern Great Barrier Reef. *Journal of Fish Biology* **56**, 774–791.

Tribble, G.W. & Nishikawa, H. 1982. An analysis of the diets of four spatially overlapping damselfishes of the genus *Chromis*. *Japanese Journal of Ichthyology* **29**, 267–271.

Unsworth, R.K.F., Garrard, S.L., Salinas De León, P., Cullen, L.C., Smith, D.J., Sloman, K.A. & Bell, J.J. 2009. Structuring of Indo-Pacific fish assemblages along the mangrove-seagrass continuum. *Aquatic Biology* **5**, 85–95.

Unsworth, R.K.F., Wylie, E., Smith, D.J. & Bell, J.J. 2007. Diel trophic structuring of seagrass bed fish assemblages in the Wakatobi Marine National Park, Indonesia. *Estuarine, Coastal and Shelf Science* **72**, 81–88.

Vincent, A.C.J. & Sadler, L.M. 1995. Faithful pair bonds in wild seahorses, *Hippocampus whitei*. *Animal Behaviour* **50**, 1557–1569.

Wantiez, L. & Kulbicki, M. 1995. Main fish populations and their relation to the benthos in a silted Bay of New Caledonia, as determined by visual censuses. *Cybium* **19**, 223–240.

Welsh, J.Q. & Bellwood, D.R. 2012a. Spatial ecology of the steephead parrotfish (*Chlorurus microrhinos*): an evaluation using acoustic telemetry. *Coral Reefs* **31**, 55–65.

Welsh, J.Q. & Bellwood, D.R. 2012b. How far do schools of roving herbivores rove? A case study using *Scarus rivulatus*. *Coral Reefs* **31**, 991–1003.

Wenger, A.S., Johansen, J.L. & Jones, G.P. 2012. Increasing suspended sediment reduces foraging, growth and condition of a planktivorous damselfish. *Journal of Experimental Marine Biology and Ecology* **428**, 43–48.

Westneat, D.F. & Stewart, I.R. 2003. Extra-pair paternity in birds: causes, correlates, and conflict. *Annual Review of Ecology, Evolution, and Systematics* **34**, 365–396.

Westneat, M.W., Alfaro, M.E., Wainwright, P.C., Bellwood, D.R., Grubich, J.R., Fessler, J.L., Clements, K.D. & Smith, L.L. 2005. Local phylogenetic divergence and global evolutionary convergence of skull function in reef fishes of the family Labridae. *Proceedings of the Royal Society B* **272**, 993–1000.

White, G.E. & Brown, C. 2013. Site fidelity and homing behaviour in intertidal fishes. *Marine Biology* doi:10.1007/s00227-013-2188-6

White, J.W. & Warner, R.R. 2007a. Behavioral and energetic costs of group membership in a coral reef fish. *Oecologia* **154**, 423–433.

White, J.W. & Warner, R.R. 2007b. Safety in numbers and the spatial scaling of density-dependent mortality in a coral reef fish. *Ecology* **88**, 3044–3054.

Whiteman, E.A. & Côté, I.M. 2003. Social monogamy in the cleaning goby *Elacatinus evelynae*: ecological constraints or net benefit? *Animal Behaviour* **66**, 281–291.

Whiteman, E.A. & Côté, I.M. 2004. Monogamy in marine fishes. *Biological Reviews* **79**, 351–375.

Wilcox, R.S. 1984. Male copulatory guarding enhances female foraging in a water strider. *Behavioral Ecology and Sociobiology* **15**, 171–174.

Williams, D.M. & Hatcher, A.I. 1983. Structure of fish communities on outer slopes of inshore, mid shelf and outer shelf reefs of the Great Barrier Reef. *Marine Ecology Progress Series* **10**, 239–250.
Williams, L.B. & Williams, E.H., Jr. 1986. Icthyological notes about fishes collected for parasite examination around Sesoko Island, Okinawa. *Galaxea* **5**, 217–221.
Wilson, S.K. 2000. Trophic status and feeding selectivity of blennies (Blenniidae: Salariini). *Marine Biology* **136**, 431–437.
Winterbottom, R. 1984. A review of the gobiid fish genus *Trimma* from the Chagos Archipelago, central Indian Ocean, with the description of seven new species. *Canadian Journal of Zoology* **62**, 695–715.
Wittenberger, J.F. & Tilson, R.L. 1980. The evolution of monogamy. *Annual Review of Ecology, Evolution, and Systematics* **11**, 197–232.
Wong, M.Y.L., Munday, P.L. & Jones, G.P. 2005. Habitat patch size, facultative monogamy and sex change in a coral-dwelling fish, *Caracanthus unipinna. Environmental Biology of Fishes* **74**, 141–150.
Woodland, D.J. 1990. *Revision of the Fish Family Siganidae with Descriptions of Two New Species and Comments on Distribution and Biology.* Honolulu, Hawaii: Bishop Museum.
Yabuta, S. 1997. Spawning migrations of the monogamous butterflyfish *Chaetodon trifasciatus. Ichthyological Research* **44**, 177–182.
Yabuta, S. 2007. Social groupings in 18 species of butterflyfish and pair bond weakening during the nonreproductive season. *Ichthyological Research* **54**, 207–210.
Yabuta, S. & Kawashima, M. 1997. Spawning behavior and haremic mating system in the corallivorous butterflyfish, *Chaetodon trifascialis*, at Kuroshima Island, Okinawa. *Ichthyological Research* **44**, 183–188.
Yahner, R.H. 1978. The adaptive nature of the social system and behavior in the eastern chipmunk, *Tamias striatus. Behavioral Ecology and Sociobiology* **3**, 397–427.
Young, L.C., Zaun, B.J. & Vander Werf, E.A. 2008. Successful same-sex pairing in Laysan albatross. *Biology Letters* **4**, 323–325.
Zagars, M., Ikejima, K., Arai, N. & Tongnunui, P. 2013. Trophic characteristics of a mangrove fish community in Southwest Thailand: important mangrove contribution and intraspecies feeding variability. *Estuarine, Coastal and Shelf Science* doi:10.1016/j.ecss.2013.01.005

Appendix

List of Indo-Pacific reef fishes and their social system, trophic affiliation, burrowing behaviour, activity patterns, size, and reproductive traits

Family	Species	Pairs	Troph.	Burr.	Act.	ML (mm)	Spa.	Gam.	Mon.	Refs.
Acanthuridae	*Acanthurus achilles*	Yes	He	No	D	200	G	Br	Yes	5
	Acanthurus albipectoralis	No	Pl	No	D	330	G	Br	No	6
	Acanthurus auranticavus	No	De	No	D	350	G	Br	No	6
	Acanthurus bariene	Yes	De	No	D	500	G	Br	No	6
	Acanthurus blochii	No	De	No	D	420	G	Br	No	6
	Acanthurus dussumieri	No	De	No	D	500	G	Br	No	7
	Acanthurus fowleri	Yes	He	No	D	450	G	Br	No	2
	Acanthurus grammoptilus	No	De	No	D	350	G	Br	No	6
	Acanthurus guttatus	No	He	No	D	280	B	Br	No	1, 3
	Acanthurus japonicus	No	He	No	D	210	G	Br	No	2
	Acanthurus leucocheilus	No	De	No	D	200	G	Br	No	2
	Acanthurus leucopareius	No	He	No	D	200	G	Br	No	1, 3
	Acanthurus leucosternon	Yes	He	No	D	380	G	Br	Yes	8
	Acanthurus lineatus	No	He	No	D	380	G	Br	No	6
	Acanthurus maculiceps	No	He	No	D	200	G	Br	No	6
	Acanthurus mata	No	Pl	No	D	500	G	Br	No	1, 2, 3
	Acanthurus nigricans	Yes	He	No	D	210	G	Br	Yes	5
	Acanthurus nigricauda	No	De	No	D	400	G	Br	No	6, 7
	Acanthurus nigrofuscus	No	He	No	D	210	B	Br	No	8, 9
	Acanthurus nigroris	No	He	No	D	250	G	Br	No	1, 2, 3
	Acanthurus nubilus	No	Pl	No	D	450	G	Br	No	2, 3, 4
	Acanthurus olivaceus	No	De	No	D	350	G	Br	No	7
	Acanthurus pyroferus	No	He	No	D	250	G	Br	No	6
	Acanthurus reversus	No	De	No	D	340	G	Br	No	3
	Acanthurus tennentii	Yes	He	No	D	310	G	Br	No	8
	Acanthurus thompsoni	No	Pl	No	D	270	G	Br	No	1, 2, 3
	Acanthurus triostegus	Yes	He	No	D	260	G	Br	No	6, 8
	Acanthurus tristis	No	He	No	D	250	G	Br	No	2
	Acanthurus xanthopterus	No	De	No	D	560	G	Br	No	1, 2, 3
	Ctenochaetus binotatus	No	De	No	D	220	G	Br	No	2, 3, 4
	Ctenochaetus cyanocheilus	No	He	No	D	200	G	Br	No	2, 3
	Ctenochaetus flavicauda	No	De	No	D	130	G	Br	No	2
	Ctenochaetus hawaiiensis	No	De	No	D	250	G	Br	No	2, 3
	Ctenochaetus marginatus	No	De	No	D	220	G	Br	No	2, 3
	Ctenochaetus striatus	No	De	No	D	260	B	Br	No	6, 8
	Ctenochaetus strigosus	No	De	No	D	180	P	Br	No	1, 8
	Ctenochaetus tominiensis	No	De	No	D	150	G	Br	No	2, 3
	Naso annulatus	No	Pl	No	D	1000	G	Br	No	1, 2, 3
	Naso brachycentron	No	He	No	D	600	G	Br	No	1, 2, 3, 4
	Naso brevirostris	No	Pl	No	D	500	G	Br	No	1, 2, 3
	Naso caeruleacauda	No	Pl	No	D	300	G	Br	No	2, 3
	Naso caesius	No	Pl	No	D	600	G	Br	No	1, 2, 3

Continued

List of Indo-Pacific reef fishes and their social system, trophic affiliation, burrowing behaviour, activity patterns, size, and reproductive traits (Continued)

Family	Species	Pairs	Troph.	Burr.	Act.	ML (mm)	Spa.	Gam.	Mon.	Refs.
	Naso hexacanthus	No	Pl	No	D	750	G	Br	No	1, 2, 3
	Naso lituratus	Yes	He	No	D	300	G	Br	No	6, 8, 9
	Naso lopezi	No	Pl	No	D	650	G	Br	No	1, 2, 3
	Naso maculatus	No	Pl	No	D	600	G	Br	No	3, 10
	Naso minor	No	Pl	No	D	190	G	Br	No	2, 4
	Naso thynnoides	No	Pl	No	D	300	G	Br	No	1, 2, 3
	Naso tonganus	No	He	No	D	600	G	Br	No	2, 3, 4
	Naso tuberosus	No	He	No	D	600	G	Br	No	1, 2, 3, 6
	Naso unicornis	No	He	No	D	700	G	Br	No	1, 2, 3, 6, 8
	Naso vlamingii	No	Pl	No	D	500	G	Br	No	1, 2, 3
	Paracanthurus hepatus	No	Pl	No	D	310	G	Br	No	1, 2, 3, 6
	Prionurus maculatus	No	He	No	D	435	G	Br	No	1, 3
	Prionurus microlepidotus	No	He	No	D	700	P	Br	No	1
	Zebrasoma desjardinii	Yes	He	No	D	400	P	Br	No	2, 4
	Zebrasoma flavescens	No	He	No	D	200	B	Br	No	2, 4
	Zebrasoma rostratum	No	He	No	D	210	P	Br	No	2, 3
	Zebrasoma scopas	Yes	He	No	D	200	B	Br	Yes	6, 8, 9
	Zebrasoma velifer (= *veliferum*)	Yes	He	No	D	400	P	Br	No	6, 8, 9
Anomalopidae	*Anomalops katoptron*	No	Pl	No	N	350	P	Br	No	4, 11
	Photoblepharon palpebratum	No	Pl	No	N	120	P	Br	No	4
Antennariidae	*Antennarius biocellatus*	No	Pi	No	D	150	P	Gel	No	2, 4
	Antennarius commerson	No	Pi	No	D	300	P	Gel	No	1, 2, 3, 4
	Antennarius maculatus	No	Pi	No	D	90	P	Gel	No	1, 2, 3, 13
	Antennarius pictus	No	Pi	No	D	160	P	Gel	No	1, 2, 3
	Antennarius randalli	No	Pi	No	D	45	P	Gel	No	1, 2, 3
	Antennarius striatus	No	Pi	No	D	220	P	Gel	No	1, 2, 3
	Antennatus (= *Antennarius*) *coccineus*	No	Pi	No	D	120	P	Gel	No	1, 2, 3, 12
	Antennatus (= *Antennarius*) *dorehensis*	No	Pi	No	D	50	P	Gel	No	2
	Antennatus (= *Antennarius*) *nummifer*	No	Pi	No	D	100	P	Gel	No	1, 2, 3
	Antennatus rosaceus	No	Pi	No	D	58	P	Gel	No	3
	Antennatus tuberosus	No	Pi	No	D	80	P	Gel	No	2
	Histiophryne cryptacanthus	No	Pi	No	D	85	P	Gel	No	1, 2, 3
	Histrio histrio	No	Ca	No	D	140	P	Gel	No	1, 2, 3, 4
	Lophiocharon trisignatus	No	Pi	No	D	180	P	Gel	No	1, 2, 4
	Tathicarpus butleri	No	Pi	No	D	120	P	Gel	No	2
Apistidae	*Apistus carinatus*	No	Pl	No	D	88	N/A	N/A	No	2, 14
Aploactinidae	*Paraploactis kagoshimensis*	No	N/A	No	D	120	N/A	N/A	No	2
Apogonidae	*Apogon amboinensis*	No	Ca	No	D	100	P	Mo	No	2, 15
	Apogon (= *Ostorhinchus*) *capricornis*	Yes	N/A	No	D	100	P	Mo	No	2

List of Indo-Pacific reef fishes and their social system, trophic affiliation, burrowing behaviour, activity patterns, size, and reproductive traits (Continued)

Family	Species	Pairs	Troph.	Burr.	Act.	ML (mm)	Spa.	Gam.	Mon.	Refs.
	Apogon (= *Ostorhinchus*) *crassiceps*	No	Ma	No	N	60	P	Mo	No	1, 2, 3, 4
	Apogon (= *Ostorhinchus*) *doederleini*	Yes	Ca	No	N	120	P	Mo	No	1, 2, 3, 16
	Apogon hyalosoma	No	Ca	No	N	200	P	Mo	No	2, 17
	Apogon norfolcensis	No	Ca	No	N	100	P	Mo	No	2, 3, 4
	Apogon (= *Ostorhinchus*) *notatus*	No	Pl	No	N	100	P	Mo	Yes	1, 2, 23
	Apogon sangiensis	No	Ca	No	N	80	P	Mo	No	1, 4
	Apogon semiornatus	No	Ca	No	N	70	P	Mo	No	1, 3, 4
	Apogon (= *Ostorhinchus*) *semiornatus*	No	N/A	No	N	70	P	Mo	No	2, 3
	Apogonichthyoides melas	No	Ca	No	D	130	P	Mo	No	2, 4
	Apogonichthyoides timorensis	No	Ca	No	N	80	P	Mo	No	2, 4
	Apogonichthyoides uninotatus	No	Ca	No	D	60	P	Mo	No	2
	Apogonichthys ocellatus	No	Ca	No	N	35	P	Mo	No	2, 4
	Archamia biguttata	No	Pl	No	N	90	P	Mo	No	1, 2, 3, 12
	Archamia bleekeri	No	Pl	No	N	90	P	Mo	No	2, 4
	Archamia fucata	No	Pl	No	N	90	P	Mo	No	1, 2, 3, 18
	Archamia leai	No	Pl	No	N	80	P	Mo	No	1, 2, 3, 16
	Archamia macroptera	No	Pl	No	D	90	P	Mo	No	2, 3,
	Archamia zosterophora	No	Pl	No	N	80	P	Mo	No	1, 2, 4
	Cercamia cladara	No	N/A	No	N	40	P	Mo	No	3, 4
	Cercamia eremia	No	N/A	No	N	40	P	Mo	No	2
	Cheilodipterus alleni	No	N/A	No	N	40	P	Mo	No	2, 4
	Cheilodipterus artus	No	Ca	No	D	120	P	Mo	No	1, 2, 3, 18
	Cheilodipterus intermedius	No	N/A	No	D	110	P	Mo	No	2, 4
	Cheilodipterus isostigmus	No	Pi	No	D	100	P	Mo	No	2, 3, 4
	Cheilodipterus macrodon	No	Pi	No	D	200	P	Mo	Yes	1, 2, 3, 18
	Cheilodipterus nigrotaeniatus	Yes	Ma	No	D	80	P	Mo	No	2, 12
	Cheilodipterus parazonatus	Yes	Ma	No	D	70	P	Mo	No	1, 2, 12
	Cheilodipterus quinquelineatus	No	Ca	No	N	120	P	Mo	No	1, 2, 3, 18
	Cheilodipterus singapurensis	No	N/A	No	N	175	P	Mo	No	2
	Cheilodipterus zonatus	Yes	Ma	No	D	70	P	Mo	No	2, 12
	Foa brachygramma	No	Pl	No	D	60	P	Mo	No	1, 4
	Foa fo	No	Pl	No	D	35	P	Mo	No	1, 2, 3
	Foa hyalina	No	Pl	No	D	50	P	Mo	No	1, 2, 3
	Fowleria marmorata	No	N/A	No	D	45	P	Mo	No	1, 2, 3
	Fowleria punctulata	No	N/A	No	N	60	P	Mo	No	1, 4
	Fowleria vaiulae	No	N/A	No	D	50	P	Mo	No	1, 2, 3
	Fowleria variegata	No	N/A	No	D	50	P	Mo	No	1, 2, 3
	Neamia octospina	No	Pl	No	D	50	P	Mo	No	1, 2, 3
	Nectamia bandanensis	No	Pl	No	N	90	P	Mo	No	2, 3, 4
	Nectamia fusca	No	Pl	No	N	100	P	Mo	No	2, 3, 4

Continued

List of Indo-Pacific reef fishes and their social system, trophic affiliation, burrowing behaviour, activity patterns, size, and reproductive traits (Continued)

Family	Species	Pairs	Troph.	Burr.	Act.	ML (mm)	Spa.	Gam.	Mon.	Refs.
	Nectamia fusca (= *Apogon guamensis*)	Yes	Pl	No	N	110	P	Mo	No	1, 2, 3, 16
	Nectamia savayensis	No	Ma	No	D	110	P	Mo	No	1, 4
	Ostorhinchus angustatus	No	Mi	No	N	90	P	Mo	No	1, 2, 3, 4
	Ostorhinchus apogonoides	No	Pl	No	D	100	P	Mo	No	1, 2, 3, 12
	Ostorhinchus aureus	No	N/A	No	D	120	P	Mo	No	1, 2, 3
	Ostorhinchus cavitensis	No	N/A	No	D	80	P	Mo	No	2
	Ostorhinchus chrysopomus	No	N/A	No	D	90	P	Mo	No	2
	Ostorhinchus chrysotaenia	No	N/A	No	N	100	P	Mo	No	2, 4
	Ostorhinchus compressus	No	Pl	No	N	120	P	Mo	No	1, 2, 3, 19
	Ostorhinchus cookii	No	N/A	No	N	100	P	Mo	No	1, 2, 3, 4
	Ostorhinchus cyanosoma	Yes	Ma	No	D	80	P	Mo	No	1, 2, 3, 16
	Ostorhinchus dispar	No	N/A	No	N	50	P	Mo	No	2, 4
	Ostorhinchus endekataenia	No	N/A	No	N	140	P	Mo	No	2, 4
	Ostorhinchus fasciatus (=*quadrifasciatus*)	No	Ma	No	N	100	P	Mo	No	2, 24
	Ostorhinchus fleurieu	No	N/A	No	D	110	P	Mo	No	2
	Ostorhinchus franssedai	No	N/A	No	D	75	P	Mo	No	2
	Ostorhinchus griffini	Yes	N/A	No	D	140	P	Mo	No	2, 4
	Ostorhinchus hartzfeldii	No	Ma	No	N	120	P	Mo	No	2, 20
	Ostorhinchus hoevenii	No	Ma	No	D	50	P	Mo	No	2, 20
	Ostorhinchus holotaenia	No	N/A	No	D	80	P	Mo	No	1, 2, 3
	Ostorhinchus kiensis	No	N/A	No	D	90	P	Mo	No	2, 3
	Ostorhinchus komodoensis	No	Pl	No	D	70	P	Mo	No	2, 4
	Ostorhinchus lateralis	No	Ma	No	N	80	P	Mo	No	2, 21
	Ostorhinchus lineomaculatus	No	N/A	No	N	65	P	Mo	No	2
	Ostorhinchus luteus	No	Pl	No	D	50	P	Mo	No	1, 2, 3, 12
	Ostorhinchus margaritophorus	No	N/A	No	N	55	P	Mo	No	2, 22
	Ostorhinchus moluccensis	No	N/A	No	N	90	P	Mo	No	2
	Ostorhinchus monospilus	Yes	N/A	No	N	80	P	Mo	No	2, 4
	Ostorhinchus multilineatus	No	N/A	No	D	100	P	Mo	No	2, 4
	Ostorhinchus nanus	No	N/A	No	D	35	P	Mo	No	2, 4
	Ostorhinchus neotes	No	N/A	No	D	30	P	Mo	No	2
	Ostorhinchus nigrofasciatus	Yes	Mi	No	N	80	P	Mo	No	1, 2, 3, 4
	Ostorhinchus novemfasciatus	Yes	Ca	No	N	90	P	Mo	No	1, 2, 3, 4
	Ostorhinchus ocellicaudus	Yes	N/A	No	D	60	P	Mo	No	2
	Ostorhinchus parvulus	No	N/A	No	D	80	P	Mo	No	2, 4
	Ostorhinchus properuptus	Yes	N/A	No	D	75	P	Mo	No	2
	Ostorhinchus rubrimacula	No	Ma	No	D	80	P	Mo	No	2, 3, 25
	Ostorhinchus rueppellii	No	N/A	No	N	120	P	Mo	No	2, 4
	Ostorhinchus sealei	No	N/A	No	D	90	P	Mo	No	2
	Ostorhinchus selas	No	N/A	No	D	55	P	Mo	No	2
	Ostorhinchus taeniophorus	N/A	N/A	No	N	100	P	Mo	No	1, 2, 3, 4
	Ostorhinchus talboti	No	N/A	No	N	100	P	Mo	No	1, 2, 3
	Ostorhinchus thermalis	No	N/A	No	D	80	P	Mo	No	2

List of Indo-Pacific reef fishes and their social system, trophic affiliation, burrowing behaviour, activity patterns, size, and reproductive traits (Continued)

Family	Species	Pairs	Troph.	Burr.	Act.	ML (mm)	Spa.	Gam.	Mon.	Refs.
	Ostorhinchus wassinki	Yes	N/A	No	D	70	P	Mo	No	2
	Pristiapogon (= Apogon) exostigma	No	Ca	No	N	110	P	Mo	No	1, 2, 16
	Pristiapogon fraenatus	No	N/A	No	N	110	P	Mo	No	1, 2, 4
	Pristiapogon kallopterus	No	Pl	No	N	150	P	Mo	No	2, 12
	Pristicon rhodopterus	No	N/A	No	N	150	P	Mo	No	2
	Pristicon trimaculatus	Yes	Pl	No	N	150	P	Mo	No	2, 3, 12
	Pseudamia gelatinosa	No	N/A	No	N	100	P	Mo	No	1, 2, 3
	Pseudamia hayashii	No	N/A	No	N	75	P	Mo	No	2
	Pseudamia zonata	No	N/A	No	N	90	P	Mo	No	2, 3
	Pseudamiops gracilicauda	No	Pl	No	N	50	P	Mo	No	3
	Pseudamiops phasma	N/A	N/A	No	D	47	P	Mo	No	3
	Pterapogon kauderni	No	Pl	No	N	65	P	Mo	No	2, 4
	Pterapogon mirifica	No	N/A	No	N	140	P	Mo	No	2, 4
	Rhabdamia cypselurus	No	Pl	No	N	60	P	Mo	No	1, 2, 3, 4
	Rhabdamia gracilis	No	Pl	No	N	60	P	Mo	No	1, 2, 3, 4
	Rhabdamia spilota	No	Pl	No	D	60	P	Mo	No	2, 4
	Siphamia corallicola	No	N/A	No	D	38	P	Mo	No	2, 4
	Siphamia elongata	No	N/A	No	D	38	P	Mo	No	2
	Siphamia fuscolineata	No	N/A	No	D	35	P	Mo	No	2
	Siphamia majimai	N/A	N/A	No	N	35	P	Mo	No	1, 4
	Siphamia tubifer	No	N/A	No	D	40	P	Mo	No	2
	Siphamia tubifer (= versicolor)	No	N/A	No	D	40	P	Mo	No	2, 3
	Sphaeramia nematoptera	No	Pl	No	N	80	P	Mo	No	1, 2, 3, 26
	Sphaeramia orbicularis	No	Pl	No	N	115	P	Mo	No	2, 3, 4
	Zapogon (= Ostorhinchus) evermanni	Yes	N/A	No	N	120	P	Mo	No	2, 4
	Zoramia fragilis	Yes	Mi	No	N	55	P	Mo	No	1, 2, 3, 18
	Zoramia gilberti	No	N/A	No	D	55	P	Mo	No	2, 4
	Zoramia leptacantha	No	N/A	No	N	60	P	Mo	No	2, 3, 4
	Zoramia perlita	No	N/A	No	N	55	P	Mo	No	2, 4
Aulostomidae	*Aulostomus chinensis*	No	Ca	No	D	800	P	Br	No	4
Balistidae	*Abalistes stellatus*	No	N/A	No	D	600	P	Bc	No	1, 2, 3
	Balistapus undulatus	No	Om	No	D	300	P	Bc	No	1, 2, 3
	Balistes polylepis	No	Ma	No	D	760	G	Bc	No	4
	Balistoides conspicillum	No	Ma	No	D	500	G	Bc	No	4
	Balistoides viridescens	No	Ma	No	D	750	G	Bc	No	1, 2, 3
	Canthidermis maculata	No	Pl	No	D	300	P	Bc	No	12
	Melichthys indicus	No	Om	No	D	240	P	Bc	No	4
	Melichthys niger	No	Om	No	D	350	P	Bc	No	1, 2, 3
	Melichthys vidua	No	Om	No	D	300	P	Bc	No	1, 2, 3
	Odonus niger	No	Pl	No	D	400	P	Bc	No	1, 2, 3
	Pseudobalistes flavimarginatus	No	Ma	No	D	600	G	Bc	No	4
	Pseudobalistes fuscus	No	Ma	No	D	550	G	Bc	Yes	1, 2, 3

Continued

List of Indo-Pacific reef fishes and their social system, trophic affiliation, burrowing behaviour, activity patterns, size, and reproductive traits (Continued)

Family	Species	Pairs	Troph.	Burr.	Act.	ML (mm)	Spa.	Gam.	Mon.	Refs.
	Rhinecanthus aculeatus	No	Om	No	D	250	P	Bc	No	27
	Rhinecanthus lunula	No	Om	No	D	280	P	Bc	No	1, 2, 3
	Rhinecanthus rectangulus	No	Om	No	D	250	P	Bc	No	4
	Rhinecanthus verrucosus	No	Om	No	D	230	P	Bc	No	28
	Sufflamen bursa	No	Om	No	D	240	P	Bc	No	4
	Sufflamen chrysopterum	No	Ma	No	D	220	P	Bc	Yes	4
	Sufflamen fraenatum	No	Om	No	D	380	P	Bc	No	4
	Xanthichthys auromarginatus	No	Pl	No	D	220	P	Bc	No	12
	Xanthichthys caeruleolineatus	No	Pl	No	D	350	P	Bc	No	12
	Xanthichthys mento	No	Pl	No	D	220	P	Bc	No	12
Batrachoididae	*Halophryne diemensis*	No	Ca	No	D	260	N/A	N/A	No	1, 2, 3, 4
Belonidae	*Platybelone argalus*	No	Pi	No	D	370	N/A	Es	No	1, 2, 3
	Strongylura incisa	No	Pi	No	D	700	N/A	Es	No	1, 2, 3
	Tylosurus acus	No	Pi	No	D	1000	N/A	Es	No	3
	Tylosurus crocodilus	No	Pi	No	D	1500	N/A	Es	No	1, 2, 4
Blenniidae	*Andamia tetradactylus*	N/A	N/A	No	D	65	P	Bc	No	2
	Aspidontus dussumieri	No	Om	No	D	120	P	Bc	No	3
	Aspidontus taeniatus	No	Ca	No	D	115	P	Bc	No	1, 2, 3, 4
	Atrosalarias fuscus	No	De	No	D	145	P	Bc	No	1, 2, 3, 29
	Blenniella caudolineata	N/A	N/A	No	D	100	P	Bc	No	3
	Blenniella chrysospilos	No	Om	No	D	140	P	Bc	No	1, 2, 3, 4
	Blenniella gibbifrons	No	N/A	No	D	100	P	Bc	No	2, 3
	Blenniella interrupta	No	N/A	No	D	80	P	Bc	No	2
	Blenniella paula	No	Om	No	D	130	P	Bc	No	1, 2, 3, 4
	Cirripectes auritus	No	Om	No	D	90	P	Bc	No	2, 3
	Cirripectes chelomatus	No	De	No	D	120	P	Bc	No	1, 2, 3, 29
	Cirripectes filamentosus	No	De	No	D	90	P	Bc	No	1, 2, 3
	Cirripectes polyzona	No	De	No	D	85	P	Bc	No	1, 2, 3, 4
	Cirripectes springeri	No	Om	No	D	80	P	Bc	No	2
	Cirripectes stigmaticus	No	De	No	D	130	P	Bc	No	2
	Cirripectes variolosus	No	De	No	D	80	P	Bc	No	2
	Cirripectus castaneus	Yes	Om	No	D	125	P	Bc	No	1, 2, 3, 4
	Crossosalarias macrospilus	No	N/A	No	D	85	P	Bc	No	1, 2, 3
	Ecsenius alleni	No	De	No	D	85	P	Bc	No	2
	Ecsenius australianus	No	De	No	D	50	P	Bc	No	1, 2
	Ecsenius axelrodi	No	De	No	D	50	P	Bc	No	2
	Ecsenius bathi	No	De	No	D	40	P	Bc	No	2
	Ecsenius bicolor	No	De	No	D	100	P	Bc	No	1, 2, 3, 29
	Ecsenius bimaculatus	No	De	No	D	50	P	Bc	No	2, 3
	Ecsenius caeruliventris	No	De	No	D	30	P	Bc	No	2
	Ecsenius collettei	No	De	No	D	40	P	Bc	No	2
	Ecsenius dilemma	No	De	No	D	50	P	Bc	No	2
	Ecsenius fijiensis	No	De	No	D	40	P	Bc	No	2, 3
	Ecsenius fourmanoiri	N/A	De	No	D	62	P	Bc	No	3
	Ecsenius isos	N/A	De	No	D	40	P	Bc	No	3

List of Indo-Pacific reef fishes and their social system, trophic affiliation, burrowing behaviour, activity patterns, size, and reproductive traits (Continued)

Family	Species	Pairs	Troph.	Burr.	Act.	ML (mm)	Spa.	Gam.	Mon.	Refs.
	Ecsenius kurti	No	De	No	D	45	P	Bc	No	2
	Ecsenius lineatus	No	De	No	D	70	P	Bc	No	2, 30
	Ecsenius lividanalis	No	De	No	D	50	P	Bc	No	2
	Ecsenius lubbocki	Yes	De	No	D	40	P	Bc	No	2
	Ecsenius mandibularis	No	De	No	D	72	P	Bc	No	29, 4
	Ecsenius melarchus	No	De	No	D	60	P	Bc	No	2, 3
	Ecsenius midas	No	Pl	No	D	130	P	Bc	No	1, 2, 3, 4
	Ecsenius monoculus	No	De	No	D	60	P	Bc	No	2
	Ecsenius namiyei	No	De	No	D	100	P	Bc	No	2, 30
	Ecsenius oculus	No	De	No	D	60	P	Bc	No	2
	Ecsenius ops	No	De	No	D	55	P	Bc	No	2
	Ecsenius pardus	N/A	De	No	D	62	P	Bc	No	3
	Ecsenius pictus	Yes	De	No	D	50	P	Bc	No	2
	Ecsenius portenoyi	N/A	De	No	D	58	P	Bc	No	3
	Ecsenius prooculis	No	De	No	D	50	P	Bc	No	2
	Ecsenius schroederi	No	De	No	D	50	P	Bc	No	2
	Ecsenius sellifer	No	De	No	D	45	P	Bc	No	2
	Ecsenius shirleyae	No	De	No	D	40	P	Bc	No	2
	Ecsenius stictus	No	De	No	D	55	P	Bc	No	2, 29
	Ecsenius stigmatura	No	De	No	D	55	P	Bc	No	2
	Ecsenius taeniatus	No	De	No	D	40	P	Bc	No	2
	Ecsenius tessera	N/A	De	No	D	40	P	Bc	No	3
	Ecsenius tigris	No	De	No	D	40	P	Bc	No	1, 2
	Ecsenius tricolor	No	De	No	D	50	P	Bc	No	2, 3
	Ecsenius trilineatus	No	De	No	D	35	P	Bc	No	2, 3
	Ecsenius yaeyamaensis	No	De	No	D	60	P	Bc	No	2
	Enchelyurus ater	N/A	N/A	No	D	53	P	Bc	No	3
	Enchelyurus kraussii	N/A	N/A	No	D	50	P	Bc	No	3
	Entomacrodus caudofasciatus	N/A	N/A	No	D	66	P	Bc	No	3
	Entomacrodus corneliae	N/A	N/A	No	D	50	P	Bc	No	3
	Entomacrodus decussatus	N/A	N/A	No	D	190	P	Bc	No	3
	Entomacrodus epalzeocheilos	N/A	N/A	No	D	128	P	Bc	No	3
	Entomacrodus macrospilus	N/A	N/A	No	D	42	P	Bc	No	3
	Entomacrodus randalli	N/A	N/A	No	D	109	P	Bc	No	3
	Entomacrodus sealei	N/A	N/A	No	D	109	P	Bc	No	3
	Entomacrodus striatus	N/A	N/A	No	D	118	P	Bc	No	3
	Entomacrodus thalassinus	N/A	N/A	No	D	60	P	Bc	No	3
	Exallias brevis	No	Co	No	D	145	P	Bc	No	1, 2, 3, 31
	Glyptoparus delicatulus	Yes	Om	No	D	50	P	Bc	No	1, 2, 3, 29
	Istiblennius bellus	No	N/A	No	D	150	P	Bc	No	2
	Istiblennius dussumieri	No	N/A	No	D	125	P	Bc	No	2
	Istiblennius edentulus	No	He	No	D	170	P	Bc	No	1, 2, 3, 32
	Istiblennius lineatus	No	He	No	D	140	P	Bc	No	2
	Meiacanthus abditus	No	Pl	No	D	65	P	Bc	No	2

Continued

List of Indo-Pacific reef fishes and their social system, trophic affiliation, burrowing behaviour, activity patterns, size, and reproductive traits (Continued)

Family	Species	Pairs	Troph.	Burr.	Act.	ML (mm)	Spa.	Gam.	Mon.	Refs.
	Meiacanthus anema	Yes	N/A	No	D	65	P	Bc	No	2, 4
	Meiacanthus atrodorsalis	Yes	Mi	No	D	110	P	Bc	No	1, 2, 3
	Meiacanthus bundoon	Yes	Pl	No	D	80	P	Bc	No	2, 3
	Meiacanthus crinitus	Yes	N/A	No	D	65	P	Bc	No	2
	Meiacanthus ditrema	No	Pl	No	D	65	P	Bc	No	1, 2, 3, 4
	Meiacanthus geminatus	No	N/A	No	D	65	P	Bc	No	2
	Meiacanthus grammistes	Yes	N/A	No	D	100	P	Bc	No	1, 2, 3
	Meiacanthus kamoharai	Yes	Om	No	D	85	P	Bc	No	2, 4
	Meiacanthus lineatus	Yes	N/A	No	D	95	P	Bc	No	1, 2
	Meiacanthus oualanensis	No	N/A	No	D	100	P	Bc	No	2
	Meiacanthus smithi	Yes	Mi	No	D	80	P	Bc	No	2, 4
	Meiacanthus urostigma	No	N/A	No	D	55	P	Bc	No	2
	Meiacanthus vicinus	Yes	N/A	No	D	65	P	Bc	No	2
	Meiacanthus vittatus	No	N/A	No	D	65	P	Bc	No	2
	Omobranchus elongatus	No	N/A	No	D	55	P	Bc	No	2
	Omobranchus germaini	No	Om	No	D	78	P	Bc	No	3
	Omobranchus obliquus	No	N/A	No	D	70	P	Bc	No	2
	Paralticus amboinensis	No	N/A	No	D	150	P	Bc	No	2
	Petroscirtes breviceps	Yes	Om	No	D	130	P	Bc	No	1, 2, 33
	Petroscirtes fallax	Yes	Om	No	D	95	P	Bc	No	1, 2
	Petroscirtes lupus	No	He	No	D	130	P	Bc	No	3, 34
	Petroscirtes mitratus	No	Om	No	D	150	P	Bc	No	1, 2, 3, 26
	Petroscirtes variabilis	No	Om	No	D	75	P	Bc	No	1, 2, 3, 26
	Petroscirtes xestus	No	N/A	No	D	95	P	Bc	No	2
	Plagiotremus flavus	No	Ca	No	D	70	P	Bc	No	2
	Plagiotremus laudandus	No	Ca	No	D	130	P	Bc	No	1, 2, 3, 4
	Plagiotremus rhinorhynchos	No	Ca	No	D	130	P	Bc	No	1, 2, 3, 4
	Plagiotremus tapeinosoma	No	Ca	No	D	130	P	Bc	No	1, 2, 3, 4
	Salarias alboguttatus	No	N/A	No	D	65	P	Bc	No	1, 2, 3
	Salarias ceramensis	No	He	No	D	140	P	Bc	No	2, 34
	Salarias fasciatus	No	De	No	D	140	P	Bc	No	1, 2, 3, 29
	Salarias guttatus	No	De	No	D	140	P	Bc	No	2, 29
	Salarias nigrocinctus	No	De	No	D	53	P	Bc	No	3
	Salarias obscurus	No	De	No	D	130	P	Bc	No	2
	Salarias patzneri	No	De	No	D	50	P	Bc	No	2, 29
	Salarias ramosus	No	De	No	D	50	P	Bc	No	2
	Salarias segmentatus	No	He	No	D	75	P	Bc	No	2, 4
	Salarias sinuosus	No	De	No	D	60	P	Bc	No	1, 2, 3
	Stanulus seychellensis	Yes	N/A	No	D	33	P	Bc	No	2
	Stanulus talboti	No	N/A	No	D	48	P	Bc	No	3
	Xiphasia matsubarai	No	Ca	No	N	300	P	Bc	No	3, 4
	Xiphasia setifer	No	Ca	No	N	530	P	Bc	No	2, 3, 4
Bothidae	*Asterorhombus fijiensis*	No	Ca	No	D	150	P	Br	No	2
	Asterorhombus filifer	No	Ca	No	D	130	P	Br	No	3
	Asterorhombus intermedius	No	Ca	No	D	160	P	Br	No	4

List of Indo-Pacific reef fishes and their social system, trophic affiliation, burrowing behaviour, activity patterns, size, and reproductive traits (Continued)

Family	Species	Pairs	Troph.	Burr.	Act.	ML (mm)	Spa.	Gam.	Mon.	Refs.
	Bothus mancus	No	Pi	No	N	420	P	Br	No	1, 2, 3, 4
	Bothus pantherinus	No	Ca	No	D	390	P	Br	No	4
	Engyprosopon grandisquama	No	Ma	No	D	110	P	Br	No	4
Caesionidae	*Caesio caerulaurea*	No	Pl	No	D	400	G	Br	No	2, 4
	Caesio cuning	No	Pl	No	D	500	G	Br	No	1, 2, 3, 12, 35
	Caesio lunaris	No	Pl	No	D	400	G	Br	No	2, 4
	Caesio teres	No	Pl	No	D	400	G	Br	No	1, 2, 3, 12
	Caesio varilineata	No	Pl	No	D	400	G	Br	No	2, 4
	Caesio xanthonota	No	Pl	No	D	400	G	Br	No	2, 4
	Dipterygonotus balteatus	No	Pl	No	D	140	G	Br	No	2, 3, 4
	Gymnocaesio gymnoptera	No	Pl	No	D	180	G	Br	No	2, 3, 4
	Pterocaesio chrysozona	No	Pl	No	D	210	G	Br	No	2, 4
	Pterocaesio digramma	No	Pl	No	D	210	G	Br	No	2, 3, 35
	Pterocaesio lativittata	No	Pl	No	D	200	G	Br	No	2, 4
	Pterocaesio marri	No	Pl	No	D	350	G	Br	No	1, 2, 3, 12
	Pterocaesio pisang	No	Pl	No	D	210	G	Br	No	2, 3, 4
	Pterocaesio randalli	No	Pl	No	D	250	G	Br	No	2, 4
	Pterocaesio tessellata	No	Pl	No	D	250	G	Br	No	2, 4
	Pterocaesio tile	No	Pl	No	D	250	G	Br	No	1, 2, 3, 12
	Pterocaesio trilineata	No	Pl	No	D	200	G	Br	No	1, 2, 3, 12
Callionymidae	*Anaora tentaculata*	No	Mi	No	D	45	P	Br	No	2, 26
	Callionymus enneactis	Yes	Mi	No	D	45	P	Br	No	2, 12
	Callionymus filamentosus	No	Mi	No	D	165	P	Br	No	3, 4
	Callionymus keeleyi	No	Mi	No	D	60	P	Br	No	2, 12
	Callionymus marquesensis	No	Mi	No	D	55	P	Br	No	3
	Callionymus simplicicornis	No	Mi	No	D	60	P	Br	No	3
	Callionymus superbus	Yes	Mi	No	D	150	P	Br	No	2, 12
	Dactylopus dactylopus	Yes	Mi	No	D	150	P	Br	No	2, 24
	Dactylopus kuiteri	Yes	Mi	No	D	150	P	Br	No	2
	Diplogrammus goramensis	No	Mi	No	D	80	P	Br	No	1, 2, 4
	Diplogrammus xenicus	Yes	Mi	No	D	70	P	Br	No	2, 36
	Synchiropus bartelsi	Yes	Mi	No	D	45	P	Br	No	2
	Synchiropus morrisoni	Yes	Mi	No	D	45	P	Br	No	1, 2, 3, 4
	Synchiropus moyeri	No	Pl	No	D	75	P	Br	No	2, 12
	Synchiropus ocellatus	No	Mi	No	D	70	P	Br	No	1, 2, 3, 12
	Synchiropus picturatus	No	Mi	No	D	60	P	Br	No	2, 4
	Synchiropus splendidus	No	Mi	No	D	60	P	Br	No	1, 2, 3, 4
	Synchiropus stellatus	No	Mi	No	D	60	P	Br	No	2, 12
Caracanthidae	*Caracanthus maculatus*	No	Ma	No	D	50	P	Gel	No	2, 12
Carangidae	*Alectis ciliaris*	No	Ca	No	D	1300	G	Br	No	1, 2, 3
	Carangoides bajad	No	Ca	No	D	610	G	Br	No	2, 37
	Carangoides chrysophrys	No	Pi	No	D	600	G	Br	No	4, 38
	Carangoides coeruleopinnatus	No	N/A	No	D	400	G	Br	No	1, 3

Continued

List of Indo-Pacific reef fishes and their social system, trophic affiliation, burrowing behaviour, activity patterns, size, and reproductive traits (Continued)

Family	Species	Pairs	Troph.	Burr.	Act.	ML (mm)	Spa.	Gam.	Mon.	Refs.
	Carangoides dinema	No	N/A	No	D	580	G	Br	No	3
	Carangoides equula	No	N/A	No	D	370	G	Br	No	3
	Carangoides ferdau	No	Ca	No	D	700	G	Br	No	1, 2, 3
	Carangoides fulvoguttatus	No	Pi	No	D	1300	G	Br	No	1, 2, 3, 39
	Carangoides gymnostethus	No	Ca	No	D	900	G	Br	No	1, 2, 3, 4
	Carangoides hedlandensis	No	Pi	No	D	320	G	Br	No	1, 3
	Carangoides humerosus	No	Ca	No	D	250	G	Br	No	1, 38
	Carangoides malabaricus	No	Ca	No	D	280	G	Br	No	1, 38
	Carangoides oblongus	No	Ca	No	D	460	G	Br	No	2, 3, 40
	Carangoides orthogrammus	No	Ca	No	D	700	G	Br	No	2, 3, 41
	Carangoides plagiotaenia	No	N/A	No	D	420	G	Br	No	2, 3
	Carangoides talamparoides	No	Pi	No	D	320	G	Br	No	1, 38
	Caranx bucculentus	No	Ca	No	D	660	G	Br	No	3, 42
	Caranx ignobilis	No	Ca	No	D	1650	G	Br	No	1, 2, 3, 43
	Caranx lugubris	No	Pi	No	N	740	G	Br	No	2, 4
	Caranx melampygus	No	Pi	No	D	1000	G	Br	No	1, 2, 3, 41, 44
	Caranx papuensis	No	Pi	No	D	800	G	Br	No	1, 2, 3 45
	Caranx sexfasciatus	No	Ca	No	N	1000	G	Br	No	2, 3
	Caranx tille	No	Ca	No	D	690	G	Br	No	3, 4
	Decapterus kurroides	No	Pl	No	D	500	G	Br	No	1, 4
	Decapterus macarellus	No	Pl	No	D	350	G	Br	No	1, 2, 3, 4
	Decapterus macrosoma	No	Pl	No	D	320	G	Br	No	3, 4
	Decapterus muroadsi	No	Pl	No	D	450	G	Br	No	3
	Decapterus russelli	No	Pl	No	D	380	G	Br	No	1, 4
	Decapterus tabl	No	Pl	No	D	500	G	Br	No	1, 4
	Elagatis bipinnulata	No	Ca	No	D	1200	G	Br	No	1, 2, 4
	Gnathanodon (= *Gnathodon*) *speciosus*	No	Ca	No	D	1400	G	Br	No	1, 2, 3
	Megalaspis cordyla	No	Ca	No	D	800	G	Br	No	3, 46
	Naucrates ductor	No	Ca	No	D	750	G	Br	No	2, 4
	Pseudocaranx dentex	No	Ca	No	D	940	G	Br	No	1, 2, 3
	Scomberoides commersonnianus	No	Pi	No	D	1200	G	Br	No	1, 2, 3, 39
	Scomberoides lysan	No	Ca	No	D	700	G	Br	No	1, 2, 3, 4
	Scomberoides tol	No	Ca	No	D	510	G	Br	No	1, 3, 4
	Selar boops	No	Ca	No	N	220	G	Br	No	1, 2, 3, 4
	Selar crumenophthalmus	No	Ca	No	N	300	G	Br	No	1, 2, 3, 4
	Selaroides leptolepis	No	Ca	No	D	220	G	Br	No	2, 47
	Seriola dumerili	No	Pi	No	D	1880	G	Br	No	1, 2, 3
	Seriola lalandi	No	Ca	No	D	1930	G	Br	No	1, 2, 3
	Seriola rivoliana	No	Pi	No	D	900	G	Br	No	1, 2, 3, 48
	Trachinotus baillonii	No	Pi	No	D	540	G	Br	No	1, 2, 3, 4
	Trachinotus blochii	No	Ma	No	D	650	G	Br	No	1, 2, 3, 4
	Trachinotus botla	No	Ma	No	D	610	G	Br	No	1, 4
	Uraspis helvola	No	N/A	No	N	500	G	Br	No	2, 4

List of Indo-Pacific reef fishes and their social system, trophic affiliation, burrowing behaviour, activity patterns, size, and reproductive traits (Continued)

Family	Species	Pairs	Troph.	Burr.	Act.	ML (mm)	Spa.	Gam.	Mon.	Refs.
Centriscidae	*Aeoliscus strigatus*	No	Pl	No	D	150	N/A	N/A	No	4
	Centriscus scutatus	No	Pl	No	D	140	N/A	N/A	No	4
Chaetodontidae	*Amphichaetodon howensis*	No	Ma	No	D	180	P	Br	No	2, 4
	Chaetodon adiergastos	Yes	N/A	No	D	160	P	Br	No	2
	Chaetodon andamanensis	Yes	Co	No	D	150	P	Br	No	1, 2, 3, 49
	Chaetodon argentatus	Yes	Om	No	D	200	P	Br	Yes	2, 49
	Chaetodon assarius	No	Pl	No	D	130	P	Br	No	2, 49
	Chaetodon aureofasciatus	Yes	Co	No	D	125	P	Br	Yes	2, 49, 50
	Chaetodon auriga	Yes	Om	No	D	230	G	Br	Yes	49, 51
	Chaetodon auripes	Yes	Om	No	D	200	P	Br	Yes	2, 52
	Chaetodon baronessa	Yes	Co	No	D	150	P	Br	Yes	1, 2, 49
	Chaetodon bennetti	Yes	Co	No	D	180	P	Br	No	1, 2, 52
	Chaetodon burgessi	Yes	N/A	No	D	140	P	Br	No	2
	Chaetodon citrinellus	Yes	Om	No	D	130	P	Br	Yes	2, 3, 52
	Chaetodon collare	Yes	Om	No	D	160	P	Br	Yes	2, 53
	Chaetodon daedalma	No	Om	No	D	150	P	Br	No	2, 52
	Chaetodon declivis	No	N/A	No	D	150	P	Br	No	4
	Chaetodon decussatus	Yes	Om	No	D	200	P	Br	No	2
	Chaetodon ephippium	Yes	Om	No	D	230	G	Br	Yes	1, 2, 3, 49
	Chaetodon falcula	Yes	Mi	No	D	200	P	Br	Yes	2, 4
	Chaetodon flavirostris	Yes	Om	No	D	200	P	Br	Yes	2, 49
	Chaetodon flavocoronatus	Yes	N/A	No	D	120	P	Br	No	2
	Chaetodon guentheri	No	N/A	No	D	130	P	Br	No	2, 4
	Chaetodon guttatissimus	Yes	N/A	No	D	120	P	Br	No	2
	Chaetodon interruptus	No	N/A	No	D	200	P	Br	No	2
	Chaetodon kleinii	Yes	Om	No	D	140	G	Br	Yes	1, 2, 3, 49
	Chaetodon lineolatus	Yes	Om	No	D	300	G	Br	Yes	1, 2, 3, 49
	Chaetodon litus	No	N/A	No	D	150	P	Br	No	2,
	Chaetodon lunula	Yes	Om	No	D	210	G	Br	Yes	1, 2, 3, 49
	Chaetodon lunulatus	Yes	Co	No	D	150	P	Br	Yes	1, 2, 3, 49
	Chaetodon melannotus	Yes	Co	No	D	150	G	Br	Yes	2, 50, 54
	Chaetodon mertensii	Yes	Om	No	D	125	P	Br	No	1, 2
	Chaetodon meyeri	Yes	Co	No	D	180	P	Br	Yes	2, 3
	Chaetodon nippon	Yes	Om	No	D	150	P	Br	No	2, 52
	Chaetodon ocellicaudus	Yes	Co	No	D	140	P	Br	No	2, 4
	Chaetodon octofasciatus	Yes	Co	No	D	120	P	Br	Yes	4
	Chaetodon ornatissimus	Yes	Co	No	D	180	P	Br	Yes	2, 49
	Chaetodon oxycephalus	Yes	Co	No	D	250	P	Br	No	1, 2
	Chaetodon pelewensis	Yes	Mi	No	D	125	P	Br	Yes	1, 2, 3, 49
	Chaetodon plebeius	Yes	Co	No	D	150	P	Br	Yes	1, 2, 3, 49
	Chaetodon punctatofasciatus	Yes	Om	No	D	120	P	Br	Yes	1, 2, 3, 49
	Chaetodon quadrimaculatus	Yes	Om	No	D	160	P	Br	Yes	2, 49
	Chaetodon rafflesii	Yes	Mi	No	D	150	G	Br	No	2, 3, 55

Continued

List of Indo-Pacific reef fishes and their social system, trophic affiliation, burrowing behaviour, activity patterns, size, and reproductive traits (Continued)

Family	Species	Pairs	Troph.	Burr.	Act.	ML (mm)	Spa.	Gam.	Mon.	Refs.
	Chaetodon rainfordi	Yes	Om	No	D	150	P	Br	Yes	1, 2, 49, 55
	Chaetodon reticulatus	Yes	Om	No	D	160	P	Br	Yes	2, 3
	Chaetodon selene	Yes	Mi	No	D	160	P	Br	No	2, 4
	Chaetodon semeion	Yes	Mi	No	D	240	G	Br	No	2, 55
	Chaetodon smithi	No	Pl	No	D	170	P	Br	No	2. 53
	Chaetodon speculum	Yes	Om	No	D	180	P	Br	Yes	2, 3, 49
	Chaetodon tinkeri	Yes	Om	No	D	150	P	Br	No	2, 53
	Chaetodon triangulum	Yes	Co	No	D	150	P	Br	Yes	2, 51
	Chaetodon trichrous	Yes	Om	No	D	120	P	Br	Yes	2, 56
	Chaetodon tricinctus	Yes	N/A	No	D	150	P	Br	Yes	2
	Chaetodon trifascialis	No	Co	No	D	180	P	Br	No	51, 57
	Chaetodon ulietensis	Yes	Om	No	D	150	P	Br	Yes	2, 3, 49
	Chaetodon unimaculatus	Yes	Om	No	D	200	P	Br	Yes	1, 2, 3, 49
	Chaetodon vagabundus	Yes	Om	No	D	230	P	Br	Yes	2, 51, 54
	Chaetodon wiebeli	Yes	He	No	D	180	P	Br	Yes	2, 4
	Chaetodon xanthurus	Yes	Om	No	D	140	P	Br	Yes	2, 3
	Chelmon marginalis	Yes	Mi	No	D	180	P	Br	No	2, 58
	Chelmon muelleri	Yes	Mi	No	D	180	P	Br	No	1, 2, 58
	Chelmon rostratus	Yes	Mi	No	D	200	P	Br	Yes	2, 49, 58
	Coradion altivelis	No	N/A	No	D	200	P	Br	No	2, 4
	Coradion chrysozonus	Yes	Sp	No	D	150	P	Br	No	2, 4
	Coradion melanopus	Yes	Sp	No	D	150	P	Br	No	2, 4
	Forcipiger flavissimus	Yes	Mi	No	D	220	P	Br	Yes	1, 2, 3
	Forcipiger longirostris	Yes	Mi	No	D	220	P	Br	Yes	1, 2, 3, 49
	Hemitaurichthys multispinosus	No	Pl	No	D	208	P	Br	No	3, 4
	Hemitaurichthys polylepis	Yes	Pl	No	D	180	P	Br	No	1, 2, 3, 49
	Hemitaurichthys thompsoni	Yes	Pl	No	D	180	P	Br	No	1, 2, 3, 49
	Hemitaurichthys zoster	Yes	Pl	No	D	160	P	Br	No	1, 2, 3, 49
	Heniochus acuminatus	Yes	Om	No	D	250	G	Br	No	1, 2, 3, 49
	Heniochus chrysostomus	Yes	Co	No	D	180	G	Br	No	1, 2, 3
	Heniochus diphreutes	No	Pl	No	D	210	G	Br	No	1, 2, 3
	Heniochus monoceros	Yes	Om	No	D	230	G	Br	No	1, 2, 3, 4
	Heniochus pleurotaenia	Yes	N/A	No	D	170	G	Br	No	2
	Heniochus singularius	Yes	Om	No	D	230	G	Br	No	1, 2, 3
	Heniochus varius	Yes	Mi	No	D	190	G	Br	No	2, 3
	Parachaetodon ocellatus	Yes	Mi	No	D	180	P	Br	No	1, 2, 59
	Roa modesta (= *Chaetodon modestus*)	No	N/A	No	D	170	P	Br	No	4
Chanidae	*Chanos chanos*	No	Om	No	D	1800	G	Br	No	1, 2, 3, 4
Cirrhitidae	*Amblycirrhitus bimacula*	Yes	Ma	No	D	85	P	Br	No	1, 2, 3, 12, 60
	Amblycirrhitus unimacula	N/A	N/A	No	D	110	P	Br	No	3
	Cirrhitichthys aprinus	No	Ca	No	D	100	P	Br	No	1, 2, 3, 61
	Cirrhitichthys aureus	No	Ca	No	D	150	P	Br	No	1, 2, 3

List of Indo-Pacific reef fishes and their social system, trophic affiliation, burrowing behaviour, activity patterns, size, and reproductive traits (Continued)

Family	Species	Pairs	Troph.	Burr.	Act.	ML (mm)	Spa.	Gam.	Mon.	Refs.
	Cirrhitichthys falco	Yes	Mi	No	D	70	P	Br	No	1, 2, 3, 12, 60
	Cirrhitichthys oxycephalus	No	Ca	No	D	95	P	Br	No	1, 2, 3, 4, 60
	Notocirrhitus (=*Cirrhitichthys*) *splendens*	No	Ca	No	D	230	P	Br	No	2
	Cirrhitops hubbardi	N/A	Ca	No	D	75	P	Br	No	3
	Cirrhitus pinnulatus	No	Ca	No	D	280	P	Br	No	1, 2, 3, 4
	Cyprinocirrhites polyactis	No	Pl	No	D	150	P	Br	No	1, 2, 3
	Isocirrhitus sexfasciatus	N/A	N/A	No	D	75	P	Br	No	3
	Itycirrhitus wilhelmi	N/A	N/A	No	D	75	P	Br	No	3
	Neocirrhites armatus	Yes	Ma	No	D	90	P	Br	Yes	1, 2, 3, 62
	Oxycirrhites typus	Yes	Mi	No	D	130	P	Br	Yes	1, 2, 3, 62
	Paracirrhites arcatus	No	Ca	No	D	130	P	Br	No	1, 2, 3, 63
	Paracirrhites forsteri	Yes	Ca	No	D	220	P	Br	No	1, 2, 3, 62
	Paracirrhites hemistictus	No	Ma	No	D	280	P	Br	No	1, 2, 3, 62
	Paracirrhites nisus	Yes	Ma	No	D	100	P	Br	No	2, 3, 4
	Paracirrhites xanthus	No	Ma	No	D	110	P	Br	No	1, 2, 3, 4, 12
Congridae	*Ariosoma anagoides*	No	N/A	No	N	400	G	Br	No	1, 2, 3
	Ariosoma fasciatum (= *Poeciloconger fasciatus*)	No	N/A	No	D	600	G	Br	No	1, 2, 3
	Ariosoma scheelei	No	N/A	No	N	200	G	Br	No	1, 2, 3
	Conger cinereus	No	Ca	No	N	1300	G	Br	No	4
	Gorgasia maculata	No	Pl	No	D	550	P	Br	No	1, 2, 3
	Gorgasia preclara	No	Pl	No	D	400	P	Br	No	4
	Heteroconger enigmaticus	No	N/A	No	D	450	G	Br	No	1, 2, 3
	Heteroconger perissodon	No	N/A	No	D	600	G	Br	No	1, 2, 3
	Heteroconger polyzona	Yes	N/A	No	D	700	G	Br	No	4
	Heteroconger taylori	No	N/A	No	D	400	G	Br	No	1, 2, 3
Coryphaenidae	*Coryphaena hippurus*	No	Ca	No	D	1620	G	Br	No	4
Dactylopteridae	*Dactyloptena orientalis*	No	Ca	No	D	380	N/A	Br	No	1, 2, 3, 4
Diodontidae	*Chilomycterus reticulatus*	No	Ma	No	D	550	P	Br	No	4, 64
	Cyclichthys orbicularis	No	Ma	No	N	150	P	Br	No	4
	Cyclichthys spilostylus	No	Ma	No	N	340	P	Br	No	4
	Diodon holocanthus	No	Ma	No	N	290	P	Br	No	4
	Diodon hystrix	No	Ma	No	N	710	P	Br	No	4
	Diodon liturosus	No	Ma	No	N	500	P	Br	No	4
	Lophodiodon calori	No	Ma	No	N	300	P	Br	No	4
Echeneidae	*Echeneis naucrates*	No	Ca	No	D	1000	N/A	N/A	No	4
Ephippidae	*Platax batavianus*	Yes	Om	No	D	500	G	Br	No	2, 4
	Platax boersii	No	Om	No	D	470	G	Br	No	2
	Platax orbicularis	Yes	Om	No	D	280	G	Br	No	2, 3
	Platax pinnatus	No	Om	No	D	370	G	Br	No	2, 65

Continued

List of Indo-Pacific reef fishes and their social system, trophic affiliation, burrowing behaviour, activity patterns, size, and reproductive traits (Continued)

Family	Species	Pairs	Troph.	Burr.	Act.	ML (mm)	Spa.	Gam.	Mon.	Refs.
	Platax teira	No	Om	No	D	410	G	Br	No	1, 2, 3, 66
	Zabidius novemaculeatus	No	N/A	No	D	450	G	Br	No	2
Fistulariidae	*Fistularia commersonii*	No	Ca	No	D	1500	P	Br	No	4
Gobiesocidae	*Diademichthys lineatus*	No	Mi	No	D	50	P	Bc	No	1, 2, 3, 67
	Discotrema (= *Lepadichthys*) *crinophilum*	Yes	Mi	No	D	30	P	Bc	No	1, 2, 3
	Lepadichthys lineatus	Yes	Mi	No	D	30	P	Bc	No	1, 2, 3, 4
Gobiidae	*Acentrogobius janthinopterus*	No	N/A	No	D	70	P	Bc	No	2
	Acentrogobius (= *Yongeichthys*) *nebulosus*	No	Pl	No	D	160	P	Bc	No	2, 17
	Amblyeleotris arcupinna	No	Pl	Yes*	D	85	P	Bc	No	2
	Amblyeleotris aurora	No	Pl	Yes*	D	90	P	Bc	No	2
	Amblyeleotris bellicauda	No	Pl	Yes*	D	70	P	Bc	No	3, 12
	Amblyeleotris biguttata	No	Pl	Yes*	D	104	P	Bc	No	3, 12
	Amblyeleotris diagonalis	No	Pl	Yes*	D	90	P	Bc	No	2, 3
	Amblyeleotris ellipse	No	Pl	Yes*	D	73	P	Bc	No	3, 12
	Amblyeleotris fasciata	Yes	Pl	Yes*	D	80	P	Bc	No	1, 2, 3, 4
	Amblyeleotris fasciata (= *katherine*)	Yes	Pl	Yes*	D	60	P	Bc	No	3, 12, 4
	Amblyeleotris fontanesii	No	Pl	Yes*	D	170	P	Bc	No	2, 3
	Amblyeleotris guttata	No	Pl	Yes*	D	90	P	Bc	No	1, 2, 3, 12
	Amblyeleotris gymnocephala	No	Pl	Yes*	D	100	P	Bc	No	2
	Amblyeleotris latifasciata	No	Pl	Yes*	D	100	P	Bc	No	2
	Amblyeleotris marquesas	No	Pl	Yes*	D	85	P	Bc	No	3, 12
	Amblyeleotris novaecaledoniae	No	Pl	Yes*	D	100	P	Bc	No	3, 12
	Amblyeleotris ogasawarensis	No	Pl	Yes*	D	110	P	Bc	No	3, 12
	Amblyeleotris periophthalma	No	Pl	Yes*	D	80	P	Bc	No	1, 2, 3
	Amblyeleotris randalli	No	Pl	Yes*	D	90	P	Bc	No	2, 3
	Amblyeleotris rhyax	No	Pl	Yes*	D	80	P	Bc	No	1, 2
	Amblyeleotris rubrimarginata	No	Pl	Yes*	D	90	P	Bc	No	2, 3
	Amblyeleotris steinitzi	No	Pl	Yes*	D	80	P	Bc	No	1, 2, 3, 12
	Amblyeleotris stenotaeniata	No	Pl	Yes*	D	95	P	Bc	No	3, 12
	Amblyeleotris wheeleri	No	Pl	Yes*	D	80	P	Bc	No	1, 2
	Amblyeleotris yanoi	No	Pl	Yes*	D	120	P	Bc	No	2
	Amblygobius buanensis	Yes	Om	No	D	60	P	Bc	No	2, 4
	Amblygobius bynoensis	Yes	He	Yes	D	80	P	Bc	No	1, 2, 3, 68
	Amblygobius decussatus	No	Om	No	D	60	P	Bc	No	1, 2, 3, 69
	Amblygobius esakiae	Yes	He	No	D	65	P	Bc	No	2
	Amblygobius nocturnus	Yes	De	Yes	D	50	P	Bc	Yes	1, 2, 3, 70
	Amblygobius phalaena	Yes	He	Yes	D	135	P	Bc	Yes	1, 2, 3, 68
	Amblygobius semicinctus	Yes	Om	Yes	D	140	P	Bc	Yes	2, 4
	Amblygobius sphynx	Yes	Mi	No	D	165	P	Bc	No	1, 2, 3
	Asterropteryx bipunctata	No	De	Yes	D	40	P	Bc	No	2, 4, 71
	Asterropteryx ensifera	No	Pl	No	D	35	P	Bc	No	1, 2, 3, 4
	Asterropteryx semipunctata	No	De	Yes	D	60	P	Bc	No	1, 2, 3, 70

List of Indo-Pacific reef fishes and their social system, trophic affiliation, burrowing behaviour, activity patterns, size, and reproductive traits (Continued)

Family	Species	Pairs	Troph.	Burr.	Act.	ML (mm)	Spa.	Gam.	Mon.	Refs.
	Asterropteryx spinosa	No	De	No	D	35	P	Bc	No	3, 71
	Asterropteryx striata	No	Pl	No	D	35	P	Bc	No	2, 72
	Bathygobius coalitus	No	De	No	D	120	P	Bc	No	1, 3, 71
	Bathygobius coalitus (= *padangensis*)	No	De	No	D	50	P	Bc	No	2, 71
	Bathygobius cocosensis	Yes	Mi	No	D	50	P	Bc	No	1, 2, 3, 73
	Bathygobius cotticeps	No	De	No	D	110	P	Bc	No	3, 71
	Bathygobius cyclopterus	No	De	No	D	70	P	Bc	No	3, 71
	Bathygobius fuscus	No	De	No	D	80	P	Bc	No	1, 3, 70
	Bathygobius laddi	No	De	No	D	50	P	Bc	No	1, 3, 71
	Bryaninops amplus	N/A	N/A	No	D	50	P	Bc	No	2
	Bryaninops dianneae	N/A	N/A	No	D	24	P	Bc	No	3
	Bryaninops erythrops	N/A	N/A	No	D	20	P	Bc	No	2
	Bryaninops loki	N/A	N/A	No	D	30	P	Bc	No	1, 2, 3
	Bryaninops natans	No	N/A	No	D	25	P	Bc	No	1, 2, 3
	Bryaninops tigris	No	Pl	No	D	55	P	Bc	No	2, 4
	Bryaninops yongei	Yes	Pl	No	D	35	P	Bc	No	2, 74
	Callogobius clitellus	No	N/A	No	D	45	P	Bc	No	2
	Callogobius hasseltii	No	N/A	No	D	60	P	Bc	No	1, 2, 3
	Callogobius maculipinnis	No	N/A	No	D	90	P	Bc	No	1, 3
	Callogobius sclateri	No	N/A	No	D	45	P	Bc	No	3
	Calumia profunda	No	N/A	No	D	23	P	Bc	No	2
	Cristatogobius lophius	No	N/A	Yes*	D	75	P	Bc	No	2, 4
	Cryptocentrus caeruleopunctatus	No	Mi	Yes*	D	80	P	Bc	No	1, 2, 75
	Cryptocentrus cinctus	Yes	Mi	Yes*	D	70	P	Bc	No	1, 2, 75
	Cryptocentrus cyanotaenia	Yes	Mi	Yes*	D	70	P	Bc	No	1, 2, 4, 75
	Cryptocentrus fasciatus	Yes	Mi	Yes*	D	80	P	Bc	No	1, 2, 3, 4, 75
	Cryptocentrus inexplicatus	No	Mi	Yes*	D	75	P	Bc	No	2, 75
	Cryptocentrus insignitus	No	Mi	Yes*	D	80	P	Bc	No	1, 2, 75
	Cryptocentrus leptocephalus	No	Mi	Yes*	D	100	P	Bc	No	2, 75
	Cryptocentrus leucostictus	No	Mi	Yes*	D	90	P	Bc	No	1, 2, 3, 75
	Cryptocentrus pavoninoides	No	Mi	Yes*	D	140	P	Bc	No	2, 75
	Cryptocentrus polyophthalmus	No	Mi	Yes*	D	90	P	Bc	No	2, 75
	Cryptocentrus strigilliceps	No	Mi	Yes*	D	60	P	Bc	No	2, 75
	Ctenogobiops aurocingulus	No	N/A	Yes*	D	55	P	Bc	No	2, 3
	Ctenogobiops feroculus	Yes	N/A	Yes*	D	55	P	Bc	No	2, 3, 4
	Ctenogobiops maculosus (= *crocineus*)	No	N/A	Yes*	D	50	P	Bc	No	2
	Ctenogobiops pomastictus	No	N/A	Yes*	D	50	P	Bc	No	1, 2, 3
	Ctenogobiops tangaroai	No	N/A	Yes*	D	50	P	Bc	No	1, 2, 3
	Ctenogobiops tongaensis	No	N/A	Yes*	D	55	P	Bc	No	3
	Discordipinna griessingeri	No	N/A	No	D	25	P	Bc	No	1, 2, 3
	Echinogobius hayashii	No	N/A	Yes	D	120	P	Bc	No	2, 4

Continued

List of Indo-Pacific reef fishes and their social system, trophic affiliation, burrowing behaviour, activity patterns, size, and reproductive traits (Continued)

Family	Species	Pairs	Troph.	Burr.	Act.	ML (mm)	Spa.	Gam.	Mon.	Refs.
	Eviota afelei	No	N/A	No	D	23	P	Bc	No	1, 3
	Eviota albolineata	No	Mi	No	D	35	P	Bc	No	1, 2, 3, 12
	Eviota bifasciata	No	Pl	No	D	30	P	Bc	No	1, 2, 4
	Eviota cometa	No	N/A	No	D	24	P	Bc	No	3
	Eviota distigma	No	N/A	No	D	27	P	Bc	No	3
	Eviota fasciola	No	N/A	No	D	24	P	Bc	No	3
	Eviota guttata	No	Mi	No	D	25	P	Bc	No	1, 2, 3, 12
	Eviota infulata	No	N/A	No	D	24	P	Bc	No	3
	Eviota lachdeberei	No	N/A	No	D	25	P	Bc	No	2
	Eviota latifasciata	No	N/A	No	D	20	P	Bc	No	1, 2, 3
	Eviota melasma	No	N/A	No	D	32	P	Bc	No	1, 2, 3
	Eviota mikiae	No	N/A	No	D	25	P	Bc	No	2
	Eviota monostigma	No	N/A	No	D	33	P	Bc	No	3
	Eviota nebulosa	No	N/A	No	D	24	P	Bc	No	3
	Eviota nigriventris	No	N/A	No	D	25	P	Bc	No	1, 2, 3
	Eviota pellucida	No	N/A	No	D	25	P	Bc	No	1, 2, 3
	Eviota prasina	No	N/A	No	D	30	P	Bc	No	1, 2
	Eviota prasites	No	Mi	No	D	25	P	Bc	No	1, 2, 3, 12
	Eviota punctulata	No	N/A	No	D	20	P	Bc	No	1, 2, 3
	Eviota queenslandica	No	Om	No	D	30	P	Bc	No	1, 2, 70
	Eviota sebreei	No	N/A	No	D	25	P	Bc	No	1, 2, 3
	Eviota sigillata	No	N/A	No	D	25	P	Bc	No	1, 2
	Eviota smaragdus	No	N/A	No	D	24	P	Bc	No	2, 3
	Eviota storthynx	No	N/A	No	D	25	P	Bc	No	2
	Eviota zebrina	No	Om	No	D	24	P	Bc	No	1, 3, 69
	Exyrias belissimus	No	Om	No	D	130	P	Bc	No	1, 2, 3
	Exyrias ferrarisi	Yes	N/A	No	D	95	P	Bc	No	2
	Exyrias puntang	No	Om	No	D	135	P	Bc	No	1, 2, 3, 15
	Favonigobius reichei	No	Ma	No	D	45	P	Bc	No	2, 15
	Fusigobius aureus	No	N/A	No	D	75	P	Bc	No	2
	Fusigobius duospilus	No	Ma	No	D	55	P	Bc	No	1, 2, 3, 12
	Fusigobius inframaculatus	No	N/A	No	D	75	P	Bc	No	2
	Fusigobius neophytus	No	Ma	No	D	70	P	Bc	No	1, 2, 3, 12
	Fusigobius signipinnis	No	N/A	No	D	30	P	Bc	No	2
	Gnatholepis anjerensis	No	N/A	No	D	55	P	Bc	No	1, 2, 3
	Gnatholepis cauerensis	No	N/A	No	D	45	P	Bc	No	1, 2, 3
	Gobiodon acicularis	No	Om	No	D	38	P	Bc	No	2, 71
	Gobiodon axillaris	N/A	N/A	No	D	40	P	Bc	No	3
	Gobiodon brochus	Yes	Om	No	D	34	P	Bc	No	1, 3, 76
	Gobiodon ceramensis	Yes	N/A	No	D	35	P	Bc	No	2
	Gobiodon citrinus	No	Co	No	D	55	P	Bc	No	1, 2, 3, 77
	Gobiodon histrio	No	Om	No	D	35	P	Bc	No	1, 3, 76
	Gobiodon okinawae	No	Om	No	D	35	P	Bc	No	1, 2, 3, 76
	Gobiodon quinquestrigatus	Yes	Om	No	D	35	P	Bc	Yes	1, 2, 3, 76
	Gobiodon spilophthalmus	No	N/A	No	D	35	P	Bc	No	2

List of Indo-Pacific reef fishes and their social system, trophic affiliation, burrowing behaviour, activity patterns, size, and reproductive traits (Continued)

Family	Species	Pairs	Troph.	Burr.	Act.	ML (mm)	Spa.	Gam.	Mon.	Refs.
	Gobiopsis exigua	No	N/A	No	D	50	P	Bc	No	2
	Heteroplopomus barbatus	No	N/A	No	D	25	P	Bc	No	2
	Istigobius decoratus	No	De	No	D	90	P	Bc	No	1, 2, 3, 70
	Istigobius goldmanni	No	De	No	D	62	P	Bc	No	3, 70
	Istigobius ornatus	No	Mi	No	D	90	P	Bc	No	1, 2, 3, 4
	Istigobius rigilius	No	Om	No	D	90	P	Bc	No	1, 2, 3, 69
	Koumansetta hectori	No	Om	No	D	50	P	Bc	No	2, 78
	Koumansetta rainfordi	No	He	No	D	55	P	Bc	No	1, 2, 3, 70
	Lotilia graciliosa	No	N/A	Yes*	D	45	P	Bc	No	1, 2, 3
	Luposicya lupus	No	N/A	No	D	35	P	Bc	No	2
	Macrodontogobius wilburi	No	Om	No	D	65	P	Bc	No	1, 2, 3, 26
	Mahidolia mystacina	No	Mi	Yes*	D	70	P	Bc	No	1, 2, 3, 79
	Myersina lachneri	No	N/A	Yes*	D	50	P	Bc	No	2
	Myersina nigrivirgata	No	N/A	Yes*	D	100	P	Bc	No	2
	Oplopomus caninoides	No	N/A	No	D	60	P	Bc	No	2
	Oplopomus oplopomus	Yes	Mi	No	D	75	P	Bc	No	1, 2, 3, 80
	Oxyurichthys notonema	N/A	N/A	Yes	D	160	P	Bc	No	3
	Oxyurichthys papuensis	No	N/A	Yes	D	200	P	Bc	No	1, 2, 3
	Pandaka pusilla	No	N/A	No	D	12	P	Bc	No	2
	Paragobiodon echinocephalus	No	N/A	No	D	35	P	Bc	Yes	1, 2, 3
	Paragobiodon lacunicolus	No	N/A	No	D	35	P	Bc	No	1, 2, 3
	Paragobiodon modestus	No	N/A	No	D	35	P	Bc	No	1, 2, 3
	Paragobiodon xanthosoma	No	N/A	No	D	40	P	Bc	Yes	1, 2, 3
	Phyllogobius platycephalops	No	N/A	No	D	25	P	Bc	No	2
	Pleurosicya bilobata	No	Mi	No	D	28	P	Bc	No	3, 4
	Pleurosicya boldinghi	Yes	N/A	No	D	35	P	Bc	No	2
	Pleurosicya coerulea	N/A	N/A	No	D	22	P	Bc	No	3
	Pleurosicya elongata	N/A	N/A	No	D	35	P	Bc	No	2
	Pleurosicya fringilla	N/A	N/A	No	D	22	P	Bc	No	3
	Pleurosicya labiata	No	N/A	No	D	35	P	Bc	No	2
	Pleurosicya micheli	No	N/A	No	D	25	P	Bc	No	2, 3
	Pleurosicya mossambica	No	N/A	No	D	35	P	Bc	No	1, 2, 3
	Priolepis aureoviridis	No	N/A	No	D	65	P	Bc	No	2,
	Priolepis cincta	No	N/A	No	D	50	P	Bc	Yes	1, 2, 3
	Priolepis compita	No	Mi	No	D	17	P	Bc	No	3
	Priolepis fallacincta	N/A	N/A	No	D	32	P	Bc	No	3
	Priolepis inhaca	No	N/A	No	D	35	P	Bc	No	1, 2, 3
	Priolepis kappa	N/A	N/A	No	D	26	P	Bc	No	3
	Priolepis nocturna	No	N/A	No	D	35	P	Bc	No	1, 2, 3
	Priolepis nuchifasciatus	Yes	Mi	No	D	40	P	Bc	No	1, 70
	Priolepis pallidicincta	N/A	N/A	No	D	36	P	Bc	No	3
	Priolepis semidoliata	N/A	N/A	No	D	36	P	Bc	No	3
	Priolepis squamogena	N/A	N/A	No	D	54	P	Bc	No	3
	Priolepis triops	N/A	N/A	No	D	26	P	Bc	No	3

Continued

List of Indo-Pacific reef fishes and their social system, trophic affiliation, burrowing behaviour, activity patterns, size, and reproductive traits (Continued)

Family	Species	Pairs	Troph.	Burr.	Act.	ML (mm)	Spa.	Gam.	Mon.	Refs.
	Signigobius biocellatus	Yes	Mi	Yes	D	65	P	Bc	Yes	1, 2, 4
	Stonogobiops larsonae	No	N/A	Yes*	D	60	P	Bc	No	2
	Stonogobiops medon	No	N/A	Yes*	D	60	P	Bc	No	2
	Stonogobiops nematodes	Yes	N/A	Yes*	D	60	P	Bc	No	1, 2, 4
	Stonogobiops xanthorhinica	Yes	N/A	Yes*	D	60	P	Bc	No	1, 2, 3, 4
	Stonogobiops yasha	Yes	N/A	Yes*	D	60	P	Bc	No	1, 2, 3, 4
	Tomiyamichthys lanceolatus	No	N/A	Yes*	D	50	P	Bc	No	2, 4
	Tomiyamichthys oni	No	N/A	Yes*	D	100	P	Bc	No	2
	Trimma anaima	No	N/A	No	D	20	P	Bc	No	2
	Trimma annosum	No	N/A	No	D	25	P	Bc	No	2
	Trimma benjamini	No	Pl	No	D	25	P	Bc	No	1, 2, 3, 4
	Trimma caesiura	No	Om	No	D	25	P	Bc	No	1, 2, 3, 70
	Trimma cana	No	N/A	No	D	25	P	Bc	No	2
	Trimma emeryi	N/A	N/A	No	D	24	P	Bc	No	3
	Trimma griffithsi	No	Pl	No	D	23	P	Bc	No	2, 81
	Trimma halonevum	No	N/A	No	D	28	P	Bc	No	2
	Trimma hoesei	Yes	N/A	No	D	25	P	Bc	No	25
	Trimma macrophthalmum	No	N/A	No	D	25	P	Bc	No	1, 2, 3
	Trimma milta	N/A	N/A	No	D	30	P	Bc	No	3
	Trimma naudei	No	Mi	No	D	30	P	Bc	No	2, 4
	Trimma okinawae	No	N/A	No	D	28	P	Bc	No	1, 2, 3
	Trimma rubromaculatum	No	N/A	No	D	22	P	Bc	No	2
	Trimma stobbsi	No	N/A	No	D	20	P	Bc	No	2
	Trimma striatum	No	Mi	No	D	35	P	Bc	No	1, 2, 3, 70
	Trimma taylori	No	Pl	No	D	25	P	Bc	No	1, 2, 3, 82
	Trimma tevegae	No	Pl	No	D	40	P	Bc	No	1, 2, 3, 4
	Tryssogobius colini	No	N/A	No	D	50	P	Bc	No	2
	Valenciennea alleni	Yes	Mi	Yes	D	65	P	Bc	No	2, 4
	Valenciennea bella	Yes	Mi	Yes	D	90	P	Bc	No	2, 4
	Valenciennea decora	Yes	Mi	Yes	D	140	P	Bc	No	1, 2, 3
	Valenciennea helsdingenii	Yes	Mi	Yes	D	145	P	Bc	Yes	1, 2, 3, 83
	Valenciennea immaculata	Yes	Mi	Yes	D	105	P	Bc	No	1, 2, 3, 4
	Valenciennea limicola	Yes	Mi	Yes	D	80	P	Bc	No	2
	Valenciennea longipinnis	Yes	Mi	Yes	D	150	P	Bc	Yes	1, 2, 3, 4
	Valenciennea muralis	Yes	Mi	Yes	D	115	P	Bc	Yes	1, 2, 3, 4, 50
	Valenciennea parva	Yes	Mi	Yes	D	65	P	Bc	Yes	1, 2, 3
	Valenciennea puellaris	Yes	Mi	Yes	D	155	P	Bc	Yes	1, 2, 3, 4
	Valenciennea randalli	No	Mi	Yes	D	90	P	Bc	No	1, 2, 3
	Valenciennea sexguttata	Yes	Mi	Yes	D	115	P	Bc	Yes	1, 2, 3
	Valenciennea strigata	Yes	Mi	Yes	D	155	P	Bc	Yes	1, 2, 3, 84
	Valenciennea wardii	Yes	Mi	Yes	D	100	P	Bc	No	1, 2, 3
	Vanderhorstia ambanoro	No	N/A	Yes*	D	120	P	Bc	No	2, 3
	Vanderhorstia flavilineata	No	N/A	Yes*	D	40	P	Bc	No	2, 4

List of Indo-Pacific reef fishes and their social system, trophic affiliation, burrowing behaviour, activity patterns, size, and reproductive traits (Continued)

Family	Species	Pairs	Troph.	Burr.	Act.	ML (mm)	Spa.	Gam.	Mon.	Refs.
	Vanderhorstia macropteryx	No	N/A	Yes*	D	60	P	Bc	No	2
	Vanderhorstia ornatissima	No	N/A	Yes*	D	60	P	Bc	No	2, 4
Haemulidae	*Diagramma melanacrum*	No	Ma	No	N	500	G	Br	No	2
	Diagramma pictum	No	Ca	No	N	940	G	Br	No	1, 2, 3, 4
	Plectorhinchus albovittatus	No	Ma	No	N	1000	G	Br	No	1, 2, 3
	Plectorhinchus chaetodonoides	No	Ca	No	N	720	G	Br	No	1, 2, 3, 4
	Plectorhinchus chrysotaenia	No	Ma	No	N	500	G	Br	No	2, 3
	Plectorhinchus flavomaculatus	No	Ca	No	N	600	G	Br	No	1, 2, 3
	Plectorhinchus gibbosus	No	Ma	No	N	600	G	Br	No	1, 2, 3
	Plectorhinchus lessonii	No	Ma	No	N	480	G	Br	No	1, 2, 3
	Plectorhinchus lineatus	No	Ma	No	N	480	G	Br	No	1, 2, 3, 4
	Plectorhinchus multivittatus	No	Ma	No	N	500	G	Br	No	2, 3
	Plectorhinchus picus	No	Ma	No	N	850	G	Br	No	1, 2, 3, 4
	Plectorhinchus polytaenia	No	Mi	No	N	400	G	Br	No	2
	Plectorhinchus unicolor	No	Ma	No	N	800	G	Br	No	1, 2, 3
	Plectorhinchus vittatus	No	Ma	No	N	850	G	Br	No	1, 2, 3
	Pomadasys argenteus	N/A	N/A	No	D	520	G	Br	No	1
Holocentridae	*Myripristis adusta*	No	Pl	No	N	320	G	Br	No	4
	Myripristis amaena	No	Pl	No	N	320	G	Br	No	4
	Myripristis berndti	No	Pl	No	N	300	G	Br	No	4
	Myripristis botche	Yes	Pl	No	N	300	G	Br	No	4
	Myripristis chryseres	No	Pl	No	N	250	G	Br	No	4
	Myripristis earlei	No	Ma	No	N	300	G	Br	No	12
	Myripristis hexagona	No	Pl	No	N	200	G	Br	No	4
	Myripristis kuntee	No	Pl	No	N	200	G	Br	No	4
	Myripristis murdjan	No	Pl	No	N	270	G	Br	No	4
	Myripristis pralinia	No	Pl	No	N	200	G	Br	No	4
	Myripristis trachyacron	No	Pl	No	N	150	G	Br	No	12
	Myripristis violacea	No	Pl	No	N	200	G	Br	No	4
	Myripristis vittata	No	Pl	No	N	200	G	Br	No	4
	Myripristis woodsi	No	Pl	No	N	210	G	Br	No	4
	Neoniphon argenteus	No	Ma	No	N	190	G	Br	No	4
	Neoniphon aurolineatus	No	Ma	No	N	250	G	Br	No	12
	Neoniphon opercularis	No	Ma	No	N	350	G	Br	No	4
	Neoniphon sammara	No	Ca	No	N	320	G	Br	No	4
	Plectrypops lima	No	Ca	No	N	160	G	Br	No	4
	Sargocentron caudimaculatum	No	Ma	No	N	250	G	Br	No	4
	Sargocentron cornutum	No	Ma	No	N	180	G	Br	No	4
	Sargocentron diadema	No	Ma	No	N	170	G	Br	No	85
	Sargocentron ensifer (= *ensiferum*)	No	Ma	No	N	250	G	Br	No	1, 2, 3
	Sargocentron iota	No	Ma	No	N	80	G	Br	No	1, 2, 3
	Sargocentron ittodai	No	Ma	No	N	170	G	Br	No	85
	Sargocentron melanospilos	No	Ma	No	N	250	G	Br	No	4

Continued

List of Indo-Pacific reef fishes and their social system, trophic affiliation, burrowing behaviour, activity patterns, size, and reproductive traits (Continued)

Family	Species	Pairs	Troph.	Burr.	Act.	ML (mm)	Spa.	Gam.	Mon.	Refs.
	Sargocentron microstoma	No	Ma	No	N	190	G	Br	No	3
	Sargocentron punctatissimum	No	Ma	No	N	200	G	Br	No	4
	Sargocentron rubrum	No	Ma	No	N	270	G	Br	No	4
	Sargocentron spiniferum	Yes	Ca	No	N	450	G	Br	No	4
	Sargocentron tiere	No	Ca	No	N	330	G	Br	No	4
	Sargocentron tiereoides	No	Ma	No	N	160	G	Br	No	4
	Sargocentron violaceum	No	Ma	No	N	250	G	Br	No	4
Kyphosidae	*Kyphosus bigibbus*	No	He	No	D	700	G	Br	No	1, 2, 3, 86
	Kyphosus cinerascens	No	He	No	D	450	G	Br	No	1, 2, 3, 86
	Kyphosus cornelii	No	He	No	D	600	G	Br	No	2, 3, 87
	Kyphosus pacificus	No	He	No	D	650	G	Br	No	3
	Kyphosus sydneyanus	No	He	No	D	600	G	Br	No	3, 86, 87
	Kyphosus vaigiensis	No	He	No	D	450	G	Br	No	1, 2, 3, 86
	Sectator ocyurus	No	He	No	D	380	G	Br	No	3, 12
Labridae	*Anampses caeruleopunctatus*	Yes	Ma	No	D	420	G	Br	No	1, 2, 3, 4
	Anampses elegans	No	Ma	No	D	300	P	Br	No	2, 4
	Anampses femininus	No	Mi	No	D	240	P	Br	No	1, 2, 3, 88
	Anampses geographicus	No	Pl	No	D	200	P	Br	No	1, 2, 3, 4
	Anampses lennardi	Yes	Mi	No	D	280	P	Br	No	2, 4
	Anampses lineatus	No	Ma	No	D	120	P	Br	No	2, 12
	Anampses melanurus	Yes	Mi	No	D	120	P	Br	No	1, 2, 3
	Anampses meleagrides	No	Ma	No	D	220	P	Br	No	1, 2, 3, 12
	Anampses neoguinaicus	No	Ma	No	D	170	P	Br	No	2, 4
	Anampses twistii	Yes	Ma	No	D	180	P	Br	No	1, 2, 3, 4
	Bodianus anthioides	No	Ma	No	D	210	P	Br	No	1, 2, 3
	Bodianus axillaris	No	Ma	No	D	200	P	Br	No	1, 2, 3, 89
	Bodianus bilunulatus	No	Ma	No	D	550	P	Br	No	2, 3
	Bodianus bimaculatus	No	Ma	No	D	100	P	Br	No	1, 2, 12
	Bodianus diana	Yes	Ma	No	D	250	P	Br	No	1, 2, 12
	Bodianus loxozonus	No	Ma	No	D	400	P	Br	No	1, 2, 3, 4
	Bodianus mesothorax	No	Ma	No	D	190	P	Br	No	1, 2, 3, 4, 12
	Bodianus neilli	No	Ma	No	D	200	P	Br	No	2, 4
	Bodianus opercularis	No	Ma	No	D	150	P	Br	No	2
	Bodianus perditio	No	Ma	No	D	150	P	Br	No	1, 2, 3
	Bodianus prognathus	No	Ma	No	D	200	P	Br	No	1, 2, 12
	Bolbometopon muricatum	No	Co	No	D	1260	B	Br	No	1, 2, 3, 90
	Calotomus carolinus	No	He	No	D	500	B	Br	No	1, 2, 3, 4
	Calotomus spinidens	No	He	No	D	190	B	Br	No	1, 2, 3, 4
	Cetoscarus bicolor	No	De	No	D	800	B	Br	No	1, 2, 3
	Cheilinus chlorourus	No	Ma	No	D	360	P	Br	No	2, 91
	Cheilinus fasciatus	No	Ma	No	D	360	P	Br	No	1, 2, 3, 91
	Cheilinus oxycephalus	Yes	Mi	No	D	170	P	Br	No	1, 2, 3, 92
	Cheilinus trilobatus	No	Ma	No	D	450	P	Br	No	1, 2, 3, 91
	Cheilinus undulatus	No	Ca	No	D	2290	G	Br	No	1, 2, 3

List of Indo-Pacific reef fishes and their social system, trophic affiliation, burrowing behaviour, activity patterns, size, and reproductive traits (Continued)

Family	Species	Pairs	Troph.	Burr.	Act.	ML (mm)	Spa.	Gam.	Mon.	Refs.
	Cheilio inermis	No	Ca	No	D	500	P	Br	No	1, 2, 3
	Chlorurus bleekeri	No	De	No	D	480	B	Br	No	1, 2, 3, 93
	Chlorurus bowersi	No	De	No	D	300	B	Br	No	2
	Chlorurus capistratoides	No	De	No	D	550	B	Br	No	2
	Chlorurus frontalis	No	De	No	D	500	B	Br	No	1, 2, 3
	Chlorurus japanensis	No	De	No	D	300	B	Br	No	2, 3
	Chlorurus microrhinos	No	De	No	D	800	B	Br	No	1, 2, 3, 93
	Chlorurus sordidus	No	De	No	D	400	B	Br	No	1, 2, 3
	Chlorurus strongylocephalus	No	De	No	D	700	B	Br	No	2
	Chlorurus troschelii	No	De	No	D	350	B	Br	No	2
	Choerodon anchorago	No	Ma	No	D	380	P	Br	No	1, 2, 3, 26
	Choerodon cephalotes	No	Ma	No	D	380	P	Br	No	1, 4
	Choerodon cyanodus	No	Ma	No	D	700	P	Br	No	1, 2
	Choerodon fasciatus	No	Ma	No	D	300	P	Br	Yes	1, 2, 3
	Choerodon graphicus	No	Ma	No	D	460	P	Br	No	1, 2, 3, 4
	Choerodon jordani	No	Ma	No	D	170	P	Br	No	2, 3
	Choerodon monostigma	No	Ma	No	D	250	P	Br	No	2
	Choerodon oligacanthus	Yes	Ma	No	D	350	P	Br	No	2
	Choerodon rubescens	No	Ma	No	D	900	P	Br	No	2, 39
	Choerodon schoenleinii	No	Ma	No	D	900	P	Br	No	1, 2
	Choerodon vitta	No	Ma	No	D	200	P	Br	No	2
	Choerodon zosterophorus	No	Ma	No	D	250	P	Br	No	2
	Cirrhilabrus aurantidorsalis	No	Pl	No	D	100	G	Br	No	2, 12
	Cirrhilabrus bathyphilus	No	Pl	No	D	110	G	Br	No	2, 12
	Cirrhilabrus condei	No	Pl	No	D	100	G	Br	No	2, 12
	Cirrhilabrus cyanopleura	No	Pl	No	D	100	G	Br	No	2, 4
	Cirrhilabrus exquisitus	No	Pl	No	D	120	G	Br	No	1, 2, 3, 12
	Cirrhilabrus filamentosus	No	Pl	No	D	120	G	Br	No	2, 4
	Cirrhilabrus flavidorsalis	No	Pl	No	D	65	G	Br	No	2, 12
	Cirrhilabrus joanallenae	No	Pl	No	D	60	G	Br	No	2, 12
	Cirrhilabrus katherinae	No	Pl	No	D	90	G	Br	No	2, 12
	Cirrhilabrus laboutei	No	Pl	No	D	110	G	Br	No	1, 2, 3, 12
	Cirrhilabrus lineatus	No	Pl	No	D	120	G	Br	No	1, 2, 3, 4
	Cirrhilabrus lubbocki	No	Pl	No	D	80	G	Br	No	2, 12, 4
	Cirrhilabrus luteovittatus	No	Pl	No	D	120	G	Br	No	2, 12
	Cirrhilabrus marjorie	No	Pl	No	D	70	G	Br	No	2, 12
	Cirrhilabrus morrisoni	No	Pl	No	D	80	G	Br	No	2, 12
	Cirrhilabrus punctatus	No	Ca	No	D	130	G	Br	No	1, 2, 3, 12
	Cirrhilabrus randalli	No	Pl	No	D	85	G	Br	No	2, 12
	Cirrhilabrus rhomboidalis	No	Pl	No	D	85	G	Br	No	2, 12
	Cirrhilabrus rubrimarginatus	No	Pl	No	D	150	G	Br	No	1, 2, 3, 12
	Cirrhilabrus rubripinnis	No	Pl	No	D	80	G	Br	No	2, 12
	Cirrhilabrus scottorum	No	Pl	No	D	120	G	Br	No	1, 2, 3, 4
	Cirrhilabrus solorensis	No	Pl	No	D	120	G	Br	No	2

Continued

List of Indo-Pacific reef fishes and their social system, trophic affiliation, burrowing behaviour, activity patterns, size, and reproductive traits (Continued)

Family	Species	Pairs	Troph.	Burr.	Act.	ML (mm)	Spa.	Gam.	Mon.	Refs.
	Cirrhilabrus temminckii	No	Pl	No	D	100	G	Br	No	2, 4, 12
	Cirrhilabrus tonozukai	No	Pl	No	D	70	G	Br	No	2, 4, 12
	Cirrhilabrus walindi	No	Pl	No	D	70	G	Br	No	2, 12
	Cirrhilabrus walshi	No	Pl	No	D	100	G	Br	No	2, 12
	Coris auricularis	No	Ma	No	D	400	P	Br	No	2, 94
	Coris aurilineata	No	Om	No	D	115	P	Br	No	2, 4, 95
	Coris aygula	No	Ma	No	D	400	P	Br	No	1, 2, 3
	Coris batuensis	No	Ma	No	D	170	P	Br	No	2, 3, 4, 12
	Coris bulbifrons	No	Ma	No	D	1400	P	Br	No	2, 4
	Coris caudimacula	No	Ma	No	D	200	P	Br	No	2, 3
	Coris centralis	No	Ma	No	D	300	P	Br	No	2, 3, 4, 12
	Coris dorsomacula	No	Ma	No	D	200	P	Br	No	1, 2, 3, 12
	Coris gaimard	No	Ma	No	D	380	P	Br	No	1, 2, 3
	Coris pictoides	No	Ma	No	D	120	P	Br	No	1, 2, 12
	Cymolutes praetextatus	No	Ma	No	D	200	P	Br	No	1, 2, 3, 12
	Cymolutes torquatus	No	N/A	No	D	120	P	Br	No	1, 2, 3
	Diproctacanthus xanthurus	No	Mi	No	D	80	P	Br	No	1, 2, 12
	Epibulus insidiator	No	Ca	No	D	350	P	Br	No	1, 2, 3
	Gomphosus varius	No	Mi	No	D	280	P	Br	No	1, 2, 3
	Halichoeres argus	No	N/A	No	D	110	P	Br	No	1, 2
	Halichoeres binotopsis	No	N/A	No	D	120	P	Br	No	2, 4
	Halichoeres biocellatus	No	N/A	No	D	120	P	Br	No	1, 2, 3
	Halichoeres chlorocephalus	No	N/A	No	D	90	P	Br	No	2
	Halichoeres chloropterus	No	Ma	No	D	190	P	Br	No	1, 2, 4
	Halichoeres chrysus	No	Mi	No	D	120	P	Br	No	1, 2, 3, 12
	Halichoeres cosmetus	No	N/A	No	D	130	P	Br	No	2
	Halichoeres hortulanus	No	Ma	No	D	270	P	Br	No	1, 2, 3, 4
	Halichoeres leucoxanthus	No	N/A	No	D	110	P	Br	No	2
	Halichoeres leucurus	Yes	N/A	No	D	120	P	Br	No	2
	Halichoeres margaritaceus	No	Ma	No	D	130	P	Br	No	1, 2, 3, 4
	Halichoeres marginatus	No	Ma	No	D	170	P	Br	No	1, 2, 3
	Halichoeres melanochir	Yes	N/A	No	D	100	P	Br	No	2
	Halichoeres melanurus	No	Mi	No	D	120	P	Br	No	1, 2, 3 4
	Halichoeres melasmapomus	No	Mi	No	D	140	P	Br	No	1, 2, 3
	Halichoeres nebulosus	No	N/A	No	D	120	P	Br	No	2, 3
	Halichoeres nigrescens	No	N/A	No	D	140	P	Br	No	2
	Halichoeres ornatissimus	No	Mi	No	D	150	P	Br	No	1, 2, 3, 4
	Halichoeres pallidus	No	Mi	No	D	80	P	Br	No	1, 2, 12
	Halichoeres papilionaceus	No	N/A	No	D	100	P	Br	No	2
	Halichoeres podostigma	No	N/A	No	D	190	P	Br	No	2
	Halichoeres prosopeion	No	N/A	No	D	130	P	Br	No	1, 2, 3
	Halichoeres richmondi	No	Mi	No	D	190	P	Br	No	2
	Halichoeres rubricephalus	No	N/A	No	D	100	P	Br	No	2
	Halichoeres scapularis	No	Mi	No	D	200	P	Br	No	1, 2, 4
	Halichoeres solorensis	No	Ma	No	D	180	P	Br	No	2, 4

List of Indo-Pacific reef fishes and their social system, trophic affiliation, burrowing behaviour, activity patterns, size, and reproductive traits (Continued)

Family	Species	Pairs	Troph.	Burr.	Act.	ML (mm)	Spa.	Gam.	Mon.	Refs.
	Halichoeres trimaculatus	No	Ma	No	D	270	P	Br	No	1, 2, 3 4
	Halichoeres trispilus	No	N/A	No	D	90	P	Br	No	2
	Halichoeres vrolikii	No	N/A	No	D	130	P	Br	No	2
	Halichoeres zeylonicus	No	N/A	No	D	150	P	Br	No	2
	Hemigymnus fasciatus	No	Ma	No	D	500	P	Br	No	1, 2, 3
	Hemigymnus melapterus	No	Ma	No	D	600	P	Br	No	1, 2, 3
	Hipposcarus harid	No	De	No	D	750	B	Br	No	2
	Hipposcarus longiceps	No	De	No	D	600	B	Br	No	1, 2, 3
	Hologymnosus annulatus	No	Pi	No	D	400	P	Br	No	1, 2, 3
	Hologymnosus doliatus	No	Ca	No	D	400	P	Br	No	1, 2, 3
	Hologymnosus longipes	No	N/A	No	D	400	P	Br	No	1, 3
	Hologymnosus rhodonotus	No	N/A	No	D	320	P	Br	No	2
	Iniistius aneitensis	No	Ma	No	D	240	P	Br	No	2, 3, 12
	Iniistius celebicus	No	Ma	No	D	250	P	Br	No	2, 3, 12
	Iniistius pavo	No	Ma	No	D	350	P	Br	No	2, 3, 4
	Iniistius pavo (= *tetrazona*)	No	Ma	No	D	250	P	Br	No	2
	Iniistius pentadactylus	No	Ma	No	D	250	P	Br	No	2, 4
	Labrichthys unilineatus	No	Co	No	D	160	P	Br	No	77
	Labroides bicolor	Yes	Mi	No	D	140	P	Br	No	1, 2, 3, 4
	Labroides dimidiatus	Yes	Mi	No	D	115	P	Br	Yes	1, 2, 3, 4
	Labroides pectoralis	No	Mi	No	D	80	P	Br	No	1, 2, 3, 4
	Labroides rubrolabiatus	Yes	Mi	No	D	90	P	Br	No	2, 3, 12
	Labropsis alleni	Yes	Co	No	D	100	P	Br	No	2, 4
	Labropsis manabei	No	Co	No	D	130	P	Br	No	77
	Labropsis micronesica	No	Mi	No	D	130	P	Br	No	2, 4, 12
	Labropsis xanthonota	No	Co	No	D	140	P	Br	No	1, 2, 3, 12
	Leptojulis cyanopleura	No	Pl	No	D	130	P	Br	No	2
	Leptoscarus vaigiensis	No	He	No	D	350	P	Br	No	1, 2, 3, 4
	Macropharyngodon choati	No	Ma	No	D	100	P	Br	No	2
	Macropharyngodon kuiteri	No	Ma	No	D	100	P	Br	No	1, 2, 3
	Macropharyngodon meleagris	No	Ma	No	D	150	P	Br	No	1, 2, 3, 4
	Macropharyngodon negrosensis	No	Ma	No	D	150	P	Br	No	1, 2, 3, 4
	Macropharyngodon ornatus	No	Ma	No	D	120	P	Br	No	2
	Novaculichthys taeniourus	No	Ma	No	D	270	P	Br	No	1, 2, 3
	Novaculoides (= *Novaculichthys*) *macrolepidotus*	No	Ma	No	D	150	P	Br	No	2, 3
	Oxycheilinus arenatus	No	Ca	No	D	190	P	Br	No	2, 1, 4
	Oxycheilinus bimaculatus	No	Ca	No	D	150	P	Br	No	2, 1, 12
	Oxycheilinus celebicus	No	N/A	No	D	240	P	Br	No	2, 1
	Oxycheilinus digramma	No	Pi	No	D	300	P	Br	No	2, 1, 12
	Oxycheilinus orientalis	No	Ca	No	D	170	P	Br	No	2, 4
	Oxycheilinus orientalis (= *rhodochrous*)	No	Ca	No	D	200	P	Br	No	2, 4
	Oxycheilinus unifasciatus	No	Ca	No	D	460	P	Br	No	2, 1, 4

Continued

List of Indo-Pacific reef fishes and their social system, trophic affiliation, burrowing behaviour, activity patterns, size, and reproductive traits (Continued)

Family	Species	Pairs	Troph.	Burr.	Act.	ML (mm)	Spa.	Gam.	Mon.	Refs.
	Oxycheilinus sp.	No	Ca	No	D	140	P	Br	No	2, 12
	Paracheilinus angulatus	No	Pl	No	D	70	P	Br	No	2
	Paracheilinus carpenteri	No	Pl	No	D	80	P	Br	No	2
	Paracheilinus filamentosus	No	Pl	No	D	80	P	Br	No	2, 4
	Paracheilinus flavianalis	No	Pl	No	D	70	P	Br	No	2
	Paracheilinus mccoskeri	No	Pl	No	D	70	P	Br	No	2
	Paracheilinus rubricaudalis	No	Pl	No	D	80	P	Br	No	2, 3
	Paracheilinus togeanensis	No	Pl	No	D	80	P	Br	No	2
	Pseudocheilinops ataenia	No	Mi	No	D	50	P	Br	No	2, 4
	Pseudocheilinus evanidus	No	Mi	No	D	80	P	Br	No	1, 2, 3, 4
	Pseudocheilinus hexataenia	No	Mi	No	D	75	P	Br	No	1, 2, 3, 4
	Pseudocheilinus ocellatus	No	Mi	No	D	85	P	Br	No	2, 1
	Pseudocheilinus octotaenia	No	Mi	No	D	135	P	Br	No	1, 2, 3, 4
	Pseudocheilinus tetrataenia	No	Mi	No	D	75	P	Br	No	2, 3, 4
	Pseudocoris aurantiofasciata	No	Pl	No	D	200	G	Br	No	2, 3
	Pseudocoris bleekeri	No	Pl	No	D	150	G	Br	No	2, 4
	Pseudocoris heteroptera	No	Pl	No	D	200	G	Br	No	2, 4
	Pseudocoris yamashiroi	No	Pl	No	D	150	G	Br	No	2, 3
	Pseudodax moluccanus	No	Ma	No	D	250	P	Br	No	1, 2, 3, 12
	Pseudojuloides atavai	No	Ma	No	D	130	P	Br	No	2, 3, 12
	Pseudojuloides cerasinus	No	Ma	No	D	120	P	Br	No	1, 2, 3, 12
	Pseudojuloides kaleidos	No	N/A	No	D	100	P	Br	No	2
	Pseudojuloides severnsi	No	Om	No	D	100	P	Br	No	2, 4
	Pteragogus cryptus	No	Mi	No	D	95	P	Br	Yes	1, 2, 3, 4
	Pteragogus enneacanthus	No	Ma	No	D	120	P	Br	No	2, 3, 12
	Pteragogus flagellifer	No	N/A	No	D	200	P	Br	No	2
	Scarus altipinnis	No	De	No	D	600	B	Br	No	1, 2, 3
	Scarus chameleon	No	De	No	D	310	B	Br	No	1, 2, 3
	Scarus dimidiatus	No	De	No	D	300	B	Br	No	1, 2, 3
	Scarus festivus	No	De	No	D	430	B	Br	No	2, 3
	Scarus flavipectoralis	No	De	No	D	410	B	Br	No	1, 2, 3
	Scarus forsteni	No	De	No	D	550	B	Br	No	1, 2, 3
	Scarus frenatus	No	De	No	D	470	B	Br	No	1, 2, 3
	Scarus ghobban	No	De	No	D	750	B	Br	No	1, 2, 3
	Scarus globiceps	No	De	No	D	270	B	Br	No	1, 2, 3
	Scarus hypselopterus	No	De	No	D	310	B	Br	No	2, 3
	Scarus koputea	N/A	De	No	D	310	B	Br	No	2, 3
	Scarus longipinnis	No	De	No	D	400	B	Br	No	1, 2, 3
	Scarus niger	No	De	No	D	350	B	Br	No	1, 2, 3
	Scarus oviceps	No	De	No	D	310	B	Br	No	1, 2, 3
	Scarus prasiognathos	No	De	No	D	700	B	Br	No	2
	Scarus psittacus	No	De	No	D	300	B	Br	No	1, 2, 3
	Scarus quoyi	No	De	No	D	210	B	Br	No	2
	Scarus rivulatus	No	De	No	D	400	B	Br	No	1, 2, 3
	Scarus rubroviolaceus	Yes	De	No	D	700	B	Br	No	1, 2, 3

List of Indo-Pacific reef fishes and their social system, trophic affiliation, burrowing behaviour, activity patterns, size, and reproductive traits (Continued)

Family	Species	Pairs	Troph.	Burr.	Act.	ML (mm)	Spa.	Gam.	Mon.	Refs.
	Scarus russelii	No	De	No	D	510	B	Br	No	2
	Scarus schlegeli	No	De	No	D	380	B	Br	No	1, 2, 3
	Scarus spinus	No	De	No	D	300	B	Br	No	1, 2, 3
	Scarus tricolor	No	De	No	D	550	B	Br	No	1, 2, 3
	Scarus viridifucatus	No	De	No	D	320	B	Br	No	2
	Scarus xanthopleura	No	De	No	D	550	B	Br	No	2, 3
	Stethojulis bandanensis	No	Mi	No	D	160	G	Br	No	1, 2, 3, 4
	Stethojulis interrupta	No	Mi	No	D	130	G	Br	No	2, 3, 4
	Stethojulis notialis	No	Mi	No	D	100	G	Br	No	2, 1
	Stethojulis strigiventer	No	Mi	No	D	140	G	Br	No	1, 2, 3
	Stethojulis trilineata	No	Mi	No	D	140	G	Br	No	2, 3
	Thalassoma amblycephalum	No	Pl	No	D	140	G	Br	No	1, 2, 3, 4
	Thalassoma hardwicke	No	Om	No	D	200	G	Br	No	1, 2, 3, 4
	Thalassoma jansenii	No	N/A	No	D	200	G	Br	No	2, 3
	Thalassoma lunare	No	Pi	No	D	250	G	Br	No	1, 2, 3
	Thalassoma lutescens	No	Ma	No	D	250	G	Br	No	1, 2, 3, 4
	Thalassoma purpureum	No	Ma	No	D	430	G	Br	No	1, 2, 3, 4
	Thalassoma quinquevittatum	No	Ma	No	D	250	G	Br	No	1, 2, 3, 4
	Thalassoma trilobatum	No	Ma	No	D	300	G	Br	No	1, 2, 3, 4
	Wetmorella albofasciata	No	Ma	No	D	60	P	Br	No	1, 2, 3
	Wetmorella nigropinnata	No	Ma	No	D	80	P	Br	No	1, 2, 3, 4
Lethrinidae	*Gnathodentex aureolineatus*	No	Ma	No	N	300	G	Br	No	2, 3, 4
	Gymnocranius euanus	No	Ma	No	D	450	G	Br	No	2, 3
	Gymnocranius frenatus	No	Ma	No	D	350	G	Br	No	2, 4
	Gymnocranius grandoculis	No	Ca	No	D	800	G	Br	No	1, 2, 3, 4
	Gymnocranius griseus	No	Ma	No	D	350	G	Br	No	2, 4
	Gymnocranius microdon	No	Ma	No	D	410	G	Br	No	2, 4
	Lethrinus amboinensis	No	Ca	No	D	570	G	Br	No	2, 4
	Lethrinus atkinsoni	No	Ca	No	D	410	G	Br	No	1, 2, 3
	Lethrinus erythracanthus	No	Ma	No	D	700	G	Br	No	1, 2, 3, 44
	Lethrinus erythropterus	No	Ca	No	D	500	G	Br	No	2, 4
	Lethrinus genivittatus	No	Ma	No	D	200	G	Br	No	1, 2, 3, 96
	Lethrinus harak	No	Ca	No	D	500	G	Br	No	1, 2, 3, 4
	Lethrinus laticaudis	No	Ca	No	D	560	G	Br	No	1, 2, 3, 4
	Lethrinus lentjan	No	Ma	No	D	500	G	Br	No	1, 2, 3
	Lethrinus microdon	No	Ca	No	D	700	G	Br	No	2, 4
	Lethrinus miniatus	No	Ca	No	D	900	G	Br	No	1, 2, 3
	Lethrinus nebulosus	No	Ca	No	D	800	G	Br	No	1, 2, 3
	Lethrinus obsoletus	No	Ma	No	D	500	G	Br	No	1, 2, 3
	Lethrinus olivaceus	No	Ca	No	D	1000	G	Br	No	1, 2, 3, 44
	Lethrinus ornatus	No	Ca	No	D	400	G	Br	No	1, 2, 4
	Lethrinus ravus	No	N/A	No	D	320	G	Br	No	3
	Lethrinus rubrioperculatus	No	Ca	No	D	500	G	Br	No	1, 2, 3
	Lethrinus semicinctus	No	Ca	No	D	290	G	Br	No	2, 4

Continued

List of Indo-Pacific reef fishes and their social system, trophic affiliation, burrowing behaviour, activity patterns, size, and reproductive traits (Continued)

Family	Species	Pairs	Troph.	Burr.	Act.	ML (mm)	Spa.	Gam.	Mon.	Refs.
	Lethrinus variegatus	No	Ma	No	D	200	G	Br	No	2, 3
	Lethrinus xanthochilus	No	Ca	No	D	600	G	Br	No	1, 2, 3, 44
	Monotaxis grandoculis	No	Ma	No	N	600	G	Br	No	1, 2, 3, 44
Lutjanidae	*Aphareus furca*	No	Ca	No	D	400	G	Br	No	1, 2, 3
	Aphareus rutilans	No	Ca	No	D	1100	G	Br	No	3, 4
	Aprion virescens	No	Ca	No	D	1000	G	Br	No	1, 2, 3, 44
	Lutjanus adetii	No	Ca	No	N	500	G	Br	No	3, 4
	Lutjanus argentimaculatus	No	Ca	No	D	1200	G	Br	No	1, 2, 3
	Lutjanus bengalensis	No	Ca	No	D	300	G	Br	No	2
	Lutjanus biguttatus	No	Ca	No	D	200	G	Br	No	2, 3, 4
	Lutjanus bohar	No	Pi	No	D	750	G	Br	No	1, 2, 3, 44
	Lutjanus boutton	No	Ca	No	D	280	G	Br	No	2, 4
	Lutjanus carponotatus	No	Ca	No	D	400	G	Br	No	1, 2, 3, 97
	Lutjanus decussatus	No	Ca	No	D	300	G	Br	No	1, 2, 98
	Lutjanus ehrenbergii	No	Ca	No	D	350	G	Br	No	1, 2, 3
	Lutjanus fulviflamma	No	Ca	No	N	350	G	Br	No	1, 2, 3, 97
	Lutjanus fulvus	No	Ca	No	D	400	G	Br	No	1, 2, 3, 44
	Lutjanus gibbus	No	Ca	No	N	500	G	Br	No	1, 2, 3
	Lutjanus johnii	No	Ca	No	D	700	G	Br	No	3, 4, 99
	Lutjanus kasmira	No	Ca	No	D	350	G	Br	No	1, 2, 3
	Lutjanus lemniscatus	No	Ca	No	D	650	G	Br	No	1, 2, 4
	Lutjanus lunulatus	No	Ca	No	D	350	G	Br	No	2
	Lutjanus lutjanus	No	Ca	No	D	300	G	Br	No	1, 2, 3, 4
	Lutjanus madras	No	Ca	No	D	300	G	Br	No	2, 24
	Lutjanus malabaricus	No	Ca	No	N	1000	G	Br	No	2, 4
	Lutjanus maxweberi	No	Ca	No	D	800	G	Br	No	2
	Lutjanus monostigma	No	Pi	No	D	600	G	Br	No	1, 2, 3, 44
	Lutjanus quinquelineatus	No	Ca	No	D	390	G	Br	No	1, 2, 3, 100
	Lutjanus rivulatus	No	Pi	No	D	800	G	Br	No	3
	Lutjanus rufolineatus	No	Ca	No	D	280	G	Br	No	2, 3, 4
	Lutjanus russelii	No	Ca	No	D	280	G	Br	No	1, 2, 3, 101
	Lutjanus sebae	No	Ca	No	D	600	G	Br	No	1, 2, 3, 4
	Lutjanus semicinctus	No	Pi	No	D	350	G	Br	No	1, 2, 4
	Lutjanus timoriensis	No	Pi	No	D	500	G	Br	No	2, 4, 102
	Lutjanus vitta	No	Ca	No	D	400	G	Br	No	1, 2, 3, 39
	Macolor macularis	No	Pl	No	N	600	G	Br	No	1, 2, 3, 4
	Macolor niger	No	Ca	No	D	600	G	Br	No	1, 2, 3, 4
	Paracaesio sordida	No	Pl	No	D	400	G	Br	No	2, 3, 4
	Paracaesio xanthura	No	Pl	No	D	400	G	Br	No	2, 3
	Pinjalo lewisi	No	Ma	No	D	500	G	Br	No	2, 3, 4
	Pinjalo pinjalo	No	Ma	No	D	500	G	Br	No	2, 3, 4
	Symphorichthys spilurus	No	Ma	No	D	600	G	Br	No	1, 2, 3
	Symphorus nematophorus	No	Pi	No	D	800	G	Br	No	1, 2, 3

List of Indo-Pacific reef fishes and their social system, trophic affiliation, burrowing behaviour, activity patterns, size, and reproductive traits (Continued)

Family	Species	Pairs	Troph.	Burr.	Act.	ML (mm)	Spa.	Gam.	Mon.	Refs.
Malacanthidae	*Hoplolatilus chlupatyi*	Yes	Pl	Yes	D	150	P	Br	Yes	2, 103
	Hoplolatilus cuniculus	Yes	Pl	Yes	D	150	P	Br	Yes	1, 2, 3, 103
	Hoplolatilus fronticinctus	Yes	Pl	Yes	D	170	P	Br	Yes	2, 103
	Hoplolatilus luteus	Yes	Pl	Yes	D	110	P	Br	Yes	2, 103
	Hoplolatilus marcosi	Yes	Pl	Yes	D	150	P	Br	Yes	2, 103
	Hoplolatilus purpureus	Yes	Pl	Yes	D	120	P	Br	Yes	2, 103
	Hoplolatilus starcki	Yes	Pl	Yes	D	150	P	Br	Yes	1, 2, 103
	Malacanthus brevirostris	Yes	Ma	Yes	D	300	P	Br	No	1, 2, 3, 104
	Malacanthus latovittatus	Yes	Ma	Yes	D	350	P	Br	Yes	1, 2, 3, 104
Microcanthidae	*Microcanthus strigatus*	No	Om	No	D	160	N/A	Br	No	1, 2, 3
	Gunnellichthys curiosus	Yes	Mi	Yes	D	120	N/A	Br	No	1, 2, 3, 4
	Gunnellichthys monostigma	Yes	Mi	Yes	D	110	N/A	Br	No	1, 2, 3, 4
	Gunnellichthys pleurotaenia	Yes	Mi	Yes	D	90	N/A	Br	No	1, 2, 3, 4
	Gunnellichthys viridescens	Yes	Mi	Yes	D	100	N/A	Br	No	1, 2, 3, 4
Molidae	*Mola mola*	No	Ca	No	D	3080	N/A	Br	No	4
Monacanthidae	*Acreichthys radiatus*	No	Co	No	D	70	P	Bc	No	4
	Acreichthys tomentosus	No	Ma	No	D	100	P	Bc	No	4
	Aluterus monoceros	No	Om	No	D	750	P	Bc	No	105
	Aluterus scriptus	No	Om	No	D	750	P	Bc	No	1, 2, 3
	Amanses scopas	Yes	Co	No	D	200	G	Br	No	106
	Anacanthus barbatus	No	N/A	No	D	350	P	Bc	No	1, 2, 3
	Cantherhines dumerilii	Yes	Co	No	D	350	P	Bc	No	77
	Cantherhines fronticinctus	No	Ma	No	D	230	P	Bc	No	4
	Cantherhines pardalis	No	Ma	No	D	250	P	Bc	No	4
	Cantherhines sandwichiensis	No	Om	No	D	194	P	Bc	No	1, 2, 3
	Chaetodermis penicilligerus	No	N/A	No	D	310	P	Bc	No	1, 2, 3
	Monacanthus chinensis	No	Om	No	D	380	P	Bc	No	4
	Oxymonacanthus longirostris	Yes	Co	No	D	90	P	Bc	Yes	77
	Paraluteres arqat	No	N/A	No	D	80	P	Bc	No	1, 2, 3
	Paraluteres prionurus	Yes	Om	No	D	100	P	Bc	No	4
	Paramonacanthus choirocephalus	No	Om	No	D	130	P	Bc	No	24
	Paramonacanthus japonicus	Yes	Mi	No	D	100	P	Bc	Yes	107
	Pervagor alternans	No	Mi	No	D	160	P	Bc	No	12
	Pervagor aspricaudus	No	Mi	No	D	120	P	Bc	No	12
	Pervagor janthinosoma	Yes	Mi	No	D	140	P	Bc	No	4
	Pervagor melanocephalus	Yes	Mi	No	D	100	P	Bc	No	12
	Pervagor nigrolineatus	No	Mi	No	D	100	P	Bc	No	12
	Pseudalutarius nasicornis	Yes	Mi	No	D	180	P	Bc	No	108
	Pseudomonacanthus macrurus	Yes	Om	No	D	240	P	Bc	No	21

Continued

List of Indo-Pacific reef fishes and their social system, trophic affiliation, burrowing behaviour, activity patterns, size, and reproductive traits (Continued)

Family	Species	Pairs	Troph.	Burr.	Act.	ML (mm)	Spa.	Gam.	Mon.	Refs.
	Rudarius excelsus	N/A	N/A	No	D	25	G	Bc	No	1, 2, 3
	Rudarius minutus	No	N/A	No	D	30	G	Bc	No	1, 2, 3
Monocentridae	*Cleidopus gloriamaris*	Yes	N/A	No	D	220	N/A	Br	No	1, 2, 3
	Monocentris japonica	Yes	N/A	No	D	170	N/A	Br	No	1, 2, 3
Mugilidae	*Crenimugil crenilabis*	No	De	No	D	400	G	Br	No	2, 3, 109
	Liza (= Chelon) macrolepis	No	De	No	D	400	G	Br	No	3, 15
	Liza (= Chelon) melinoptera	No	De	No	D	300	G	Br	No	3, 15
	Liza (= Chelon) subviridis	No	De	No	D	400	G	Br	No	3, 4
	Liza (= Ellochelon) vaigiensis	No	Om	No	D	520	G	Br	No	2, 3, 15
	Moolgarda seheli	No	De	No	D	500	G	Br	No	3, 4
	Mugil cephalus	No	De	No	D	300	G	Br	No	3, 15
	Neomyxus leuciscus	No	Om	No	D	460	G	Br	No	2, 3, 4
	Valamugil (= Moolgarda) engeli	No	De	No	D	150	G	Br	No	2, 109
Mullidae	*Mulloidichthys flavolineatus*	No	Ma	No	N	400	G	Br	No	1, 2, 3, 4
	Mulloidichthys mimicus	No	Ma	No	N	300	G	Br	No	1, 2, 3
	Mulloidichthys pfluegeri	Yes	Ma	No	D	480	G	Br	No	1, 2, 3,
	Mulloidichthys vanicolensis	No	Ma	No	N	380	G	Br	No	1, 2, 3, 4
	Parupeneus barberinoides	No	Ma	No	D	250	G	Br	No	1, 2, 3, 4
	Parupeneus barberinus	No	Ma	No	D	500	G	Br	No	1, 2, 3, 4
	Parupeneus ciliatus	No	Ma	No	N	380	G	Br	No	1, 2, 3, 4
	Parupeneus crassilabris	No	Ma	No	D	350	G	Br	No	1, 2, 3
	Parupeneus cyclostomus	Yes	Ca	No	D	500	P	Br	No	1, 2, 3, 4
	Parupeneus heptacanthus	No	Ma	No	D	360	P	Br	No	1, 2, 3, 110
	Parupeneus indicus	No	Ma	No	N	350	P	Br	No	1, 2, 3, 4
	Parupeneus insularis	No	Ma	No	D	330	P	Br	No	1, 2, 3, 111
	Parupeneus macronemus	No	Ma	No	D	320	P	Br	No	1, 2, 3, 112
	Parupeneus multifasciatus	No	Ma	No	D	300	P	Br	No	1, 2, 3, 110
	Parupeneus pleurostigma	No	Ma	No	D	330	P	Br	No	1, 2, 3, 110
	Parupeneus rubescens	No	Ma	No	D	430	P	Br	No	1, 2, 3, 113
	Parupeneus spilurus	No	Ma	No	N	350	P	Br	No	1, 2, 3
	Parupeneus trifasciatus	No	Ca	No	N	350	P	Br	No	1, 2, 3, 4
	Upeneus moluccensis	No	Ca	No	D	200	P	Br	No	114
	Upeneus taeniopterus	No	Ca	No	D	360	P	Br	No	2
	Upeneus tragula	No	Ca	No	D	300	P	Br	No	115
	Upeneus vittatus	No	Ca	No	D	280	P	Br	No	115
Muraenidae	*Channomuraena vittata*	No	N/A	No	N	1500	P	Br	No	4
	Echidna delicatula	No	Ma	No	N	650	P	Br	No	116
	Echidna nebulosa	No	Ma	No	N	750	P	Br	No	116
	Echidna polyzona	No	Ma	No	N	600	P	Br	No	116

List of Indo-Pacific reef fishes and their social system, trophic affiliation, burrowing behaviour, activity patterns, size, and reproductive traits (Continued)

Family	Species	Pairs	Troph.	Burr.	Act.	ML (mm)	Spa.	Gam.	Mon.	Refs.
	Enchelycore bayeri	No	Pi	No	N	600	P	Br	No	116
	Enchelycore lichenosa	No	Pi	No	N	925	P	Br	No	4
	Enchelycore pardalis	No	Pi	No	N	800	P	Br	No	12
	Enchelycore schismatorhynchus	No	Pi	No	N	1200	P	Br	No	4
	Enchelynassa canina	No	Ca	No	N	1540	P	Br	No	4
	Gymnomuraena zebra	No	Ma	No	D	1540	P	Br	No	117
	Gymnothorax albimarginatus	No	N/A	No	N	1000	P	Br	No	4
	Gymnothorax breedeni	No	Pi	No	N	750	P	Br	No	116
	Gymnothorax buroensis	No	N/A	No	N	330	P	Br	No	1, 2, 3
	Gymnothorax chilospilus	No	N/A	No	D	500	P	Br	No	1, 2, 3
	Gymnothorax chlamydatus	No	N/A	No	D	800	P	Br	No	1, 2, 3
	Gymnothorax cribroris	No	N/A	No	N	470	P	Br	No	1, 2, 3
	Gymnothorax enigmaticus	No	N/A	No	N	580	P	Br	No	1, 2, 3
	Gymnothorax eurostus	No	Ca	No	D	650	P	Br	No	3
	Gymnothorax favagineus	No	Ca	No	D	1800	P	Br	No	4
	Gymnothorax fimbriatus	No	Ca	No	N	800	P	Br	No	4
	Gymnothorax flavimarginatus	No	Ca	No	N	1500	P	Br	No	3
	Gymnothorax gracilicauda	No	N/A	No	N	320	P	Br	No	1, 2, 3
	Gymnothorax herrei	No	N/A	No	D	300	G	Br	No	1, 2, 3
	Gymnothorax isingteena	No	N/A	No	D	1800	P	Br	No	1, 2, 3
	Gymnothorax javanicus	No	Pi	No	N	2390	P	Br	No	3
	Gymnothorax melatremus	No	Ma	No	N	200	P	Br	No	12
	Gymnothorax meleagris	No	Pi	No	N	1200	P	Br	No	3
	Gymnothorax minor	No	N/A	No	D	600	P	Br	No	1, 2, 3
	Gymnothorax nudivomer	No	N/A	No	D	1800	P	Br	No	1, 2, 3
	Gymnothorax pictus	No	Ca	No	D	1200	P	Br	No	4
	Gymnothorax richardsonii	No	N/A	No	D	320	P	Br	No	1, 2, 3
	Gymnothorax rueppelliae	No	Ca	No	N	800	P	Br	No	12
	Gymnothorax thyrsoideus (= *Siderea thysoidea*)	Yes	N/A	No	D	650	P	Br	No	4
	Gymnothorax tile	No	N/A	No	D	530	P	Br	No	1, 2, 3
	Gymnothorax undulatus	No	Ca	No	N	1500	P	Br	No	4
	Gymnothorax zonipectis	No	N/A	No	N	800	P	Br	No	1, 2, 3
	Pseudechidna brummeri	No	N/A	No	N	1030	P	Br	No	1, 2, 3
	Rhinomuraena quaesita	Yes	Pi	No	D	850	P	Br	No	116
	Scuticaria okinawae	No	N/A	No	D	930	P	Br	No	1, 2, 3
	Scuticaria tigrina	No	N/A	No	N	1200	P	Br	No	1, 2, 3
	Strophidon sathete	No	Ca	No	D	3750	P	Br	No	4
	Uropterygius fasciolatus	No	Pi	No	D	530	P	Br	No	117
	Uropterygius macrocephalus	No	Pi	No	D	400	P	Br	No	117
Nemipteridae	*Nemipterus furcosus*	No	Ca	No	D	300	G	Br	No	2, 3, 101
	Nemipterus peronii	No	Ca	No	D	280	G	Br	No	3, 4
	Nemipterus zysron	No	Ca	No	D	250	G	Br	No	3, 4
	Pentapodus aureofasciatus	Yes	Ca	No	D	250	G	Br	No	2, 3

Continued

List of Indo-Pacific reef fishes and their social system, trophic affiliation, burrowing behaviour, activity patterns, size, and reproductive traits (Continued)

Family	Species	Pairs	Troph.	Burr.	Act.	ML (mm)	Spa.	Gam.	Mon.	Refs.
	Pentapodus bifasciatus	No	Ca	No	D	200	G	Br	No	2, 4
	Pentapodus caninus	No	Ca	No	D	250	G	Br	No	2, 3
	Pentapodus emeryii	No	Ca	No	D	350	G	Br	No	2, 4
	Pentapodus nagasakiensis	No	Ma	No	D	250	G	Br	No	2, 4
	Pentapodus paradiseus	No	Ca	No	D	350	G	Br	No	1, 2, 3
	Pentapodus porosus	No	Ca	No	D	300	G	Br	No	2
	Pentapodus setosus	No	Ma	No	D	250	G	Br	No	2, 4
	Pentapodus trivittatus	Yes	Ca	No	D	300	G	Br	No	2, 4
	Pentapodus (= *Scaevius*) *vitta*	No	Ca	No	D	400	G	Br	No	2, 3
	Scaevius milii	No	Ca	No	D	300	G	Br	No	2, 4
	Scolopsis affinis	No	Ma	No	D	300	G	Br	No	2, 118
	Scolopsis aurata	No	N/A	No	D	300	G	Br	No	2
	Scolopsis bilineata	No	Ma	No	N	250	G	Br	No	1, 2, 3, 119
	Scolopsis ciliata	No	Ca	No	D	250	G	Br	No	2, 3, 4
	Scolopsis ghanam	No	Ca	No	D	250	G	Br	No	2, 4
	Scolopsis lineata	No	Ca	No	D	250	G	Br	No	1, 2, 3, 4
	Scolopsis margaritifera	No	Ca	No	D	250	G	Br	No	1, 2, 120
	Scolopsis monogramma	No	Ca	No	D	380	G	Br	No	2, 4
	Scolopsis taenioptera	No	Ca	No	D	300	G	Br	No	2, 4
	Scolopsis trilineata	No	Ma	No	D	250	G	Br	No	1, 2, 3, 20
	Scolopsis vosmeri	No	Ma	No	D	250	G	Br	No	2, 4
	Scolopsis xenochrous	No	Ma	No	D	250	G	Br	No	2, 4
Ophichthidae	*Apterichtus klazingai*	No	Ca	No	D	400	G	Br	No	4
	Brachysomophis cirrocheilos	No	Ca	No	D	1250	G	Br	No	4
	Brachysomophis crocodilinus	No	Ca	No	N	820	G	Br	No	121
	Brachysomophis henshawi	No	Ca	No	N	1060	G	Br	No	4
	Callechelys catostoma	No	Ca	No	D	850	G	Br	No	1, 2, 3
	Callechelys marmorata	No	Ca	No	D	900	G	Br	No	1, 2, 3
	Leiuranus semicinctus	No	Ca	No	N	600	G	Br	No	4
	Leiuranus versicolor	No	N/A	No	D	520	G	Br	No	1, 2, 3
	Myrichthys colubrinus	No	N/A	No	D	900	G	Br	No	1, 2, 3
	Myrichthys maculosus	No	N/A	No	N	1000	G	Br	No	4
	Ophichthus altipennis (= *melanochir*)	No	N/A	No	D	800	G	Br	No	4
	Ophichthus bonaparti	No	N/A	No	N	750	G	Br	No	4
	Ophichthus cephalozona	No	N/A	No	N	1080	G	Br	No	1, 2, 3
	Ophichthus polyophthalmus	No	N/A	No	D	350	G	Br	No	4
	Pisodonophis cancrivorus	Yes	Ma	No	D	750	G	Br	No	122
Opistognathidae	*Opistognathus darwiniensis*	No	Pl	No	D	450	P	Mo	No	2
	Opistognathus dendriticus	No	Pl	No	D	450	P	Mo	No	1, 2
	Opistognathus papuensis	No	Pl	No	D	450	P	Mo	No	1, 2
	Opistognathus solorensis	No	Pl	No	D	450	P	Mo	No	1, 2
Ostraciidae	*Lactoria cornuta*	No	Ma	No	D	460	P	Br	No	1, 2, 3, 4
	Lactoria diaphana	No	Ma	No	D	250	P	Br	No	2, 123

List of Indo-Pacific reef fishes and their social system, trophic affiliation, burrowing behaviour, activity patterns, size, and reproductive traits (Continued)

Family	Species	Pairs	Troph.	Burr.	Act.	ML (mm)	Spa.	Gam.	Mon.	Refs.
	Lactoria fornasini	No	Ma	No	D	150	P	Br	No	1, 2, 3, 123
	Ostracion cubicus	No	Om	No	D	450	P	Br	No	1, 2, 3, 4
	Ostracion meleagris	Yes	Om	No	D	180	P	Br	No	1, 2, 3, 4
	Ostracion rhinorhynchos	Yes	Ma	No	D	300	P	Br	No	2, 4
	Ostracion solorensis	Yes	N/A	No	D	110	P	Br	No	1, 2, 3
	Ostracion whitleyi	Yes	Ma	No	D	155	P	Br	No	1, 2, 3, 12
	Rhynchostracion nasus	No	N/A	No	D	300	P	Br	No	2
	Tetrosomus concatenatus	No	N/A	No	D	300	P	Br	No	2
	Tetrosomus gibbosus	No	Ma	No	D	300	P	Br	No	1, 2, 3, 4
Pegasidae	*Eurypegasus draconis*	Yes	Mi	No	D	70	P	Br	Yes	1, 2, 3, 4
	Pegasus volitans	Yes	Mi	No	D	110	P	Br	No	1, 2, 3, 4
Pempheridae	*Parapriacanthus ransonneti*	No	Pl	No	N	100	G	Br	No	4
	Pempheris adusta	No	Pl	No	N	170	G	Br	No	4
	Pempheris oualensis	No	Pl	No	N	220	G	Br	No	4
	Pempheris schwenkii	No	Pl	No	N	150	G	Br	No	4
	Pempheris vanicolensis	No	Pl	No	N	200	G	Br	No	4
Pinguipedidae	*Parapercis clathrata*	No	Ma	No	D	175	P	Br	No	1, 2, 3, 12
	Parapercis cylindrica	No	Ma	No	D	230	P	Br	No	2, 89
	Parapercis hexophtalma	No	Ma	No	D	280	P	Br	No	1, 2, 3
	Parapercis lata	No	Ma	No	D	260	P	Br	No	2, 3,
	Parapercis lineopunctata	No	Ma	No	D	120	P	Br	No	2
	Parapercis maculata	No	Ma	No	D	200	P	Br	No	2
	Parapercis millepunctata	No	Ma	No	D	260	P	Br	No	1, 2, 3
	Parapercis multiplicata	No	Ma	No	D	120	P	Br	No	1, 2, 3
	Parapercis schauinslandii	No	Pl	No	D	130	P	Br	No	1, 2, 3, 4
	Parapercis snyderi	No	Ma	No	D	110	P	Br	No	1, 2, 3
	Parapercis tetracantha	No	Ma	No	D	260	P	Br	No	2
	Parapercis xanthozona	No	Ma	No	D	230	P	Br	No	1, 2, 3
Platycephalidae	*Cociella punctata*	No	Pl	No	D	350	P	Br	No	2, 14
	Cymbacephalus beauforti	No	Pi	No	D	470	P	Br	No	2, 124
	Inegocia japonica	No	N/A	No	D	250	P	Br	No	2
	Onigocia spinosa	No	N/A	No	N	130	P	Br	No	2
	Rogadius patriciae	No	N/A	No	D	270	P	Br	No	2
	Rogadius pristiger	No	N/A	No	D	210	P	Br	No	2
	Rogadius welanderi	No	N/A	No	D	130	P	Br	No	2
	Sunagocia (= *Thysanophrys*) *arenicola*	No	Ca	No	D	370	P	Br	No	1, 2, 3, 4
	Sunagocia (= *Thysanophrys*) *carbunculus*	No	Pi	No	D	400	P	Br	No	2, 125
	Sunagocia (= *Thysanophrys*) *otaitensis*	No	Ca	No	D	250	P	Br	No	1, 2, 3
	Thysanophrys chiltonae	No	Ca	No	D	220	P	Br	No	1, 2, 3, 4

Continued

List of Indo-Pacific reef fishes and their social system, trophic affiliation, burrowing behaviour, activity patterns, size, and reproductive traits (Continued)

Family	Species	Pairs	Troph.	Burr.	Act.	ML (mm)	Spa.	Gam.	Mon.	Refs.
Plesiopidae	*Assessor flavissimus*	No	Pl	No	D	55	G	Mo	No	1, 2
	Assessor macneilli	No	Om	No	D	60	G	Mo	No	1, 2, 3, 70
	Assessor randalli	No	Pl	No	D	60	G	Mo	No	2
	Belonepterygion fasciolatum	No	N/A	No	D	50	G	Bc	No	2
	Calloplesiops altivelis	No	N/A	No	D	130	G	Bc	No	1, 2, 3
	Paraplesiops poweri	No	Ma	No	D	149	G	Bc	No	1, 126
	Plesiops coeruleolineatus	No	Ca	No	N	80	G	Bc	No	1, 2, 3, 4
	Plesiops corallicola	No	Ma	No	N	160	G	Bc	No	1, 2, 3, 126
	Plesiops insularis	No	Ma	No	D	149	G	Bc	No	1, 126
	Plesiops verecundus	No	Ma	No	D	149	G	Bc	No	1, 126
Pleuronectidae	*Pseudorhombus dupliciocellatus*	No	Ca	No	D	400	P	Br	No	2
	Samaris cristatus	No	Ma	No	D	220	P	Br	No	1, 2, 3, 4
	Samariscus triocellatus	No	Ma	No	N	90	P	Br	No	1, 2, 3, 4
Plotosidae	*Paraplotosus albilabris*	No	Ma	No	N	1300	P	Bc	No	4
	Plotosus lineatus	No	Ca	No	N	320	P	Bc	No	4
Pomacanthidae	*Apolemichthys griffisi*	Yes	Sp	No	D	250	P	Bc	No	2, 4
	Apolemichthys trimaculatus	Yes	Sp	No	D	250	P	Br	No	1, 2, 3
	Apolemichthys xanthopunctatus	No	Sp	No	D	250	P	Br	No	2, 4
	Centropyge aurantia	No	N/A	No	D	100	P	Br	No	2
	Centropyge bicolor	Yes	Om	No	D	150	P	Br	No	2, 127
	Centropyge bispinosa	No	He	No	D	100	P	Br	No	2, 4, 53
	Centropyge boylei	N/A	He	No	D	70	P	Br	No	3, 4
	Centropyge colini	No	He	No	D	90	P	Br	No	1, 2, 3
	Centropyge eibli	Yes	He	No	D	110	P	Br	No	2, 4
	Centropyge ferrugata	No	He	No	D	100	P	Br	No	2, 53
	Centropyge fisheri	No	He	No	D	76	P	Br	No	2
	Centropyge flavipectoralis	No	He	No	D	100	P	Br	No	2
	Centropyge flavissima	No	He	No	D	100	P	Br	No	2, 128
	Centropyge heraldi	No	He	No	D	100	P	Br	No	2, 128
	Centropyge hotumatua	No	He	No	D	80	P	Br	No	2, 3
	Centropyge interrupta	No	He	No	D	150	P	Br	Yes	128
	Centropyge loricula	No	He	No	D	100	P	Br	No	2, 3, 4
	Centropyge multicolor	No	He	No	D	90	P	Br	No	2
	Centropyge multispinis	No	He	No	D	100	P	Br	No	2
	Centropyge narcosis	Yes	He	No	D	72	P	Br	No	3, 4
	Centropyge nigriocella	No	He	No	D	60	P	Br	No	2, 4
	Centropyge nox	Yes	He	No	D	90	P	Br	No	2, 4
	Centropyge shepardi	No	He	No	D	120	P	Br	No	2, 4, 53
	Centropyge tibicen	No	He	No	D	180	P	Br	No	2, 128
	Centropyge venusta	Yes	He	No	D	120	P	Br	No	129
	Centropyge vrolikii	No	Om	No	D	120	P	Br	Yes	2, 53
	Chaetodontoplus ballinae	No	Sp	No	D	200	P	Br	No	2, 53
	Chaetodontoplus chrysocephalus	No	Sp	No	D	220	P	Br	No	2, 4, 53
	Chaetodontoplus conspicillatus	Yes	Sp	No	D	250	P	Br	No	1, 2, 3, 53

List of Indo-Pacific reef fishes and their social system, trophic affiliation, burrowing behaviour, activity patterns, size, and reproductive traits (Continued)

Family	Species	Pairs	Troph.	Burr.	Act.	ML (mm)	Spa.	Gam.	Mon.	Refs.
	Chaetodontoplus dimidiatus	Yes	Sp	No	D	200	P	Br	No	2
	Chaetodontoplus duboulayi	Yes	Sp	No	D	250	P	Br	No	2, 53
	Chaetodontoplus melanosoma	Yes	Sp	No	D	200	P	Br	No	2, 4, 53
	Chaetodontoplus meredithi	Yes	Sp	No	D	250	P	Br	No	2, 53
	Chaetodontoplus mesoleucus	Yes	Om	No	D	180	P	Br	No	2, 53
	Chaetodontoplus personifer	N/A	N/A	No	D	350	P	Br	No	2
	Chaetodontoplus septentrionalis	No	Sp	No	D	200	P	Br	No	2, 4, 53
	Genicanthus bellus	No	Pl	No	D	180	P	Br	No	1, 2, 3, 53
	Genicanthus caudovittatus	No	Pl	No	D	200	P	Br	No	2, 3, 53
	Genicanthus lamarck	No	Pl	No	D	230	P	Br	No	2, 53
	Genicanthus melanospilos	Yes	Pl	No	D	180	P	Br	No	2, 3, 53
	Genicanthus semicinctus	No	Pl	No	D	250	P	Br	No	2, 53
	Genicanthus semifasciatus	No	Pl	No	D	210	P	Br	No	2, 53
	Genicanthus spinus	No	Pl	No	D	375	P	Br	No	3, 53
	Genicanthus watanabei	No	Pl	No	D	150	P	Br	No	2, 53
	Paracentropyge (= Centropyge) multifasciata	No	He	No	D	100	P	Br	No	1, 2, 3
	Pomacanthus annularis	Yes	Sp	No	D	450	P	Br	No	2, 53, 129
	Pomacanthus imperator	Yes	Sp	No	D	380	P	Br	No	2, 3, 53
	Pomacanthus navarchus	No	Sp	No	D	380	P	Br	No	2, 53
	Pomacanthus semicirculatus	No	Om	No	D	350	P	Br	No	2, 3, 53
	Pomacanthus sexstriatus	Yes	Om	No	D	460	P	Br	No	2, 53, 130
	Pomacanthus xanthometopon	No	Sp	No	D	380	P	Br	No	2, 53
	Pygoplites diacanthus	Yes	Sp	No	D	250	P	Br	No	2, 3, 53
Pomacentridae	*Abudefduf bengalensis*	No	Om	No	D	170	G	Bc	No	1, 2, 4
	Abudefduf conformis	N/A	N/A	No	D	170	G	Bc	No	3
	Abudefduf lorenzi	No	He	No	D	150	G	Bc	No	2, 131
	Abudefduf notatus	No	N/A	No	D	150	G	Bc	No	2, 4
	Abudefduf septemfasciatus	No	Om	No	D	190	G	Bc	No	1, 2, 3
	Abudefduf sexfasciatus	No	Pl	No	D	150	G	Bc	No	1, 2, 3, 132
	Abudefduf sordidus	No	Om	No	D	190	G	Bc	No	1, 2, 3
	Abudefduf vaigiensis	No	Pl	No	D	190	G	Bc	No	1, 2, 3, 132
	Abudefduf whitleyi	No	Pl	No	D	150	G	Bc	No	1, 2, 3, 133
	Acanthochromis polyacanthus	Yes	Pl	No	D	130	P	Bc	Yes	1, 2, 138
	Altrichthys azurelineatus	Yes	Pl	No	D	60	P	Bc	No	2, 4
	Altrichthys curatus	Yes	Pl	No	D	60	P	Bc	No	2, 4
	Amblyglyphidodon aureus	Yes	Pl	No	D	150	P	Bc	No	1, 2, 3
	Amblyglyphidodon batunai	Yes	Om	No	D	100	P	Bc	No	2, 4
	Amblyglyphidodon curacao	No	Pl	No	D	115	P	Bc	No	1, 2, 3, 133
	Amblyglyphidodon leucogaster	No	Om	No	D	130	P	Bc	No	1, 2, 4

Continued

List of Indo-Pacific reef fishes and their social system, trophic affiliation, burrowing behaviour, activity patterns, size, and reproductive traits (Continued)

Family	Species	Pairs	Troph.	Burr.	Act.	ML (mm)	Spa.	Gam.	Mon.	Refs.
	Amblyglyphidodon melanopterus	No	Om	No	D	140	P	Bc	No	3
	Amblyglyphidodon orbicularis	Yes	Om	No	D	130	P	Bc	No	2, 3
	Amblyglyphidodon ternatensis	No	Om	No	D	120	P	Bc	No	2
	Amblypomacentrus breviceps	No	N/A	No	D	80	P	Bc	No	1, 2, 4
	Amphiprion akallopisos	No	Pl	No	D	100	P	Bc	No	2, 4, 132, 134
	Amphiprion akindynos	No	Pl	No	D	120	P	Bc	No	3, 134
	Amphiprion chrysopterus	No	Om	No	D	170	P	Bc	No	1, 2, 3, 4
	Amphiprion clarkii	No	Om	No	D	120	P	Bc	Yes	1, 2, 3, 4
	Amphiprion ephippium	Yes	Pl	No	D	120	P	Bc	No	2, 4
	Amphiprion frenatus	No	Pl	No	D	60	P	Bc	Yes	2, 4, 134
	Amphiprion latezonatus	No	Pl	No	D	140	P	Bc	No	2, 4, 134
	Amphiprion leucokranos	No	Pl	No	D	130	P	Bc	No	2, 4, 134
	Amphiprion mccullochi	No	Pl	No	D	120	P	Bc	No	2, 4, 134
	Amphiprion melanopus	Yes	Om	No	D	130	P	Bc	No	1, 2, 3, 4
	Amphiprion ocellaris	No	Pl	No	D	90	P	Bc	No	2, 135
	Amphiprion percula	No	Pl	No	D	90	P	Bc	No	1, 2, 4
	Amphiprion perideraion	No	Pl	No	D	100	P	Bc	Yes	1, 2, 3, 4, 134
	Amphiprion polymnus	No	Pl	No	D	120	P	Bc	No	2, 4, 134
	Amphiprion rubrocinctus	No	Pl	No	D	120	P	Bc	No	2, 4
	Amphiprion sandaracinos	Yes	Pl	No	D	130	P	Bc	No	2, 4, 134
	Amphiprion sebae	No	Pl	No	D	140	P	Bc	No	2, 4, 134
	Amphiprion tricinctus	No	Om	No	D	130	P	Bc	No	2, 4
	Cheiloprion labiatus	No	Co	No	D	80	P	Bc	No	1, 2
	Chromis abrupta	No	Pl	No	D	80	P	Bc	No	3
	Chromis acares	No	Pl	No	D	55	P	Bc	No	1, 2, 3
	Chromis agilis	No	Pl	No	D	90	P	Bc	No	1, 2, 3, 136
	Chromis albomaculata	No	Pl	No	D	180	P	Bc	No	2
	Chromis alleni	No	Pl	No	D	80	P	Bc	No	2
	Chromis alpha	No	Pl	No	D	100	P	Bc	No	2
	Chromis amboinensis	No	Pl	No	D	75	P	Bc	No	1, 2, 3
	Chromis analis	No	Pl	No	D	150	P	Bc	No	1, 2, 3
	Chromis atripectoralis	No	Pl	No	D	100	P	Bc	No	1, 2, 3
	Chromis atripes	No	Pl	No	D	70	G	Bc	No	1, 2
	Chromis bami	No	Pl	No	D	85	P	Bc	No	3
	Chromis caudalis	No	Pl	No	D	90	P	Bc	No	2
	Chromis chrysura	No	Pl	No	D	160	P	Bc	No	1, 2, 3
	Chromis cinerascens	No	Pl	No	D	130	P	Bc	No	2
	Chromis delta	No	Pl	No	D	65	P	Bc	No	1, 2, 3
	Chromis dimidiata	No	Pl	No	D	70	P	Bc	No	2
	Chromis elerae	No	Pl	No	D	70	P	Bc	No	1, 2, 3
	Chromis fatuhivae	No	Pl	No	D	67	P	Bc	No	3
	Chromis flavapicis	No	Pl	No	D	135	P	Bc	No	3

List of Indo-Pacific reef fishes and their social system, trophic affiliation, burrowing behaviour, activity patterns, size, and reproductive traits (Continued)

Family	Species	Pairs	Troph.	Burr.	Act.	ML (mm)	Spa.	Gam.	Mon.	Refs.
	Chromis flavipectoralis	No	Pl	No	D	70	P	Bc	No	2
	Chromis flavomaculata	No	Pl	No	D	150	P	Bc	No	2, 137
	Chromis fumea	No	Pl	No	D	130	P	Bc	No	1, 2, 3
	Chromis iomelas	No	Pl	No	D	70	P	Bc	No	1, 2, 3
	Chromis lepidolepis	No	Pl	No	D	80	P	Bc	No	1, 2, 3, 139
	Chromis leucura	No	Pl	No	D	55	P	Bc	No	2, 3
	Chromis lineata	No	Pl	No	D	50	P	Bc	No	2
	Chromis margaritifer	No	Pl	No	D	80	P	Bc	No	2, 4
	Chromis nitida	No	Pl	No	D	90	P	Bc	No	1, 2, 3, 140
	Chromis notata	No	Pl	No	D	160	P	Bc	No	2
	Chromis opercularis	No	Pl	No	D	160	P	Bc	No	2
	Chromis ovatiformis	No	Pl	No	D	90	P	Bc	No	2
	Chromis pamae	No	Pl	No	D	137	P	Bc	No	3
	Chromis retrofasciata	No	Pl	No	D	55	P	Bc	No	1, 2, 3, 141
	Chromis scotochiloptera	No	Pl	No	D	150	P	Bc	No	2, 4
	Chromis ternatensis	No	Pl	No	D	70	P	Bc	No	1, 2, 3, 132
	Chromis vanderbilti	No	Pl	No	D	60	P	Bc	No	1, 2, 3
	Chromis viridis	No	Pl	No	D	80	P	Bc	No	1, 2, 3
	Chromis weberi	No	Pl	No	D	120	P	Bc	No	1, 2, 3
	Chromis westaustralis	No	Pl	No	D	100	P	Bc	No	2
	Chromis xanthochira	No	Pl	No	D	140	P	Bc	No	1, 2, 139
	Chromis xanthura	No	Pl	No	D	150	P	Bc	No	1, 2, 3
	Chrysiptera albata	No	N/A	No	D	46	P	Bc	No	3
	Chrysiptera biocellata	No	Om	No	D	70	P	Bc	No	1, 2, 3
	Chrysiptera bleekeri	Yes	N/A	Yes	D	80	P	Bc	No	2, 4
	Chrysiptera brownriggii	No	Om	No	D	85	P	Bc	No	2, 3, 4
	Chrysiptera caeruleolineata	No	Pl	No	D	55	P	Bc	No	1, 2, 3
	Chrysiptera cyanea	No	Pl	No	D	80	P	Bc	No	1, 2, 4
	Chrysiptera cymatilis	No	Pl	No	D	60	P	Bc	No	2
	Chrysiptera flavipinnis	No	N/A	No	D	80	P	Bc	No	1, 2
	Chrysiptera galba	No	N/A	No	D	95	P	Bc	No	3
	Chrysiptera glauca	No	He	No	D	80	P	Bc	No	1, 2, 3
	Chrysiptera hemicyanea	No	Pl	No	D	60	P	Bc	No	2, 4
	Chrysiptera kuiteri	Yes	Pl	Yes	D	60	P	Bc	No	2, 142
	Chrysiptera oxycephala	Yes	Pl	No	D	80	P	Bc	No	2, 4
	Chrysiptera parasema	No	N/A	No	D	60	P	Bc	No	2, 4
	Chrysiptera rex	No	He	No	D	80	P	Bc	No	1, 2, 3, 4
	Chrysiptera rollandi	No	Pl	No	D	50	P	Bc	No	1, 2, 3, 4
	Chrysiptera sinclairi	No	N/A	No	D	60	P	Bc	No	2
	Chrysiptera springeri	No	Om	No	D	60	P	Bc	No	2, 139

Continued

List of Indo-Pacific reef fishes and their social system, trophic affiliation, burrowing behaviour, activity patterns, size, and reproductive traits (Continued)

Family	Species	Pairs	Troph.	Burr.	Act.	ML (mm)	Spa.	Gam.	Mon.	Refs.
	Chrysiptera starcki	No	N/A	No	D	90	P	Bc	No	1, 2, 3
	Chrysiptera talboti	No	Pl	No	D	60	P	Bc	No	1, 2, 3, 4
	Chrysiptera taupou	No	N/A	No	D	80	P	Bc	No	1, 2, 3
	Chrysiptera traceyi	No	Om	No	D	60	P	Bc	No	2, 4
	Chrysiptera tricincta	Yes	N/A	No	D	60	P	Bc	No	2, 142
	Chrysiptera unimaculata	No	He	No	D	80	P	Bc	No	1, 2, 3, 132
	Dascyllus aruanus	No	Pl	No	D	80	P	Bc	No	1, 2, 3
	Dascyllus auripinnis	No	Pl	No	D	140	P	Bc	No	2, 3, 139
	Dascyllus carneus	No	Pl	No	D	70	P	Bc	No	2, 4
	Dascyllus flavicaudus	No	Pl	No	D	110	P	Bc	No	2, 3, 139
	Dascyllus melanurus	No	Pl	No	D	80	P	Bc	No	1, 2, 3, 143
	Dascyllus reticulatus	No	Pl	No	D	80	P	Bc	No	1, 2, 3, 143
	Dascyllus strasburgi	No	Om	No	D	105	P	Bc	No	3, 143
	Dascyllus trimaculatus	No	Om	No	D	140	P	Bc	No	1, 2, 3, 4, 143
	Dischistodus chrysopoecilus	No	De	No	D	150	P	Bc	No	1, 2
	Dischistodus darwiniensis	No	De	No	D	130	P	Bc	No	2
	Dischistodus fasciatus	No	De	No	D	130	P	Bc	No	2
	Dischistodus melanotus	No	De	No	D	150	P	Bc	No	1, 2, 144
	Dischistodus perspicillatus	No	De	No	D	190	P	Bc	No	1, 2
	Dischistodus prosopotaenia	No	De	No	D	180	P	Bc	No	1, 2
	Dischistodus pseudochrysopoecilus	No	De	No	D	160	P	Bc	No	1, 2
	Hemiglyphidodon plagiometopon	No	De	No	D	180	P	Bc	No	1, 2, 145
	Lepidozygus tapeinosoma	No	Pl	No	D	90	P	Bc	No	1, 2, 3
	Neoglyphidodon bonang	No	N/A	No	D	130	P	Bc	No	2, 4
	Neoglyphidodon carlsoni	No	N/A	No	D	120	P	Bc	No	2, 3
	Neoglyphidodon crossi	No	N/A	No	D	130	P	Bc	No	2, 4
	Neoglyphidodon melas	Yes	Co	No	D	150	P	Bc	No	1, 2, 4, 144
	Neoglyphidodon nigroris	No	Pl	No	D	110	P	Bc	No	1, 2, 3, 4
	Neoglyphidodon oxyodon	No	N/A	No	D	140	P	Bc	No	2, 4
	Neoglyphidodon polyacanthus	No	N/A	No	D	140	P	Bc	No	1, 2, 3
	Neoglyphidodon thoracotaeniatus	No	Pl	No	D	100	P	Bc	No	2, 4
	Neopomacentrus aquadulcis	No	Pl	No	D	100	P	Bc	No	2
	Neopomacentrus azysron	No	Pl	No	D	80	G	Bc	No	1, 2, 35
	Neopomacentrus bankieri	No	Pl	No	D	70	G	Bc	No	1, 2, 146
	Neopomacentrus cyanomos	No	Pl	No	D	90	G	Bc	No	1, 2, 133
	Neopomacentrus filamentosus	No	Pl	No	D	80	G	Bc	No	2, 3, 118
	Neopomacentrus metallicus	No	Pl	No	D	80	G	Bc	No	3, 4
	Neopomacentrus nemurus	No	Pl	No	D	80	G	Bc	No	2, 3, 4
	Neopomacentrus taeniurus	No	Pl	No	D	100	G	Bc	No	2

List of Indo-Pacific reef fishes and their social system, trophic affiliation, burrowing behaviour, activity patterns, size, and reproductive traits (Continued)

Family	Species	Pairs	Troph.	Burr.	Act.	ML (mm)	Spa.	Gam.	Mon.	Refs.
	Neopomacentrus violascens	No	Pl	No	D	70	G	Bc	No	2, 4
	Parma oligolepis	No	N/A	No	D	200	P	Bc	No	1
	Parma polylepis	No	De	No	D	220	P	Bc	No	1, 147
	Plectroglyphidodon dickii	No	Om	No	D	110	P	Bc	No	1, 2, 3, 4
	Plectroglyphidodon flaviventris	No	N/A	No	D	80	P	Bc	No	3, 4
	Plectroglyphidodon imparipennis	No	Om	No	D	60	P	Bc	No	1, 2, 3, 4
	Plectroglyphidodon johnstonianus	No	Co	No	D	90	P	Bc	No	1, 2, 3, 4
	Plectroglyphidodon lacrymatus	No	Om	No	D	100	P	Bc	No	1, 2, 3, 145
	Plectroglyphidodon leucozonus	No	He	No	D	100	P	Bc	No	1, 2, 3, 4
	Plectroglyphidodon phoenixensis	No	De	No	D	90	P	Bc	No	1, 2, 3, 4
	Plectroglyphidodon sagmarius	No	N/A	No	D	64	P	Bc	No	3
	Pomacentrus adelus	No	De	No	D	85	P	Bc	No	2, 3, 145
	Pomacentrus albimaculus	No	He	No	D	90	P	Bc	No	2, 4
	Pomacentrus alexanderae	No	Om	No	D	100	P	Bc	No	2, 4
	Pomacentrus alleni	No	N/A	No	D	65	P	Bc	No	2
	Pomacentrus amboinensis	No	Om	No	D	100	P	Bc	No	1, 2, 3, 148
	Pomacentrus armillatus	No	N/A	No	D	70	P	Bc	No	2
	Pomacentrus aurifrons	No	Pl	No	D	70	P	Bc	No	2, 3
	Pomacentrus auriventris	No	N/A	No	D	70	P	Bc	No	2
	Pomacentrus australis	No	He	No	D	80	P	Bc	No	1, 2, 139
	Pomacentrus azuremaculatus	No	N/A	No	D	100	P	Bc	No	2
	Pomacentrus bankanensis	No	De	No	D	100	P	Bc	No	1, 2, 3, 139, 145
	Pomacentrus bipunctatus	No	N/A	No	D	100	P	Bc	No	2
	Pomacentrus brachialis	No	Om	No	D	100	P	Bc	No	2, 3, 139
	Pomacentrus burroughi	No	De	No	D	80	P	Bc	No	2, 145
	Pomacentrus callainus	No	N/A	No	D	95	P	Bc	No	3
	Pomacentrus chrysurus	No	De	No	D	90	P	Bc	No	1, 2, 3, 145
	Pomacentrus coelestis	No	Pl	No	D	70	P	Bc	No	1, 2, 3, 4
	Pomacentrus colini	No	N/A	No	D	90	P	Bc	No	2
	Pomacentrus cuneatus	No	N/A	No	D	90	P	Bc	No	2
	Pomacentrus geminospilus	No	N/A	No	D	75	P	Bc	No	2
	Pomacentrus grammorhynchus	No	De	No	D	120	P	Bc	No	1, 2, 149
	Pomacentrus imitator	No	Pl	No	D	100	P	Bc	No	1, 2, 3
	Pomacentrus javanicus	No	N/A	No	D	80	P	Bc	No	2
	Pomacentrus komodoensis	No	N/A	No	D	80	P	Bc	No	2
	Pomacentrus lepidogenys	No	Pl	No	D	90	P	Bc	No	1, 2, 139
	Pomacentrus limosus	No	N/A	No	D	70	P	Bc	No	2
	Pomacentrus littoralis	No	N/A	No	D	100	P	Bc	No	2
	Pomacentrus melanochir	No	Pl	No	D	70	P	Bc	No	2, 4

Continued

List of Indo-Pacific reef fishes and their social system, trophic affiliation, burrowing behaviour, activity patterns, size, and reproductive traits (Continued)

Family	Species	Pairs	Troph.	Burr.	Act.	ML (mm)	Spa.	Gam.	Mon.	Refs.
	Pomacentrus milleri	No	De	No	D	90	P	Bc	No	2, 150
	Pomacentrus moluccensis	No	Om	No	D	70	P	Bc	No	1, 2, 3, 69, 139
	Pomacentrus nagasakiensis	No	Pl	No	D	120	P	Bc	No	1, 2, 3, 4
	Pomacentrus nigromanus	No	Om	No	D	90	P	Bc	No	2, 4
	Pomacentrus nigromarginatus	No	Om	No	D	90	P	Bc	No	1, 2, 3, 139
	Pomacentrus opisthostigma	No	N/A	No	D	80	P	Bc	No	2
	Pomacentrus pavo	No	Pl	No	D	110	P	Bc	No	1, 2, 3
	Pomacentrus philippinus	No	Pl	No	D	100	P	Bc	No	1, 2, 3
	Pomacentrus polyspinus	No	N/A	No	D	100	P	Bc	No	2
	Pomacentrus proteus	No	N/A	No	D	100	P	Bc	No	2
	Pomacentrus reidi	No	N/A	No	D	110	P	Bc	No	2
	Pomacentrus saksonoi	No	N/A	No	D	90	P	Bc	No	2
	Pomacentrus similis	No	Om	No	D	70	P	Bc	No	2, 148
	Pomacentrus simsiang	No	De	No	D	90	P	Bc	No	2, 139
	Pomacentrus smithi	No	Pl	No	D	70	P	Bc	No	2, 139
	Pomacentrus spilotoceps	No	N/A	No	D	80	P	Bc	No	2
	Pomacentrus stigma	No	N/A	No	D	120	P	Bc	No	2
	Pomacentrus taeniometopon	No	De	No	D	100	P	Bc	No	2, 4
	Pomacentrus tripunctatus	No	De	No	D	100	P	Bc	No	1, 2, 145
	Pomacentrus vaiuli	No	Om	No	D	100	P	Bc	No	2, 3
	Pomacentrus wardi	No	De	No	D	100	P	Bc	No	1, 2, 139, 145
	Pomacentrus yoshii	No	N/A	No	D	80	P	Bc	No	2, 3
	Pomachromis fuscidorsalis	No	Pl	No	D	85	P	Bc	No	3
	Pomachromis guamensis	No	Pl	No	D	55	P	Bc	No	2
	Pomachromis richardsoni	No	N/A	No	D	55	P	Bc	No	1, 2, 3
	Premnas biaculeatus	Yes	Om	No	D	80	P	Bc	No	2, 4
	Pristotis obtusirostris	No	Pl	No	D	100	P	Bc	No	2, 3
	Stegastes albifasciatus	No	De	No	D	110	P	Bc	No	1, 2, 3
	Stegastes altus	No	De	No	D	150	P	Bc	No	2, 151
	Stegastes apicalis	No	De	No	D	130	P	Bc	No	1, 2, 145
	Stegastes aureus	No	De	No	D	110	P	Bc	No	1, 2, 3, 4
	Stegastes emeryi	No	N/A	No	D	90	P	Bc	No	3
	Stegastes fasciolatus	No	De	No	D	110	P	Bc	Yes	1, 2, 3, 152
	Stegastes gascoynei	No	De	No	D	150	P	Bc	No	1, 2, 3, 4
	Stegastes lividus	No	De	No	D	150	P	Bc	No	1, 2, 3
	Stegastes nigricans	No	De	No	D	140	P	Bc	No	1, 2, 3
	Stegastes obreptus	No	De	No	D	150	P	Bc	No	2, 150
	Stegastes punctatus	No	De	No	D	150	P	Bc	No	3
Priacanthidae	*Heteropriacanthus cruentatus*	No	Ca	No	N	320	G	Br	No	4
	Priacanthus blochii	No	Pl	No	N	350	G	Br	No	12
	Priacanthus hamrur	No	Ca	No	N	400	G	Br	No	4

List of Indo-Pacific reef fishes and their social system, trophic affiliation, burrowing behaviour, activity patterns, size, and reproductive traits (Continued)

Family	Species	Pairs	Troph.	Burr.	Act.	ML (mm)	Spa.	Gam.	Mon.	Refs.
	Pristigenys niphonia	No	N/A	No	N	350	G	Br	No	1, 2, 3
Pseudochromidae	*Amsichthys knighti*	No	N/A	No	D	45	P	Bc	No	1
	Congrogadus subducens	No	Ca	No	D	450	P	Bc	No	1, 2, 3, 126
	Cypho purpurascens	Yes	N/A	No	D	75	P	Bc	No	1, 2
	Labracinus cyclophthalmus	Yes	Ca	No	D	200	P	Bc	No	1, 2, 89
	Labracinus lineatus	No	N/A	No	D	250	P	Bc	No	2
	Lubbockichthys multisquamatus	No	Mi	No	D	75	P	Bc	No	1, 2, 153
	Manonichthys (= *Pseudochromis*) *paranox*	No	Ma	No	D	70	P	Bc	No	1, 2
	Manonichthys polynemus	No	N/A	No	D	120	P	Bc	No	2, 4
	Manonichthys splendens	Yes	N/A	No	D	130	P	Bc	No	2, 4
	Ogilbyina novaehollandiae	Yes	N/A	No	D	100	P	Bc	No	1, 2
	Ogilbyina queenslandiae	No	N/A	No	D	150	P	Bc	No	1, 2
	Ogilbyina salvati	N/A	N/A	No	D	80	P	Bc	No	3
	Oxycercichthys veliferus	No	N/A	No	D	120	P	Bc	No	1, 2
	Pictichromis coralensis	No	N/A	No	D	70	P	Bc	No	3
	Pictichromis (= *Pseudochromis*) *diadema*	No	N/A	No	D	60	P	Bc	No	2
	Pictichromis paccagnellae	No	N/A	No	D	70	P	Bc	No	1, 2
	Pictichromis porphyrea	No	Ma	No	D	70	P	Bc	No	2, 3
	Pseudochromis andamanensis	No	N/A	No	D	70	P	Bc	No	2
	Pseudochromis bitaeniatus	No	Ma	No	D	70	P	Bc	No	1, 2
	Pseudochromis cyanotaenia	Yes	Mi	No	D	60	P	Bc	No	1, 2, 4
	Pseudochromis elongatus	No	Ma	No	D	35	P	Bc	No	2
	Pseudochromis flammicauda	Yes	Ma	No	D	55	P	Bc	No	1, 2
	Pseudochromis fuscus	No	Ca	No	D	90	P	Bc	No	1, 2, 3, 154
	Pseudochromis jamesi	No	Ma	No	D	55	P	Bc	No	1, 3
	Pseudochromis marshallensis	No	Ma	No	D	70	P	Bc	No	2
	Pseudochromis moorei	Yes	Ma	No	D	100	P	Bc	No	2, 4
	Pseudochromis perspicillatus	No	Ma	No	D	120	P	Bc	No	1, 2
	Pseudochromis quinquedentatus	No	Ma	No	D	95	P	Bc	No	1
	Pseudochromis ransonneti	No	Ma	No	D	120	P	Bc	No	2
	Pseudochromis steenei	Yes	Ma	No	D	120	P	Bc	No	2
	Pseudochromis wilsoni	No	Ma	No	D	80	P	Bc	No	1, 2
	Pseudoplesiops immaculatus	No	N/A	No	D	50	P	Bc	No	1, 2
	Pseudoplesiops rosae	No	Ma	No	D	23	P	Bc	No	1, 4
	Pseudoplesiops typus	No	N/A	No	D	70	P	Bc	No	1
Ptereleotridae	*Aioliops megastigma*	No	N/A	No	D	30	P	Bc	No	2
	Nemateleotris decora	Yes	Pl	Yes	D	85	P	Bc	Yes	1, 2, 3, 4
	Nemateleotris helfrichi	Yes	Pl	Yes	D	65	P	Bc	Yes	2, 3
	Nemateleotris magnifica	Yes	Pl	Yes	D	80	P	Bc	Yes	1, 2, 3, 4
	Oxymetopon compressus	No	N/A	No	D	200	P	Bc	No	2

Continued

List of Indo-Pacific reef fishes and their social system, trophic affiliation, burrowing behaviour, activity patterns, size, and reproductive traits (Continued)

Family	Species	Pairs	Troph.	Burr.	Act.	ML (mm)	Spa.	Gam.	Mon.	Refs.
	Oxymetopon cyanoctenosum	Yes	N/A	Yes	D	200	P	Bc	No	2, 4
	Parioglossus formosus	No	N/A	No	D	40	P	Bc	No	2
	Parioglossus interruptus	No	N/A	No	D	30	P	Bc	No	2
	Parioglossus nudus	No	N/A	No	D	20	P	Bc	No	2
	Parioglossus raoi	No	N/A	No	D	30	P	Bc	No	2
	Ptereleotris evides	Yes	Pl	Yes	D	135	P	Bc	Yes	1, 2, 3, 12
	Ptereleotris grammica	Yes	Pl	Yes	D	100	P	Bc	No	1, 2, 12
	Ptereleotris hanae	Yes	Pl	Yes	D	120	P	Bc	No	1, 2, 3, 12
	Ptereleotris heteroptera	Yes	Pl	Yes	D	120	P	Bc	Yes	1, 2, 3, 4
	Ptereleotris microlepis	Yes	Pl	Yes	D	150	P	Bc	Yes	1, 2, 3, 4
	Ptereleotris monoptera	Yes	Pl	Yes	D	150	P	Bc	Yes	1, 2, 3, 4
	Ptereleotris uroditaenia	No	Pl	No	D	100	P	Bc	No	1, 2
	Ptereleotris zebra	No	Pl	No	D	120	P	Bc	No	1, 2, 3, 4
Rachycentridae	*Rachycentron canadum*	No	Ca	No	D	2000	G	Br	No	4
Scatophagidae	*Scatophagus argus*	No	Om	No	D	300	N/A	Bc	No	2, 3
	Selenotoca multifasciata	No	N/A	No	D	280	N/A	Bc	No	2
Scombridae	*Acanthocybium solandri*	No	Pi	No	D	2100	G	Br	No	2
	Grammatorcynus bilineatus	No	Ca	No	D	1000	G	Br	No	1, 2, 3
	Gymnosarda unicolor	No	Pi	No	D	1800	G	Br	No	2, 3, 145
	Rastrelliger kanagurta	No	Pl	No	D	380	G	Br	No	3
	Sarda orientalis	No	Ca	No	D	1020	G	Br	No	2, 4
	Scomberomorus commerson	No	Pi	No	D	2350	G	Br	No	1, 2, 3
	Thunnus albacares	No	Ca	No	D	2100	G	Br	No	2, 4
Scorpaenidae	*Brachypterois serrulata*	No	N/A	No	N	120	P	Gel	No	2
	Dendrochirus biocellatus	No	Pi	No	N	100	P	Gel	No	2, 12
	Dendrochirus brachypterus	No	Ma	No	N	170	P	Gel	No	1, 2, 3, 155
	Dendrochirus zebra	No	Ma	No	N	180	P	Gel	No	1, 2, 3, 155
	Ebosia bleekeri	No	N/A	No	N	220	P	Gel	No	2
	Iracundus signifer	No	Pi	No	D	130	P	Gel	No	2, 4, 156
	Parapterois heterura	No	N/A	No	N	230	P	Gel	No	2
	Parascorpaena mossambica	No	Ma	No	N	100	P	Gel	No	1, 2, 3, 157
	Parascorpaena picta	No	Ma	No	N	150	P	Gel	No	2, 155
	Pteroidichthys amboinensis	No	N/A	No	N	120	P	Gel	No	2
	Pterois antennata	No	Ma	No	N	200	P	Gel	No	1, 2, 3, 4
	Pterois miles	No	Ca	No	N	380	P	Gel	No	1, 2, 3, 158
	Pterois mombasae	No	Ma	No	N	160	P	Gel	No	2, 155
	Pterois radiata	No	Ma	No	N	240	P	Gel	No	2, 155
	Pterois russelii (= kodipungi)	No	N/A	No	N	350	P	Gel	No	2
	Pterois volitans	No	Ca	No	N	380	P	Gel	No	1, 2, 3, 158
	Rhinopias aphanes	No	N/A	No	D	240	P	Gel	No	2

List of Indo-Pacific reef fishes and their social system, trophic affiliation, burrowing behaviour, activity patterns, size, and reproductive traits (Continued)

Family	Species	Pairs	Troph.	Burr.	Act.	ML (mm)	Spa.	Gam.	Mon.	Refs.
	Rhinopias eschmeyeri	No	N/A	No	D	190	P	Gel	No	2
	Rhinopias frondosa	No	Ca	No	N	230	P	Gel	No	2, 4
	Scorpaenodes guamensis	No	Ma	No	N	140	P	Gel	No	1, 2, 3, 155
	Scorpaenodes hirsutus	No	Ma	No	N	70	P	Gel	No	1, 2, 3, 155
	Scorpaenodes kelloggi	No	Ma	No	D	48	P	Gel	No	3, 159
	Scorpaenodes littoralis	No	Ma	No	N	80	P	Gel	No	2
	Scorpaenodes minor	N/A	Ma	No	D	52	P	Gel	No	3
	Scorpaenodes parvipinnis	No	Ma	No	N	130	P	Gel	No	1, 2, 3, 155
	Scorpaenodes quadrispinosus	N/A	Ma	No	D	97	P	Gel	No	3
	Scorpaenodes varipinnis	No	Ma	No	N	70	P	Gel	No	2, 155
	Scorpaenopsis diabolus	Yes	Pi	No	D	280	P	Gel	No	1, 2, 3, 12
	Scorpaenopsis macrochir	Yes	N/A	No	D	150	P	Gel	No	2, 3, 4
	Scorpaenopsis neglecta	No	Pi	No	D	150	P	Gel	No	2, 160
	Scorpaenopsis oxycephala	No	Ma	No	D	360	P	Gel	No	2, 12
	Scorpaenopsis papuensis	No	N/A	No	D	220	P	Gel	No	2
	Scorpaenopsis possi	No	N/A	No	D	220	P	Gel	No	2, 3
	Scorpaenopsis venosa	No	N/A	No	D	200	P	Gel	No	2
	Sebastapistes cyanostigma	No	Ma	No	N	80	P	Gel	No	1, 2, 3, 4
	Sebastapistes mauritiana	No	Pi	No	N	80	P	Gel	No	2
	Sebastapistes strongia	No	Pi	No	N	95	P	Gel	No	3, 4
	Taenianotus triacanthus	Yes	Ca	No	D	100	P	Gel	No	1, 2, 3, 4
Serranidae	*Aethaloperca rogaa*	No	Pi	No	D	600	P	Br	No	1, 3, 155
	Anyperodon leucogrammicus	No	Pi	No	D	520	P	Br	No	1, 2, 3
	Aporops bilinearis	N/A	N/A	No	D	115	P	Br	No	1, 3
	Belonoperca chabanaudi	No	Ma	No	D	150	P	Br	No	1, 2, 3, 12
	Belonoperca pylei	N/A	N/A	No	D	80	P	Br	No	1, 3
	Cephalopholis argus	No	Pi	No	D	550	G	Br	No	1, 2, 3, 155
	Cephalopholis boenak	No	Ca	No	D	260	P	Br	No	1, 2, 3, 4
	Cephalopholis cyanostigma	No	Ca	No	D	300	P	Br	No	1, 2, 4
	Cephalopholis formosa	No	N/A	No	D	340	P	Br	No	1, 2
	Cephalopholis leopardus	No	Pi	No	D	200	P	Br	No	1, 2, 3, 12
	Cephalopholis microprion	No	Ca	No	D	240	P	Br	No	1, 2, 3, 40, 161
	Cephalopholis miniata	No	Pi	No	D	410	G	Br	No	1, 2, 3, 12
	Cephalopholis polleni	No	N/A	No	D	430	P	Br	No	2, 3
	Cephalopholis sexmaculata	No	Pi	No	D	480	G	Br	No	1, 2, 3, 12
	Cephalopholis sonnerati	No	Ma	No	D	570	G	Br	No	1, 2, 3, 155
	Cephalopholis spiloparaea	No	Pi	No	D	220	P	Br	No	1, 2, 3, 12
	Cephalopholis urodeta	No	Pi	No	D	270	G	Br	No	1, 2, 3, 12

Continued

List of Indo-Pacific reef fishes and their social system, trophic affiliation, burrowing behaviour, activity patterns, size, and reproductive traits (Continued)

Family	Species	Pairs	Troph.	Burr.	Act.	ML (mm)	Spa.	Gam.	Mon.	Refs.
	Cromileptes altivelis	No	Ca	No	D	700	P	Br	No	1, 2, 3, 4
	Diploprion bifasciatum	No	Ca	No	D	250	P	Br	No	1, 2, 3, 89, 162
	Epinephelus areolatus	No	Ca	No	D	400	G	Br	No	1, 2, 3, 4
	Epinephelus areolatus (= *waandersii*)	No	N/A	No	D	600	G	Br	No	2
	Epinephelus bilobatus	No	N/A	No	D	330	G	Br	No	2
	Epinephelus bleekeri	No	Ca	No	D	750	G	Br	No	2, 24
	Epinephelus bontoides	No	N/A	No	D	300	G	Br	No	2
	Epinephelus chlorostigma	No	Ca	No	D	750	G	Br	No	2, 4
	Epinephelus coeruleopunctatus	No	Ca	No	D	600	G	Br	No	1, 2, 3, 4
	Epinephelus coioides	No	Ca	No	D	950	G	Br	No	1, 2, 3, 4
	Epinephelus corallicola	No	N/A	No	D	490	G	Br	No	1, 2
	Epinephelus cyanopodus	No	Pi	No	D	120	G	Br	No	1, 2, 3, 12
	Epinephelus erythrurus	No	N/A	No	D	430	G	Br	No	2
	Epinephelus fasciatus	No	Ca	No	D	400	G	Br	No	1, 2, 3, 155
	Epinephelus fuscoguttatus	No	Ca	No	D	1000	G	Br	No	1, 2, 3, 44, 155
	Epinephelus hexagonatus	No	Ca	No	D	260	G	Br	No	1, 2, 3, 155
	Epinephelus howlandi	No	Pi	No	D	440	G	Br	No	1, 2, 3, 12
	Epinephelus irroratus	No	N/A	No	D	340	G	Br	No	3
	Epinephelus lanceolatus	No	Ca	No	D	2340	G	Br	No	1, 2, 3, 4
	Epinephelus longispinis	No	Ma	No	D	500	G	Br	No	2, 4
	Epinephelus macrospilos	No	Ca	No	D	430	G	Br	No	1, 2, 3, 155
	Epinephelus maculatus	No	Ca	No	D	600	G	Br	No	1, 2, 3, 4
	Epinephelus malabaricus	No	Ca	No	D	2340	G	Br	No	1, 2, 3, 4
	Epinephelus melanostigma	No	Pi	No	D	330	G	Br	No	2, 3, 12
	Epinephelus merra	No	Ca	No	D	320	G	Br	No	1, 2, 3
	Epinephelus miliaris	No	Ma	No	D	530	G	Br	No	2, 3, 4
	Epinephelus multinotatus	No	Ca	No	D	1000	G	Br	No	2, 4
	Epinephelus ongus	No	Ca	No	D	350	G	Br	No	1, 2, 3, 163
	Epinephelus polyphekadion	No	Ca	No	D	610	G	Br	No	1, 2, 3
	Epinephelus quoyanus	No	Ca	No	D	380	G	Br	No	1, 2, 4
	Epinephelus retouti	No	N/A	No	D	470	G	Br	No	3
	Epinephelus rivulatus	No	Ca	No	D	450	G	Br	No	1, 2, 3, 4
	Epinephelus socialis	No	Pi	No	D	420	G	Br	No	2, 3, 12
	Epinephelus spilotoceps	No	Ca	No	D	310	G	Br	No	2, 4
	Epinephelus tauvina	No	Pi	No	D	330	G	Br	No	1, 2, 3, 12
	Epinephelus tukula	No	Ca	No	D	2000	G	Br	No	1, 2, 3, 4
	Gracila albomarginata	No	Pi	No	D	500	N/A	Br	No	1, 2, 3, 12
	Grammistes sexlineatus	No	Pi	No	D	270	N/A	Br	No	1, 2, 3, 4
	Grammistops ocellatus	Yes	N/A	No	D	130	N/A	Gel	No	1, 2, 3

List of Indo-Pacific reef fishes and their social system, trophic affiliation, burrowing behaviour, activity patterns, size, and reproductive traits (Continued)

Family	Species	Pairs	Troph.	Burr.	Act.	ML (mm)	Spa.	Gam.	Mon.	Refs.
	Liopropoma flavidum	N/A	N/A	No	D	61	N/A	Br	No	3
	Liopropoma latifasciatum	N/A	N/A	No	D	160	N/A	Br	No	2
	Liopropoma mitratum	N/A	N/A	No	D	82	N/A	Br	No	3
	Liopropoma multilineatum	N/A	N/A	No	D	77	N/A	Br	No	3
	Liopropoma pallidum	N/A	N/A	No	D	78	N/A	Br	No	3
	Liopropoma susumi	N/A	N/A	No	D	91	N/A	Br	No	3
	Liopropoma tonstrinum	N/A	N/A	No	D	80	N/A	Br	No	3
	Luzonichthys earlei	No	Pl	No	D	44	N/A	Br	No	2
	Luzonichthys waitei	No	Pl	No	D	70	N/A	Br	No	1, 2, 3, 12
	Luzonichthys whitleyi	No	Pl	No	D	60	N/A	Br	No	2, 3
	Luzonichthys williamsi	No	N/A	No	D	58	N/A	Br	No	3
	Mycteroperca acutirostris (= *Epinephelus undulosus*)	No	Ca	No	D	500	G	Br	No	2, 4
	Plectranthias inermis	No	N/A	No	D	350	N/A	Br	No	2, 4
	Plectranthias longimanus	No	N/A	No	D	350	N/A	Br	No	1, 2, 3
	Plectranthias nanus	No	N/A	No	D	50	N/A	Br	No	1, 2, 3
	Plectranthias winniensis	No	N/A	No	D	350	N/A	Br	No	1, 2, 3
	Plectropomus areolatus	No	Pi	No	D	800	G	Br	No	1, 2, 3, 12
	Plectropomus laevis	No	Pi	No	D	1250	G	Br	No	1, 2, 3, 12
	Plectropomus leopardus	No	Pi	No	D	700	G	Br	No	1, 2, 3, 12
	Plectropomus maculatus	No	Pi	No	D	1250	G	Br	No	1, 2, 3
	Plectropomus oligacanthus	No	Ca	No	D	750	G	Br	No	1, 2, 3, 4
	Plectropomus pessuliferus	No	Pi	No	D	1200	G	Br	No	1, 2, 3, 12
	Pogonoperca punctata	Yes	Pi	No	D	350	N/A	Gel	No	1, 2, 3, 12
	Pseudanthias aurulentus	No	Pl	No	D	60	P	Br	No	2
	Pseudanthias bartlettorum	No	Pl	No	D	90	P	Br	No	2, 3
	Pseudanthias bicolor	No	Pl	No	D	130	P	Br	No	1, 2, 3, 12
	Pseudanthias bimaculatus	No	Pl	No	D	90	P	Br	No	2
	Pseudanthias carlsoni	No	Pl	No	D	100	P	Br	No	3
	Pseudanthias cooperi	No	Pl	No	D	140	P	Br	No	1, 2, 3
	Pseudanthias dispar	No	Pl	No	D	95	P	Br	No	1, 2, 3
	Pseudanthias engelhardi	No	Pl	No	D	100	P	Br	No	1, 2, 3
	Pseudanthias evansi	No	Pl	No	D	100	P	Br	No	2
	Pseudanthias fasciatus	No	Pl	No	D	210	P	Br	No	1, 2
	Pseudanthias flavoguttatus	No	Pl	No	D	110	P	Br	No	2
	Pseudanthias hiva	No	Pl	No	D	140	P	Br	No	3
	Pseudanthias huchtii	No	Pl	No	D	120	P	Br	No	1, 2
	Pseudanthias hutomoi	No	Pl	No	D	120	P	Br	No	2
	Pseudanthias hypselosoma	No	Pl	No	D	70	P	Br	No	1, 2
	Pseudanthias ignitus	No	Pl	No	D	80	P	Br	No	2, 4
	Pseudanthias lori	No	Pl	No	D	120	P	Br	No	1, 2, 3
	Pseudanthias luzonensis	No	Pl	No	D	145	P	Br	No	1, 2
	Pseudanthias mooreanus	N/A	Pl	No	D	72	P	Br	No	3
	Pseudanthias olivaceus	No	Pl	No	D	120	P	Br	No	2, 3

Continued

List of Indo-Pacific reef fishes and their social system, trophic affiliation, burrowing behaviour, activity patterns, size, and reproductive traits (Continued)

Family	Species	Pairs	Troph.	Burr.	Act.	ML (mm)	Spa.	Gam.	Mon.	Refs.
	Pseudanthias parvirostris	No	Pl	No	D	100	P	Br	No	2, 4
	Pseudanthias pascalus	No	Pl	No	D	170	P	Br	No	2, 3
	Pseudanthias pictilis	No	Pl	No	D	135	P	Br	No	2
	Pseudanthias pleurotaenia	No	Pl	No	D	200	P	Br	No	1, 2, 3, 12
	Pseudanthias randalli	No	Pl	No	D	70	P	Br	No	2
	Pseudanthias regalis	No	Pl	No	D	62	P	Br	No	3
	Pseudanthias rubrizonatus	No	Pl	No	D	100	P	Br	No	2, 4
	Pseudanthias sheni	No	Pl	No	D	200	P	Br	No	2
	Pseudanthias smithvanizi	No	Pl	No	D	95	P	Br	No	1, 2
	Pseudanthias squamipinnis	No	Pl	No	D	150	P	Br	No	1, 2, 3
	Pseudanthias tuka	No	Pl	No	D	120	P	Br	No	1, 2
	Pseudanthias venator	No	Pl	No	D	70	P	Br	No	2
	Pseudanthias ventralis	No	Pl	No	D	70	P	Br	No	1, 2, 3, 12
	Pseudogramma polyacantha	No	Mi	No	D	75	N/A	Br	No	1, 2, 3
	Serranocirrhitus latus	No	N/A	No	D	350	N/A	Br	No	1, 2, 3
	Suttonia lineata	N/A	N/A	No	D	96	N/A	Br	No	3
	Variola albimarginata	No	Pi	No	D	550	G	Br	No	1, 2, 3
	Variola louti	No	Pi	No	D	550	G	Br	No	1, 2, 3, 44
Siganidae	*Siganus argenteus*	No	He	No	D	420	G	Es	No	1, 2, 3, 6, 164
	Siganus canaliculatus	No	He	No	D	290	G	Es	No	1, 2, 3, 164, 165
	Siganus corallinus	Yes	He	No	D	250	G	Es	No	1, 2, 3, 6, 164
	Siganus doliatus	Yes	He	No	D	250	G	Es	No	1, 2, 3
	Siganus guttatus	No	He	No	D	350	G	Es	No	1, 2, 3, 164
	Siganus javus	Yes	He	No	D	530	G	Es	No	1, 2, 3, 164, 166
	Siganus lineatus	No	De	No	D	350	G	Es	No	1, 2, 3, 164, 167
	Siganus magnificus	Yes	He	No	D	230	G	Es	No	2, 4, 164
	Siganus puelloides	Yes	He	No	D	310	G	Es	No	2, 164
	Siganus puellus	Yes	Sp	No	D	380	G	Es	No	164, 168
	Siganus punctatissimus	Yes	He	No	D	280	G	Es	No	2, 6, 164
	Siganus punctatus	Yes	He	No	D	300	G	Es	No	1, 2, 3, 6, 164
	Siganus randalli	No	He	No	D	250	G	Es	No	2, 164
	Siganus spinus	No	He	No	D	200	G	Es	No	1, 2, 3, 164
	Siganus stellatus	Yes	He	No	D	350	G	Es	No	2, 164
	Siganus unimaculatus	Yes	He	No	D	240	G	Es	No	2, 4, 164
	Siganus uspi	Yes	He	No	D	220	G	Es	No	2, 4, 164
	Siganus vermiculatus	No	He	No	D	370	G	Es	No	1, 2, 3, 164

List of Indo-Pacific reef fishes and their social system, trophic affiliation, burrowing behaviour, activity patterns, size, and reproductive traits (Continued)

Family	Species	Pairs	Troph.	Burr.	Act.	ML (mm)	Spa.	Gam.	Mon.	Refs.
	Siganus virgatus	No	He	No	D	300	G	Es	No	2, 164
	Siganus vulpinus	Yes	He	No	D	240	G	Es	No	1, 2, 3, 164
Soleidae	*Aseraggodes kaianus*	No	Ma	No	D	140	N/A	Br	No	1, 2, 3, 4
	Aseraggodes melanostictus	No	Ma	No	D	40	N/A	Br	No	2
	Liachirus melanospilos	No	Ma	No	D	150	N/A	Br	No	2
	Pardachirus pavoninus	No	Ma	No	N	220	N/A	Br	No	2
	Soleichthys heterorhinos	No	Ma	No	N	150	N/A	Br	No	2
	Synaptura marginata	No	Ma	No	N	300	N/A	Br	No	1, 2, 3, 4
	Zebrias fasciatus	No	Ma	No	N	250	N/A	Br	No	2
Solenostomidae	*Solenostomus cyanopterus*	Yes	Mi	No	D	160	P	Po	Yes	4
	Solenostomus halimeda	Yes	Mi	No	D	70	P	Po	No	1, 2, 3
	Solenostomus paegnius	Yes	Mi	No	D	120	P	Po	No	4
	Solenostomus paradoxus	Yes	Mi	No	D	110	P	Po	No	1, 2, 3
Sparidae	*Acanthopagrus berda*	No	Ca	No	D	500	G	Br	No	2, 4
Sphyraenidae	*Sphyraena barracuda*	No	Pi	No	D	1800	G	Br	No	1, 2, 3, 169
	Sphyraena flavicauda	No	N/A	No	N	500	G	Br	No	1, 2, 3, 4
	Sphyraena forsteri	No	Ca	No	N	650	G	Br	No	1, 2, 3, 4
	Sphyraena helleri	No	N/A	No	N	850	G	Br	No	1, 2, 3, 4
	Sphyraena jello	No	Ca	No	D	1500	G	Br	No	1, 2, 3, 170
	Sphyraena qenie	No	Pi	No	N	1000	G	Br	No	1, 2, 3, 40
Synanceiidae	*Choridactylus multibarbus*	No	N/A	No	D	120	P	Gel	No	2
	Dampierosa daruma	No	N/A	No	D	130	P	Gel	No	2
	Inimicus caledonicus	No	N/A	No	D	250	P	Gel	No	2, 3
	Inimicus didactylus	Yes	Ca	No	D	180	P	Gel	No	1, 2, 3, 4
	Inimicus sinensis	No	N/A	No	D	180	P	Gel	No	2
	Minous trachycephalus	No	N/A	No	N	90	P	Gel	No	2, 4
	Synanceia horrida	No	Ca	No	N	300	P	Gel	No	1, 2, 3
	Synanceia verrucosa	No	Ca	No	D	350	P	Gel	No	2, 4
Syngnathidae	*Acentronura tentaculata (= breviperula)*	Yes	Mi	No	D	50	P	Po	No	1, 2, 3
	Choeroichthys brachysoma	No	Mi	No	D	65	P	Po	No	1, 2, 3
	Choeroichthys cinctus	No	Mi	No	D	80	P	Po	No	1, 2, 3
	Corythoichthys amplexus	Yes	Mi	No	D	95	P	Po	Yes	171
	Corythoichthys flavofasciatus	Yes	Mi	No	D	180	P	Po	Yes	172
	Corythoichthys haematopterus	Yes	Mi	No	D	180	P	Po	Yes	1, 2, 3
	Corythoichthys intestinalis	Yes	Mi	No	D	180	P	Po	Yes	171
	Corythoichthys nigripectus	Yes	Mi	No	D	110	P	Po	Yes	1, 2, 3
	Corythoichthys ocellatus	Yes	Mi	No	D	110	P	Po	No	1, 2, 3
	Corythoichthys polynotatus	Yes	Mi	No	D	160	P	Po	No	1, 2, 3
	Corythoichthys schultzi	Yes	Mi	No	D	160	P	Po	Yes	1, 2, 3
	Doryrhamphus excisus	Yes	Mi	No	D	70	P	Po	Yes	4

Continued

List of Indo-Pacific reef fishes and their social system, trophic affiliation, burrowing behaviour, activity patterns, size, and reproductive traits (Continued)

Family	Species	Pairs	Troph.	Burr.	Act.	ML (mm)	Spa.	Gam.	Mon.	Refs.
	Doryrhamphus janssi	Yes	Mi	No	D	130	P	Po	No	4
	Doryrhamphus japonicus	Yes	Mi	No	D	70	P	Po	Yes	4
	Doryrhamphus negrosensis	Yes	Mi	No	D	55	P	Po	No	1, 2, 3
	Dunckerocampus boylei	Yes	Mi	No	D	160	P	Po	No	1, 2, 3
	Dunckerocampus dactyliophorus	Yes	Mi	No	D	180	P	Po	No	173
	Dunckerocampus multiannulatus	Yes	Mi	No	D	180	P	Po	No	1, 2, 3
	Dunckerocampus pessuliferus	Yes	Mi	No	D	160	P	Po	No	173
	Halicampus brocki	No	N/A	No	D	115	P	Po	No	1, 2, 3
	Halicampus macrorhynchus	Yes	N/A	No	D	160	P	Po	No	4
	Halicampus mataafae	No	N/A	No	D	150	P	Po	No	1, 2, 3
	Halicampus nitidus	No	N/A	No	D	75	P	Po	No	1, 2, 3
	Hippichthys cyanospilos	No	N/A	No	D	160	P	Po	No	1, 2, 3
	Hippocampus barbouri	No	N/A	No	D	150	P	Po	No	1, 2, 3
	Hippocampus bargibanti	Yes	Mi	No	D	20	P	Po	No	174, 175
	Hippocampus colemani	No	Mi	No	D	10	P	Po	No	1, 2, 3
	Hippocampus comes	Yes	Mi	No	N	160	P	Po	Yes	4
	Hippocampus denise	No	Mi	No	D	15	P	Po	No	175
	Hippocampus histrix	No	Mi	No	D	150	P	Po	Yes	175
	Hippocampus kuda	Yes	Mi	No	D	150	P	Po	No	4
	Hippocampus trimaculatus	No	Mi	No	D	220	P	Po	No	175
	Micrognathus andersonii	No	N/A	No	D	60	P	Po	No	1, 2, 3
	Micrognathus brevirostris pygmaeus	No	N/A	No	D	60	P	Po	No	1, 2, 3
	Phoxocampus tetrophthalmus	No	N/A	No	D	80	P	Po	No	1, 2, 3
	Siokunichthys nigrolineatus	No	N/A	No	D	400	P	Po	No	1, 2, 3
	Syngnathoides biaculeatus	Yes	Mi	No	D	280	P	Po	No	176
	Trachyrhamphus bicoarctatus	No	N/A	No	D	400	P	Po	No	1, 2, 3
	Trachyrhamphus longirostris	No	N/A	No	D	400	P	Po	No	1, 2, 3
Synodontidae	*Saurida gracilis*	No	Pi	No	N	280	P	Br	No	1, 2, 3, 4
	Saurida nebulosa	No	Pi	No	D	200	P	Br	No	1, 2, 3
	Synodus binotatus	Yes	Pi	No	D	170	P	Br	No	1, 2, 3, 12
	Synodus dermatogenys	Yes	Pi	No	D	200	P	Br	No	1, 2, 3, 161
	Synodus jaculum	Yes	Pi	No	D	140	P	Br	No	1, 2, 3, 12
	Synodus rubromarmoratus	No	Pi	No	D	85	P	Br	No	1, 2, 3
	Synodus variegatus	Yes	Pi	No	D	240	P	Br	No	1, 2, 3, 12
	Trachinocephalus myops	No	Ca	No	D	250	P	Br	No	1, 2, 3, 177
Tetraodontidae	*Arothron caeruleopunctatus*	No	Om	No	D	700	G	Es	No	1, 2, 3
	Arothron hispidus	No	Om	No	D	480	G	Es	No	4
	Arothron immaculatus	No	Om	No	D	280	G	Es	No	176
	Arothron manilensis	No	Om	No	D	310	G	Es	No	15
	Arothron mappa	No	Om	No	D	600	G	Es	No	1, 2, 3
	Arothron meleagris	No	Co	No	D	500	G	Es	No	77
	Arothron nigropunctatus	Yes	Co	No	D	330	G	Es	No	77

List of Indo-Pacific reef fishes and their social system, trophic affiliation, burrowing behaviour, activity patterns, size, and reproductive traits (Continued)

Family	Species	Pairs	Troph.	Burr.	Act.	ML (mm)	Spa.	Gam.	Mon.	Refs.
	Arothron reticularis	No	Ma	No	D	300	G	Es	No	4
	Arothron stellatus	No	Om	No	D	900	G	Es	No	1, 2, 3
	Canthigaster amboinensis	No	Om	No	D	140	P	Bc	No	178
	Canthigaster bennetti	Yes	He	No	D	100	P	Bc	Yes	4, 179
	Canthigaster callisterna	Yes	N/A	No	D	240	P	Bc	No	1, 2, 3
	Canthigaster compressa	Yes	Mi	No	D	100	P	Bc	No	4, 108
	Canthigaster coronata	No	Om	No	D	135	P	Bc	No	4
	Canthigaster epilampra	Yes	Om	No	D	110	P	Bc	No	4
	Canthigaster janthinoptera	Yes	Om	No	D	90	P	Bc	No	4
	Canthigaster leoparda	Yes	N/A	No	D	70	P	Bc	No	1, 2, 3
	Canthigaster ocellicincta	Yes	N/A	No	D	65	P	Bc	No	4
	Canthigaster papua	Yes	Mi	No	D	90	P	Bc	No	108
	Canthigaster rivulata	No	N/A	No	D	180	P	Bc	No	1, 2, 3
	Canthigaster smithae	No	N/A	No	D	130	P	Bc	No	1, 2, 3
	Canthigaster solandri	Yes	Om	No	D	105	P	Bc	No	4, 179
	Canthigaster tyleri	No	Om	No	D	80	P	Bc	No	4
	Canthigaster valentini	Yes	Om	No	D	90	P	Bc	Yes	4, 179
	Chelonodon patoca	No	Om	No	D	330	P	Bc	No	180
	Lagocephalus sceleratus	No	Ca	No	D	850	P	Bc	No	181, 182
	Torquigener brevipinnis	No	N/A	No	D	140	P	Bc	No	1, 2, 3
Tetrarogidae	*Ablabys macracanthus*	Yes	N/A	No	D	150	P	Br	No	2
	Ablabys taenianotus	Yes	Ca	No	N	150	P	Br	No	1, 2, 3, 20
	Paracentropogon longispinis	No	N/A	No	D	120	P	Br	No	2
	Richardsonichthys leucogaster	No	N/A	No	N	100	P	Br	No	2, 4
	Tetraroge barbata	Yes	Ca	No	D	110	P	Br	No	2, 4
	Tetraroge niger	Yes	N/A	No	D	135	P	Br	No	2
Trichonotidae	*Trichonotus elegans*	No	Ma	No	D	180	N/A	N/A	No	1, 2, 3, 183
	Trichonotus halstead	No	Ma	No	D	150	N/A	N/A	No	2, 183
	Trichonotus setiger	No	Ma	No	D	180	N/A	N/A	No	1, 2, 3, 183
Tripterygiidae	*Ceratobregma helenae*	No	N/A	No	D	45	P	Bc	No	1, 2, 3
	Enneapterygius atrogulare	Yes	Om	No	D	53	P	Bc	No	3, 73
	Enneapterygius elegans	N/A	N/A	No	D	35	P	Bc	No	3
	Enneapterygius flavoccipitis	Yes	N/A	No	D	35	P	Bc	No	2
	Enneapterygius hemimelas	N/A	N/A	No	D	48	P	Bc	No	3
	Enneapterygius mirabilis	No	N/A	No	D	35	P	Bc	No	1, 2
	Enneapterygius nanus	N/A	N/A	No	D	28	P	Bc	No	3
	Enneapterygius niger	N/A	N/A	No	D	35	P	Bc	No	3
	Enneapterygius nigricauda	N/A	N/A	No	D	35	P	Bc	No	3
	Enneapterygius pallidoserialis	Yes	N/A	No	D	35	P	Bc	No	2
	Enneapterygius paucifasciatus	N/A	N/A	No	D	35	P	Bc	No	3
	Enneapterygius philippinus	Yes	N/A	No	D	35	P	Bc	No	2
	Enneapterygius pyramis	N/A	N/A	No	D	34	P	Bc	No	3

Continued

List of Indo-Pacific reef fishes and their social system, trophic affiliation, burrowing behaviour, activity patterns, size, and reproductive traits (Continued)

Family	Species	Pairs	Troph.	Burr.	Act.	ML (mm)	Spa.	Gam.	Mon.	Refs.
	Enneapterygius randalli	N/A	N/A	No	D	34	P	Bc	No	3
	Enneapterygius rhabdotus	N/A	N/A	No	D	32	P	Bc	No	3
	Enneapterygius rhothion	N/A	N/A	No	D	37	P	Bc	No	3
	Enneapterygius rufopileus	No	Mi	No	D	45	P	Bc	No	3, 184
	Enneapterygius similis	N/A	N/A	No	D	39	P	Bc	No	3
	Enneapterygius triserialis	N/A	N/A	No	D	45	P	Bc	No	3
	Enneapterygius williamsi	N/A	N/A	No	D	33	P	Bc	No	3
	Helcogramma capidata	No	N/A	No	D	41	P	Bc	No	3
	Helcogramma chica	No	Om	No	D	40	P	Bc	No	3, 185
	Helcogramma rhinoceros	Yes	N/A	No	D	40	P	Bc	No	2
	Helcogramma striata	No	Pl	No	D	50	P	Bc	No	1, 2, 3, 4
	Helcogramma vulcana	Yes	N/A	No	D	40	P	Bc	No	2
	Norfolkia brachylepis	N/A	N/A	No	D	73	P	Bc	No	3
	Norfolkia squamiceps	N/A	N/A	No	D	66	P	Bc	No	3
	Norfolkia thomasi	N/A	N/A	No	D	50	P	Bc	No	3
	Springerichthys kulbickii	N/A	N/A	No	D	35	P	Bc	No	3
	Ucla xenogrammus	No	N/A	No	D	45	P	Bc	No	1, 2, 3
Uranoscopidae	*Uranoscopus sulphureus*	No	N/A	No	D	350	N/A	N/A	No	2
Zanclidae	*Zanclus cornutus*	Yes	Om	No	D	160	G	Br	No	1, 2, 3

Note: Asterisks mark burrowing species that associate with alpheid shrimps. Species names are those listed as currently accepted in FishBase (Froese & Pauly 2013); where those used in original articles were different, they are given in parentheses.

Act. = activity period; B = both; Bc = benthic clutch; Br = broadcast; Burr. = burrowing; Ca = carnivore; Co = corallivore; D = diurnal; De = detritivore; Es = egg scattering; G = group; Gam. = gamete release; Gel = gelatinous egg mass; He = herbivore; Ma = macroinvertevore; Mi = microinvertevore; ML = maximum length; Mo = mouth brooding; Mon. = monogamy; N = nocturnal; N/A = no available information. Om = omnivore; P = pair; Pi = Piscivore; Pl = planktivore; Po = pouch brooding; Refs. = references; Sp = spongivore; Spa. = spawning mode; Troph. = trophic group.

Appendix references: 1: Randall et al. 1997, 2: Allen et al. 2003, 3: Randall 2005, 4: Froese & Pauly 2013, 5: Barlow 1974, 6: Cheal et al. 2012, 7: Russ 1984, 8: Robertson et al. 1979, 9: Brandl & Bellwood 2013b, 10: Randall & Struhsaker 1981, 11: Meyer-Rochow 1976, 12: Sandin & Williams 2010, 13: Pietsch & Grobecker 1987, 14: Nakane et al. 2011, 15: Nanjo et al. 2008, 16: Marnane & Bellwood 2002, 17: Zagars et al. 2013, 18: Barnett et al. 2006, 19: Job & Shand 2001, 20: Unsworth et al. 2007, 21: Unsworth et al. 2009, 22: Schmitz & Wainwright 2011, 23: Fukumori et al. 2008, 24: Hajisamae 2009, 25: Longenecker & Langston 2006, 26: Nakamura et al. 2003, 27: Kuwamura 1991, 28: Chen et al. 2001, 29: Wilson 2000, 30: Ho et al. 2007, 31: Carlson 2012, 32: Roberts 1987, 33: Kwak et al. 2004, 34: Townsend & Tibbetts 2000, 35: Hamner et al. 1988, 36: Fricke & Zaiser 1982, 37: Grandcourt et al. 2004, 38: Salini et al. 1994, 39: Farmer & Wilson 2011, 40: Blaber et al. 1990, 41: Meyer et al. 2001, 42: Brewer et al. 1989, 43: Smith & Parrish 2002, 44: Randall 1980, 45: Blaber & Cyrus 1983, 46: Sreenivasan 1978, 47: Tandon 1960, 48: Barreiros et al. 2003, 49: Roberts & Ormond 1992, 50: Pratchett et al. 2006, 51: Reese 1975, 52: Sano 1989, 53: Allen et al. 1998, 54: Claydon 2004, 55: Pratchett 2005, 56: Reavis & Copus 2011, 57: Yabuta & Kawashima 1997, 58: Ferry-Graham et al. 2001, 59: Bellwood et al. 2010, 60: Donaldson 1990, 61: Kadota et al. 2011, 62: Donaldson 1989, 63: Kane et al. 2009, 64: Clemente et al. 2010, 65: Bellwood et al. 2006b, 66: Taquet et al. 2007, 67: Sakashita 1992, 68: Hernaman et al. 2009, 69: Kramer et al. 2012b, 70: Depczynski & Bellwood 2003, 71: Herler et al. 2011, 72: Allen & Munday 1995, 73: White & Brown 2013, 74: Munday et al. 2002, 75: Karplus 1979, 76: Brooker et al. 2010, 77: Cole et al. 2008, 78: Randall & Goren 1993, 79: Daroonchoo 1991, 80: Sano et al. 1984, 81: Hagiwara & Winterbottom 2007, 82: Winterbottom 1984, 83: Clark et al. 2000, 84: Reavis & Barlow 1998, 85: Randall 1998, 86: Clements & Choat 1997, 87: Rimmer & Wiebe 1987, 88: DiSalvo et al. 2007, *(Continued)*

List of Indo-Pacific reef fishes and their social system, trophic affiliation, burrowing behaviour, activity patterns, size, and reproductive traits (Continued)

Appendix references (Continued): 89: Williams & Williams 1986, 90: Bellwood et al. 2003, 91: Muñoz et al. 2006, 92: Ferry-Graham et al. 2002, 93: Choat et al. 2004, 94: Lek et al. 2011, 95: Pratchett et al. 2001, 96: Debenay et al. 2011, 97: Connell 1998, 98: Nanami & Yamada 2008, 99: Kiso & Mahyam 2003, 100: Sweatman 1993, 101: Salini et al. 1990, 102: Bonin et al. 2011, 103: Clark et al. 1998, 104: Clark & Pohle 1992, 105: López-Peralta & Arcila 2002, 106: Harmelin-Vivien 1979, 107: Kokita & Mizota 2002, 108: Wantiez & Kulbicki 1995, 109: Blaber 1977, 110: McCormick 1995, 111: Randall & Myers 2002, 112: Randall 2004, 113: Letourneur 1996, 114: Golani & Galil 1991, 115: Sabrah & El-Ganainy 2009, 116: Mehta 2009, 117: Reece et al. 2010, 118: Mequila & Campos 2007, 119: Boaden & Kingsford 2012, 120: Russell 1990, 121: Hiatt & Strasburg 1960, 122: Tamaki et al. 1992, 123: Moyer & Sano 1987, 124: Metian et al. 2013, 125: Sivakumar & Ramaiyan 1987, 126: Mooi & Gill 2004, 127: Ang & Manica 2010, 128: Moyer & Nakazono 1978, 129: Pyle 2003, 130: Mantyka & Bellwood 2007, 131: Hensley & Allen 1977, 132: Frédérich et al. 2009, 133: Williams & Hatcher 1983, 134: Fautin & Allen 1992, 135: Fricke & Fricke 1977, 136: Hobson 1991, 137: Tribble & Nishikawa 1982, 138: Wenger et al. 2012, 139: Allen 1991, 140: Sackley & Kaufman 1996, 141: Gluckmann & Vandewalle 1998, 142: Allen & Rajasuriya 1995, 143: Randall & Allen 1977, 144: Elliott & Bellwood 2003, 145: Ceccarelli 2007, 146: Bellwood et al. 2006a, 147: Horn 1989, 148: Öhman et al. 1998, 149: Lewis 1998, 150: Hamilton & Dill 2003, 151: Kohda 1988, 152: Russ 1987, 153: Allen 1987, 154: Feeney et al. 2012, 155: Harmelin-Vivien & Bouchon 1976, 156: Shallenberger & Madden 1973, 157: Marguillier et al. 1997, 158: Muñoz et al. 2011, 159: Longenecker 2007, 160: Barros et al. 2008, 161: Holmes & McCormick 2011, 162: Stewart & Jones 2001, 163: Craig 2007, 164: Woodland 1990, 165: Fox & Bellwood 2008, 166: Cvitanovic & Bellwood 2009, 167: Fox et al. 2009, 168: Hoey et al. 2013, 169: Grubich et al. 2008, 170: Hajisamae et al. 2003, 171: Gronell 1984, 172: Iyer 2012, 173: Leysen et al. 2011, 174: Lourie & Randall 2003, 175: Foster & Vincent 2004, 176: Horinouchi et al. 2012, 177: Kizhakudan & Gomathy 2007, 178: Sikkel & Sikkel 2012, 179: Sikkel 1990, 180: Beumer 1978, 181: Sabrah et al. 2006, 182: Aydin 2011, 183: Clark & Pohle 1996, 184: Silberschneider & Booth 2001, 185: Hadley Hansen 1986.

Oceanography and Marine Biology: An Annual Review, 2014, **52**, 81-132

Taylor & Francis

THE ECOSYSTEM ROLES OF PARROTFISHES ON TROPICAL REEFS

ROBERTA M. BONALDO[1], ANDREW S. HOEY[2] & DAVID R. BELLWOOD[2,3]

[1]*Departamento de Ecologia, Instituto de Biociências, Universidade de São Paulo, São Paulo, Brazil*
E-mail:robertabonaldo@gmail.com (corresponding author)
[2]*Australian Research Council Centre of Excellence for Coral Reef Studies, James Cook University, Townsville, Queensland 4811, Australia*
[3]*School of Marine and Tropical Biology, James Cook University, Townsville, Queensland 4811, Australia*

Global reductions in biodiversity and the accelerating loss and degradation of many of the world's ecosystems have intensified research into the roles of species in ecosystem processes. Within coral-reef systems, the parrotfishes (Labridae) are widely viewed as a key functional group in facilitating the recovery of reefs from recurrent disturbances. Although parrotfishes are commonly viewed as herbivores, exerting top-down control of algal communities, their unique jaws allow them to feed on almost all coral-reef substratum types. Consequently, parrotfish are the primary agents responsible for a number of key ecological processes on coral reefs, namely, bioerosion, sediment production and transport, provision of space for coral settlement, and predation of live coral colonies. The parrotfishes, however, cannot be considered a uniform group. Their functional roles are highly dependent on species feeding mode (scrapers, excavators, and browsers) and body size, with larger individuals having a disproportionately greater effect on the dynamics of benthic communities than smaller conspecifics. Parrotfish are ubiquitous on tropical reefs worldwide, yet there is strong spatial structuring in the taxonomic and functional composition of the group. This spatial variation has been shaped by their biogeographic history, the productivity of their environment, and the habitat requirements of individual taxa. Over recent decades, increasing fishing pressure and habitat destruction have had a dramatic impact on the structure of parrotfish assemblages, and as a consequence, on many reefs, normal ecosystem processes have been disrupted. Indeed, reef systems with severely depleted parrotfish communities are less resilient to anthropogenic or natural environmental perturbations. Management strategies for the protection of this unique and critical functional group are urgently needed if we are to maintain the diversity, resilience, and structure of coral-reef ecosystems.

Introduction

Global reductions in biodiversity and the accelerating loss and degradation of many of the world's ecosystems have intensified research into the relationship between biodiversity and ecosystem function and the roles of individual species in ecosystem processes (Loreau & Hector 2001, Worm et al. 2006). Consequently, species are increasingly viewed in terms of their function within an ecosystem (Duffy 2003, Bellwood et al. 2004, Johansson et al. 2013), as opposed to their taxonomic labels or affinities. In this context, ecosystem function (or processes) refers to the movement or storage of energy through trophic or bioconstructional pathways (i.e., the processes leading to the formation of biogenic reefs and calcareous sediments derived from or otherwise associated with them). A functional group is a collection of organisms that are primarily responsible for carrying out or modifying particular ecosystem functions, irrespective of their taxonomic affinities (following Done et al. 1996). Viewing taxa from a functional perspective allows not only the impact of a species to be quantified but also the likely consequences of reductions or complete loss of the species from the system (Bellwood et al. 2003, 2012a, Petchey & Gaston 2006, O'Gorman et al. 2011). This extensive and growing body of work has shown that in many ecosystems there are key (or critical) functional groups that perform a disproportionately large role in the maintenance of ecosystem functions (Steneck et al. 2002, Smith & Knapp 2003). Establishing how ecosystems will respond to future biodiversity loss and degradation requires a clearer, quantitative understanding of the roles of key functional groups and the locations in which these roles are fulfilled.

Consequently, much of the early research into relationships between biodiversity and ecosystem function focused on experimentally manipulated assemblages of primary producers in terrestrial or aquatic ecosystems (e.g., Tilman 1999, Balvanera et al. 2006). This early research was driven, at least in part, by the strong and direct links between the traits (or functional characteristics) of plants and ecosystem processes such as primary production (Duffy 2002). Consumers, however, can have a much more pervasive effect on a range of ecosystem processes, often having a strong influence on the productivity, biomass, and composition of primary-producer communities (Estes & Duggins 1995, Hughes et al. 2007); the biomass and composition of lower-order consumers (Terborgh et al. 2001, Estes et al. 2011); the physical structure of the habitat (McNaughton et al. 1988, Bruno & Bertness 2001); energy flow through trophic pathways (Odum 1968); and the production and movement of abiotic material (Bellwood 1985).

Coral reefs are one of the world's most diverse and productive ecosystems, yet they are also one of the most threatened. The effects of overfishing, pollution, sedimentation, eutrophication, habitat modification, and diseases are greatly compounded by climate change and are having a marked effect on the structure and function of coral-reef ecosystems globally (Jackson et al. 2001, Pandolfi et al. 2003, Bellwood et al. 2004). In particular, reductions in, or the total loss of, consumer functional groups are undermining critical ecosystem functions and limiting the capacity of reefs to regenerate following disturbances (Hughes et al. 2005, Mumby et al. 2006). Fish are the dominant consumers within coral-reef systems and contribute to a range of biological and physical processes. Much of the energy transferred between trophic groups passes through fish, and collectively, fish influence the biomass, productivity, and composition of algal communities (Williams & Polunin 2001, Russ 2003, Hughes et al. 2007), as well as the composition and morphology of coral assemblages (Cole et al. 2008, Rotjan & Lewis 2008, Bonaldo & Bellwood 2011a). Fish also facilitate the settlement of benthic organisms by clearing patches of substratum (Mumby et al. 2006) and are major agents of bioerosion and the production, reworking, and transport of sediments on coral reefs (Bellwood 1996, Bruggemann et al. 1996, Goatley & Bellwood 2012). Critically, the parrotfishes, more than any other group of fishes, are major contributors to all of these processes on coral reefs (Figure 1) and, if not the most important group, are arguably one of the most important groups of fishes on coral reefs.

The parrotfishes (family Labridae, tribe Scarini) are a monophyletic group of approximately 100 species (Bellwood 1994, Westneat & Alfaro 2005, Cowman et al. 2009, Parenti & Randall 2011, Comeros-Raynal et al. 2012). They are ubiquitous components of fish communities on tropical reefs around the world, where they graze in multispecies schools over biogenic and non-biogenic substrata (Bellwood & Choat 1990, Bonaldo et al. 2006, Hoey & Bellwood 2008). It is their specialized feeding morphology, however, that makes the parrotfishes unique: They have well-developed pharyngeal jaw apparatus that allows them to effectively grind calcareous material (Gobalet 1989, Bellwood 1994), and their oral teeth are fused to form strong, beak-like jaws (Bellwood & Choat 1990). As a result of this unique jaw morphology, parrotfishes are able to feed on almost all types

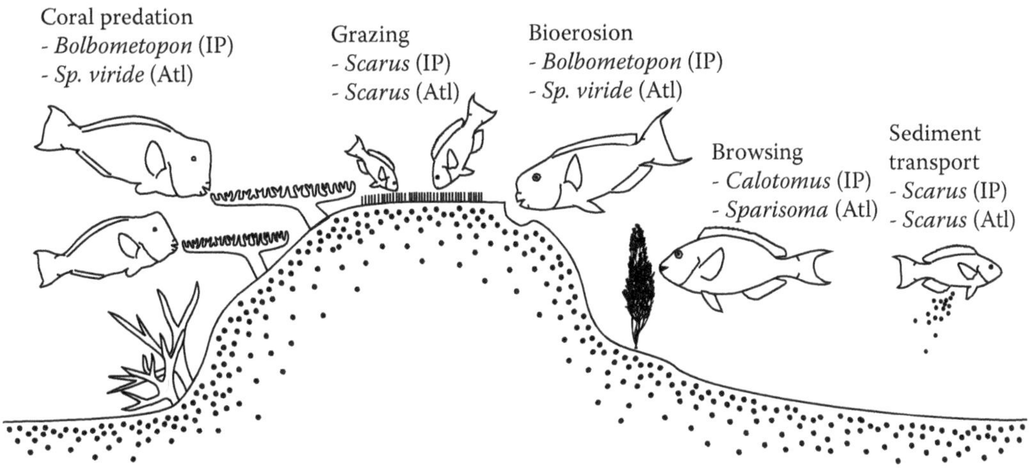

Figure 1 Schematic representation of the primary functional roles of parrotfishes on tropical reefs: bioerosion, grazing (scraping), coral predation, browsing, and sediment transport. The dominant taxa responsible for each function on tropical reefs in the Atlantic (Atl) and in the Indo-Pacific (IP) are given.

of coral-reef substratum, including live coral colonies, algal turfs covering dead coral surfaces or non-biogenic rock, macroalgae, and seagrass (Bellwood & Choat 1990, Bruggemann et al. 1994b,c, Bonaldo et al. 2006, Cole et al. 2008, Hoey & Bellwood 2008).

Although all parrotfishes have modified oral and pharyngeal jaws, there are marked differences in jaw morphology and feeding behaviour among species (Bellwood & Choat 1990), which directly affect the nature and intensity of the impact of these species on the reef benthos (Bruggemann et al. 1994b, Hoey & Bellwood 2008, Bonaldo & Bellwood 2009, Bonaldo et al. 2011). Based on the osteology and myology of the jaws, parrotfishes are usually classified into three main feeding modes or functional groups: browsers, scrapers, and excavators (Bellwood 1994; Figures 2 and 3). Both morphological and molecular phylogenies of parrotfish genera have identified two major clades that correspond to genera that are associated with either seagrass beds (i.e., *Sparisoma* clade, formerly Sparisomatinae; Figure 3A, 3C, 3D) or reefs (i.e., *Scarus* clade, formerly Scarinae; Figure 3B, 3E, 3F) (Figure 2).

Owing to their abundance and high feeding rates (Table 1), parrotfish facilitate high rates of removal and turnover of the benthic communities covering reef surfaces (Choat 1991, Bruggemann et al. 1994a,b, Bellwood et al. 2003, Floeter et al. 2005, Bonaldo et al. 2006, Fox & Bellwood 2007). By primarily feeding on surfaces covered by the epilithic algal matrix (EAM; *sensu* Wilson et al. 2003), parrotfish consume large amounts of turf algae and associated trapped material (primarily detritus, microbes, and sediments) and as such are considered as one of the main groups of 'nominally herbivorous' tropical fishes (Lewis & Wainwright 1985, Bellwood et al. 2004, Mumby 2006, 2009a). Several parrotfish species also feed directly on live coral colonies and have been identified as important coral predators (e.g., Bruckner & Bruckner 1998, Bellwood et al. 2003, Rotjan & Lewis 2008, Bonaldo & Bellwood 2011a). Furthermore, some of the larger parrotfish species are among the most important agents of bioerosion on coral reefs (Bellwood 1995a, Bruggemann et al. 1996), as portions of consolidated substratum ingested when feeding are triturated in the pharyngeal apparatus before discharge as sediment onto the reef and nearby areas (Frydl & Stearn 1978, Bellwood 1995b, Bruggemann et al. 1996, Bellwood et al. 2003, Alwany et al. 2009). Because of this processing of reef carbonates, parrotfish are also one of the most important producers of sediments and contribute to the reworking and transport of sediments on tropical reefs (Bellwood 1996, Bruggemann et al. 1996).

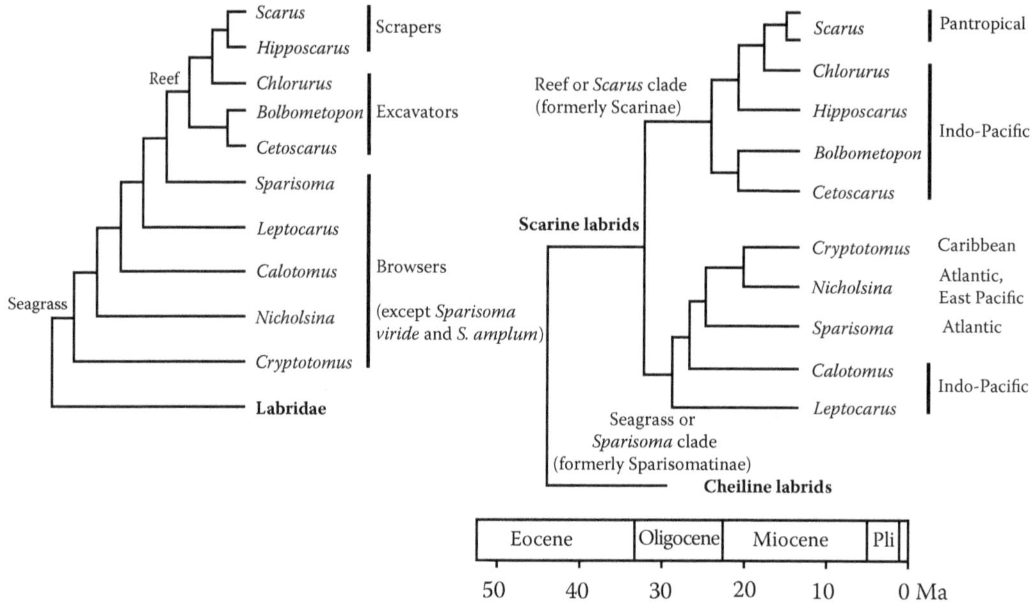

Figure 2 Cladograms depicting relationships among parrotfish genera. (A) Cladogram derived from comparative morphology of 143 characters from 69 species of parrotfish (redrawn from D.R. Bellwood, *Records of the Australian Museum Supplement* **20**, 1994, 86pp). The predominant feeding mode (browser, scraper, excavator) and habitat (seagrass, reef) are indicated. (B) Chronogram derived from molecular data showing the ages of origin and diversification of the 10 parrotfish genera (adapted from P.F. Cowman, D.R. Bellwood, & L. van Herwerden, *Molecular Phylogenetics and Evolution* **52**, 621–631, 2009). Analyses were based on two mitochondrial and two nuclear gene fragments for each species. The geographic distribution and estimated ages of the two clades and ten genera are shown. Pli, Pliocene.

Figure 3 (See also colour figure in the insert) Photographs of representative species from the three functional groups of parrotfishes. From the West Atlantic: (A) the browser *Sparisoma axillare*, (B) the scraper *Scarus vetula*, and (C) the excavator *Sparisoma viride*. From the Indo-Pacific: (D) the browser *Calotomus carolinus*, (E) the scraper *Scarus rivulatus*, and (F) the large excavator *Bolbometopon muricatum*. (Photos: João Paulo Krajewski.)

Table 1 Species-specific feeding rates (mean bites per minute) and foray sizes (mean number of bites) of parrotfishes on reef systems around the world

Species[a]	Location	Mean bites min^{-1} (SE)[b]	Foray size (bites) (SE)[b]	Source
Excavators				
Bolbometopon muricatum	Lizard Island, Australia	2.1 (0.8)	1.1 (0.12)	Bellwood & Choat 1990
Cetoscarus bicolor	South Sinai, Red Sea, Egypt	5.88 (0.8)	—	Alwany et al. 2009
C. ocellatus	Lizard Island, Australia	3.9 (0.4)	1.3 (0.1)	Bellwood & Choat 1990
Chlorurus bleekeri	Lizard Island, Australia	15.6 (3.2)	4.8 (1.3)	Bellwood & Choat 1990
C. gibbus	South Sinai, Red Sea, Egypt	6.38 (2.0)	—	Alwany et al. 2009
C. microrhinos	Lizard Island, Australia	7.9 (1.1)	2.3 (0.2)	Bellwood & Choat 1990
	Heron Island, Australia	9.6 (1.1)	3.1 (0.3)	Bellwood & Choat 1990
C. perspicillatus	Oahu, Hawaii	12.89–15.59[c]	—	Ong & Holland 2010
C. spilurus	Lizard Island, Australia	13.8 (2.2)	1.5 (0.3)	Bellwood & Choat 1990
	Heron Island, Australia	14.8 (2.1)	2.9 (0.5)	Bellwood & Choat 1990
C. sordidus	South Sinai, Red Sea, Egypt	15.3 (2.6)	—	Alwany et al. 2009
C. strongylocephalus	Zanzibar Island, Tanzania	12.9	—	Lokrantz et al. 2008
Sparisoma amplum	Abrolhos, Brazil	7.0 (0.7) to 6.1 (0.9)	1.7 (0.1) to 1.3 (0.1)	Francini-Filho et al. 2008
	Abrolhos, Brazil	0 (0) to 7 (5.4)	4.56 (0.4)	Francini-Filho et al. 2010
	Fernando de Noronha, Brazil	0.8 (0.2) to 5.9 (1.6)	—	Bonaldo et al. 2006
S. viride	Karpata, Bonaire	—	3.6	Bruggemann et al. 1994b
Scrapers/browsers				
Scarus chameleon	Lizard Island, Australia	9.6 (1.6)	3.6 (0.3)	Bellwood & Choat 1990
	Heron Island, Australia	7.1 (1.2)	2.4 (0.3)	Bellwood & Choat 1990
S. ferrugineus	South Sinai, Red Sea, Egypt	11.88 (2.4)	—	Alwany et al. 2009
	Sheikh Said Island, Red Sea, Eritrea	—	1–35	Afeworki et al. 2011
S. flavipectoralis	Heron Island, Australia	15.5 (1.9)	4.0 (0.3)	Bellwood & Choat 1990
S. frenatus	Heron Island, Australia	12.2 (1.6)	3.7 (0.5)	Bellwood & Choat 1990
	South Sinai, Red Sea, Egypt	10.72 (1.1)	—	Alwany et al. 2009
S. ghobban	Heron Island, Australia	13.5 (2.2)	2.8 (0.2)	Bellwood & Choat 1990
	South Sinai, Red Sea, Egypt	10.92 (1.7)	—	Alwany et al. 2009
S. globiceps	Heron Island, Australia	39.3 (5.8)	8.7 (1.1)	Bellwood & Choat 1990
S. iseri	Bocas del Toro, Panama	40.4 (2.98)	—	S.R. Floeter & C.E.L. Ferreira unpublished data, 2002
S. niger	Heron Island, Australia	22.9 (2.7)	5.8 (0.9)	Bellwood & Choat 1990
	South Sinai, Red Sea, Egypt	19.78 (1.5)	—	Alwany et al. 2009
	Zanzibar Island, Tanzania	11.3–18.4	—	Lokrantz et al. 2008
S. oviceps	Heron Island, Australia	12.9 (2.2)	4.1 (0.7)	Bellwood & Choat 1990
S. psittacus	Lizard Island, Australia	21.6 (2.5)	3.9 (0.3)	Bellwood & Choat 1990
	Heron Island, Australia	23.6 (3.3)	5.9 (1.2)	Bellwood & Choat 1990
S. rivulatus	Lizard Island, Australia	17.6 (1.7)	4.4 (0.4)	Bellwood & Choat 1990
	Heron Island, Australia	17.3 (2.9)	6.1 (1.1)	Bellwood & Choat 1990
	Orpheus Island, Australia	20 (2.1) to 32 (2.2)	—	Bonaldo & Bellwood 2008

Continued

Table 1 (Continued) Species-specific feeding rates (mean bites per minute) and foray sizes (mean number of bites) of parrotfishes on reef systems around the world

Species[a]	Location	Mean bites min^{-1} (SE)[b]	Foray size (bites) (SE)[b]	Source
S. rubroviolaceus	Lizard Island, Australia	10.1 (1.6)	2.7 (0.2)	Bellwood & Choat 1990
	Oahu, Hawaii	11.3–15.84[c]	—	Ong & Holland 2010
S. schlegeli	Lizard Island, Australia	20.1 (2.4)	4.5 (0.4)	Bellwood & Choat 1990
	Heron Island, Australia	16.7 (2.7)	6.3 (0.5)	Bellwood & Choat 1990
S. spinus	Heron Island, Australia	45 (6.7)	12.9 (2.9)	Bellwood & Choat 1990
S. trispinosus	Abrolhos, Brazil	17.8 (1.8)	3.0 (0.2)	Francini-Filho et al. 2008
		9.8 (1.5) to 17.22 (1.3)	5.46 (0.22)	Francini-Filho et al. 2010
S. vetula	Karpata, Bonaire	—	4.2	Bruggemann et al. 1994b
S. zelindae	Abrolhos, Brazil	11–19.42	5.56 (0.31)	Francini-Filho et al. 2010
Sparisoma axillare	Fernando de Noronha, Brazil	1 (0.92) to 5 (2.3)	—	Bonaldo et al. 2006
	Abrolhos, Brazil	8.04 (1.5) to 9.72 (1.4)	4.9 (0.2)	Francini-Filho et al. 2010
S. cretense	Azores Archipelago	6.86 (1.67)	—	J.P. Barreiros unpublished data, 2002
S. frondosum	Fernando de Noronha, Brazil	3 (1.0) to 5 (2.6)	—	Bonaldo et al. 2006
	Abrolhos, Brazil	5.45 (1.6) to 7.11 (1.8)	4.22 (0.2)	Francini-Filho et al. 2010
S. tuiupiranga	Arraial do Cabo, Brazil	1.83 (0.2) to 5.38 (0.3)	—	Ferreira et al. 1998b

[a] Where necessary, species names used in original articles have been replaced by the currently valid name (Froese & Pauly 2013).

[b] SE, standard error. Values given as ranges usually refer to different size classes, ontogenetic phases, or times of the day.

[c] Original values presented as daily rates.

The unique importance of the parrotfishes and number of roles they perform in reef ecosystems (Figure 1) has led to the widely held view that information on the ecology of this group is fundamental to our understanding of the structure, dynamics, and resilience of coral reefs (Bellwood et al. 2004, Mumby et al. 2006, 2007, Hughes et al. 2007, Ledlie et al. 2007). In this context, there has been an increased focus on patterns of feeding, distribution, resource use, and functional ecology of parrotfishes over the last few decades, helping to build an extensive literature on this subject. These studies have been timely given the current losses and future threats facing reef ecosystems around the world (Nyström et al. 2000, Baird & Marshall 2002, Gardner et al. 2003, Hughes et al. 2003, 2005, Worm et al. 2006, Hoegh-Guldberg et al. 2007, De'ath et al. 2009, Baird et al. 2013). Globally, coral reefs have suffered from massive habitat destruction and species depletion, with significant reductions in numbers and diversity of herbivorous fishes (e.g., Bellwood et al. 2003, Aswani & Hamilton 2004, Lokrantz et al. 2010, Graham et al. 2011, Comeros-Raynal et al. 2012). This potential loss of species and functional groups has highlighted the need to better understand the ecological role of herbivorous species and functional groups if we wish to assess the degree of resilience of coral-dominated reef systems (Bellwood et al. 2004, Nyström et al. 2008).

Clearly, parrotfish perform an important role in the functioning of healthy reef systems, but there is considerable variation among parrotfishes in the functions they perform and their relative contribution to each function. Species-specific differences in jaw morphologies, feeding modes, bite rates, and habitat associations influence the functional role of individual taxa (Bellwood & Choat 1990, Hoey & Bellwood 2008). Many other factors, such as body size, may also directly determine the nature and intensity of the impact of parrotfish feeding on the reef substratum (Bruggemann et al. 1994b, 1996, Bonaldo & Bellwood 2008, Francini-Filho et al. 2008, Lokrantz et al. 2008, Alwany et al. 2009, Jayewardene 2009). Together with this variation among taxa, the functional roles of parrotfish can also vary across a range of spatial scales, from small scales, such as among habitats or reefs (Fox & Bellwood 2007, Hoey & Bellwood 2008), to larger biogeographical scales (Bellwood & Wainwright 2002). As a consequence, assessing the functional roles of parrotfish in a system is a complex task that requires careful consideration of the multiple characteristics of the parrotfish community in a given location.

This review provides the first comprehensive assessment of the functional roles of parrotfish on tropical reefs and a detailed discussion of the factors that influence these roles. Herbivory has long been recognized as a key process determining the benthic community structure on tropical reefs (Stephenson & Searles 1960, Randall 1965), with large, mobile, nominally herbivorous fish playing a predominant role. Foremost among these are the parrotfishes. While parrotfish are often viewed in terms of their role as herbivores, exerting top-down control on algal communities, they are the primary agents for a number of key ecological processes on coral reefs (e.g., Bellwood 1985, Bruggemann 1995, Bonaldo et al. 2011). This review identifies the main ecosystem roles of parrotfishes on tropical reefs: grazing (scraping algae-covered substratum); browsing (biting off parts of erect macroalgae); coral predation; bioerosion; and the production, reworking, and transport of sediments. The review also critically evaluates the importance of particular parrotfish taxa in performing these roles and their relative contribution in comparison to other reef organisms. The main factors affecting parrotfish functional roles on tropical reefs are then discussed—in particular how feeding mode and body size influence functional affinities, the extent of functional redundancy both within the parrotfishes and among other groups of herbivorous fishes, and how spatial and temporal variation in parrotfish assemblages influence the nature and intensity of the impacts of these fishes on benthic reef communities. The importance of the roles performed by parrotfish for the 'healthy' functioning of coral reefs is then discussed, and this information is used to predict the likely effects of current and ongoing exploitation of parrotfish and identify strategies to protect parrotfish assemblages and conserve their functions on tropical reefs globally. Finally, the review identifies current gaps in knowledge of the biology, ecology, and functioning of parrotfishes on tropical reefs and highlights future research to gain a better understanding of this key component of coral-reef ecosystems.

Ecosystem roles of parrotfishes on tropical reefs

Grazing

Herbivory is considered one of the most important ecological processes on coral reefs, maintaining a healthy balance between corals and benthic algae. Coral-reef degradation has been frequently associated with declines in reef-building corals and a concomitant proliferation of benthic algae, especially macroalgae and algal turfs. These changes are often termed 'phase shifts' (e.g., Hughes et al. 2003, Bellwood et al. 2004, Ledlie et al. 2007, Norström et al. 2009). Reductions in the abundance or cover of scleractinian corals on degraded reefs are usually accompanied by decreases in the overall structural complexity of the reef and in the availability of shelter and food for many reef species (Hughes 1994, Gardner et al. 2003, Hughes et al. 2003, 2007, Graham et al. 2006, Alvarez-Filip et al. 2009, Norström et al. 2009). As a consequence, the balance between coral and algae in coral-reef systems is a major determinant of benthic community structure and has direct influence on food web dynamics, topographic complexity of the habitat, biodiversity, productivity, and ecosystem function (McCook 1999, Bellwood et al. 2004, Mumby et al. 2006, Birrell et al. 2008; reviewed by Wilson et al. 2006, Pratchett et al. 2008, 2011).

Coral-algae interactions in reef ecosystems are mediated by the feeding activity of herbivores (McCook 1997, Bellwood et al. 2004, Mumby et al. 2006, Burkepile & Hay 2008). On reefs with undisturbed fish communities, herbivores remove up to 90% of the net algal production (Polunin & Klumpp 1992, Ferreira et al. 1998b), with the vast majority consumed by large mobile, or 'roving', herbivorous fishes. This intense grazing maintains algal communities in a cropped state, dominated by highly productive algal turfs. On tropical reefs, these large roving herbivorous fishes are predominantly parrotfishes, surgeonfishes (Acanthuridae), and drummers or sea chubs (Kyphosidae), with an additional group, the rabbitfishes (Siganidae), present in the Indo-Pacific but absent from the tropical Atlantic. Of these groups, the parrotfishes and the surgeonfishes are typically the most abundant roving herbivorous fishes on tropical reefs (Lewis & Wainwright 1985, Fox & Bellwood 2007, Hoey & Bellwood 2008, Cheal et al. 2012, Afeworki et al. 2013, Johansson et al. 2013). Together with their high abundance, parrotfishes have one of the highest feeding rates recorded among coral-reef fishes. For example, *Scarus spinus* and *S. globiceps* in the Indo-Pacific and *S. iseri* in the Caribbean have been recorded to take up to 40 bites per minute (Bellwood & Choat 1990; Table 1). Although these rates are the extremes for the group, they are markedly higher than those recorded for the rabbitfishes (ca. 3–12 bites min^{-1}; Fox et al. 2009, Fox & Bellwood 2013) and surgeonfishes (ca. 9–17 bites min^{-1}; Goatley & Bellwood 2010). Such intense grazing activity and high abundances, coupled with their feeding preference for reef surfaces covered by algal material (Bellwood & Choat 1990, Bruggemann et al. 1994a,b, Bonaldo et al. 2006, Alwany et al. 2009, Jayewardene 2009, Afeworki et al. 2011; Tables 1, 2), highlight the potential importance of parrotfish in the transfer of energy and nutrients to consumers in food webs of reef ecosystems (Choat 1991). The intense feeding of parrotfish has been widely recognized as a major factor in controlling fast-growing algae and in influencing the composition and distribution of the benthos in these systems (Carpenter 1986, Bellwood et al. 2004, Ceccarelli et al. 2005, Mumby et al. 2006, Fox & Bellwood 2007), including the invertebrate cryptofauna (Kramer et al. 2013). However, it is their feeding mode that sets parrotfish apart from other herbivorous fishes.

Parrotfishes have a unique feeding mode that differentiates them from other groups of roving herbivorous fishes, especially in terms of their role in reef processes. The majority of surgeonfishes, drummers, and rabbitfishes tend to bite or crop the upper portions of algae (Choat et al. 2002, Hoey & Bellwood 2008), thereby allowing rapid regrowth from the basal portions of the algae. In contrast, the beak-like jaw of parrotfishes allows them to 'scrape' the substratum, removing both the upper and lower portions of the algae along with portions of the substratum (Bellwood & Choat 1990, Bruggemann et al. 1994b, 1996, Hoey & Bellwood 2008, Bonaldo & Bellwood 2009). This unique feeding mode not only has a greater impact on algal communities than other herbivorous fish groups but also may facilitate the settlement of other benthic taxa (e.g., corals, sponges, algae) by providing areas of bare substratum. On tropical reefs, the densities and biomass of parrotfish are often broadly comparable to those of herbivorous surgeonfish and usually exceed those of herbivorous rabbitfish and drummers (Bruggemann et al. 1995, Krajewski & Floeter 2011, Afeworki et al. 2013, Johansson et al. 2013). Indeed, it has been estimated that parrotfish represent 40–60% of the total biomass of roving herbivorous fish assemblages on the Great Barrier Reef (GBR; Fox & Bellwood 2007, Hoey & Bellwood 2010a,b); up to 72% in the Red Sea (Afeworki et al. 2013); 54–98% in the Caribbean (Bruggemann et al. 1996, Williams & Polunin 2001); and over 50% in the South-Western Atlantic province (*sensu* Kulbicki et al. 2013) off Brazil (Krajewski & Floeter 2011). Irrespective of the relative abundances of the other groups of herbivorous fish, they are not functional equivalents of parrotfish (i.e., do not fulfil the same role) in reef processes. As grazers of reef substrata, parrotfishes perform a role that is fundamentally different from that of other herbivorous taxa, highlighting the critical importance of parrotfishes to the functioning of healthy reef systems.

Of all plant and algal material available on tropical reefs, the EAM is the primary feeding substratum of parrotfish worldwide (Table 2). The EAM is a conglomeration of short, turf-forming filamentous algae (<1 cm high), macroalgal propagules, microalgae, sediment, detritus, and associated fauna (Purcell 2000, Purcell & Bellwood 2001, Wilson et al. 2003, Bruno et al. 2009, Bonaldo & Bellwood 2011b, Kramer et al. 2012). Because of the complex composition of the EAM, it has

Table 2 Summary of dominant feeding substrata of parrotfishes on tropical reefs around the world, presented as either the mean or range of bites (%) on each substratum type

		Proportion of bites (%)					
Species[a]	Location	EAM	Macroalgae	Live coral	CCA[b]	Seagrass	Source
Excavators							
Bolbometopon muricatum	Northern GBR, Australia	37.11	0	48.17	14.71	0	Hoey & Bellwood 2008
Chlorurus perspicillatus	Oahu, Hawaii	61	—	1	38	—	Ong & Holland 2010
C. sordidus	Zanzibar Island, Tanzania	91.9–100	0–2.7	0–6.3	0–13.3	—	Lokrantz et al. 2008
C. strongylocephalus	Zanzibar Island, Tanzania	80–100	0–10	0–26.7	0–10.5	—	Lokrantz et al. 2008
Sparisoma amplum	Abrolhos, Brazil	48.77	0.89 (0.9)	4.33 (1.4)	35.82 (3.8)	0.53	Francini-Filho et al. 2010
	Fernando de Noronha, Brazil	60–65	7	0	0	0	Bonaldo et al. 2006
S. viride	Karpata, Bonaire	79–97	0	0–5.1	—	—	Bruggemann et al. 1994b
	San Blas, Panama	91	5	—	—	5	McAfee & Morgan 1996
Scrapers/browsers							
Scarus ferrugineus	Sheikh Said Island, Red Sea, Eritrea	92–97	—	2–8	0	0	Afeworki et al. 2011
S. iseri	San Blas, Panama	8.5	23	—	—	—	McAfee & Morgan 1996
S. niger	Zanzibar, Tanzania	92.3–100	0	0–9.1	0–7.7	—	Lokrantz et al. 2008
S. rivulatus	Orpheus Island, Australia	97.17	0	2.82	0	0	Bonaldo & Bellwood 2008
S. rubroviolaceus	Oahu, Hawaii	55	—	1	40	—	Ong & Holland 2010
S. trispinosus	Abrolhos, Brazil	50.99 (3.9)	4.32 (1.5)	1.3 (0.4)	34.74 (3.7)	7.15 (1.8)	Francini-Filho et al. 2010
S. vetula	Karpata, Bonaire	67–99.8	0	0–0.2	—	—	Bruggemann et al. 1994b
S. zelindae	Abrolhos, Brazil	51.65 (3.8)	2.60 (1.4)	2.73 (1.2)	33.0 (3.7)	5.9 (1.5)	Francini-Filho et al. 2010
Sparisoma aurofrenatum	San Blas, Panama	26	23	—	—	23	McAfee & Morgan 1996
S. axillare	Abrolhos, Brazil	64.39 (3.3)	6.2 (1.7)	1.24 (0.5)	13.88 (1.9)	13.17 (2.64)	Francini-Filho et al. 2010
	Fernando de Noronha, Brazil	45–55	60–65	0	0	0	Bonaldo et al. 2006

Continued

Table 2 (Continued) Summary of dominant feeding substrata of parrotfishes on tropical reefs around the world, presented as either the mean or range of bites (%) on each substratum type

		Proportion of bites (%)					
Species[a]	Location	EAM	Macroalgae	Live coral	CCA[b]	Seagrass	Source
S. chrysopterum	San Blas, Panama	11	11	—	—	43	McAfee & Morgan 1996
S. frondosum	Abrolhos, Brazil	53.54 (3.9)	5.7 (1.6)	1.98 (0.9)	16.98 (2.7)	18.79 (3.2)	Francini-Filho et al. 2010
	Fernando de Noronha, Brazil	49–55	33–36	0	0	0	Bonaldo et al. 2006
S. rubripinne	San Blas, Panama	17	14	—	—	23	McAfee & Morgan 1996

Note: Numbers in parenthesis are standard errors.

[a] Where necessary, species names used in original articles have been replaced by the currently valid name (Froese & Pauly 2013).

[b] CCA, crustose coralline algae.

been questioned if parrotfish actually target the algal material or other components of the EAM that are of similar or even higher nutritional value, such as detritus and associated fauna (Choat et al. 2002, 2004, Wilson et al. 2003, Crossman et al. 2005, Kramer et al. 2013). Parrotfish ingest large amounts of detritus, bacterial complexes, meiofauna, coral tissue, sessile invertebrates, and algal material when feeding on the EAM (Bellwood & Choat 1990, Comeros-Raynal et al. 2012). This is consistent with studies examining the gastrointestinal short-chain fatty acid (SCFA) profiles of three parrotfishes (*Chlorurus microrhinos*, *C. sordidus*, and *Scarus schlegeli*), which reported low levels of SCFA with an unusually high proportion of isovalerate, indicative of a diet rich in detritus, bacteria, and meiofauna (Choat et al. 2002, 2004). Although detailed analyses of SCFA have been restricted to these three species, similarities in feeding mode suggest that detritivory may be widespread among parrotfishes, especially those that feed predominantly from the EAM. Further detailed studies are still necessary to verify the food items responsible for providing nutrition to different species in the group.

Irrespective of the nutritional target of parrotfish when feeding on the EAM, their constant feeding on this substratum clearly removes large amounts of benthic algae, clearing spaces on the reef surface than can be occupied by or promote growth of other sessile organisms (Bruggemann et al. 1994b, 1996, Fox & Bellwood 2007, Hughes et al. 2007, Alwany et al. 2009, Bonaldo & Bellwood 2009, Mumby & Harborne 2010). For instance, on the GBR estimated rates of grazing by parrotfish range from 250–450% of the reef surface area per year on outer-shelf reefs to 500–600% per year and 500–1700% per year on mid- and inner-shelf reefs, respectively (Fox & Bellwood 2007, Hoey & Bellwood 2008). On inner-shelf reefs, this is the equivalent of each square metre of the reef crest being grazed every 18 days (Fox & Bellwood 2007). The spaces cleared by parrotfish may remain largely free of algae and sediments for over 8 days, thus representing potential sites for the settlement of corals and other sessile organisms in the reef (Bonaldo & Bellwood 2009). Additionally, the effects of parrotfish grazing on young macroalgae are still not well known, and as these fish deliver frequent and relatively large bites on the feeding substratum, they may be able to reduce the proliferation of macroalgae by consuming it during vulnerable early stages (Diaz-Pulido & McCook 2003). In marine reserves in the Caribbean, for example, high parrotfish biomass and density were correlated with higher cover of scleractinian corals, higher density of coral recruits, and lower cover of macroalgae (Mumby 2006, Mumby et al. 2006, 2007). Even by targeting components other than algal material when feeding on the EAM, parrotfish may perform important roles as reef herbivores, having a large impact on the distribution, abundance, and succession of algal communities.

Browsing

Grazing fishes feeding on the EAM clearly play an important role on coral reefs, maintaining algal assemblages in a cropped state and aiding in the prevention of macroalgal establishment and expansion. The removal of mature fleshy macroalgae (i.e., browsing), however, represents a separate process on coral reefs and is becoming increasingly important because of recent increases in macroalgal biomass in many regions. Following large-scale coral mortality, the dead coral skeletons are rapidly colonized by algal turfs and macroalgal propagules (Diaz-Pulido & McCook 2002, Mumby & Steneck 2008). While reefs with relatively intact herbivore communities seem to be able to compensate for the increased algal production (Adam et al. 2011, Gilmour et al. 2013), reductions in herbivorous fishes in many regions have compromised the capacity of reefs to absorb the increased algal production, thereby releasing macroalgal propagules from top-down control. Such scenarios are becoming increasingly common on reefs worldwide and may ultimately lead to shifts to macroalgal dominance (e.g., Smith et al. 1981, Done 1992, Knowlton 1992, Hughes 1994, McManus & Polsenberg 2004, Graham et al. 2006, Cheal et al. 2010).

Transitions in the benthic composition of coral reefs, from coral to macroalgal dominance, represent a fundamental shift in the physical structure and functioning of these systems. Once established, these shifts are difficult to reverse as interactions among elements of the new state form positive feedbacks, reinforcing and maintaining the state (Hughes et al. 2005, Mumby & Steneck 2008, Scheffer et al. 2009, Hoey & Bellwood 2011). Grazing fishes feeding on the EAM may readily eliminate macroalgal propagules (Diaz-Pulido & McCook 2003); however, these fishes appear to have a limited capacity to remove larger, fleshy (predominantly brown) macroalgae (Choat et al. 2002, 2004, Bellwood et al. 2006a, Hoey & Bellwood 2009, Burkepile & Hay 2011, Vergés et al. 2012). The removal of larger, mature, fleshy macroalgae appears to be restricted to a smaller suite of species, the macroalgal browsers. This distinction is important and highlights the difference between those species that may prevent (i.e., grazers) and those that may potentially reverse (i.e., browsers) shifts to macroalgal dominance on tropical reefs (Bellwood et al. 2006a, Hoey & Bellwood 2011).

Within the parrotfishes, macroalgal browsing taxa are almost exclusively from the seagrass or *Sparisoma* clade (Figures 1, 2), which are most speciose and abundant in the tropical Atlantic and have long been recognized as one of the most important browsers of macroalgae and seagrass on reefs in this region (Randall 1967, Lobel & Ogden 1981, Targett & Targett 1990, Bonaldo et al. 2006, Ferreira & Gonçalves 2006, Mumby et al. 2006). An early study quantifying the gut content of five species of *Sparisoma* revealed that the diet of four of these species (*S. aurofrenatum*, *S. chrysopterum*, *S. rubripinne*, and *S. viride*) was dominated by filamentous and fleshy macroalgae (83–98%, including *Polysiphonia*, *Laurencia*, *Dictyota*, *Halimeda*), while the diet of *Sparisoma radians*, the smallest species of the genus, was dominated by seagrass (88%; Randall 1967). Subsequent studies relating the loss of transplanted macroalgae and seagrass (i.e., bioassays) to the distribution of herbivorous fishes among habitats and reefs and direct observations of feeding on macroalgal bioassays highlighted the importance of *Sparisoma* spp. as browsers of macrophytes (Hay 1981, 1984, Lewis 1985, 1986, Reinthal & Macintyre 1994). For example, *S. rubripinne* and *S. chrysopterum* were almost solely responsible for the consumption of assays of the brown macroalgae *Sargassum polyceratium*, *Turbinaria turbinata* and *Padina jamaicensis*, and the seagrass *Thalassia testudinum* on a Belizean reef (Lewis 1985). More recently, video observations of fishes feeding on algal assemblages generated by herbivore manipulations identified *Sparisoma aurofrenatum*, *S. chrysopterum*, and *S. viride* as the dominant consumers of fleshy brown (*Dictyota* and *Lobophora*) and jointed calcareous macroalgae (*Amphiroa*, *Jania*, and *Halimeda*) in the Florida

Keys (Burkepile & Hay 2011). Large *Scarus* species and *Kyphosus* (Kyphosidae) have also been recorded feeding on jointed calcareous and fleshy brown macroalgae, respectively (Randall 1967, Paddack et al. 2006, Burkepile & Hay 2011); however, they are typically less abundant and appear to play a relatively minor role compared to that of *Sparisoma* spp.

In contrast to the situation in the tropical Atlantic, the *Sparisoma* clade is less abundant and speciose on Indo-Pacific reefs, being represented by *Calotomus* spp. (five species) and *Leptoscarus vaigiensis* in the Indo-West Pacific and *Nicholsina denticulata* in the Tropical East Pacific (TEP; Figure 2). Parrotfish assemblages in these regions are instead dominated by members of the *Scarus* clade (e.g., Russ 2003, Fox & Bellwood 2007, Hoey & Bellwood 2008, 2011, Alwany et al. 2009; Figure 4). Analyses of stomach content and direct observations indicate that, like the Atlantic members of the *Sparisoma* clade, *Calotomus* spp. and *Leptoscarus vaigiensis* have a diet dominated by fleshy brown macroalgae (primarily *Sargassum* spp.) or seagrass (McClanahan et al. 1999, Ohta & Tachihara 2004, Hoey & Bellwood 2010c). Although relatively widespread across the Indo-Pacific, these species are typically not abundant and have a limited impact on the removal of macroalgae on most coral-dominated reefs (Hoey & Bellwood 2008, 2009, Michael et al. 2013). The only exceptions may be in macroalgae- (*Sargassum*) and seagrass-dominated habitats, which typically support higher densities of both juvenile and adult *Leptoscarus vaigiensis* (McClanahan et al. 1999, 2010, Hoey & Bellwood 2010b, Wilson et al. 2010).

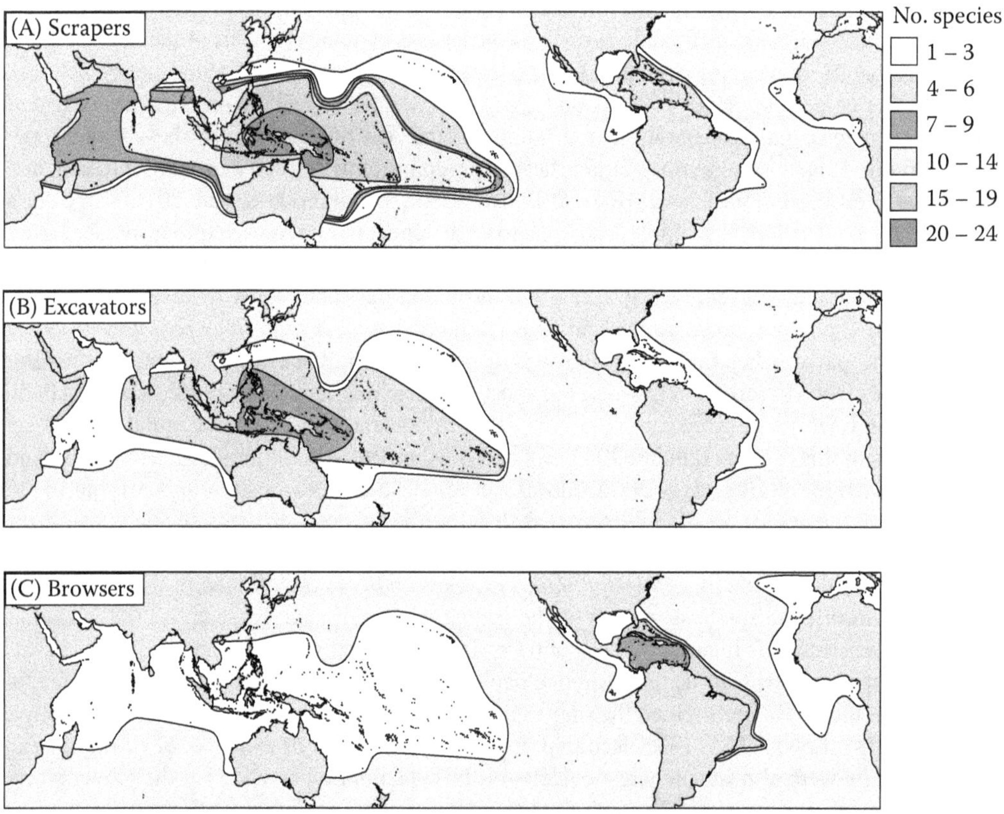

Figure 4 (See also colour figure in the insert) Biogeographic patterns of species richness within the three functional groups of parrotfishes: (A) scrapers, (B) excavators, and (C) browsers. The distributions highlight the Indo-Australian Archipelago and Caribbean as the centres of diversity within the Indo-Pacific and Atlantic Oceans, respectively, and that the Indo-Pacific supports a higher diversity of scraping and excavating species, and the Atlantic supports a higher diversity of browsing species.

On Indo-Pacific reefs, macroalgal browsing is predominantly carried out by other (i.e., non-parrotfish) reef fishes. Analyses of gut contents and video observations of feeding on macroalgal assays have shown that the dominant browsers of brown fleshy macroalgae on Indo-Pacific reefs include unicornfishes (Acanthuridae: *Naso unicornis*, *N. lituratus*, *N. brachycentron*, and *N. tonganus*; Robertson & Gaines 1986, Choat et al. 2002, Hoey & Bellwood 2009, Vergés et al. 2012, Rasher et al. 2013); drummers or sea chubs (Kyphosidae: *Kyphosus vaigiensis*, *K. cinerescens*, and *K. bigibbus*; Choat et al. 2002, Hoey 2010, Hoey & Bellwood 2011, Lefèvre & Bellwood 2011, Michael et al. 2013); rabbitfishes (Siganidae: *Siganus canaliculatus*, and *S. javus*; Fox & Bellwood 2008, Cvitanovic & Bellwood 2009, Bennett & Bellwood 2011, Hoey et al. 2013); and a batfish (Ephippidae: *Platax pinnatus*; Bellwood et al. 2006a).

Despite the predominance of other fish taxa in the consumption of fleshy macroalgae, some parrotfishes may still contribute to the removal of macroalgal biomass on Indo-Pacific reefs. Grazing fishes, including *Scarus* spp., have frequently been observed taking bites from the surface of large macroalgae, such as *Sargassum* (Fox & Bellwood 2008, Hoey & Bellwood 2009, Vergés et al. 2012, Michael et al. 2013). Although these species are likely to be targeting surface epibiota or detrital material rather than the macroalga itself, they may contribute to the removal of macroalgal biomass through the dislodgement or incidental ingestion of macroalgal material when feeding. This mechanism of removal is especially relevant when the epiphytic load of the macroalgae is high and the macroalgae are senescent (Lefèvre & Bellwood 2011). In addition, several Indo-Pacific parrotfishes have been identified as the dominant consumers of jointed calcareous macroalgae. Three parrotfishes (*Chlorurus microrhinos*, *Hipposcarus longiceps*, and *Scarus rivulatus*) were the dominant consumers of species of *Halimeda* and *Amphiroa* on an inshore reef on the GBR (Mantyka & Bellwood 2007a,b); *Chlorurus spilurus* was the dominant consumer of *Amphiroa* and *Galaxaura* on a Fijian reef (Rasher et al. 2013). This ability to consume calcareous macroalgae appears to be related to the strong jaws and lack of acidic gastric secretions in the parrotfishes (Bellwood & Choat 1990, Choat et al. 2004). Although parrotfish are the dominant browsers of macroalgae on tropical Atlantic reefs, their potential importance in the removal of macroalgae from Indo-Pacific reefs should not be discounted.

Coral predation

It is axiomatic that scleractinian (hard) corals constitute one of the dominant benthic components of tropical reef ecosystems, yet coral predation is one of the most specialized feeding habits among reef animals, with less than 5% of reef fish species known to consume live coral (Cole et al. 2008). Although approximately 10 reef fish families contain corallivorous representatives, only 1, the butterflyfishes (Chaetodontidae), is predominantly corallivorous, having more than 50% of its species feeding mainly, if not exclusively, on corals (Cole et al. 2008, Rotjan & Lewis 2008, Bellwood et al. 2010).

Parrotfishes are generally viewed as herbivores or detritivores, but a number of species also feed directly on live coral colonies (e.g., Tables 2, 3). Corallivorous parrotfishes have been recorded in most tropical reef systems around the world, including the Indo-Pacific (Bellwood & Choat 1990, Bellwood et al. 2003, Bonaldo & Bellwood 2011a); the Caribbean (Bruggemann et al. 1996, Bruckner & Bruckner 1998, Rotjan & Lewis 2005, 2008); the Red Sea (Alwany et al. 2009); and Brazilian coastal waters (Francini-Filho et al. 2008, 2010). For the majority of parrotfishes, corallivory represents only a small proportion of their feeding (less than 1%), especially in comparison to their feeding on dead coral surfaces covered by the EAM (Table 2). However, in a few species, especially those that attain larger body sizes and are able to bite deeper into the substratum, up to 50% of the diet may consist of live coral (e.g., Table 2). The most corallivorous of all parrotfishes is the bumphead wrasse, *Bolbometopon muricatum*, the largest species in the group (up to 130 cm total length, TL), which has been reported to take approximately half of its bites from live corals

Table 3 Parrotfishes known to feed on live coral colonies and corals targeted by each species (if known)

Parrotfishes	Location	Corals	Source
Excavators			
Bolbometopon muricatum	Lizard Island, Australia	*Acropora cytherea*, *A. divaricata*, *A. hyacinthus*, *A. latistella*, *A. listeria*, *A. millepora*, *A. nasuta*, *A. secale*, *Anamastrea* sp., *Favites* sp., *Porites cylindrica*	Bellwood 1985
	Lizard Island, Australia	*Acropora* spp., massive *Porites* spp.	Bonaldo & Bellwood 2011a
	Indo-Pacific	*Acropora* spp., *Montipora* spp., *Pocillopora verrucosa*	Bellwood et al. 2003, Hoey & Bellwood 2008
Cetoscarus bicolor	South Sinai, Red Sea, Egypt	*Porites* spp.	Alwany et al. 2009
C. ocellatus	GBR, Australia	*Montipora* sp., *Platygyra* spp., encrusting species	Bellwood 1985
Chlorurus gibbus	South Sinai, Red Sea, Egypt	Not specified	Alwany et al. 2009
C. microrhinos	GBR, Australia	Massive *Porites* spp.	Bonaldo & Bellwood 2011a, Bonaldo et al. 2012
C. perspicillatus	Oahu, Hawaii	Not specified	Ong & Holland 2010
C. spilurus	GBR, Australia	*Porites australiensis*, *P. lobata*, *P. lutea*	Bellwood 1985
C. strongylocephalus	Zanzibar, Tanzania	Not specified	Lokrantz et al. 2008
Sparisoma amplum	Abrolhos, Brazil	*Mussismilia braziliensis*, *Montastrea cavernosa*, *Siderastrea* spp.	Francini-Filho et al. 2008
S. viride	Karpata, Bonaire	Not specified	Bruggemann et al. 1994a,c
	Carrie Bow Cay, Belize	*Montastrea annularis*, *M. cavernosa*, *M. franksi*, *M. faveolata*, *Siderastrea siderastrea*, *Agaricia agaricites*, *Porites porites*, *P. astreoides*, *Diploria strigosa*	Rotjan & Lewis 2006
	Belize	*Porites porites*	Littler et al. 1989
	Carrie Bow Cay, Belize	*Porites astreoides*	Rotjan & Lewis 2005
	Florida Keys	*Porites divaricata*	Miller & Hay 1998
	Rosario Islands, Colombia	*Montastrea annularis*	Sánchez et al. 2004
	Puerto Rico	*Colpophyllia natans*	Bruckner & Bruckner 1998
	St Croix, US Virgin Islands	*Montastrea annularis*	Bruckner et al. 2000
Scrapers or browsers			
Scarus ferrugineus	South Sinai, Red Sea, Egypt	*Porites* spp.	Alwany et al. 2009
S. flavipectoralis	Lizard Island, Australia	Massive *Porites* spp.	Bonaldo & Bellwood 2011a
S. frenatus	South Sinai, Red Sea, Egypt		Alwany et al. 2009
	Lizard Island, Australia	Faviidae	Bonaldo & Bellwood 2011a
	Heron Island, Australia	*Porites* spp.	Bellwood & Choat 1990
S. ghobban	South Sinai, Red Sea, Egypt	*Porites* spp.	Alwany et al. 2009

Table 3 (Continued) Parrotfishes known to feed on live coral colonies and corals targeted by each species (if known)

Parrotfishes	Location	Corals	Source
S. coelestinus	Caribbean	—	Randall 1967, 1974
S. guacamaia	Caribbean	—	Glynn 1997
S. niger	Lizard Island, Australia	Massive *Porites* spp.	Bonaldo & Bellwood 2011a
	Zanzibar, Tanzania	—	Lokrantz et al. 2008
S. rivulatus	Lizard Island, Australia	*Porites* spp.	Bellwood 1985
S. rubroviolaceus	Oahu, Hawaii	—	Ong & Holland 2010
S. trispinosus	Abrolhos, Brazil	*Montastrea cavernosa, Mussismilia braziliensis, M. hartii, Favia gravida, Porites astreoides, Siderastrea* spp.	Francini-Filho et al. 2008
Sparisoma aurofrenatum	Florida Keys	*Porites divaricata*	Miller & Hay 1998

Note: Where necessary, original species names where replaced by the currently valid nomenclature.

(Bellwood et al. 2003). It should be noted, however, that this species appears to feed non-selectively on live coral colonies and dead coral surfaces covered by EAM, as these substrata are bitten largely in proportion to their availability (Bellwood et al. 2003, Hoey & Bellwood 2008). Other large parrotfishes, such as *Cetoscarus bicolor* (to 80 cm TL) and *Chlorurus gibbus* (to 70 cm TL) in the Red Sea (Alwany et al. 2009) and *C. strongylocephalus* (to 70 cm TL) in Zanzibar (Lokrantz et al. 2008), may deliver 20–30% of their bites on live coral colonies. In contrast, on the GBR *Cetoscarus ocellatus* (to 80 cm TL), the sister taxon of *C. bicolor*, and *Chlorurus microrhinos* (to 70 cm TL), sister taxon of both *C. gibbus* and *C. strongylocephalus*, have been reported to take fewer than 10% of their bites from live coral colonies (Bellwood & Choat 1990). Such variation among regions suggests that the extent of corallivory within the parrotfishes may be dependent, at least to some extent, on local environmental conditions, such as the species of coral present (Bonaldo & Bellwood 2011a).

Of the remaining species for which data are available, the smaller species of *Chlorurus* and *Scarus* appear to take relatively few bites from live coral (<5%; Table 2); however, data are lacking for many species. It is possible that other parrotfishes worldwide, especially larger species and individuals, may feed on corals to a greater extent than is currently known. One of the difficulties may be that feeding by larger individuals is often difficult to observe in areas where parrotfish are targeted by fisheries, especially those using spearfishing (Bellwood et al. 2003, 2012a, Aswani & Hamilton 2004, Bejarano et al. 2013), as these individuals are likely to be highly averse to divers (Dickens et al. 2011, Feary et al. 2011, Januchowski-Hartley et al. 2012, Welsh & Bellwood 2012a).

Overall, the limited feeding by most parrotfishes on live coral colonies, especially compared to their extensive feeding on surfaces covered by EAM (Bruggemann et al. 1996, Ferreira & Gonçalves 2006, Bonaldo & Bellwood 2008, Francini-Filho et al. 2008, Lokrantz et al. 2008), may suggest that corallivory by parrotfish is not an important factor in shaping the distribution or survivorship of coral colonies in reef ecosystems. There is increasing evidence, however, that in areas of high parrotfish abundance, they may have a significant impact on the distribution, reproduction, survivorship, and overall health of highly selected coral species (e.g., Littler et al. 1989, Miller & Hay 1998, Rotjan & Lewis 2008, Rotjan & Dimond 2010, Bonaldo & Bellwood 2011a, Bonaldo et al. 2011, 2012). On Lizard Island, a midshelf reef in the northern GBR, it has been estimated that in areas of high parrotfish biomass density (i.e., the reef crest), the total surface area of massive *Porites* colonies is completely grazed every 270 days by parrotfish (Bonaldo & Bellwood 2011a). Such intense rates of corallivory suggest that parrotfish may be controlling massive *Porites* populations in some reef zones. This conclusion was supported by a negative correlation between the areal extent of massive *Porites* spp. colonies and the population density of parrotfish or the number of parrotfish bite marks per unit area on live *Porites* colonies (Bonaldo & Bellwood 2011a). Likewise, parrotfish corallivory has been considered one of the determinants of the distribution of *Porites astreoides* and *P. porites* f. *furcata* on Caribbean reefs (Littler et al. 1989, Miller & Hay 1998).

In addition to adult coral colonies, several studies have suggested that parrotfish cause damage to or mortality of recently settled or juvenile corals (Sammarco 1980, Raimondi & Morse 2000, Box & Mumby 2007, Rotjan & Lewis 2008, Mumby 2009b, Penin et al. 2010, 2011, Trapon et al. 2013b). The impact of parrotfish grazing on coral recruits, however, is not straightforward. As described previously, through their feeding activities parrotfishes may provide areas of clear substratum and thereby facilitate the settlement of corals, but they may also contribute to the mortality of small corals through incidental predation. On reefs at Lizard Island on the GBR, the exclusion of herbivorous fishes, including parrotfishes, increased the survivorship of coral recruits on artificial substrata (tiles) by 50% on the reef crest but had no detectable effects on the back reef (Trapon et al. 2013b). As the reef crest has higher herbivorous fish densities and grazing rates (Fox & Bellwood 2007, Hoey & Bellwood 2008), it appears that fish grazing might be a major source of mortality for recently settled corals. Indeed, survivorship on the reef crest was negatively correlated with the density of parrotfish feeding scars on the experimental tiles after 4 weeks (Trapon et al. 2013b). However, several studies have found that the densities of juvenile corals (<50 mm in diameter) are positively related to the population density of parrotfish, presumably because moderate increases in grazing activity provide suitable settlement sites and prevent smothering of newly settled corals by algae and sediment on the algae (Hughes et al. 2007, Hoey et al. 2011, Penin et al. 2011). At higher parrotfish densities, these positive effects may be outweighed by the effects of increased mortality through incidental grazing, with the population density of juvenile corals either negatively correlated or uncorrelated with the abundance or biomass of parrotfish (Trapon et al. 2013a,c).

Corallivorous parrotfishes have a fundamentally different impact on coral colonies in comparison to other corallivorous taxa that typically only remove the living tissue when feeding and leave the underlying coral skeleton largely intact (e.g., crown-of-thorns starfish: Keesing & Lucas 1992; butterflyfish: Harmelin-Vivien & Bouchon-Navaro 1983, Cole et al. 2008). Many corallivorous fishes, such as butterflyfishes and filefishes (Monacanthidae), have small and highly modified jaws capable of removing individual coral polyps (Cole et al. 2008). In contrast, the robust jaws of the parrotfishes, alongside some pufferfishes (Tetraodontidae) and triggerfishes (Balistidae), allow them to bite deeper into coral colonies, removing not only the living tissue but also portions of the underlying coral skeleton (Glynn 1997, Jayewardene & Birkeland 2006, Jayewardene et al. 2009, Bonaldo et al. 2011, 2012). As a consequence, although areas of live coral colonies preyed on by butterflyfishes usually recover relatively quickly, predation by excavating parrotfishes and pufferfishes usually results in the death of part or all of the colony and possible changes in colony shape (Bonaldo et al. 2011, 2012).

Corallivory by parrotfishes, and other groups that bite deep into coral skeletons, is likely to have a greater impact on coral populations than feeding by other corallivores that bite or pick individual polyps when feeding. Feeding by butterflyfishes and other corallivores that remove individual polyps typically elicits the retraction of adjacent polyps and tentacles into the coral skeleton (Gochfeld 2004). This withdrawal of polyps is likely to render them inaccessible to these 'nipping' corallivores and will reduce feeding on the adjacent polyps and across the entire colony as the wave of polyp retraction spreads from the initial site of attack. Such short-term retraction of polyps, however, offers little defence against corallivorous parrotfishes that remove large areas of coral tissue and the underlying skeleton with a single bite (Bonaldo et al. 2011). The difference in the extent of coral tissue loss is likely to influence the morphology, physiology, and regeneration of affected colonies. For example, the tissue damage caused by the feeding of a butterflyfish (i.e., loss of a single polyp) may regenerate within 7–10 days, whereas parrotfish and pufferfish grazing scars can take from 40 days to 3 years to heal (Bak & Steward-Van Es 1980, Bruckner & Bruckner 1998, Sánchez et al. 2004, Jayewardene & Birkeland 2006, Bonaldo et al. 2011).

Parrotfishes and butterflyfishes are the two main groups of corallivorous fishes on coral reefs (Cole et al. 2008, Rotjan & Lewis 2008). Both groups are widespread and abundant, are characterized by high feeding rates (Table 1; Gregson et al. 2008, Pratchett et al. 2013), and have been shown to influence the distribution of coral colonies in coral reefs (Littler et al. 1989, Miller & Hay 1998, Cole et al. 2008, Bonaldo & Bellwood 2011a). On the GBR, estimates of daily tissue removal of massive *Porites* by parrotfish range from 0.57–6.84 mg m^{-2} on an inshore reef to 1.19–7.53 mg m^{-2} on a midshelf reef (Bonaldo & Bellwood 2011a, Bonaldo et al. 2012). These estimates are generally lower than those for butterflyfish on Indo-Pacific reefs (e.g., Harmelin-Vivien & Bouchon-Navaro 1983, Cole et al. 2011, 2012). On Lizard Island, for instance, butterflyfish consume *Acropora* tissue at rates of 73–137 mg m^{-2} d^{-1} (Cole et al. 2011, 2012). Similarly, butterflyfish have been estimated to remove coral tissue at rates of approximately 10 and 63 mg m^{-2} d^{-1} in the Gulf of Aqaba and Moorea, respectively (Harmelin-Vivien & Bouchon-Navaro 1982, 1983). Thus, overall removal of *Porites* tissue by parrotfish per unit area on inner- and midshelf reefs on the GBR was 2–15 times lower than the rates of *Acropora* tissue removal by butterflyfish on Indo-Pacific reefs. However, rates of coral predation by parrotfish on outer-shelf reefs of the GBR (35.62–41.09 g m^{-2} d^{-1}; Hoey & Bellwood 2008) are over 1000 times greater than estimates for inner- and midshelf reefs. Although these last estimates include the mass of carbonate skeleton removed when feeding, it would appear that coral predation by parrotfish on the outer-shelf GBR reefs, and indeed anywhere that *Bolbometopon muricatum* is abundant, is likely to greatly exceed that of butterflyfishes and other corallivorous fishes (Bellwood et al. 2003, 2012a, Hoey & Bellwood 2008).

Corallivorous parrotfishes and butterflyfishes differ not only in terms of feeding mode and the intensity of coral predation but also in the coral species they target. Corallivorous butterflyfishes feed preferentially on fast-growing branching corals, such as *Acropora* and *Pocillopora* (Pratchett 2005, Cole et al. 2008, 2012), which grows at rates of approximately 10–15 cm yr^{-1}. In contrast, the majority of corallivorous parrotfishes prey predominantly on slow-growing massive and submassive corals, such as *Porites* and *Montastrea* (with the exception of *Bolbometopon muricatum*; Bellwood et al. 2003, Hoey & Bellwood 2008; Table 3). Massive *Porites* are among the slowest-growing coral species worldwide, with growth rates as low as 9 mm yr^{-1}, about 10–15 times lower than growth rates of *Acropora* spp. (Veron 2000). Moreover, as the deeper marks left by parrotfish remove parts of the underlying skeleton, especially when delivered in clusters (Bruckner & Bruckner 1998, Bonaldo et al. 2011, 2012), individual parrotfish bite marks appear to have a greater impact on coral colonies than those of butterflyfish.

Although parrotfish may have negative effects on corals (Littler et al. 1989, Miller & Hay 1998, Rotjan & Lewis 2005, Mumby 2009b), records of total colony mortality caused by parrotfish are relatively rare (e.g., Bruckner & Bruckner 1998, Rotjan & Lewis 2005, Francini-Filho et al. 2008). In some extreme cases, coral predation by parrotfish has been reported to affect the growth form and survivorship of adult coral colonies (e.g., Bruckner & Bruckner 1998, Rotjan & Lewis 2005, 2008, Francini-Filho et al. 2008, Bonaldo et al. 2011, 2012). For instance, *Sparisoma viride*, an excavating parrotfish species that delivers concentrated bites on coral colonies when feeding ('focused biting' *sensu* Bruckner et al. 2000), has been reported occasionally to cause total mortality of *Colpophyllia natans* colonies in the Caribbean (Bruckner & Bruckner 1998). Similarly, portions of massive *Porites* colonies on the GBR are rapidly colonized by algal turfs and appear unable to recover from intensive feeding by excavating parrotfishes, such as *Chlorurus microrhinos* (Bonaldo et al. 2011). Thus, the negative effects of parrotfish on corals appear to be restricted to particular coral species and within specific reef habitats (Littler et al. 1989, Rotjan & Lewis 2005, Mumby 2009b, Bonaldo & Bellwood 2011a). Although this predation may result in local declines, it may also support coral species richness within reef zones by decreasing the abundance of dominant coral species (e.g., *Porites* spp.) in areas with high parrotfish abundance (e.g., Littler et al. 1989,

Rotjan & Lewis 2005, Mumby 2009b, Bonaldo & Bellwood 2011a). Overall, the effects of parrotfish on corals are complex. They enhance settlement but may increase juvenile mortality (Trapon et al. 2013b). Parrotfish may change the growth rates, growth forms, and survival of coral species yet may maintain diversity by removing competitively dominant species. Regardless of the exact nature of the interactions, parrotfish undoubtedly play an important role in coral demography.

Bioerosion

Coral reefs by their very nature are dynamic growing structures. The reef grows as a result of carbonate accretion primarily because of deposition by corals and coralline algae (Glynn 1997, Hopley et al. 2012). However, infilling by sediment from invertebrates and algae and subsequent diagenesis also contribute to the production of a solid reef matrix (Glynn 1997, Mallela & Perry 2007, Hopley et al. 2012). The opposite side of the equation in the carbonate budget (*sensu* Scoffin et al. 1980) is the loss of carbonate. The relative magnitudes of the rates of accretion and erosion will determine if the reef grows or shrinks. Most work on carbonate budgets has been in the Caribbean and TEP, where the limited species diversity of corals, fishes, and other reef organisms facilitates quantification. For many years, it has been clear that the balance has been changing, with some exceptional areas eroding to the extent that reefs have been completely eroded back to the base rock (e.g., Eakin 1996). In recent years, the Caribbean has seen widespread declines in carbonate accretion to the extent that many areas may now be in negative carbonate status (Perry et al. 2013).

The loss of carbonate from a reef may be physical, chemical, or biological. Chemical loss through dissolution is limited, although it may increase somewhat as ocean acidification progresses (Hoegh-Guldberg et al. 2007). Physical loss may be considerably greater, although empirical evidence of its magnitude is limited. Our best understanding of physical loss is associated with extreme weather events (i.e., tropical storms), which periodically dislodge corals from the reef surface (Madin & Connolly 2006). However, background levels of carbonate loss may also be high. Using a series of mesh traps to quantify loss of corals by fragmentation, Hughes (1999) estimated that on Lizard Island on the GBR, 1.87 kg of corals per metre of reef front per annum were lost as fragments, a value that approaches other sources of carbonate loss. Most of this loss is probably through wave action on dead colonies or parts of colonies. However, pieces of coral (*Acropora* spp.) and whole colonies (*Acropora hyacinthus*) may also be dislodged during feeding by *Bolbometopon muricatum*. Thus, parrotfishes may not ingest all of the material they detach from the reef, and some physical losses may be biologically mediated.

By far the greatest source of carbonate loss from reefs is bioerosion. Bioerosion is the removal of carbonate by living organisms. This may be through dissolution by acids (boring polychaetes, gastropods) or direct physical action, whether boring (sponges, bivalves) or surface excavation (fish, urchins). Several studies have investigated the relative contribution of internal (boring) versus external (excavating) erosion on coral reefs (Hutchings 1986, Kiene & Hutchings 1994, Peyrot-Clausade et al. 2000). The results were highly consistent: Erosion on intact offshore reefs was overwhelmingly dominated by external erosion. Based mainly on the rates of erosion of premeasured blocks of coral, the rates of external erosion could be two orders of magnitude higher than internal erosion (e.g., Kiene & Hutchings 1994). The only exception to this general trend is in highly productive, polluted, inshore, degraded, or marginal reefs (i.e., reefs that exist near or beyond normal environmental limits of reef distribution), where internal bioerosion (principally by bivalves and clionid sponges) may be considerably higher and in some instances may exceed external erosion (e.g., Risk et al. 1995, Hutchings et al. 2005). Intact reefs may also differ from damaged reefs in the taxa primarily responsible for external erosion. On healthy reefs, external erosion is almost invariably dominated by parrotfishes; however, in degraded, overfished, or marginal reefs, external erosion is typically dominated by echinoids (McClanahan & Muthiga 1988, Eakin 1996, Carreiro-Silva & McClanahan 2001). Erosion by echinoids is important in some exceptional locations, such as on the slope of Ningaloo Reef, Western Australia (Johansson et al. 2010), and on some reefs in the Caribbean (Brown-Saracino et al. 2007).

The earliest studies of parrotfish bioerosion were in the Caribbean, where carbonate budgets constructed by palaeontologists investigating reef processes identified parrotfishes (especially *Sparisoma viride*) as the key agent of external erosion (Gygi 1969, 1975, Frydl & Stearn 1978, Scoffin et al. 1980). The feeding of *S. viride* was described in detail and quantified by Bruggemann et al. (1994a,c), who subsequently examined the relative contribution of two parrotfishes (*Scarus vetula* and *S. viride*) to patterns of bioerosion on the fringing reef of Bonaire in the Netherlands Antilles (Bruggemann et al. 1994b, 1996). Again, *S. viride* was by far the greatest contributor to bioerosion, removing reef material at a rate of 0.55–5.38 kg m^{-2} yr^{-1} (Table 4).

The similarities in the patterns of bioerosion by parrotfishes in the Caribbean and Indo-Pacific were highlighted by Bellwood & Wainwright (2002). The main difference was in the taxonomic composition of the eroders (and their evolutionary history). There is no overlap in the species or genera

Table 4 Variation in the production (i.e., by bioerosion) and reworking of sediments among parrotfishes and geographic locations

Species[a]	Maximum fish length (TL, cm)	Location	Bioerosion or sediment production (kg individual^{-1} yr^{-1})	Sediment reworked (% of total)	Source
Excavators					
Bolbometopon muricatum	130	Indo-Pacific, GBR, Australia	5690	—	Bellwood et al. 2003, Hoey & Bellwood 2008
Chlorurus gibbus	70	South Sinai, Red Sea, Egypt	290.3	—	Alwany et al. 2009
C. microrhinos	70	Lizard Island, Australia	1017.7	—	Bellwood 1995a
		Lizard Island, Australia	—	1.4	Bellwood 1996
C. perspicillatus	61	Oahu, Hawaii	21–301	—	Ong & Holland 2010
C. spilurus	40	Lizard Island, Australia	23.6	—	Bellwood 1995a
		Lizard Island, Australia	—	19	Bellwood 1996
C. sordidus		South Sinai, Red Sea, Egypt	42.3	—	Alwany et al. 2009
Sparisoma viride	64	Karpata, Caribbean	0–142	—	Bruggemann et al. 1996
		Barbados	—	<34	Scoffin et al. 1980
		Bermuda	—	30	Gygi 1975
		Karpata, Bonaire	—	10–15	Bruggemann 1995
Scrapers					
Scarus ferrugineus	41	South Sinai, Red Sea, Egypt	43[b]	—	Alwany et al. 2009
S. frenatus	40	South Sinai, Red Sea, Egypt	44[b]	—	Alwany et al. 2009
S. ghobban	75	South Sinai, Red Sea, Egypt	232[b]	—	Alwany et al. 2009
S. niger	40	South Sinai, Red Sea, Egypt	15[b]	—	Alwany et al. 2009
S. rubroviolaceus	70	Oahu, Hawaii	14–380	—	Ong & Holland 2010
S. vetula	61	Karpata, Bonaire	0–51	—	Bruggemann et al. 1996
			—	61–79	Bruggemann 1995

[a] When necessary, original species names were replaced by the currently valid nomenclature (Froese & Pauly 2013).
[b] Approximate values calculated from original values given in kilograms per hour.

of key bioeroders between the two regions. The dominant genus in the Caribbean, *Sparisoma*, is restricted to the Atlantic and was probably separated from its Indo-Pacific relatives around 20 million years before present (Ma; Figure 2). The evolutionary history of the two areas has been largely independent for at least the past 19 million years (Bellwood & Wainwright 2002, Floeter et al. 2008, Cowman et al. 2009, Cowman & Bellwood 2013a,b). Despite their evolutionary separation, the functional processes in the Atlantic and Indo-Pacific are remarkably consistent. There are four main similarities: Firstly, regardless of the local species richness, the bioeroding functional group is invariably composed of relatively few species, usually with just one or two species predominating, for example, one in the Caribbean (*Sparisoma viride*), two to three in the Indo-Pacific (such as *Chlorurus microrhinos* and *C. spilurus* (formerly *C. sordidus*) on the GBR) and three in the Red Sea (*C. gibbus*, *C. sordidus,* and *Cetoscarus bicolor*) (Bruggemann et al. 1994b, 1996, Bellwood & Wainwright 2002, Hoey & Bellwood 2008). Wherever present, *Bolbometopon muricatum* typically dominates bioerosion in Indo-Pacific locations (Bellwood et al. 2012a; Table 4).

Secondly, in all locations, erosion is dominated by one relatively large and uncommon species of parrotfish. On coral reefs, open areas predominate, and the convex surfaces targeted by excavating forms may only represent a small fraction of the surfaces available for grazing (Choat & Bellwood 1985, Goatley & Bellwood 2011). This suggests that there may be constraints on excavating feeding resulting from the limited availability of suitable surfaces.

Thirdly, in both Atlantic and Indo-Pacific regions, bioeroding species are functionally excavators (*sensu* Bellwood & Choat 1990) with well-developed musculature and robust jaws with uneven, crenulated edges (Bellwood 1994). Given the markedly different evolutionary histories of parrotfishes in the two regions, these similarities appear to have arisen as a result of functional convergence. Excavating Indo-Pacific parrotfishes (*Bolbometopon*, *Cetoscarus*, and *Chlorurus*) probably shared a common excavating ancestor; excavating *Sparisoma* probably evolved later from a seagrass-feeding lineage (Streelman et al. 2002, Robertson et al. 2006, Choat et al. 2012; Figure 2).

Finally, in both regions, estimated rates of erosion by parrotfishes closely approximate local calcification rates (Bellwood 1995b, Bruggemann et al. 1994b, 1996, Bellwood et al. 2003). This suggests that a large proportion or even the majority of carbonate produced in an area is removed by parrotfish and that, in a dynamic carbonate budget, parrotfish are major agents of bioerosion on both Caribbean and Indo-Pacific coral reefs.

Although parrotfishes are, by far, the most important of the fishes implicated in bioerosion, other records of potential bioerosion by fish include tetraodontiforms, such as pufferfishes, *Arothron* spp. (e.g., Jayewardene & Birkeland 2006, Jayewardene et al. 2009), and the surgeonfish, *Ctenochaetus striatus* (Krone et al. 2006). Although important in affecting distributions of corals (Jayewardene et al. 2009), erosion by pufferfish is likely to constitute only a minor loss of carbonate, as they are usually less abundant and have lower feeding rates than parrotfish. For instance, a species of pufferfish in Panama that erodes about 20 g of coral per day results in a total carbonate loss of only 30 g $CaCO_3$ m^{-2} yr^{-1} because of a relatively low population density of 40 individuals per hectare (Glynn 1997). Furthermore, the study on *C. striatus* is based on inferences from soft anatomy, with no laboratory or field validation. Rather than bioeroding, *C. striatus* appears to be primarily involved in sediment transfer (Choat 1991, Goatley & Bellwood 2010; see 'Production, reworking and transport of sediment' below).

Although erosion in the Indo-Pacific is primarily caused by excavating parrotfish genera, there may be some erosion as a result of the grazing activities of larger scraping *Scarus* species. The initial separation of parrotfishes into two functional groups, 'excavators' and 'scrapers', reflected behavioural, anatomical, and evolutionary divisions (Bellwood & Choat 1990, Bellwood 1994). However, there is some overlap in function, with some large scrapers capable of excavating to some extent, largely as a result of their size. In the Caribbean, this includes *S. vetula* (Bruggemann et al. 1996) and *S. guacamaia* (Gygi 1975); in Brazilian waters, *S. trispinosus* (Francini-Filho et al. 2008, 2010); in the Indo-Pacific, *S. rubroviolaceus*, *S. ghobban*, and *S. altipinnis* (Bellwood &

Choat 1990, Alwany et al. 2009, Ong & Holland 2010). Among the excavators, there is also a size-dependent variation in erosion rates, with large fishes having a disproportionately large influence on erosion rates. Thus, although *Chlorurus spilurus* is, based on behaviour and morphology, an excavating species, its contribution is usually small and only dominates erosion when abundant and in the absence of other, larger, species (Alwany et al. 2009, Johansson et al. 2010). When present, it is the larger *Chlorurus* species (e.g., *C. microrhinos*, *C. gibbus*) and especially *Bolbometopon muricatum* that overwhelmingly dominate erosion. Per capita erosion rates of *Chlorurus sordidus* are probably comparable to those of smaller *Chlorurus* species (e.g., *C. atrilunula*, *C. genazonatus*, *C. japanensis*), with estimated mean values of 23.6 kg per individual per year (kg yr^{-1}; Bellwood 1995a). The slightly larger species, *C. perspicillatus*, is estimated to remove 380 kg yr^{-1} (Ong & Holland 2010), while the largest species of *Chlorurus*, *C. microrhinos*, is estimated to remove 1.02 t yr^{-1} (Bellwood 1995a). In comparison, *Bolbometopon muricatum* individuals remove an estimated 5.4 t yr^{-1} (Bellwood et al. 2003; Table 4).

It is this overwhelming effect of the larger individuals that determines bioerosion patterns across the GBR, from 0.9–3.8 kg m^{-2} yr^{-1} on inshore reefs to 23.1–32.3 kg m^{-2} yr^{-1} on outer reefs—a difference that is caused largely by the distribution of *B. muricatum* (Hoey & Bellwood 2008). This is also the underlying factor that drives global patterns of erosion across the Indo-Pacific, with the loss of *B. muricatum* reflecting the intensity of human exploitation of reef fishes (Bellwood et al. 2003, 2012a). Indeed, the rates of erosion across much of the Indo-Pacific are less than half the rates expected from historical estimates and appear to reflect a collapse of this process as a result of human activity within the last 50 years (Bellwood et al. 2012a). With the effective functional loss of *B. muricatum*, these reefs may be experiencing an erosional environment that has not existed since before species of *Bolbometopon* appeared around 28 Ma (Cowman et al. 2009, Cowman & Bellwood 2011, 2013a; Figure 2).

High rates of parrotfish bioerosion reflect the abundance of larger parrotfishes, which is greatest on reefs exposed to the open ocean and in the shallower areas, especially on exposed reef crests (Bruggemann et al. 1996, Hoey & Bellwood 2008, Ong & Holland 2010). This is where the productivity of the EAM is greatest (Williams et al. 1986) and where detrital aggregates are the most nutritious (with the lowest proportion of inorganic material) (Crossman et al. 2001, Wilson et al. 2003). In these areas, eroding parrotfishes feed predominantly on convex (raised) surfaces (Bellwood & Choat 1990), and unlike echinoids, rates of erosion will decline as convex surfaces become increasingly flattened. Although the diets of several parrotfishes do include corals (Rotjan & Lewis 2006, Cole et al. 2008, Bonaldo & Bellwood 2011a; see the previous section on coral predation), only *B. muricatum* feeds primarily on corals (largely in proportion to availability; Bellwood et al. 2003, Hoey & Bellwood 2008). In all cases, whether corals are ingested alive or as part of the consolidated reef matrix, the carbonate that is ingested must be discharged, and as a result the rate of erosion is correlated with the rate of sediment production.

Production, reworking, and transport of sediment

When feeding on the EAM, large scraping parrotfishes and excavators remove part of the underlying reef matrix in addition to the algae and particulate material in the EAM. Because of their unusual feeding mode and pharyngeal apparatus (Gobalet 1989, Bellwood & Choat 1990, Bellwood 1994), ingested carbonate is passed into the intestine as fine sediment. This is then voided in the faeces (Bellwood 1996, Bruggemann et al. 1996). Thus, the sediment discharged by parrotfish is a combination of sediment production (the material eroded from the reef matrix; primary sediment production) and reworking, by which sediment is ingested and processed along with other material before discharge, probably with a smaller particle size (secondary production or reworking; Scoffin et al. 1980, Bellwood 1985, 1995b, 1996). The resulting modal particle size of sediment reworked by parrotfish is typically less than 250 µm (fine sand) and may be less than 63 µm (mud) (Bellwood 1996).

Parrotfishes vary in the extent to which they produce or rework sediment. In general, the more delicate the feeding anatomy, the greater the reworking. Thus, the vast majority of smaller *Scarus* species are predominantly sediment reworkers (Bellwood 1996; Table 4). In contrast, larger species, especially excavating species, release a higher proportion of primary sediment in the faeces. Thus, the extent of primary sediment production is a reflection of a species' individual bioerosion rate, ranging from almost zero in *S. schlegeli* to several tonnes per year in *Bolbometopon muricatum* (Bellwood et al. 2003, Hoey & Bellwood 2008). However, sediment transport and reworking occurs in all parrotfishes that feed on soft sediments or reef surfaces that are covered with EAM (with associated carbonate sediments; Bellwood 1996). Rates of sediment reworking are comparable to those of other major sediment removers, such as the surgeonfish *Ctenochaetus striatus* (Goatley & Bellwood 2010), and sediment can be transported over considerable distances (Bellwood 1996).

The source of transported sediment reflects the main parrotfish feeding areas, which are predominantly shallow, low-sediment areas, especially on the reef crest (e.g., Bruggemann et al. 1996, Fox & Bellwood 2007, Bonaldo & Bellwood 2011b). However, the location of sediment release is the key to the transport. Although many parrotfish species defecate close to their feeding areas, others, like the surgeonfish *Ctenochaetus striatus*, which feeds in shallow areas and defecates in deeper zones of the reef (Goatley & Bellwood 2010), have specific defecation sites away from feeding sites (Bellwood 1985, 1996). In these cases, parrotfish promote a net transport off the reef as they a) remove sediment trapped within the epilithical algal matrix, b) actively transport sediment away from the reef, c) reduce particle size (increasing the chances of hydrodynamic transport off the reef), and d) change the algal configuration of the EAM (reducing further sediment retention; Bellwood 1996, Purcell & Bellwood 2001, Wilson et al. 2003). Although sediment reworking by parrotfish may be comparable to that of surgeonfish, evidence of a significant reduction in sediment size is largely restricted to parrotfish species (Bellwood 1996, Hoey & Bellwood 2008, Goatley & Bellwood 2010). Overall, of all fishes on tropical reefs, parrotfishes are by far the main producers and transporters of sediments.

Even though parrotfish produce and ingest large quantities of sediment, there is increasing evidence that they avoid ingesting sediment if given the opportunity. The suggestion of Purcell & Bellwood (2001) that herbivorous fish prefer high detrital quality and low-particulate (sediment) areas has been supported by recent field experiments (Bellwood & Fulton 2008, Goatley & Bellwood 2012), which showed increases in feeding by parrotfish and other reef herbivores on EAM with lowered sediment loads.

The high proportion of sediments in the faeces of parrotfishes, especially in comparison to other reef fish groups, is believed to decrease their quality as a food source for other reef organisms. Robertson (1982) found that among 88 species from 11 families in Palau, the parrotfishes were the only group whose faeces were not consumed by other reef fishes. There is only one study, to our knowledge, that reports consumption of parrotfish faeces, in an oceanic archipelago off northeastern Brazil. The species feeding on parrotfish faeces, the wrasse *Thalassoma noronhanum*, is an opportunistic feeder, exhibiting a wide variety of feeding habits. It frequently follows grazing parrotfish and feeds on invertebrates disturbed by them (Sazima et al. 2005). Given the limited quantity of carbonate sediments and diverse diets of parrotfishes in this area, the faeces might have been of slightly higher nutritional quality than in other areas.

Factors influencing parrotfish functional roles

Feeding mode

The presence of oral jaws and a second set of independent powerful jaws in the throat of parrotfishes (Gobalet 1989, Bellwood & Choat 1990) have been suggested as the main characteristics driving diversification within the group. The pharyngeal jaws have effectively freed the oral jaws from the constraints of food processing and allowed them to specialize in food procurement (i.e., 'functional decoupling'; Liem & Sanderson 1986). However, recent work has emphasized the importance of their unusual intermandibular articulation (between the articular and the dentary), rather than the pharyngeal jaws, in promoting diversification within the group (Price et al. 2010).

The division of parrotfishes into seagrass- and reef-associated clades corresponds closely with the functional classification of parrotfish species. The seagrass (*Sparisoma*) clade consists of five genera (*Calotomus*, *Cryptotomus*, *Leptoscarus*, *Nicholsina*, and *Sparisoma*), and with the exception of a few species of *Sparisoma*, all members of the clade are browsers (Figure 2). These taxa inhabit both reefal and non-reef habitats, such as seagrass beds and rocky coasts, and feed predominantly on seagrasses and macroalgae (Bellwood 1994, Streelman et al. 2002). The oral jaws typically possess pointed and protruding teeth, which are individually positioned or partly coalesced on the jaws, often with little or no external cementation (with the exception of some *Sparisoma* species; see the following discussion). The muscle that closes the lower jaw (i.e., adductor mandibulae) and the bones of the lower jaw (i.e., angular and dentary) exhibit typical labrid form (Bellwood 1994). When feeding, these browsing parrotfishes use their teeth to remove EAM, macroalgae, or seagrass from the substratum without scarring it (Bellwood 1994).

In contrast, all members of the reef clade are scrapers or excavators and usually inhabit reefal environments. In this group, the teeth are found in vertical alternating rows, and the outer surface of the jaws is usually covered with cement (Bellwood 1985, 1994). A similar pattern is seen in *Sparisoma aurofrenatum*, *S. viride*, and *S. amplum*, which show strong similarities to the Indo-Pacific genus *Cetoscarus*. Members of the reef clade also have highly modified adductor mandibulae and, with the exception of *Bolbometopon* and *Cetoscarus*, possess an unusual intermandibular articulation, between the dentary and angular bones in the lower jaw. Although there are common features of the morphology of feeding apparatus within the clade, there is also considerable variation, which highlights the diversity of feeding modes. Excavating taxa (*Bolbometopon*, *Cetoscarus*, *Chlorurus*, and *Sparisoma amplum* and *S. viride* of the seagrass clade) have robust jaws with simple and strong articulations among elements. This, coupled with a relatively large muscle mass and modified adductor mandibulae in which the A3 component of the adductor mandibulae inserts directly on the dentary in *Bolbometopon*, *Cetoscarus*, and *Chlorurus*, generates a forceful bite (Bellwood & Choat 1990, Bellwood 1994). In comparison, scraping species (*Hipposcarus* and *Scarus*) have a gracile feeding apparatus with relatively mobile articulations in the oral jaw and smaller musculature. In addition, the A1 and A2 components of the adductor mandibulae are fused in *Hipposcarus* and *Scarus* (Bellwood 1994). Interestingly, some members of the genus *Sparisoma* exhibit several characteristic features of both scraping (*S. aurofrenatum*) and excavating parrotfishes (*S. amplum* and *S. viride*) from the reef-associated lineage (Bruggemann et al. 1994b, Bernardi et al. 2000). These similarities appear to have arisen as a result of functional convergence (Streelman et al. 2002, Robertson et al. 2006). Overall, scraping taxa take small bites, have high feeding rates, and show no particular affinities for the angle or curvature of feeding substrata; excavating taxa take larger bites, have lower feeding rates, and have a preference for convex feeding substrata (Bellwood & Choat 1990, Bruggemann et al. 1994b).

Scraping and excavating parrotfishes usually leave distinctive bite marks or 'feeding scars' on the surfaces on which they feed. Differences in the jaw structure and morphology have a direct influence on size and depth of bite marks left on the substratum (Figure 5). Feeding scars of excavating species are approximately twice as long as they are wide, with four or more deep grooves running parallel to the major axis, whereas feeding scars of scraping species are characterized by two parallel feeding marks with the length approximately 10 times as long as they are wide (Bellwood & Choat 1990, Hoey & Bellwood 2008; Figure 5).

The different feeding modes of parrotfishes have direct implications for how each group impacts benthic communities. The deeper and relatively larger feeding scars of excavating parrotfishes remain open and free of algae and sediments for longer periods of time compared to the shallow bite

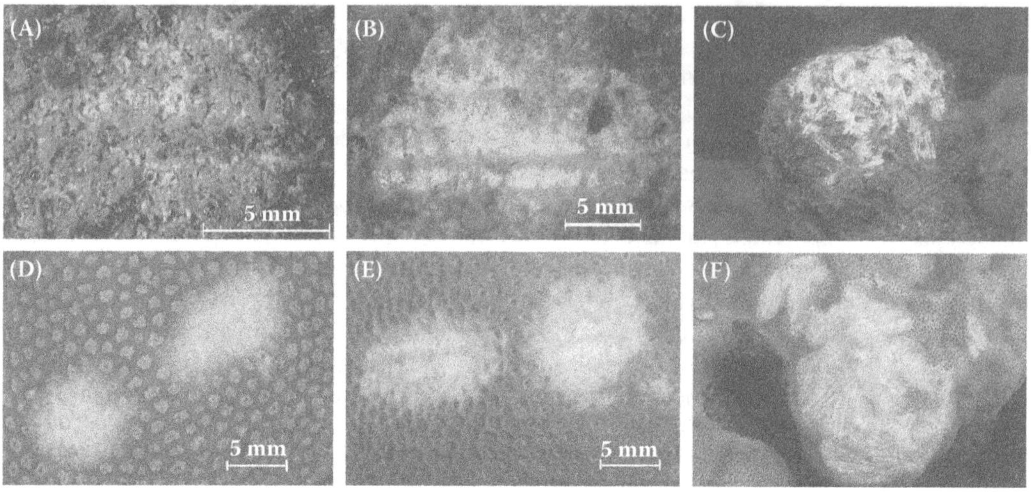

Figure 5 (See also colour figure in the insert) Bite marks of parrotfishes on dead coral surfaces covered by the EAM: individual marks of scraping and excavating parrotfishes (A and B, respectively) and multiple marks of scraping and excavating parrotfishes (C and D, respectively). Bite marks of parrotfishes on live coral colonies (massive *Porites* sp.): individual marks of scraping and excavating parrotfishes (E and F, respectively) and multiple marks of scraping and excavating parrotfishes (G and H, respectively). (Photos: João Paulo Krajewski.)

marks of scraping parrotfishes (Bonaldo & Bellwood 2009, Bonaldo et al. 2011). Thus, reef areas bitten by parrotfishes with different feeding modes usually have different recovery and recolonization rates. Scraping parrotfish bite marks on dead coral colonies covered by the EAM (Figure 5A, 5C), for instance, are rapidly reoccupied by reef algae, with only 10% of the feeding scar remaining after 4 days. Bite marks of excavating parrotfishes (Figure 5B, 5D) are recolonized at a much slower rate, with almost 40% of the scar area remaining after 4 days (Bonaldo & Bellwood 2009). As a consequence, the feeding scars of excavating parrotfishes probably represent better potential sites for the settlement of organisms with lower growth rates, such as corals. Live coral colonies bitten by scraping parrotfish species also have faster recovery, as polyps usually remain intact (Bonaldo et al. 2011; Figure 5E, 5G).

Although differences in jaw morphology and their implications for parrotfish feeding are reasonably well understood (Bellwood & Choat 1990, Bruggemann et al. 1994b, Bonaldo & Bellwood 2009, Bonaldo et al. 2011), little is known about differences in the intake of nutrients and energy. As excavating parrotfishes usually have lower feeding rates and are able to reach larger body sizes in comparison to most scraping species (Bellwood & Choat 1990, Bruggemann et al. 1994b, Alwany et al. 2009), it would be expected that they obtain more nutrients and energy per bite than scraping taxa. Indeed, studies of coral predation suggest that excavating parrotfish species are able to exploit different food items. In the Caribbean, for instance, the excavator *Sparisoma viride* preferentially targets *Montastrea annularis* colonies with a higher number of mature gonads (Rotjan & Lewis 2009). Similarly, on an inshore reef in the central GBR, bite marks of excavating parrotfishes on massive *Porites* colonies increased during the period that coincided with coral spawning, while the number of scraping parrotfish bites remained unchanged between seasons (Bonaldo et al. 2012). These studies suggest that excavating parrotfishes are able to reach coral eggs and gonads better than their scraping counterparts as a result of their deeper bites. Also, they appear able to differentiate the nutritional content of individual coral colonies or even regions within a single colony. The fact that, worldwide, almost all parrotfish species for which live corals are a major component of the diet are excavators (Bruggemann et al. 1994b,c, 1996, Bellwood et al. 2003, Francini-Filho et al. 2008, Lokrantz et al. 2008, Alwany et al. 2009) reinforces the present suggestions that this group is able to feed on this substratum more efficiently than scraping parrotfishes. Further studies are necessary, however, to verify if excavating parrotfish species have any nutritional and energetic advantages of biting deeper into the substratum in comparison to scraping species.

Body size and ontogeny

In addition to the importance of jaw morphology and feeding mode in influencing parrotfish functional roles, body size is an important factor in influencing the nature and intensity of feeding of these fish and their subsequent impacts on benthic assemblages, even within a species (Bruggemann et al. 1994a,b, 1996, Mumby et al. 2006, Bonaldo & Bellwood 2008, Lokrantz et al. 2008, Jaywardene et al. 2009, Ong & Holland 2010, Afeworki et al. 2011). Parrotfish undergo marked changes as they grow, which directly affect their impact on the substratum. Juvenile parrotfish (15 mm TL) feed predominately on crustaceans and foraminiferans, and once this carnivorous phase has passed (Bellwood 1988, Chen 2002), small individuals usually only crop the upper portions of algal material and do not have an impact on or scrape the underlying carbonate substratum (Bonaldo & Bellwood 2008). As parrotfish grow, the enlargement and development of the oral jaws and associated musculature allow them to bite deeper into the EAM, effectively scraping and then, in some species, excavating the substratum (Bellwood & Choat 1990, Bruggemann et al. 1994a,b, 1996, Francini-Filho et al. 2008, Lokrantz et al. 2008, Bonaldo & Bellwood 2009, Ong & Holland 2010). In Nyange, Zanzibar, for example, a non-linear relationship between scraping rate and body length of *Chlorurus sordidus* implied that a 50% reduction in the population mean length, through fishing, for example, could result in a 90% decrease in the amount of substratum cleared during feeding (Lokrantz et al. 2008). Therefore, larger parrotfish make a disproportionately high contribution to removal of algae and bioerosion in comparison to smaller individuals as, regardless of species, the rate of substratum removal per unit biomass tends to be higher for larger-bodied parrotfish (e.g., Bruggemann et al. 1994a,b, 1996, Bonaldo & Bellwood 2008, Lokrantz et al. 2008, Ong & Holland 2010).

Feeding changes associated with ontogeny and body size may also be related to the social aspects of some parrotfish species. Parrotfishes have complex social and sexual systems, with many species having dichromatic adult phases and sequential protogynous hermaphroditism (Warner 1984, van Rooij et al. 1996a,b). Several species are also territorial, with large males and even females defending feeding and reproductive territories (van Rooij et al. 1996a,b). Differences in feeding rates among juvenile and adult phases of parrotfishes may reflect this complex behaviour, with higher bite rates in juveniles (e.g., Lokrantz et al. 2008) and lower feeding rates in territorial terminal-phase males (e.g., Bruggemann et al. 1994c, Bonaldo et al. 2006). The social structures of some species have also been shown to vary among habitats, with males exhibiting different feeding rates and behaviour in different areas of the reef (e.g., van Rooij et al. 1996b, Girolamo et al. 1999). Terminal males and large females of some *Sparisoma* species usually have lower feeding rates and spend a greater amount of time patrolling and defending territories in comparison to smaller, non-territorial conspecifics (van Rooij et al. 1996b, Girolamo et al. 1999, Bonaldo et al. 2006). In contrast, other parrotfishes, such as *Scarus rivulatus*, seem to display low intensity or no territorial behaviour, and in these cases, feeding behaviour does not seem to be affected by social systems or ontogenetic phase (e.g., Bonaldo & Bellwood 2008). However, there have been few studies of parrotfish social systems, and to date they have focused on species in the Atlantic Ocean and Mediterranean Sea (e.g., van Rooij et al. 1996b, Girolamo et al. 1999). Clearly, further information is necessary to evaluate the effects of social systems on the feeding activity and associated ecosystem roles of different size parrotfish individuals on tropical reefs. Parrotfishes exhibit marked changes in their ecological roles during ontogeny, changing from carnivores, to scrapers or browsers, and then excavators in some of the larger species. In all cases, feeding behaviour appears to be labile, with non-linear relationships between body size and the rates of ecosystem functions.

Among-habitat distributions

Parrotfishes display marked variation in their distribution among reef habitats and, like many species of herbivorous fish, generally attain their highest abundance and biomass in shallow areas of exposed seaward reefs, decreasing both across the reef flat and down the reef slope (Bruggemann et al. 1996, Wilson et al. 2003, Fox & Bellwood 2007, Hoey & Bellwood 2008, Ong & Holland 2010, Afeworki et al. 2013). Although there is some variation in the habitat preferences/associations of individual species (e.g., Russ 1984, McAfee & Morgan 1996, Hoey & Bellwood 2008), the among-habitat distribution of parrotfish biomass is remarkably consistent across broad geographic scales, from the Red Sea to the GBR and the Caribbean (Figure 6). Such similarities suggest that among-habitat distributions are not influenced by the taxonomic or functional composition of the assemblage and point toward generalities in the mechanisms structuring these assemblages.

The distribution of parrotfishes among reef habitats may be influenced by a range of physical and biological processes, including wave energy, habitat structure, availability of preferred food sources, sedimentation, and fishing. Within coral-reef systems, hydrodynamic forces, principally caused by wave action, are typically highest in shallow exposed habitats and decrease substantially with water depth, degree of shelter from prevailing winds (i.e., leeward vs. windward), and distance from the crest (Bellwood & Wainwright 2001, Fulton & Bellwood 2005). Such variation in wave-induced water motion among habitats has important implications for fish swimming performance (Webb 2004). Fish inhabiting locations with high wave energy must be able to cope with high and turbulent water flow while still performing daily activities (e.g., foraging, predator avoidance).

Swimming performance has been shown to play an important role in shaping among-habitat distributions of fishes, including parrotfishes, on both coral and rocky reefs (Bellwood & Wainwright 2001, Bellwood et al. 2002, Fulton & Bellwood 2004). In general, fishes with stronger swimming abilities inhabit exposed high-wave-energy habitats, and weaker swimmers occupy more benign habitats with reduced wave energy and water flow (Fulton et al. 2001, 2005, Bellwood et al. 2002). Over half of coral-reef fishes, including parrotfishes, use labriform locomotion (i.e., primarily use their pectoral fins for thrust), with the shape of the pectoral fin closely linked with swimming performance (Wainwright et al. 2002, Fulton & Bellwood 2005). The majority of parrotfishes have high-aspect-ratio (i.e., tapered) pectoral fins, indicative of efficient swimming at high speeds (Bellwood & Wainwright 2001). The swimming performance of parrotfishes may be further enhanced by their large body size (Fulton & Bellwood 2004), especially in exposed habitats. While swimming performance may explain the absence of weaker/poorer swimming species from high-wave-energy environments, it does not explain the converse. The swimming performances of parrotfishes may allow them to occupy shallow exposed reef habitats, but enhanced swimming performance does not exclude them from more sheltered environments.

The distribution of fishes among habitats may also be influenced by the availability of their preferred foods. The cover and growth rates of scleractinian corals are often highest in shallow exposed habitats (Done 1982, Barnes & Chalker 1990, Wismer et al. 2009); however, corals are only consumed in significant quantities by two species, *Bolbometopon muricatum* in the Indo-Pacific (Bellwood et al. 2003, Hoey & Bellwood 2008) and *Sparisoma viride* in the Caribbean (Bruggemann et al. 1996, Bruckner & Bruckner 1998, Rotjan & Lewis 2005, 2008). Neither of these species displays particular affinities for live coral, largely consuming corals in proportion to their availability (Bellwood et al. 2003, Hoey & Bellwood 2008, Rotjan & Lewis 2006, 2008).

The availability and productivity of surfaces covered by EAM, the main feeding substratum for most parrotfishes (Table 2), may directly influence the distribution of parrotfishes. Relationships between parrotfish abundance or biomass and the availability of EAM are varied. Numerous studies have reported positive relationships between the biomass of parrotfish (or herbivorous fish collectively) and the cover or standing biomass of EAM (e.g., Williams & Polunin 2001, Russ 2003,

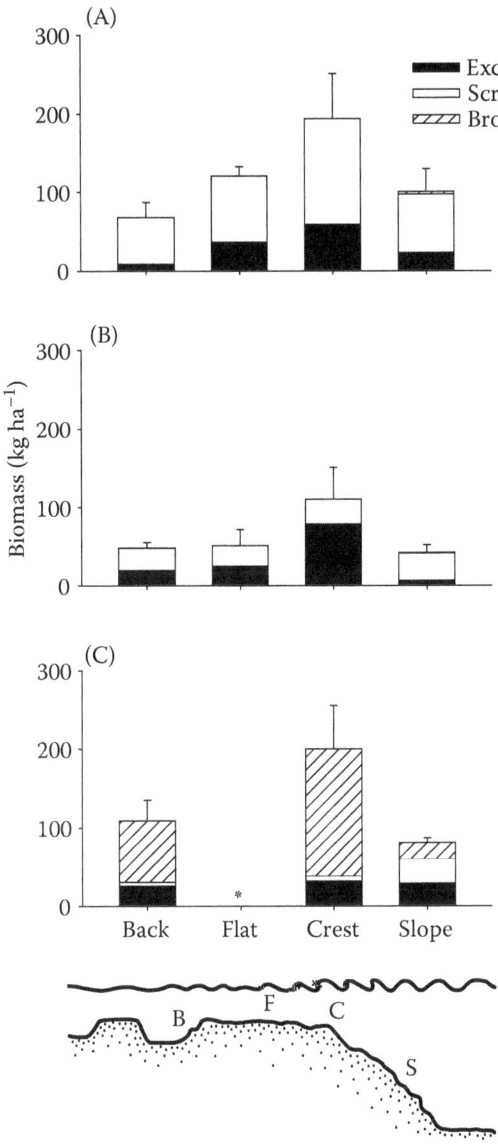

Figure 6 Among-habitat variation in the distribution of excavating, scraping, and browsing parrotfish biomass density (kg ha^{-1}, mean + SE) in (A) the Great Barrier Reef (GBR), Australia; (B) the Red Sea; and (C) the Caribbean. Mean values are based on four replicate 20-min timed surveys in each location in which all parrotfishes within a 5-m wide belt were recorded (GBR: Lizard Island, Hoey & Bellwood 2010b; Red Sea: Saudi Arabia, A.S. Hoey unpublished data 2012; Caribbean: Little Cayman Island, D.R. Bellwood unpublished data 2007). *No surveys were conducted in reef flat habitat in the Caribbean.

Hoey et al. 2011); others have reported increases in parrotfish biomass following extensive coral loss, presumably as a result of the increased availability of algal turfs that rapidly colonize dead coral skeletons (Pratchett et al. 2008, Adam et al. 2011, Gilmour et al. 2013). Yet other studies have failed to detect any relationship between parrotfish biomass and EAM availability (Howard et al. 2009, Wismer et al. 2009). Irrespective of the relationship, it is difficult to know whether parrotfish may be responding to or contributing to the cover of EAM. It appears most likely that parrotfish would be responding to the nutritional content or productivity, as opposed to standing biomass, of epilithic algal communities. In support of this, Russ (2003) showed that the biomass of herbivorous fish, including parrotfish, correlated more strongly with the productivity than the standing biomass of EAM among habitats on an offshore reef on the GBR.

The detrital and sediment loads of the EAM also differ among habitats and may influence the distribution of fishes that feed on the EAM. Detritus is an important source of nutrients for fishes that feed on the EAM (Choat et al. 2002, Wilson et al. 2003, Crossman et al. 2005), while sediments generally suppress feeding rates of fishes on EAM-covered substratum (Bellwood & Fulton 2008, Goatley & Bellwood 2012). The detrital and sediment loads within the EAM are directly related to wave energy, with the mass of both detritus and sediment decreasing with increasing wave exposure (Crossman et al. 2001, Purcell & Bellwood 2001, Wilson et al. 2003, Bonaldo & Bellwood 2011b). Habitats with higher parrotfish biomass, such as exposed reef crest, tend to have lower detrital and sediment mass; conversely, habitats with lower parrotfish biomass, such as sheltered back reef, have higher detrital and sediment mass. However, the key aspect appears to be the detritus-to-sediment ratio. Regardless of total availability, parrotfishes appear to prefer areas with higher ratios of detritus to sediment. Thus, the crest is preferred as, although there is little detritus or sediment there, the particulate material has a higher proportion of detritus (Purcell & Bellwood 2001).

The physical structure of a habitat has long been recognized as an important determinant of fish assemblages on coral reefs, with structurally complex habitats supporting a greater number of species and individuals than less-complex habitats (Luckhurst & Luckhurst 1978, Friedlander & Parrish 1998). The relationship between structural complexity and biomass and feeding rates of parrotfish assemblages, however, is complex. The structure provided by corals is generally positively related to the biomass or grazing intensity of parrotfish (Randall 1965, Fox & Bellwood 2007). In contrast, Graham et al. (2006) recorded no change in parrotfish biomass, despite dramatic declines in the biomass of other groups, following the structural collapse of reefs in the Seychelles. The structure provided by stands of tall, fleshy macroalgae has also been shown to suppress feeding by scraping and excavating parrotfishes (Hoey & Bellwood 2011), yet appears to be an important habitat for some browsing taxa, including the parrotfish *Leptoscarus vaigiensis* (Wilson et al. 2010, Hoey et al. 2013). Clearly, habitat structure may have an important influence on the structure of parrotfish assemblages. However, the complexities of the relationship (Nash et al. 2012, 2013, Brandl & Bellwood 2013), coupled with regional variation in disturbance regimes (e.g., coral bleaching, severe storms, outbreaks of crown-of-thorns starfish), preclude any generalizations regarding the role of structural complexity in influencing the distribution of parrotfishes among zones.

In addition to differences in parrotfish distribution among different reef zones, individual parrotfish tend to occupy a given area on the reef (Welsh & Bellwood 2012a, Howard et al. 2013). For many years, our knowledge of the spatial and temporal extent of site fidelity by parrotfish has been limited to direct visual observations on the home ranges of a few species (Bellwood 1985, van Rooij et al. 1996a,b, Mumby & Wabnitz 2002, Chateau & Wantiez 2009). Recently, new technologies for tracking animals in the field have allowed not only the spatial range of individual fishes to be quantified over longer timescales but also habitat utilization patterns within a fish's home range to be examined. Collectively, these studies have shown that parrotfish and other roving herbivorous fish have smaller home ranges than previously assumed. For instance, at Orpheus Island (GBR), *Chlorurus microrhinos*, one of the largest parrotfish species in the area, has a mean home range of about 8000 m^2, within which over half of the feeding activity is concentrated in a relatively small proportion (22%) of the total area (Welsh & Bellwood 2012a). Similarly, *Scarus rubroviolaceus* exhibited strong site fidelity on reefs in Hawaii, with individual movement patterns and home ranges (mean size 4144 m^2) relatively stable throughout a 2-month study period (Howard et al. 2013). Even for schooling parrotfishes, such as *S. rivulatus*, which are usually regarded as the most wide ranging of roving herbivores (Myers 1989), individuals only rove over a limited reef area, leaving the school once they reach the edge of their individual home range (Welsh & Bellwood 2012b). Schools may cover large areas, yet individuals do not. The home ranges of these schooling *S. rivulatus* (ca. 24,000 m^2) were only two to three times larger than those of similar-size solitary or haremic species. In contrast to these coral-reef species, *Sparisoma cretense*, a parrotfish of rocky reefs in the Mediterranean and northern Africa, seems to exhibit greater mobility, with estimates of home ranges in the waters around Lampedusa, Italy, varying from 70,387 to 256,398 m^2 (La Mesa et al. 2012). Overall, the limited spatial range of parrotfish individuals, especially in coral-reef systems, suggests that the impact of individual parrotfish is spatially restricted, and as a consequence, the ecosystem role of parrotfishes is dependent on the local density of parrotfishes on the reef (Welsh & Bellwood 2012b, Howard et al. 2013).

Superimposed on all of the aforementioned physical and biological factors is fishing pressure. Parrotfish are the targets of nocturnal spearfishers throughout much of the Indo-Pacific and Caribbean and are extremely susceptible to this form of fishing (Bellwood et al. 2003, 2012a, Aswani & Hamilton 2004, Bejarano et al. 2013). This fishing practice not only explains the rapid declines in parrotfish populations in some regions but also likely causes a fundamental change in their distribution among habitats (Hay 1984). Fishing effort, and hence yield, of spearfishers is initially concentrated in the shallow waters, and while they may gradually fish deeper areas of the reef, the greatest impact will always be in the shallow habitats. The removal of shallow-water populations is likely to be compounded by a tendency of surviving fish to move into deeper areas to avoid spearfishers (Feary et al. 2011), ultimately leading to a system in which the ecological impact of parrotfish is concentrated in deeper, as opposed to shallower, habitats.

Temporal variation in parrotfish feeding and distributions

Biomass and nutritional quality of reef algae vary considerably across a range of temporal scales. The primary productivity of reef algae may be determined by several factors, such as day length and light intensity, which directly affect rates of photosynthesis and the amount and quality of nutrients available for herbivores (Klumpp & McKinnon 1989, Afeworki et al. 2011, 2013). Indeed, a number of studies on feeding by parrotfishes and other nominally herbivorous fishes have found temporal variation in feeding rates (Bruggemann et al. 1994a,b, McAfee & Morgan 1996, Ferreira et al. 1998b), selection of feeding substrata (Afeworki et al. 2011), and distribution of fish species (Paddack et al. 2006) on reef ecosystems.

One of the most widespread temporal patterns in the feeding behaviour of herbivorous fishes is variation in feeding rates throughout the day. Nominally herbivorous fishes, including several parrotfishes, tend to exhibit higher feeding rates in the middle of the day or in the afternoon (Polunin & Klumpp 1989, Bellwood 1996, Fox & Bellwood 2007, Alwany et al. 2009). It has been hypothesized that this feeding pattern is related to diel patterns in the nutritional quality of the algae, with a noticeable increase in algal nutriment from the morning to a midday maximum and remaining relatively constant throughout the afternoon. The highest feeding rates coincide with periods of greatest nutritional value of the algae (Zemke-White et al. 2002).

Feeding by herbivorous fishes may also vary seasonally (e.g., Paddack et al. 2006, Lefèvre & Bellwood 2011, Afeworki et al. 2013). This variation may be reflected in differences in fish distribution among reefs or reef zones (Paddack et al. 2006), in feeding rates (Bellwood 1995a) or in selection of food and feeding substratum (Afeworki et al. 2011, Lefèvre & Bellwood 2011). Overall, seasonal differences in feeding patterns of parrotfish usually reflect variations in biomass, productivity, and availability of algal resources throughout the year (e.g., Bruggemann et al. 1995, Ateweberhan et al. 2006, Lefèvre & Bellwood 2011, Afeworki et al. 2013). In Bonaire, the higher primary production on shallow reefs during summer is associated with higher densities of parrotfish and surgeonfish in this zone than during winter, when primary production is lower (Bruggemann et al. 1994a,b, 1995). Similarly, on rocky reefs in south-eastern Brazil, *Sparisoma atomarium* exhibits higher feeding rates and has greater gut content volume during the summer than in winter. This higher feeding intensity in summer coincides with periods when there are more frequent blooms of algae that are more palatable and nutritious (Guimaraens & Coutinho 1996, Ferreira et al. 1998a,b). Overall, temporal differences in feeding patterns of parrotfish suggest that their effects on benthic communities are likely to vary temporally, although variation appears to be predominantly in the intensity of feeding rather than in their qualitative effects.

Regional variation in the distribution of parrotfishes

There is considerable regional variation in the taxonomic and functional composition of parrotfishes that has been shaped by a complex biogeographic history (Bellwood 1994, Bernardi et al. 2000, Streelman et al. 2002, Choat et al. 2012) and, more recently, by human exploitation of these fish and their habitats (Bellwood et al. 2003, 2012a, Hawkins & Roberts 2003, Mumby 2006). The most pronounced division in the composition of parrotfish assemblages occurs between the Indo-Pacific and Atlantic Oceans, with no overlap in species and limited overlap in genera between these two regions (Parenti & Randall 2000, 2011, Streelman et al. 2002, Bellwood et al. 2004). Of the 100 species of parrotfish that are currently recognized, the majority (73 species) are confined to the Indo-Pacific (Parenti & Randall 2011). Furthermore, of the ten parrotfish genera, six are restricted to the Indo-Pacific (*Bolbometopon*, *Calotomus*, *Cetoscarus*, *Chlorurus*, *Hipposcarus*, and *Leptoscarus*); two to the Atlantic (*Cryptotomus* and *Sparisoma*); and one genus to the Atlantic and East Pacific (*Nicholsina*; Figure 2B). The remaining genus, *Scarus*, is the most speciose (52 species) and has a pantropical distribution, with 43 species in the Indo-Pacific and nine in the Atlantic (Figure 2B). These marked differences point toward largely independent evolutionary histories of parrotfishes within these two regions following the closure of the Tethyan connection between the Atlantic and Indo-Pacific basins, at approximately 14 Ma (Choat et al. 2012, Cowman & Bellwood 2013a,b).

Phylogenetic reconstructions based on morphological and molecular data have provided considerable insights into the evolutionary relationships within the parrotfishes and how these have influenced the present distribution of the group. Bellwood (1994), using a phylogeny of genera based on morphology (Figure 2A), suggested that parrotfish were likely to have arisen in the tropical Tethys Sea and differentiated into eastern (i.e., proto-Indo-Pacific) and western (i.e., proto-Atlantic) populations, prior to and following the closure of the Tethys. Differentiation prior to the closure of the Tethys is supported by a lack of pantropical early diverging taxa, a fossil parrotfish (*Calotomus preisli*) that confirms the presence of the parrotfishes in the mid-Miocene (~14 Ma) in the key central region of the Tethys (Bellwood & Schultz 1991), and the estimated age of the genus *Sparisoma* (14–35 million years) from molecular data (Bernardi et al. 2002). Molecular phylogenies of parrotfish genera offer further support for this early differentiation (Figure 2B; Streelman et al. 2002, Cowman et al. 2009, Wainwright et al. 2012), identifying a clear break between typically browsing, seagrass-associated genera (*Cryptotomus*, *Nicholsina*, *Sparisoma*, *Calotomus*, and *Leptoscarus*) and scraping and excavating reef-associated genera (*Bolbometopon*, *Cetoscarus*, *Chlorurus*, *Hipposcarus*, and *Scarus*). The presence of early diverging Indo-Pacific genera in each group and the estimated date of divergence (~30 Ma) suggest that the separation of reef and seagrass clades occurred before the closure of the Tethys (Streelman et al. 2002). Following the closure of the Tethys, the eastern and western lineages were largely isolated. While both were influenced by sea-level changes, there was considerable variation in the availability of coral-reef habitat, with continuity of coral-reef development and habitat in the Indo-Pacific, and extensive loss of coral-reef habitats in the Atlantic (Bellwood 1994, Bellwood & Wainwright 2002). Subsequently, present-day parrotfish assemblages in the Atlantic are dominated by seagrass-associated taxa and in the Indo-Pacific by reef-associated taxa.

The pantropical distribution of *Scarus* is in marked contrast to all other parrotfish genera. Given the age of the genus (4.4–7.8 million years), it is unlikely that *Scarus* was present in the Atlantic prior to the closure of the Tethys (Bellwood 1994, Choat et al. 2012). There have been some suggestions that *Scarus* arose in the Indo-Pacific and crossed the East Pacific Barrier before the closure of the Isthmus of Panama (IOP), implying colonization of the Atlantic from the west (Bellwood 1994, Smith et al. 2008). A recent comprehensive molecular phylogeny of *Scarus* and *Chlorurus* (Choat et al. 2012) provides some support for this hypothesis, with *Scarus compressus* separating from *S. ghobban* in the East Pacific after the closure of the IOP (presumably as a result of a founder effect from an early colonization of the East Pacific by a '*ghobban*' lineage). The age of the other East Pacific species (*S. perrico*) and its Atlantic sister species (*S. hoefleri*) at about 5 million years is equivocal as these species are sisters to a broad Atlantic clade, which is in turn sister to a large Indo-Pacific clade. As noted by Choat et al. (2012), it is also possible (although perhaps less parsimonious) that ancestral *Scarus* entered the Atlantic from the Indian Ocean (via southern Africa) and moved westward to colonize the Caribbean and then eastern Pacific prior to the closure of the IOP (Figure 7).

The evolutionary history of the parrotfishes has led not only to marked regional variation in the taxonomic composition but also to the functional composition of parrotfish assemblages (Figures 2, 3, 4, 6, 8A). The relative contribution of the seagrass clade versus reef clade, in particular, appears to have shaped the functional composition of Atlantic and Indo-Pacific parrotfish faunas. Given the dominance of the seagrass clade in the Atlantic and the reef clade in the Indo-Pacific, it is hardly surprising that scraping and excavating taxa are dominant in the Indo-Pacific and browsing taxa are dominant in the Atlantic (Figure 4). Excavating parrotfishes, in particular, are largely restricted to the Indo-Pacific, with 95% of excavating species and all reef-based excavating genera (*Bolbometopon*, *Cetoscarus*, and *Chlorurus*) exclusive to this region; only two excavating species, *Sparisoma viride* and its sister taxon *S. amplum*, are present in the Atlantic (Bruggemann et al. 1994a,b). Despite having arisen from the seagrass lineage, *S. viride* shares a suite of features characteristic of the reef lineage (robust jaws with simple and strong articulations among elements and well-developed musculature to generate a forceful bite; Bellwood 1994) and appears to be an example of functional convergence (Streelman et al. 2002, Robertson et al. 2006). In marked contrast to the distribution of excavating and scraping taxa, browsing taxa account for over half of all Atlantic parrotfishes (16 of 26 species) and less than 10% of Indo-Pacific parrotfish species (seven of 73 species; Figures 2B, 4, 7, 8A).

Within the Indo-Pacific and Atlantic Oceans, there are marked gradients in species richness of parrotfishes, with each of the three functional groups (scrapers, excavators, and browsers) displaying similar patterns (Figures 4, 8A). The Indo-Australian Archipelago and Caribbean Sea support the highest number of parrotfish species in the Indo-Pacific and Atlantic, respectively, with species richness declining with distance from these areas. Such gradients in diversity are widespread among marine taxa (e.g., Hughes et al. 2002, 2003, Connolly et al. 2003, reviewed in Bellwood et al. 2012b). The roles of parrotfishes in reef processes are, however, not related to the presence (or absence) of component species but rather to the abundance or biomass of taxa within each functional group (e.g., Hoey & Bellwood 2008, Alwany et al. 2009, Bonaldo & Bellwood 2009). Overall, the distribution of biomass among groups broadly reflects the species composition of each biogeographic region (i.e., Atlantic reefs characterized by browsers, Indo-Pacific reefs by scrapers and excavators, and East Pacific reefs by scrapers; Figures 4, 8A); however, the consistency of function across large gradients in species richness within each region is striking. On low-latitude Indo-Pacific reefs with relatively intact herbivorous fish assemblages, the biomass of scraping and excavating parrotfishes shows little variation from the Red Sea (17 species) to the GBR (30 species; Figures 6, 8A). Browsing taxa, while widespread across the Indo-Pacific, contribute little to the overall biomass of parrotfish assemblages (Hoey & Bellwood 2008, 2010b; Figures 6, 8A). This consistency suggests that the processes performed by parrotfishes may be maintained irrespective of local species richness. The only exception to this appears to be high-latitude, or subtropical, reefs where the overall parrotfish biomass, and that of excavators in particular, is greatly reduced (Feary et al. 2010, Hoey et al. 2011).

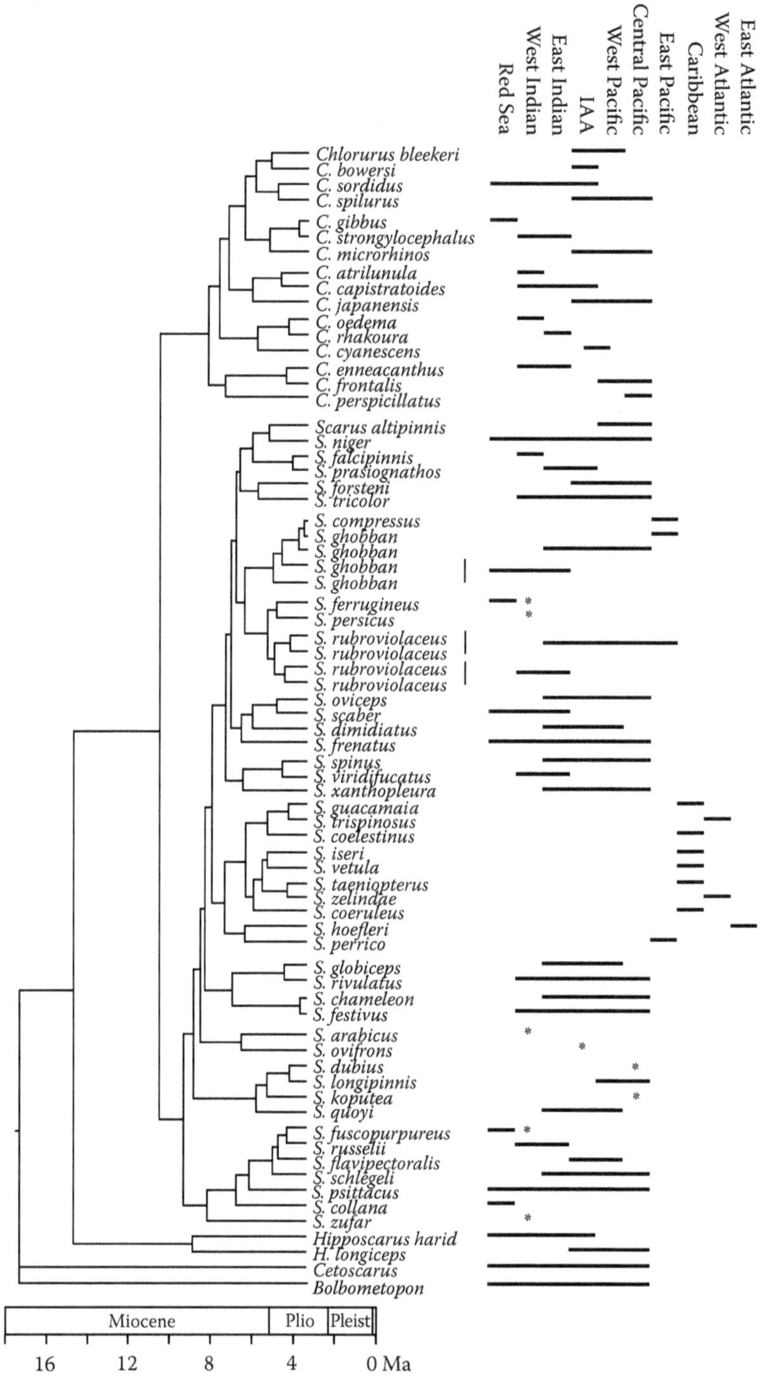

Figure 7 Chronogram derived from molecular data showing the ages of origin and diversification of 65 parrotfish species from the reef or *Scarus* clade. (Adapted from J.H. Choat, O.S. Klanten, L. van Herwerden, D.R. Robertson, & K.D. Clements, *Biological Journal of the Linnean Society* **107**, 529–557, 2012.) Phylogenetic analysis was based on two mitochondrial and one nuclear gene fragments for each species. The geographic distribution of each species across 10 biogeographic regions is illustrated. IAA, Indo-Australian Archipelago. *Limited distributions within the region indicated. Plio, Pliocene; Pleist, Pleistocene.

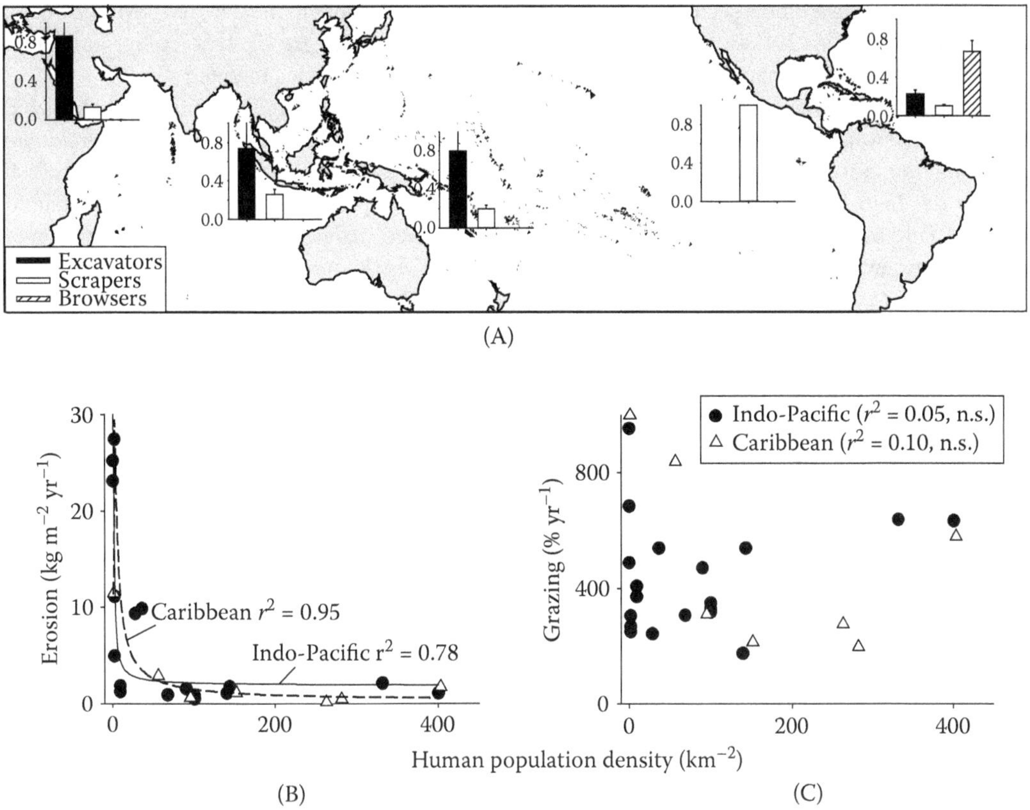

Figure 8 Influence of biogeographic setting and human population density on the structure and function of parrotfish assemblages. (A) Biogeographic variation in the functional composition of parrotfish assemblages. Values are the proportion of total parrotfish biomass within each of the three functional groups and are based on four replicate surveys (ca. 1200 m^2) in each of four habitats (reef slope, crest, flat, and back reef) at each location. Locations are (from left to right) Red Sea (A.S. Hoey unpublished 2012); Cocos (Keeling) Island; East Indian Ocean (Bellwood et al. 2003); northern GBR (Bellwood et al. 2003); Panama, East Pacific (D.R. Bellwood unpublished 2013); and Little Cayman Island, Caribbean (D.R. Bellwood unpublished 2007). Effect of human population density on rates of (B) bioerosion and (C) grazing by parrotfish on Indo-Pacific and Caribbean reefs. Data for Indo-Pacific reefs from Bellwood et al. (2012a) and Caribbean reefs from Bruggemann (1995) and Hawkins & Roberts (2003). The fitted trend lines are inverse polynomials (Indo-Pacific: $r^2 = 0.78$, $p < 0.0001$; Caribbean $r^2 = 0.95$, $p = 0.0001$).

Among the anthropogenic effects on coral reefs, fishing is arguably one of the oldest and most pressing, with a long history of exploitation by coastal human populations (Jackson et al. 2001, Pandolfi et al. 2003, O'Connor et al. 2011). Fishing is typically selective, removing some species and individuals faster than others, and may lead to distorted food webs, altered ecosystem function, and increased vulnerability to other disturbances (Jennings & Polunin 1996, Pandolfi et al. 2003, Mumby et al. 2006).

Commercial, artisanal, and subsistence fishers target parrotfish using a diversity of gears throughout much of their range, and fishing has led to marked declines in the biomass of parrotfish in many regions (e.g., McClanahan 1994, Aswani & Hamilton 2004, Floeter et al. 2006, Tuya et al. 2006, Rhodes et al. 2008, Campbell et al. 2012). On reefs with low local human populations and high levels of effective protection from fishing, the biomass of parrotfish can exceed 500 kg ha^{-1} (averaged across shallow and deep habitats, Caribbean: Bruggemann et al. 1995; Indo-Pacific: Bellwood et al. 2012a). In contrast, on reefs with high local human population densities and limited protection, parrotfish biomass can be rapidly reduced and is often less than 50 kg ha^{-1} (Mumby et al. 2006, Bellwood et al. 2012a). Such reductions in parrotfish biomass are likely to compromise reef processes, but the exact nature of the changes will depend on the species targeted.

Not surprisingly it is the larger species and individuals of parrotfish that are typically targeted by fishers (Aswani & Hamilton 2004, Bellwood et al. 2012a, Bejarano et al. 2013). The removal of these large individuals has had a dramatic effect on the magnitude of reef erosion by parrotfish in both the Indo-Pacific and Caribbean, with both systems displaying congruent and exponential declines (Figure 8B). Given the disproportionate importance of large parrotfish in bioerosion—both large excavating species (Indo-Pacific: *Bolbometopon muricatum*, *Cetoscarus* spp., *Chlorurus microrhinos*, *C. gibbus*, *C. strongylocephalus*; Caribbean: *Sparisoma viride*) and large individuals of scraping species (Indo-Pacific: *Scarus rubroviolaceus*, *S. ghobban*, *S. altipinnis*; Caribbean: *S. vetula*, *S. guacamaia*)—the fact that their removal leads to reduced erosion is not surprising. The nature of the relationship, however, highlights the vulnerability to fishing of excavating fishes and the role they perform. In both the Indo-Pacific and the Caribbean, erosion by parrotfish effectively disappears when human population densities reach around 20 km^{-2} (Figure 8B; Bellwood et al. 2012a). Although targeted by a range of gears, nocturnal quiescence and a tendency of some species to aggregate while sleeping makes these large fishes extremely susceptible to nocturnal spearfishing, facilitates profitable catches, and likely explains the rapid declines in their populations (Johannes 1981, Aswani & Hamilton 2004, Bellwood et al. 2012a).

In contrast to erosion, grazing by parrotfish appears to be less sensitive to fishing and shows no clear response to variation in human population densities in either the Indo-Pacific or Caribbean (Figure 8C). This lack of response appears to be related to the resilience of smaller-bodied, primarily scraping species to fishing pressure. Comparing 18 reefs from across the Indo-Pacific, Bellwood et al. (2012a) reported that the population density of scraping taxa (i.e., *Scarus* and *Hipposcarus*) showed no relationship to human population density, while the density of small excavating species (i.e., *Chlorurus bleekeri*, *C. japanensis*, *C. sordidus*) was actually positively correlated. Furthermore, heavily fished reefs had almost double the number of small parrotfishes (<25 cm in length) than lightly fished reefs. Remarkably similar responses of small-bodied parrotfishes to fishing have been reported in the Caribbean, with small-bodied scraping species (*Scarus iseri* and *S. taeniopterus*) showing no clear response to fishing and the abundance of smaller size classes increasing with increasing fishing pressure (Hawkins & Roberts 2004). This apparent resilience of parrotfish grazing to fishing may, however, conceal dangerous processes and predispose reefs to future ecological surprises. Indeed, it was the overfishing of parrotfish that probably caused subsequent phase shifts to macroalgal dominance on Caribbean reefs (Hughes 1994). While few phase shifts have been documented in the Indo-Pacific, the increasing prevalence of small-bodied, and presumably faster-growing, species may lead to compromised ecosystem processes and increased vulnerability to future disturbances (Bellwood et al. 2012a).

Functional redundancy among the parrotfishes

The literature on the ecology of parrotfishes on tropical reefs worldwide indicates that while some species play an almost exclusive role (e.g., Bellwood et al. 2003, Francini-Filho et al. 2008, Burkepile & Hay 2011), others seem to be part of larger groups of functionally equivalent species and may be considered to be 'functionally redundant' (Johansson et al. 2013), that is, the loss of any one species can potentially be compensated for by the actions of another (Bellwood et al. 2004). The feeding modes of parrotfishes are so distinctive that recent studies have concluded that they perform different ecological roles on coral reefs than other nominally herbivorous groups (e.g., Bellwood et al. 2004, Hoey & Bellwood 2011, Johansson et al. 2013). For instance, the largest parrotfish, the excavating *Bolbometopon muricatum*, plays such a unique role on coral reefs (Mouillot et al. 2013, in press), as a result of its larger and stronger jaws and higher feeding preference for live coral colonies in comparison to other parrotfishes, that its feeding impact cannot be replaced by any other species within the clade (Bellwood & Choat 1990, Bellwood et al. 2003, Hoey & Bellwood 2008). As a consequence, the functioning of Indo-Pacific reef systems is extremely vulnerable to the loss of this single species. Areas in which populations of these species are reduced or absent, as a result of intensive overfishing (Aswani & Hamilton 2004), have experienced marked declines in rates of external bioerosion, sediment production, and coral predation (Bellwood et al. 2003, 2012a, Hoey & Bellwood 2008). Similarly, a single species is largely responsible for external bioerosion, sediment production, and coral predation by parrotfish across most of the tropical Atlantic. *Sparisoma viride* in the Caribbean and its sister taxon *S. amplum* on Brazilian reefs differ from all other parrotfish species in these regions because they are the only excavating species and the only parrotfish species that feed frequently on live scleractinian corals (Bruggemann et al. 1994b, Bruckner & Bruckner 1998, Ferreira & Gonçalves 2006, Francini-Filho et al. 2008, 2010, Rotjan & Lewis 2009). Given this limited functional redundancy, there is a pressing need to look beyond the preservation of species diversity and focus on the protection of key species in critical functional groups if we are to protect and manage coral reefs into the future (Bellwood et al. 2004).

Differences in the impact of scraping and excavating species are also visible among parrotfishes in the Caribbean (Bruggemann et al. 1994b, 1996, Burkepile & Hay 2008). An experimental study in the Florida Keys, for instance, used inclusion cages to manipulate the composition of herbivorous fishes in small areas (4 m^2) of reef to determine the effects of these fishes on the benthic communities (Burkepile & Hay 2008). The results showed that reef surfaces exposed to feeding by the browsing *Sparisoma aurofrenatum* only, the scraping *Scarus taeniopterus* only, or the two species collectively showed clear differences in the algal communities that established over a 7-month period (Burkepile & Hay 2008). In particular, the absence of *Sparisoma aurofrenatum* (i.e., the '*Scarus taeniopterus*–only' treatment) led to a marked increase in the density of erect macroalgae, *Lobophora*, *Sargassum*, and *Codium*. Conversely, the cover of turf algae was greatest in the '*Sparisoma aurofrenatum*–only' treatment. These results clearly demonstrate the different impacts of these two parrotfish species on benthic communities. Similar results may be expected for other species across functional groups; however, the potential differences between species within a single functional group may be limited.

For some functional groups of parrotfishes, there appears to be a relatively high level of redundancy, with member species appearing to perform similar roles in coral-reef ecosystems. This seems to be the case for scraping parrotfishes on high-diversity reefs in the Indo-Pacific, in which most species possess broadly similar body sizes, feeding rates, and selection of feeding substratum (high preference for dead coral colonies covered by the EAM) (Tables 1, 2; Bellwood & Choat 1990, Johansson et al. 2013). Such similarities are indicative of a level of functional redundancy within this group, suggesting that the loss of one species may be compensated by the remaining species to maintain the function. In summary, there appears to be limited redundancy between functional groups of parrotfishes or within the relatively species-poor excavating group. However, there appears to be considerable scope for functional redundancy within the species-rich scraping parrotfishes in the Indo-Pacific and the browsing parrotfishes in the Caribbean.

Practical implications

The identification of the critical functional roles played by reef species is essential if we wish to obtain a better understanding of the resilience of natural systems (Tilman et al. 1997, Loreau et al. 2001, Bellwood et al. 2004, Loreau 2004). The body of literature on the feeding ecology of parrotfishes shows consistent patterns in the impacts of parrotfish grazing on benthic communities, with parrotfishes playing distinct roles in the consumption of algae (Paddack et al. 2006, Fox & Bellwood 2007, Mumby 2009b); predation of live coral colonies (Rotjan & Lewis 2005, 2006, 2008, Bonaldo & Bellwood 2011a); bioerosion (Bellwood 1995a, Bellwood et al. 2003, Alwany et al. 2009); and sediment reworking and production (Bellwood 1996, Bruggemann et al. 1996) on tropical reefs around the world. This increasing recognition of parrotfish as key drivers of the benthic composition and functioning of coral-reef ecosystems emphasizes the critical need to conserve not only parrotfish populations but also key species and larger individuals that contribute disproportionately to ecosystem processes.

The overall grazing impact of parrotfish on algal turfs and live coral colonies presents a conundrum. Parrotfishes, especially excavating species, seem to play conflicting roles for the perceived 'health' of a reef system that is manifest through their impacts on the benthic communities. When their feeding impact is delivered on surfaces covered by algal material, they clear space and may increase the potential settlement of corals on the substratum. However, when their feeding activity is focused on coral colonies, they are likely to cause loss of coral tissue in the reef habitat (Rotjan & Lewis 2008, Mumby 2009a). Moreover, incidental grazing by parrotfish appears to be a potential source of early post-settlement mortality in corals (Trapon et al. 2013b,c). Similar, potentially conflicting functions have also been observed for some sea urchins, such as *Diadema antillarum*, which appears to function as both a herbivore and a corallivore (Rotjan & Lewis 2008). However, a large body of literature has shown clear evidence that the predominant role of parrotfish and sea urchins in a number of reef systems around the world is the removal of algal material (e.g., Carpenter 1993, Bellwood et al. 2004, Mumby 2006, 2009a,b, Paddack et al. 2006, Bonaldo & Bellwood 2009). These findings demonstrate that parrotfish have a demonstrable contribution to the control of algal growth and, as a consequence, may be considered as key agents in the maintenance of coral cover on tropical reefs (Choat 1991, Paddack et al. 2006, Fox & Bellwood 2007, Bonaldo & Bellwood 2008, Mumby 2009b).

That is not to say that the importance of corallivory by some parrotfishes should be discounted. It is clear that corallivory must be balanced by coral growth if coral populations are to be maintained (Cole et al. 2008). If this balance is altered by environmental degradation and rates of coral predation exceed rates of coral growth, it is possible that parrotfish could negatively impact the remaining corals and threaten the persistence of local coral assemblages (Rotjan et al. 2006, Rotjan & Lewis 2008). Overall, this scenario highlights the vulnerability of coral reefs to unnatural disturbances, as even one of the most important sources of resilience on coral reefs, the grazing by parrotfish, could become deleterious under abnormal conditions.

Parrotfishes remain one of the few groups able to feed on and alter the structure of almost all coral-reef benthic substratum types. In a changing world, where tropical reefs are experiencing a significant depletion of herbivorous fishes and reduction in coral cover (e.g., Bellwood et al. 2003, 2004, Gardner et al. 2003, Hughes et al. 2003, 2007, Graham et al. 2006, Paddack et al. 2009), it is likely that parrotfish will become even more important for maintaining coral-reef processes. In particular, two parrotfishes, regarded as critically important to their host ecosystems when they are present (*Bolbometopon muricatum* in the Indo-Pacfic and *Scarus trispinosus* in Brazilian waters), are qualified for listing as 'threatened' under current IUCN (International Union for Conservation of Nature) categorization (Comeros-Raynal et al. 2012). Also, in some regions, the overall reduction in size and biomass of parrotfish individuals imposes serious constraints on the functioning of these fishes on coral reefs; in other areas, overfishing and habitat degradation have caused local extinction of some species. The latter is the case for several areas in the Coral Triangle, where the large human populations living adjacent to the coast, intense fishing, and high dependencies on marine resources have caused marked depletions in a number of reef fish species and the local extinction of some species (Comeros-Raynal et al. 2012).

In spite of the visible reduction of parrotfish populations worldwide, the relatively low proportion of species globally listed in threatened categories is either encouraging or concerning (Comeros-Raynal et al. 2012): encouraging in that some species may be more resilient than previously thought (Bellwood et al. 2012a); concerning in that serious declines, even when documented, do not trigger a management/conservation response. Most protection is aimed at preventing the loss of single species, not the loss of a function. For example, although *Bolbometopon muricatum* is present across most of the Indo-Pacific, it is in near ecologically irrelevant densities over most of its geographic range. Conservation actions, such as the establishment and enforcement of marine reserves and the protection of herbivorous reef fish species, including parrotfishes (Bellwood et al. 2003, Mumby 2006, Mumby et al. 2006, Comeros-Raynal et al. 2012), have been successful in some regions. The implementation of marine protected areas (MPAs) with high levels of compliance seems to be a valid strategy in protecting some parrotfish and surgeonfish species, as visible recoveries in the biomass of these species have been observed in a number of systems (e.g., McClanahan & Muthiga 1988, Bellwood et al. 2003, Mumby et al. 2006). Although the adequate protection of parrotfishes and their habitats is still a long way off (less than 5% of the geographical range of parrotfishes is currently encompassed by reserves; Comeros-Raynal et al. 2012), the impacts of protection in marine reserves are promising and provide an important first step in establishing management strategies for these important species in tropical reefs around the world. However, the prospects for successful conservation of parrotfishes and their ecological functions will depend to a greater extent on environmental and fisheries management outside MPAs.

Future directions in research

Despite a large body of literature on the biology and ecology of parrotfishes on tropical reefs, many important questions remain. For instance, it is possible that parrotfish may limit the distribution of reef components other than benthic algae (Hixon & Brostoff 1996, Burkepile & Hay 2008, Bonaldo & Bellwood 2009, Mumby 2009b) and coral colonies (Bruckner & Bruckner 1998, Miller & Hay 1998, Bonaldo & Bellwood 2011a). Studies in Panama and Florida, for instance, reported that several species of *Sparisoma* were the main consumers of sponges in bioassay experiments (Wulff 1997, Dunlap & Pawlik 1998). Furthermore, it has been estimated that two parrotfishes (*Chlorurus microrhinos* and *Scarus rivulatus*) consume 5.28% of the available harpacticoid copepods each day on an inshore reef in the central GBR, which corresponds to 27 times more than the consumption of these organisms by all of the benthic carnivores observed at the site combined (Kramer et al. 2013). As a consequence, in addition to the well-recognized consumption of algal material and live coral colonies, parrotfish probably influence the distribution and abundance of other benthic components, such as sponges, organisms living in association with the EAM, and other invertebrates.

As the majority of studies of the functional roles of parrotfishes on coral reefs have been carried out on the GBR (e.g., Hoey & Bellwood 2008, Cheal et al. 2010) and in the Caribbean (e.g., Bruggemann et al. 1994a,b,c, 1996, Burkepile & Hay 2008, 2011), relatively little is known about the functional roles of parrotfish feeding in other coral-reef regions (Berumen et al. 2013). There are also relatively few studies of parrotfish on tropical rocky reefs. On coral reefs, the balance between live coral colonies and benthic algae is fundamentally important, as corals provide the basis of the structure and dynamics of these systems. For this reason, parrotfishes, along with other reef herbivores, are considered key species in mediating coral-algae interactions, as they directly influence the structure of benthic communities in these ecosystems (Miller & Hay 1998, Fox & Bellwood 2007, Burkepile & Hay 2008). However, on reefs where coral colonies are not the primary basis of the ecosystem, and where the benthic community is naturally characterized by a low abundance of live coral colonies and a high abundance of other organisms, such as algae (e.g., Bonaldo et al. 2006, Ferreira et al. 2005, Francini-Filho et al. 2008), the importance of parrotfish in shaping the structure and dynamics of the reef benthos remains largely unknown.

Given the differences in the feeding ecology of parrotfishes and the geographical distribution of species in the group, caution should be exercised in extrapolating patterns and processes derived from studies of coral reefs in the Caribbean or Indo-Pacific to other areas. For example, although parrotfish are usually a dominant component in the overall biomass and abundance of herbivorous fish assemblages on tropical reefs, this does not seem to be the case for most tropical rocky reefs in the

TEP (Rosenblatt & Hobson 1969, Dominici-Arosemena & Wolff 2006). The low coral abundance on reefs in this region, however, cannot be considered the only explanation for the low proportion of parrotfish in herbivorous fish assemblages in the TEP, as other regions with few species of parrotfish and low coral cover, such as some reefs in the Caribbean and along the Brazilian coast, have higher proportions of parrotfish in the biomass and abundance of reef fish communities (e.g., Francini-Filho et al. 2010, Krajewski & Floeter 2011). One possible explanation may be that, while reefs in the West Atlantic are dominated by parrotfishes in the seagrass clade (especially *Sparisoma*), the TEP has only one species in this clade (*Nicholsina denticulata*; Rosenblatt & Hobson 1969). The area is dominated by the five reef-clade parrotfishes that are usually regarded as having a closer, perhaps obligate relationship with coral-reef habitats (Streelman et al. 2002, Comeros-Raynal et al. 2012). Indeed, *N. denticulata* has the widest latitudinal and habitat distribution range within the TEP, while the five other species are abundant only in a few locations within the entire region. These marked differences in the structure (identity, abundance) of parrotfish assemblages between coral and rocky reefs, and even among reefs with similar formation, probably affect reef functioning in each region, with direct implications for the dynamics and energy flow in the respective communities. Further research on parrotfish feeding in these less-studied systems is important for a better understanding of the overall impact of parrotfish feeding on reefs with different structural basis and for a general view of the importance of parrotfishes globally.

Despite the body of work on the biology and ecology of parrotfishes on tropical reefs, there is still a paucity of information on the nutritional ecology of the parrotfishes. Parrotfishes are a clade within the Labridae, in which the majority of species are represented by wrasses, known for feeding predominantly on invertebrates (Westneat et al. 2005, Bellwood et al. 2006b). Parrotfishes have, for a long time, been classified as reef herbivores, but there is increasing evidence that these fishes do not gain their nutrition primarily from the algae (Choat et al. 2002, 2004, Crossman et al. 2005, Ferreira & Gonçalves 2006, Rotjan & Lewis 2008, Bonaldo et al. 2012, Kramer et al. 2013). There is still a great deal to be learned about the specific sources of nutrients and energy required by parrotfishes, including possible variations among species and functional groups. *Cetoscarus* spp., for example, take less than a third of the number of bites of similar-size *Chlorurus microrhinos* (Bellwood & Choat 1990). Clearly, different nutritional strategies are in operation.

These aspects are just part of a long and exciting list of questions about parrotfish that have yet to be fully explored. There is, for instance, little known about how their larval biology compares with that of other reef species (Leis & McCormick 2006). The role of colour patterns during ontogeny and at maturity is still poorly understood (cf. Crook 1997, Streelman et al. 2002, Choat et al. 2012), few parrotfish predators have been identified, and regional variations in demographic patterns remain to be explained (Choat 1991, Bellwood & Choat 2011). Recently, more-detailed studies have begun on the factors determining foraging ranges (Nash et al. 2012, 2013), the use of shelter (Kerry & Bellwood 2012), and the response of parrotfish to sediments (Goatley & Bellwood 2012), but these issues are still far from well understood. Likewise, the biomechanics of the intermandibular articulation and of pharyngeal jaw motion offer a promising avenue for future research (cf. Wainwright et al. 2012), while the function of the unusual sacculated intestine has only been preliminarily examined (Bellwood 1994). Understanding the role of the mucous cocoon secreted by parrotfish remains an exciting challenge (cf. Grutter et al. 2010), while the role of thermoregulation in parrotfish so far defies explanation (Welsh & Bellwood 2012c).

However, the greatest challenge in the study of parrotfishes will be to understand their role on coral reefs in the future. Through overfishing, habitat degradation, and now climate change, humans have fundamentally altered the world's coral reefs. In these new conditions, parrotfishes may no longer fulfil the roles they once did. Many reefs around the world lack the large parrotfishes that supported key ecosystem functions in the recent past (Bellwood et al. 2012a). These reefs are now reliant on small, fast-growing parrotfishes to fulfil some of the remaining functions of parrotfishes, such as the role of scraping grazers. These altered conditions on coral reefs open novel challenges in terms of the increased risk of unanticipated ecosystem changes and the potential for a reduction in the ecosystem goods and services provided by coral reefs (cf. Bellwood et al. 2004). The challenge ahead is to better understand the roles of parrotfishes in a human-altered world and to enable us to manage these new reef systems for the good of both reefs and the humans that depend on them. Parrotfishes have been key components of coral-reef ecosystems in the past and will probably be critically important in shaping the future of the world's coral reefs.

Acknowledgements

We thank C.H.R. Goatley for help with Figure 1 and the manuscript; J.P. Krajewski for photographs; J.P. Barreiros, C.E.L. Ferreira, R.B. Francini-Filho, and S.R. Floeter for use of their raw data; S.R. Floeter, C.E.L. Ferreira, and I.P. Smith for commenting on the manuscript; and J.H. Choat for helpful discussions.

References

Adam, T.C., Schmitt, R.J., Holbrook, S.J., Edmunds, P.J., Carpenter, R.C. & Bernardi, G. 2011. Herbivory, connectivity, and ecosystem resilience: response of a coral reef to a large-scale perturbation. *PLoS One* **6**, e23717.

Afeworki, Y.A., Bruggemann, J.H. & Videler, J.J. 2011. Limited flexibility in resource use in a coral reef grazer foraging on seasonally changing algal communities. *Coral Reefs* **30**, 109–122.

Afeworki, Y.A., Videler, J.J. & Bruggemann, J.H. 2013. Seasonally changing habitat use patterns among roving herbivorous fishes in the southern Red Sea: the role of temperate and algal community structure. *Coral Reefs* **32**, 475–485.

Alvarez-Filip, L., Dulvy, N.K., Gill, J.A., Côté, I.M. & Watkinson, A.R. 2009. Flattening of Caribbean coral reefs: region-wide declines in architectural complexity. *Proceedings of the Royal Society B* **276**, 3019–3025.

Alwany, M.A., Thaler, E. & Stachowitsch, M. 2009. Parrotfish bioerosion on Egyptian Red Sea reefs. *Journal of Experimental Marine Biology and Ecology* **371**, 170–176.

Aswani, S. & Hamilton, R.J. 2004. Integrating indigenous ecological knowledge and customary sea tenure with marine and social science for conservation of bumphead parrotfish (*Bolbometopon muricatum*) in the Roviana Lagoon, Solomon Islands. *Environmental Conservation* **31**, 69–83.

Ateweberhan, M., Bruggemann, J.H. & Breeman, A.M. 2006. Effects of extreme seasonality on community structure and functional group dynamics of coral reef algae in the southern Red Sea (Eritrea). *Coral Reefs* **25**, 391–406.

Baird, A.H., Hoey, A.S., Campbell, S.J., Herdiana, Y. & Pratchett, M.S. 2013. *Acanthaster planci* is a major cause of coral mortality in Indonesia. *Coral Reefs* **32**, 803–812.

Baird, A.H. & Marshall, P.A. 2002. Mortality, growth and reproduction in scleractinian corals following bleaching on the Great Barrier Reef. *Marine Ecology Progress Series* **237**, 133–141.

Bak, R.P.M. & Steward-Van Es, Y. 1980. Regeneration of superficial damage in the scleractinian corals *Agaricia agaricites* f. *purpurea* and *Porites astreoides*. *Bulletin of Marine Science* **30**, 883–887.

Balvanera, P., Pfisterer, A.B., Buchmann, N., He, J.-S., Nakashizuka, T., Raffaelli, D. & Schmid, B. 2006. Quantifying the evidence for biodiversity effects on ecosystem functioning and services. *Ecology Letters* **9**, 1146–1156.

Barnes, D.D.J. & Chalker, B.B.E. 1990. Calcification and photosynthesis in reef-building corals and algae. In *Coral Reefs. Ecosystems of the World*. Volume 25, D.D.J. Barnes & B.B.E. Chalker (eds). Amsterdam: Elsevier, 109–131.

Bejarano, S., Golbuu, Y., Sapolu, T. & Mumby, P.J. 2013. Ecological risk and the exploitation of herbivorous reef fish across Micronesia. *Marine Ecology Progress Series* **482**, 197–215.

Bellwood, D.R. 1985. *The functional morphology, systematic and behavioural ecology of parrotfishes (Family Scaridae)*. PhD thesis, James Cook University, Australia.

Bellwood, D.R. 1988. Ontogenetic changes in the diet of early post-settlement *Scarus* species (Pisces: Scaridae). *Journal of Fish Biology* **33**, 213–219.

Bellwood, D.R. 1994. A phylogenetic study of the parrotfishes: family Scaridae (Pisces: Labroidei), with a revision of genera. *Records of the Australian Museum Supplement* **20**, 1–86.
Bellwood, D.R. 1995a. Direct estimate of bioerosion by two parrotfish species, *Chlorurus gibbus* and *C. sordidus*, on the Great Barrier Reef, Australia. *Marine Biology* **121**, 419–429.
Bellwood, D.R. 1995b. Carbonate transport and within-reef patterns of bioerosion and sediment release by parrotfishes (family Scaridae) on the Great Barrier Reef. *Marine Ecology Progress Series* **117**, 127–136.
Bellwood, D.R. 1996. Production and reworking of sediment by parrotfishes (family Scaridae) on the Great Barrier Reef, Australia. *Marine Biology* **125**, 795–800.
Bellwood, D.R. & Choat, J.H. 1990. A functional-analysis of grazing in parrotfishes (family Scaridae): the ecological implications. *Environmental Biology of Fishes* **28**, 189–214.
Bellwood, D.R. & Choat, J.H. 2011. Dangerous demographics: the lack of juvenile humphead parrotfishes *Bolbometopon muricatum* on the Great Barrier Reef. *Coral Reefs* **30**, 549–554.
Bellwood, D.R. & Fulton, C.J. 2008. Sediment-mediated suppression of herbivory on coral reefs: decreasing resilience to rising sea-levels and climate change? *Limnology and Oceanography* **53**, 2695–2701.
Bellwood, D.R., Hoey, A.S. & Choat, J.H. 2003. Limited functional redundancy in high diversity systems: resilience and ecosystem function on coral reefs. *Ecology Letters* **6**, 281–285.
Bellwood, D.R., Hoey, A.S. & Hughes, T.P. 2012a. Human activity selectively impacts the ecosystem roles of parrotfishes on coral reefs. *Proceedings of the Royal Society B* **279**, 1621–1629.
Bellwood, D.R., Hughes, T.P., Folke, C. & Nyström, M. 2004. Confronting the coral reef crisis: supporting biodiversity, functional groups and resilience. *Nature* **429**, 827–833.
Bellwood, D.R., Hughes, T.P. & Hoey, A.S. 2006a. Sleeping functional group drives coral reef recovery. *Current Biology* **16**, 2434–2439.
Bellwood, D.R., Klanten, S., Cowman, P.F., Pratchett, M.S., Konow, N. & van Herwerden, L. 2010. Evolutionary history of the butterflyfishes (f: Chaetodontidae) and the rise of coral feeding fishes. *Journal of Evolutionary Biology* **23**, 335–349.
Bellwood, D.R., Renema, W. & Rosen, B.R. 2012b. Biodiversity hotspots, evolution and coral biogeography: a review. In *Biotic Evolution and Environmental Change in Southeast Asia*, D.J. Gower et al. (eds). Cambridge, UK: Cambridge University Press, 216–245.
Bellwood, D.R. & Schultz, O. 1991. A review of the fossil record of the parrotfishes (Labroidei: Scaridae) with a description of a new *Calotomus* species from the Middle Miocene (Badenian) of Austria. *Annals Naturhistorisches Museum Wien* **92**, 55–71.
Bellwood, D.R. & Wainwright, P.C. 2001. Locomotion in labrid fishes: implications for habitat use and cross-shelf biogeography on the Great Barrier Reef. *Coral Reefs* **20**, 139–150.
Bellwood, D.R. & Wainwright, P.C. 2002. The history and biogeography of fishes on coral reefs. In *Coral Reef Fishes: Dynamics and Diversity in a Complex Ecosystem*, P.F. Sale (ed.). San Diego, California: Academic Press, 5–32.
Bellwood, D.R., Wainwright, P.C., Fulton, C.J. & Hoey, A. 2002. Assembly rules and functional groups at global biogeographic scales. *Functional Ecology* **16**, 557–562.
Bellwood, D.R., Wainwright, P.C., Fulton, C.J. & Hoey, A.S. 2006b. Functional versatility supports coral reef biodiversity. *Proceedings of the Royal Society B* **273**, 101–107.
Bennett, S. & Bellwood, D.R. 2011. Latitudinal variation in macroalgal consumption by fishes on the Great Barrier Reef. *Marine Ecology Progress Series* **426**, 241–252.
Bernardi, G., Holbrook, S.J., Schmitt, R.J., Crane, N.L. & DeMartini, E. 2002. Species boundaries, populations, and colour morphs in the coral reef three-spot damselfish (*Dascyllus trimaculatus*) species complex. *Proceedings of the Royal Society of London B* **269**, 599–605.
Bernardi, G., Robertson, D.R., Clifton, K.E. & Azzurro, E. 2000. Molecular systematics, zoogeography, and evolutionary ecology of the Atlantic parrotfish genus *Sparisoma*. *Molecular Phylogenetics and Evolution* **15**, 292–300.
Berumen, M.L., Hoey, A.S., Bass, W.H., Bouwmeester, J., Catania, D., Cochran, J.E.M., Khalil, M.T., Miyake, S., Mughal, M.R., Spaet, J.L.Y. & Saenz-Agudelo, P. 2013. The status of coral reef ecology research in the Red Sea. *Coral Reefs* **32**, 737–748.
Birrell, C., McCook, L., Willis, B. & Diaz-Pulido, G. 2008. Effects of benthic algae on the replenishment of corals and the implications for the resilience of coral reefs. *Oceanography and Marine Biology: An Annual Review* **46**, 25–65.

Bonaldo, R.M. & Bellwood, D.R. 2008. Size-dependent variation in the functional role of the parrotfish *Scarus rivulatus* on the Great Barrier Reef, Australia. *Marine Ecology Progress Series* **360**, 237–244.
Bonaldo, R.M. & Bellwood, D.R. 2009. Dynamics of parrotfish grazing scars. *Marine Biology* **156**, 771–777.
Bonaldo, R.M. & Bellwood, D.R. 2011a. Parrotfish predation on massive *Porites* on the Great Barrier Reef. *Coral Reefs* **30**, 259–269.
Bonaldo, R.M. & Bellwood, D.R. 2011b. Spatial variation on the response of epilithic algal turfs to grazing pressure on coral reefs. *Coral Reefs* **30**, 381–390.
Bonaldo, R.M., Krajewski, J.P. & Bellwood, D.R. 2011. Relative impact of parrotfish grazing scars on massive *Porites* on Lizard Island, Great Barrier Reef. *Marine Ecology Progress Series* **423**, 223–233.
Bonaldo, R.M., Krajewski, J.P., Sazima, C. & Sazima, I. 2006. Foraging activity and resource use by three parrotfish species at Fernando de Noronha Archipelago, tropical West Atlantic. *Marine Biology* **149**, 423–433.
Bonaldo, R.M., Welsh, J.Q. & Bellwood, D.R. 2012. Spatial and temporal variation in coral predation by parrotfishes on the GBR: evidence from an inshore reef. *Coral Reefs* **31**, 263–272.
Box, S.J. & Mumby, P.J. 2007. The effect of macroalgal competition on the growth and survival of juvenile Caribbean corals. *Marine Ecology Progress Series* **342**, 139–149.
Brandl, S.J. & Bellwood, D.R. 2013. Individual-based analyses reveal limited functional overlap in a coral reef fish community. *Journal of Animal Ecology* doi: 10.1111/1365-2656.12171
Brown-Saracino, J., Peckol, P., Curran, H.A. & Robbart, M.L. 2007. Spatial variation in sea urchins, fish predators, and bioerosion rates on coral reefs of Belize. *Coral Reefs* **26**, 71–78.
Bruckner, A.W. & Bruckner, R.J. 1998. Destruction of coral by *Sparisoma viride*. *Coral Reefs* **17**, 350 only.
Bruckner, A.W., Bruckner, R.J. & Sollins, P. 2000. Parrotfish predation on live coral: 'spot biting' and 'focused biting'. *Coral Reefs* **19**, 50 only.
Bruggemann, J.H. 1995. *Parrotfish grazing on coral reefs: a trophic novelty*. PhD thesis, University of Groningen, the Netherlands.
Bruggemann, J.H., Begeman, J., Bosma, E.M., Verburg, P. & Breeman, A.M. 1994a. Foraging by the stoplight parrotfish *Sparisoma viride* II. Intake and assimilation of food, protein and energy. *Marine Ecology Progress Series* **106**, 57–71.
Bruggemann, J.H., Kuyper, M.W.M. & Breeman, A.M. 1994b. Comparative analysis of foraging and habitat use by the sympatric Caribbean parrotfish *Scarus vetula* and *Sparisoma viride*. *Marine Ecology Progress Series* **112**, 51–66.
Bruggemann, J.H., van Kessel, A.M., van Rooij, J.M. & Breeman, A.M. 1996. Bioerosion and sediment ingestion by the Caribbean parrotfish *Scarus vetula* and *Sparisoma viride*: implications of fish size, feeding mode and habitat use. *Marine Ecology Progress Series* **134**, 59–71.
Bruggemann, J.H., van Oppen, M.J.H. & Breeman, A.M. 1994c. Foraging by the stoplight parrotfish *Sparisoma viride* I. Food selection in different, socially determined habitats. *Marine Ecology Progress Series* **106**, 41–55.
Bruggemann, J.H., van Rooij, J., Videler, J. & Breeman, A. 1995. Dynamics and limitations of herbivore populations on a Caribbean coral reef. In *Parrotfish grazing on coral reefs: a trophic novelty*. PhD thesis, University of Groningen, the Netherlands, 153–178.
Bruno, J. & Bertness, M.D. 2001. Positive interactions, facilitations and foundation species. In *Marine Community Ecology*, M.D. Bertness et al. (eds). Sunderland, Massachusetts: Sinauer Associates, 201–220.
Bruno, J.F., Sweatman, H., Precht, W.F., Selig, E.R. & Schutte, V.G.W. 2009. Assessing evidence of phase shifts from coral to macroalgal dominance on coral reefs. *Ecology* **90**, 1478–1484.
Burkepile, D.E. & Hay, M.E. 2008. Herbivore richness and feeding complementarity affect community structure and function on a coral reef. *Proceedings of the National Academy of Sciences of the United States of America* **105**, 16201–16206.
Burkepile, D.E. & Hay, M.E. 2011. Feeding complimentarity versus redundancy among herbivorous fishes on a Caribbean reef. *Coral Reefs* **30**, 351–362.
Campbell, S.J., Hoey, A.S., Maynard, J., Kartawijaya, T., Cinner, J., Graham, N.A.J. & Baird, A.H. 2012. Weak compliance undermines the success of no-take zones in a large government-controlled marine protected area. *PLoS ONE* **7**, e50074.
Carpenter, R.C. 1986. Partitioning herbivory and its effects on coral reef algal communities. *Ecological Monographs* **56**, 345–364.

Carpenter, R.C. 1993. Effects of algal canopy height and microscale substratum topography on profiles of flow speed in a coral forereef environment. *Limnology and Oceanography* **38**, 687–694.
Carreiro-Silva, M. & McClanahan, T.R. 2001. Echinoid bioerosion and herbivory on Kenyan coral reefs: the role of protection from fishing. *Journal of Experimental Marine Biology and Ecology* **262**, 133–153.
Ceccarelli, D.M., Jones, G.P. & McCook, L.J. 2005. Foragers versus farmers: contrasting effects of two behavioural groups of herbivores on coral reefs. *Oecologia* **145**, 445–453.
Chateau, O. & Wantiez, L. 2009. Movement patterns of four coral reef fish species in a fragmented habitat in New Caledonia: implications for the design of marine protected area networks. *ICES Journal of Marine Science* **66**, 50–55.
Cheal, A., Emslie, M., Miller, I. & Sweatman, H. 2012. The distribution of herbivorous fishes on the Great Barrier Reef. *Marine Biology* **159**, 1143–1154.
Cheal, A.J., MacNeil, M.A., Cripps, E., Emslie, M.J., Jonker, M., Schaffelke, B. & Sweatman, H. 2010. Coral–macroalgal phase shifts or reef resilience: links with diversity and functional roles of herbivorous fishes on the Great Barrier Reef. *Coral Reefs* **29**, 1005–1015.
Chen, L.S. 2002. Post-settlement diet shift of *Chlorurus sordidus* and *Scarus schlegeli* (Pisces: Scaridae). *Zoological Studies* **41**, 47–58.
Choat, J.H. 1991. The biology of herbivorous fishes on coral reefs. In *The Ecology of Fishes on Coral Reefs*, P.F. Sale (ed.). San Diego, California: Academic Press, 120–155.
Choat, J.H. & Bellwood, D.R. 1985. Interactions amongst herbivorous reef fishes on a coral reef: influence of spatial variation. *Marine Biology* **89**, 221–234.
Choat, J.H., Clements, K.D. & Robbins, W.D. 2002. The trophic status of herbivorous fishes on coral reefs. I: Dietary analysis. *Marine Biology* **140**, 613–623.
Choat, J.H., Klanten, O.S., van Herwerden, L., Robertson, D.R. & Clements, K.D. 2012. Patterns and processes in the evolutionary history of parrotfishes (Family Labridae). *Biological Journal of the Linnean Society* **107**, 529–557.
Choat, J.H., Robbins, W.D. & Clements, K.D. 2004. The trophic status of herbivorous fishes on coral reefs. II. Food processing modes and trophodynamics. *Marine Biology* **145**, 445–454.
Cole, A.J., Lawton, R.J., Pratchett, M.S. & Wilson, S.K. 2011. Chronic coral consumption by butterflyfishes. *Coral Reefs* **30**, 85–93.
Cole, A.J., Lawton, R.J., Wilson, S.K. & Pratchett, M.S. 2012. Consumption of tabular acroporid corals by reef fishes: a comparison with plant-herbivore interactions. *Functional Ecology* **26**, 307–316.
Cole, A.J., Pratchett, M.S. & Jones, G.P. 2008. Diversity and functional importance of coral-feeding fishes on tropical coral reefs. *Fish and Fisheries* **9**, 286–307.
Comeros-Raynal, M.T., Choat, J.H., Polidoro, B.A., Clements, K.D., Abesamis, R., Craig, M.T., Lazuardi, M.E., McIlwain, J., Muljadi, A., Myers, R.F., Nañola, C.L., Jr, Pardede, S., Rocha, L.A., Russel, B., Sanciangco, J.C., Stockwell, B., Harwell, H. & Carpenter, K.E. 2012. The likelihood of extinction of iconic and dominant herbivores and detritivores of coral reefs: the parrotfishes and surgeonfishes. *PLoS One* **7**, e39825.
Connolly, S.R., Bellwood, D.R. & Hughes, T.P. 2003. Indo-Pacific biodiversity of coral reefs: deviations from a mid-domain model. *Ecology* **84**, 2178–2190.
Cowman, P.F. & Bellwood, D.R. 2011. Coral reefs as drivers of cladogenesis: expanding coral reefs, cryptic extinction events and the development of biodiversity hotspots. *Journal of Evolutionary Biology* **24**, 2543–2562.
Cowman, P.F. & Bellwood, D.R. 2013a. The historical biogeography of coral reef fishes: global patterns of origination and dispersal. *Journal of Biogeography* **40**, 209–224.
Cowman, P.F. & Bellwood, D.R. 2013b. Vicariance across major marine biogeographic barriers: temporal concordance and the relative intensity of hard versus soft barriers. *Proceedings of the Royal Society B* **280**, 20131541. doi: 10.1098/rspb.2013.1541
Cowman, P.F., Bellwood, D.R. & van Herwerden, L. 2009. Dating the evolutionary origins of the wrasses (Labridae) and the rise of trophic novelty on coral reefs. *Molecular Phylogenetics and Evolution* **52**, 621–631.
Crook, A.C. 1997. Colour patterns in a coral reef fish: is background complexity important? *Journal of Experimental Marine Biology and Ecology* **217**, 237–252.
Crossman, D.J., Choat, J.H. & Clements, K.D. 2001. Detritus as food for grazing fishes on coral reefs. *Limnology and Oceanography* **46**, 1596–1605.

Crossman, D.J., Choat, J.H. & Clements, K.D. 2005. Nutritional ecology of nominally herbivorous fishes on coral reefs. *Marine Ecology Progress Series* **296**, 129–142.
Cvitanovic, C. & Bellwood, D.R. 2009. Local variation in herbivore feeding activity on an inshore reef of the Great Barrier Reef. *Coral Reefs* **28**, 127–133.
De'ath, G., Lough, J.M. & Fabricius, K.E. 2009. Declining coral calcification on the Great Barrier Reef. *Science* **323**, 116–119.
Diaz-Pulido, G. & McCook, L.J. 2002. The fate of bleached corals: patterns and dynamics of algal recruitment. *Marine Ecology Progress Series* **232**, 115–128.
Diaz-Pulido, G. & McCook, L.J. 2003. Relative roles of herbivory and nutrients in the recruitment of coral-reef seaweeds. *Ecology* **84**, 2026–2033.
Dickens, L.C., Goatley, C.H.R., Tanner, J.K. & Bellwood, D.R. 2011. Quantifying relative diver effects in underwater visual censuses. *PLoS ONE* **6**, e18965.
Dominici-Arosemena, A. & Wolff, M. 2006. Reef fish community in the Tropical Eastern Pacific (Panamá): living on a relatively stable rocky reef environment. *Helgoland Marine Research* **60**, 287–305.
Done, T.J. 1982. Patterns in the distribution of coral communities across the central Great Barrier Reef. *Coral Reefs* **1**, 95–107.
Done, T.J. 1992. Phase shifts in coral reef communities and their ecological significance. *Hydrobiologia* **247**, 121–132.
Done, T.J., Ogden, J.C. & Wiebe, W.J. 1996. Biodiversity and ecosystem function of coral reefs. In *Functional Roles of Biodiversity: a Global Perspective*, A. Mooney et al. (eds). Chichester, UK: Wiley, 393–429.
Duffy, J.E. 2002. Biodiversity and ecosystem function: the consumer connection. *Oikos* **99**, 201–219.
Duffy, J.E. 2003. Biodiversity loss, trophic skew and ecosystem functioning. *Ecology Letters* **6**, 680–687.
Dunlap, M. & Pawlik, J.R. 1998. Spongivory by parrotfish in Florida mangrove and reef habitats. *Marine Ecology* **19**, 325–337.
Eakin, C.M. 1996. Where have all the carbonates gone? A model comparison of calcium carbonate budgets before and after the 1982–1983 El Niño at Uva Island in the eastern Pacific. *Coral Reefs* **15**, 109–119.
Estes, J.A. & Duggins, D.O. 1995. Sea otters and kelp forests in Alaska: generality and variation in a community ecological paradigm. *Ecological Monographs* **65**, 75–100.
Estes, J.A., Terborgh, J., Brashares, J.S., Power, M.E., Berger, J., Bond, W.J., Carpenter, S.R., Essington, T.E., Holt, R.D., Jackson, J.B.C., Marquis, R.J., Oksanen, L., Oksanen, T., Paine, R.T., Pikitch, E.K., Ripple, W.J., Sandin, S.A., Scheffer, M., Schoener, T.W., Shurin, J.B., Sinclair, A.R.E., Soulé, M.E., Virtanen, R. & Wardle, D.A. 2011. Trophic downgrading of Planet Earth. *Science* **333**, 301–306.
Feary, D.A., Burt, J.A., Bauman, A.G., Usseglio, P., Sale, P.F. & Cavalcante, G.H. 2010. Fish communities in the world's warmest reefs: what can they tell us about the effects of climate change in the future? *Journal of Fish Biology* **77**, 1931–1947.
Feary, D.A., Cinner, J.E., Graham, N.A.J. & Januchowski-Hartley, F.A. 2011. Effects of customary marine closures on fish behaviour, spear-fishing success, and underwater visual surveys. *Conservation Biology* **25**, 341–349.
Ferreira, C.E.L., Gasparini, J.L., Carvalho-Filho, A. & Floeter, S.R. 2005. A recently extinct parrotfish species from Brazil. *Coral Reefs* **24**, 128 only.
Ferreira, C.E.L. & Gonçalves, J.E.A. 2006. Community structure and diet of roving herbivorous reef fishes in the Abrolhos Archipelago, south-western Atlantic. *Journal of Fish Biology* **69**, 1533–1551.
Ferreira, C.E.L., Gonçalves, J.E.A., Coutinho, R. & Peret, A.C.P. 1998a. Herbivory by the Dusky Damselfish, *Stegastes fuscus* (Cuvier, 1830) in a tropical rocky shore: effects on the benthic community. *Journal of Experimental Marine Biology and Ecology* **229**, 241–264.
Ferreira, C.E.L., Peret, A.C. & Coutinho, R. 1998b. Seasonal grazing rates and food processing by tropical herbivorous fishes. *Journal of Fish Biology* **53**, 222–235.
Floeter, S.R., Behrens, M.D., Ferreira, C.E.L., Paddack, M.J. & Horn, M.H. 2005. Latitudinal gradients of marine herbivorous fishes: patterns and processes. *Marine Biology* **147**, 1435–1447.
Floeter, S.R., Halpern, B.S. & Ferreira, C.E.L. 2006. Effects of fishing and protection on Brazilian reef fishes. *Biological Conservation* **128**, 391–402.
Floeter, S.R., Rocha, L.A., Robertson, D.R., Joyeux, J.C., Smith-Vaniz, W.F., Wirtz, P., Edwards, A.J., Barreiros, J.P., Ferreira, C.E.L., Gasparini, J.L., Brito, A., Falcón, J.M., Bowen, B.W. & Bernardi, G. 2008. Atlantic reef fish biogeography and evolution. *Journal of Biogeography* **35**, 22–47.

Fox, R.J. & Bellwood, D.R. 2007. Quantifying herbivory across a coral reef depth gradient. *Marine Ecology Progress Series* **339**, 49–59.

Fox, R.J. & Bellwood, D.R. 2008. Remote video bioassays reveal the potential feeding impact of the rabbitfish *Siganus canaliculatus* (f: Siganidae) on an inner-shelf reef of the Great Barrier Reef. *Coral Reefs* **27**, 605–615.

Fox, R.J. & Bellwood, D.R. 2013. Niche partitioning creates a unique ecosystem function for rabbitfishes (Perciformes, Siganidae) on coral reefs. *Coral Reefs* **32**, 13–23.

Fox, R.J., Sunderland, T.L., Hoey, A.S. & Bellwood, D.R. 2009. Estimating functional roles in coral reef ecosystems: behaviour drives contrasting roles in herbivorous fishes (f: Siganidae). *Marine Ecology Progress Series* **385**, 261–269.

Francini-Filho, R.B., Ferreira, C.M., Coni, E.O.C., Moura, R.L. & Kaufman, L. 2010. Foraging activity of roving herbivorous reef fish (Acanthuridae and Scaridae) in eastern Brazil: influence of resource availability and interference competition. *Journal of the Marine Biological Association of the United Kingdom* **90**, 481–492.

Francini-Filho, R.B., Moura, R.L., Ferreira, C.M. & Coni, E.O.C. 2008. Live coral predation by parrotfishes (Perciformes: Scaridae) in the Abrolhos Bank, eastern Brazil, with comments on the classification of species into functional groups. *Neotropical Ichthyology* **6**, 191–200.

Friedlander, A.M. & Parrish, J.D. 1998. Habitat characteristics affecting fish assemblages on a Hawaiian coral reef. *Journal of Experimental Marine Biology and Ecology* **224**, 1–30.

Frydl, P. & Stearn, C.W. 1978. Rates of bioerosion by parrotfish in Barbados reef environments. *Journal of Sedimentary Research* **48**, 1149–1157.

Fulton, C.J. & Bellwood, D.R. 2004. Wave exposure, swimming performance, and the structure of tropical and temperate reef fish assemblages. *Marine Biology* **144**, 429–437.

Fulton, C.J. & Bellwood, D.R. 2005. Wave-induced water motion and the functional implications for coral reef fish assemblages. *Limnology and Oceanography* **50**, 255–264.

Fulton, C.J., Bellwood, D.R. & Wainwright, P.C. 2001. The relationship between swimming ability and habitat use in wrasses (Labridae). *Marine Biology* **139**, 25–33.

Fulton, C.J., Bellwood, D.R. & Wainwright, P.C. 2005. Wave energy and swimming performance shape coral reef fish assemblages. *Proceedings of the Royal Society B* **272**, 827–832.

Gardner, T.A., Côtè, I.M., Gill, J.A., Grant, A. & Watkinson, A.R. 2003. Long-term region-wide declines in Caribbean corals. *Science* **301**, 958–960.

Gilmour, J.P., Smith, L.D., Heyward, A.J., Baird, A.H. & Pratchett, M.S. 2013. Recovery of an isolated coral reef system following severe disturbance. *Science* **340**, 69–71.

Girolamo, M., Scaggiante, M. & Rasotto, M.B. 1999. Social organization and sexual pattern in the Mediterranean parrotfish *Sparisoma cretense* (Teleostei: Scaridae). *Marine Biology* **135**, 353–360.

Glynn, P.W. 1997. Bioerosion and coral reef growth: a dynamic balance. In *Life and Death of Coral Reefs*, C. Birkeland (ed.). New York: Chapman and Hall, 68–95.

Goatley, C.H.R. & Bellwood, D.R. 2010. Biologically mediated sediment fluxes on coral reefs: sediment removal and off-reef transportation by the surgeonfish *Ctenochaetus striatus*. *Marine Ecology Progress Series* **415**, 237–245.

Goatley, C.H.R. & Bellwood, D.R. 2011. The role of dimensionality, canopies and complexity in ecosystem monitoring. *PLoS ONE* **6**, e27307.

Goatley, C.H.R. & Bellwood, D.R. 2012. Sediment suppresses herbivory across a coral reef depth gradient. *Biology Letters* **8**, 1016–1018.

Gobalet, K.W. 1989. Morphology of the parrotfish pharyngeal jaw apparatus. *Integrative and Comparative Biology* **29**, 319–331.

Gochfeld, D.J. 2004. Predation-induced morphological and behavioural defenses in a hard coral: implications for foraging behavior of coral-feeding butterflyfishes. *Marine Ecology Progress Series* **267**, 145–158.

Graham, N.A.J., Chabanet, P., Evans, R.D., Jennings, S., Letourneur, Y., MacNeil, M.A., McClanahan, T.R., Öhman, M.C., Polunin, N.V.C. & Wilson, S.K. 2011. Extinction vulnerability of coral reef fishes. *Ecology Letters* **14**, 341–348.

Graham, N.A.J., Wilson, S.K., Jennings, S., Polunin, N.V.C., Bijoux, J.P. & Robinson, J. 2006. Dynamic fragility of oceanic coral reef ecosystems. *Proceedings of the National Academy of Sciences of the United States of America* **22**, 8425–8429.

Gregson, M.A., Pratchett, M.S., Berumen, M.L. & Goodman, B.A. 2008. Relationships between butterflyfish (Chaetodontidae) feeding rates and coral consumption on the Great Barrier Reef. *Coral Reefs* **27**, 583–591.

Grutter, A.S., Rumney, J.G., Sinclair-Taylor, T., Waldie, P. & Franklin, C.E. 2010. Fish mucous cocoons: the 'mosquito nets' of the sea. *Biology Letters* **7**, 292–294.

Guimaraens, M.A. & Coutinho, R. 1996. Spatial and temporal variation of benthic marine algae at the Cabo Frio upwelling region, Rio de Janeiro, Brazil. *Aquatic Botany* **52**, 283–299.

Gygi, R.A. 1969. An estimate of the erosional effect of *Sparisoma viride* (Bonnaterre), the green parrotfish, on some Bermuda reefs. *Special Publication of Bermuda Biological Station for Research* **2**, 137–143.

Gygi, R.A. 1975. *Sparisoma viride* (Bonnaterre), the stoplight parrotfish, a major sediment producer on coral reefs of Bermuda? *Eclogae Geologicae Helvetiae* **68**, 327–359.

Harmelin-Vivien, M.L. & Bouchon-Navaro, Y. 1982. Trophic relationships among Chaetodontid fishes in the Gulf of Aqaba (Red Sea). *Proceedings of the 4th International Coral Reef Symposium* **2**, E.D. Gomez et al. (eds). Manila: Marine Science Center, University of the Philippines, 537–544.

Harmelin-Vivien, M.L. & Bouchon-Navaro, Y. 1983. Feeding rates and significance of coral feeding among chaetodontid fishes in Moorea (French Polynesia). *Coral Reefs* **2**, 119–127.

Hawkins, J.P. & Roberts, C.M. 2003. Effects of fishing on sex-changing Caribbean parrotfishes. *Biological Conservation* **115**, 213–226.

Hawkins, J.P. & Roberts, C.M. 2004. Effects of artisanal fishing on Caribbean coral reefs. *Conservation Biology* **18**, 215–226.

Hay, M.E. 1981. Spatial patterns of grazing intensity on a Caribbean barrier reef: herbivory and algal distribution. *Aquatic Botany* **11**, 97–109.

Hay, M.E. 1984. Predictable spatial escapes from herbivory: how do these affect the evolution of herbivore resistance in tropical marine communities? *Oecologia* **64**, 396–407.

Hixon, M.A. & Brostoff, W.N. 1996. Succession and herbivory: effects of differential fish grazing on Hawaiian coral-reef algae. *Ecological Monographs* **66**, 67–90.

Hoegh-Guldberg, O., Mumby, P.J., Hooten, A.J., Steneck, R.S., Greenfield, P., Gomez, E., Harvell, C.D., Sale, P.F., Edwards, A.J., Caldeira, K., Knowlton, N., Eakin, C.M., Iglesias-Prieto, R., Muthiga, N., Bradbury, R.H., Dubi, A. & Hatziolos, M.E. 2007. Coral reefs under rapid climate change and ocean acidification. *Science* **318**, 1737–1742.

Hoey, A.S. 2010. Size matters: macroalgal height influences the feeding response of coral reef herbivores. *Marine Ecology Progress Series* **411**, 299–301.

Hoey, A.S. & Bellwood, D.R. 2008. Cross shelf variation in the role of parrotfishes on the Great Barrier Reef. *Coral Reefs* **27**, 37–47.

Hoey, A.S. & Bellwood, D.R. 2009. Limited functional redundancy in a high diversity system: single species dominates key ecological process on coral reefs. *Ecosystems* **12**, 1316–1328.

Hoey, A.S. & Bellwood, D.R. 2010a. Among-habitat variation in herbivory on *Sargassum* spp. on a mid-shelf reef in the northern Great Barrier Reef. *Marine Biology* **157**, 189–200.

Hoey, A.S. & Bellwood, D.R. 2010b. Cross-shelf variation in browsing intensity on the Great Barrier Reef. *Coral Reefs* **29**, 499–508.

Hoey, A.S. & Bellwood, D.R. 2010c. Damselfish territories as a refuge for macroalgae on coral reefs. *Coral Reefs* **29**, 107–118.

Hoey, A.S. & Bellwood, D.R. 2011. Suppression of herbivory by macroalgal density: a critical feedback on coral reefs? *Ecology Letters* **14**, 267–273.

Hoey, A.S., Brandl, S.J. & Bellwood, D.R. 2013. Diet and cross-shelf distribution of rabbitfishes (f. Siganidae) on the northern Great Barrier Reef: implications for ecosystem function. *Coral Reefs* **32**, 973–984.

Hoey, A.S., Pratchett, M.S. & Cvitanovic, C. 2011. High macroalgal cover and low coral recruitment undermines the potential resilience of the world's southernmost coral reef assemblage. *PLoS ONE* **6**, e25824.

Hopley, D., Smithers, S.G. & Parnell, K. 2012. *The Geomorphology of the Great Barrier Reef: Development, Diversity and Change*. Cambridge, UK: Cambridge University Press.

Howard, K.G., Claisse, J.T., Clark, T.B., Boyle, K. & Parrish, J.D. 2013. Home range and movement patterns of the redlip parrotfish (*Scarus rubroviolaceus*) in Hawaii. *Marine Biology* **160**, 1583–1595.

Howard, K.G., Schumacher, B.D. & Parrish, J.D. 2009. Community structure and habitat associations of parrotfishes on Oahu, Hawaii. *Environmental Biology of Fishes* **89**, 175–186.

Hughes, T.P. 1994. Catastrophes, phase shifts, and large-scale degradation of a Caribbean coral reef. *Science* **265**, 1547–1551.
Hughes, T.P. 1999. Off-reef transport of coral fragments at Lizard Island, Australia. *Marine Geology* **157**, 1–6.
Hughes, T.P., Baird, A.H., Bellwood, D.R., Card, M., Connolly, S.R., Folke, C., Grosberg, R., Hoegh-Guldberg, O., Jackson, J.B.C., Kleypas, J., Lough, J., Marshall, P., Nyström, M., Palumbi, S.R., Pandolfi, J.M., Rosen, B. & Roughgarden, J. 2003. Climate change, human impacts, and the resilience of coral reefs. *Science* **301**, 929–933.
Hughes, T.P., Bellwood, D.R. & Connolly, S. 2002. Biodiversity hotspots, centres of endemicity, and the conservation of coral reefs. *Ecology Letters* **5**, 775–784.
Hughes, T.P., Bellwood, D.R., Folke, C., Steneck, R.S. & Wilson, J. 2005. New paradigms for supporting the resilience of marine ecosystems. *Trends in Ecology and Evolution* **20**, 380–386.
Hughes, T.P., Rodrigues, M.J., Bellwood, D.R., Ceccarelli, D., Hoegh-Guldberg, O., McCook, L., Moltschaniwsky, N., Pratchett, M.S., Steneck, R.S. & Willis, B. 2007. Phase shifts, herbivory, and the resilience of coral reefs to climate change. *Current Biology* **17**, 360–365.
Hutchings, P.A. 1986. Biological destruction of coral reefs. *Coral Reefs* **4**, 239–252.
Hutchings, P.A., Peyrot-Clausade, M. & Osorno, A. 2005. Influence of land runoff on rates and agents of bioerosion of coral substrates. *Marine Pollution Bulletin* **51**, 438–447.
Jackson, J.B.C., Kirby, M.X., Berger, W.H., Bjorndal, K.A., Botsford, L.W., Bourque, B.J., Bradbury, R.H., Cooke, R., Erlandson, J., Estes, J.A., Hughes, T.P., Kidwell, S., Lange, C.B., Lenihan, H.S., Pandolfi, J.M., Peterson, C.H., Steneck, R.S., Tegner, M.J. & Warner, R.R. 2001. Historical overfishing and the recent collapse of coastal ecosystems. *Science* **293**, 629–637.
Januchowski-Hartley, F.A., Nash, K.L. & Lawton, R.J. 2012. Influence of spear guns, dive gear and observers on estimating fish flight initiation distance on coral reefs. *Marine Ecology Progress Series* **469**, 113–119.
Jayewardene, D. 2009. A factorial experiment quantifying the influence of parrotfish density and size on algal reduction on Hawaiian coral reefs. *Journal of Experimental Marine Biology and Ecology* **375**, 64–69.
Jayewardene, D. & Birkeland, C. 2006. Fish predation on Hawaiian corals. *Coral Reefs* **25**, 499–506.
Jayewardene, D., Donahue, M.J. & Birkeland, C. 2009. Effects of frequent fish predation on corals in Hawaii. *Coral Reefs* **28**, 499–506.
Jennings, S. & Polunin, N.V.C. 1996. Impacts of fishing on tropical reef ecosystems. *Ambio* **25**, 44–49.
Johannes, R.E. 1981. *Words of the Lagoon: Fishing and Marine Lore in the Palau District of Micronesia.* Berkeley, California: University of California Press.
Johansson, C.L., Bellwood, D.R. & Depczynski, M. 2010. Sea urchins, macroalgae and coral reef decline: a functional evaluation of an intact reef system, Ningaloo, Western Australia. *Marine Ecology Progress Series* **414**, 65–74.
Johansson, C.L., van de Leemput, I.A., Depczynski, M., Hoey, A.S. & Bellwood, D.R. 2013. Key herbivores reveal limited functional redundancy on inshore coral reefs. *Coral Reefs* **32**, 963–972.
Keesing, J.K. & Lucas, J.S. 1992. Field measurement of feeding and movement rates of the crown-of-thorns starfish *Acanthaster planci* (L.). *Journal of Experimental Marine Biology and Ecology* **156**, 89–104.
Kerry, J.T. & Bellwood, D.R. 2012. The effect of coral morphology on shelter selection by coral reef fishes. *Coral Reefs* **31**, 415–424.
Kiene, W.E. & Hutchings, P.A. 1994. Bioerosion experiments at Lizard Island, Great Barrier Reef. *Coral Reefs* **13**, 91–98.
Klumpp, D.W. & McKinnon, A.D. 1989. Temporal and spatial patterns in primary production of a coral-reef epilithic algal community. *Journal of Experimental Marine Biology and Ecology* **131**, 1–22.
Knowlton, N. 1992. Thresholds and multiple stable states in coral reef community dynamics. *American Zoologist* **32**, 674–682.
Krajewski, J.P. & Floeter, S.R. 2011. Reef fish community structure of the Fernando de Noronha Archipelago (Equatorial Western Atlantic): the influence of exposure and benthic composition. *Environmental Biology of Fishes* **92**, 25–40.
Kramer, M.J., Bellwood, D.R. & Bellwood, O. 2012. Cryptofauna of the epilithic algal matrix on an inshore coral reef, Great Barrier Reef. *Coral Reefs* **31**, 1007–1015.
Kramer, M.J., Bellwood, O. & Bellwood, D.R. 2013. The trophic importance of algal turfs for coral reef fishes: the crustacean link. *Coral Reefs* **32**, 575–583.
Krone, R., van Treeck, P., Schuhmacher, H., Nebel, H. & Epple, M. 2006. A special palate structure of *Ctenochaetus striatus*—a hidden tool for bioerosion. *Coral Reefs* **25**, 645 only.

Kulbicki, M., Parravicini, V., Bellwood, D.R., Arias-Gonzàlez, E., Chabanet, P., Floeter, S.R., Friedlander, A., McPherson, J., Myers, R.E., Vigliola, L. & Mouillot, D. 2013. Global biogeography of reef fishes: a hierarchical quantitative delineation of regions. *PLoS ONE* **8**, e81847. doi:10.1371/journal.pone.0081847

La Mesa, G., Consalvo, I., Annunziatellis, A. & Canese, S. 2012. Movement patterns of the parrotfish *Sparisoma cretense* in a Mediterranean marine protected area. *Marine Environmetal Research* **82**, 59–68.

Ledlie, M.H., Graham, N.A.J., Bythell, J.C., Wilson, S.K., Jennings, S., Polunin, N.V.C. & Hardcastle, J. 2007. Phase shifts and the role of herbivory in the resilience of coral reefs. *Coral Reefs* **26**, 641–653.

Lefèvre, C.D. & Bellwood, D.R. 2011. Temporal variation in coral reef ecosystem processes: herbivory of macroalgae by fishes. *Marine Ecology Progress Series* **422**, 239–251.

Leis, J.M. & McCormick, M.I. 2006. The biology, behavior and ecology of the pelagic, larval stage of coral reef fishes. In *Coral Reef Fishes: Dynamics and Diversity in a Complex Ecosystem*, P.F. Sale (ed.). San Diego, California: Elsevier Academic Press, 171–200.

Lewis, S.M. 1985. Herbivory on coral reefs: algal susceptibility to herbivorous fishes. *Oecologia* **65**, 370–375.

Lewis, S.M. 1986. The role of herbivorous fishes in the organization of a Caribbean reef community. *Ecological Monographs* **56**, 183–200.

Lewis, S.M. & Wainwright, P.C. 1985. Herbivore abundance and grazing intensity on a Caribbean coral reef. *Journal of Experimental Marine Biology and Ecology* **87**, 215–228.

Liem, K.F. & Sanderson, S.L. 1986. The pharyngeal jaw apparatus of labrid fishes: a functional morphological perspective. *Journal of Morphology* **187**, 143–158.

Littler, M.M., Taylor, P.R. & Littler, D.S. 1989. Complex interactions in the control of coral zonation on a Caribbean reef flat. *Oecologia* **80**, 331–340.

Lobel, P.S. & Ogden, J.C. 1981. Foraging by the herbivorous parrotfish *Sparisoma radians*. *Marine Biology* **64**, 173–183.

Lokrantz, J., Nyström, M., Norström, A.V., Folke, C. & Cinner, J.E. 2010. Impacts of artisanal fishing on key functional groups and the potential vulnerability of coral reefs. *Environmental Conservation* **36**, 327–337.

Lokrantz, J., Nyström, M., Thyresson, M. & Johansson, C. 2008. The non-linear relationship between body size and function in parrotfishes. *Coral Reefs* **150**, 1145–1152.

Loreau, M. 2004. Does functional redundancy exist? *Oikos* **104**, 606–611.

Loreau, M. & Hector, A. 2001. Partitioning selection and complementarity in biodiversity experiments. *Nature* **412**, 72–76.

Loreau, M., Naeem, S., Inchausti, P., Bengtsson, J., Grime, J.P., Hector, A., Hooper, D.U., Huston, M.A., Raffaelli, D., Schmid, B., Tilman, D. & Wardle, D.A. 2001. Biodiversity and ecosystem functioning: current knowledge and future challenges. *Science* **294**, 804–808.

Luckhurst, B.E. & Luckhurst, K. 1978. Analysis of the influence of substrate variables on coral reef fish communities. *Marine Biology* **49**, 317–323.

Madin, J.S. & Connolly, S.R. 2006. Ecological consequences of major hydrodynamic disturbances on coral reefs. *Nature* **444**, 477–480.

Mallela, J. & Perry, C.T. 2007. Calcium carbonate budgets for two coral reefs affected by different terrestrial runoff regimes, Rio Bueno, Jamaica. *Coral Reefs* **26**, 129–145.

Mantyka, C.S. & Bellwood, D.R. 2007a. Direct evaluation of macroalgal removal by herbivorous reef fishes. *Coral Reefs* **26**, 435–442.

Mantyka, C.S. & Bellwood, D.R. 2007b. Macroalgal grazing selectivity among herbivorous coral reef fishes. *Marine Ecology Progress Series* **325**, 177–185.

McAfee, S.T. & Morgan, S.G. 1996. Resource use by five sympatric parrotfishes in the San Blas Archipelago, Panama. *Marine Biology* **125**, 427–437.

McClanahan, T.R. 1994. Kenyan coral reef lagoon fish: effects of fishing, substrate complexity, and sea urchins. *Coral Reefs* **13**, 231–241.

McClanahan, T.R., Hendrick, V., Rodrigues, M.J. & Polunin, N.V.C. 1999. Varying responses of herbivorous and invertebrate-feeding fishes to macroalgal reduction on a coral reef. *Coral Reefs* **18**, 195–203.

McClanahan, T.R., Kaunda-Arara, B. & Omukoto, J.O. 2010. Composition and diversity of fish and fish catches in closures and open-access fisheries of Kenya. *Fisheries Management and Ecology* **17**, 63–76.

McClanahan, T.R. & Muthiga, N.A. 1988. Changes in Kenyan coral reef community structure and function due to exploitation. *Hydrobiologia* **166**, 269–276.

McCook, L.J. 1997. Effects of herbivory on zonation of *Sargassum* spp. within fringing reefs of the central Great Barrier Reef. *Marine Biology* **129**, 713–722.

McCook, L.J. 1999. Macroalgae and phase shifts on coral reefs of the Great Barrier Reef and other regions: a perspective on scientific issues and management consequences. *Coral Reefs* **18**, 357–367.

McManus, J.M. & Polsenberg, J.F. 2004. Coral-algal phase shifts on coral reefs: ecological and environmental aspects. *Progress in Oceanography* **60**, 263–279.

McNaughton, S.J., Ruess, R.W. & Seagle, S.W. 1988. Large mammals and process dynamics in African ecosystems. *BioScience* **38**, 794–800.

Michael, P.J., Hyndes, G.A., Vanderklift, M.A. & Vergés, A. 2013. Identity and behavior of herbivorous fish influence large-scale spatial patterns of macroalgal herbivory in a coral reef. *Marine Ecology Progress Series* **482**, 227–240.

Miller, M.W. & Hay, M.E. 1998. Effects of fish predation and seaweed competition on the survival and growth of corals. *Oecologia* **113**, 231–238.

Mouillot, D., Bellwood, D.R., Baraloto, C., Chave, J., Galzin, R., Harmelin-Vivien, M., Kulbicki, M., Lavergne, S., Lavorel, S., Mouquet, N., Paine, C.E.T., Renaud, J. & Thuiler, W. 2013. Rare species support vulnerable functions in high-diversity ecosystems. *PLoS ONE* **11**, e1001569. doi: 10.1371/journal.pbio.1001569

Mouillot, D., Villéger, S., Parravicini, V., Kulbicki, M., Arias-Gonzales, J.E., Bender, M.G., Chabanet, P., Floeter, S.R., Friedlander, A., Vigliola, L. & Bellwood, D.R. in press. Functional over-redundancy and high functional vulnerability in global fish faunas of tropical reefs. *Proceedings of the National Academy of Sciences of the United States of America.*

Mumby, P.J. 2006. The impact of exploiting grazers (Scaridae) on the dynamics of Caribbean coral reefs. *Ecological Applications* **16**, 747–769.

Mumby, P.J. 2009a. Phase shifts and the stability of macroalgal communities on Caribbean coral reefs. *Coral Reefs* **28**, 761–773.

Mumby, P.J. 2009b. Herbivory versus corallivory: are parrotfish good or bad for Caribbean coral reefs? *Coral Reefs* **28**, 683–690.

Mumby, P.J., Dahlgren, C.P., Harborne, A.R., Kappel, C.V., Micheli, F., Brumbaugh, D.R., Holmes, K.E., Mendes, J.M., Broad, K., Sanchirico, J.N., Buch, K., Box, S., Stoffle, R.W. & Gill, A.B. 2006. Fishing, trophic cascades and the process of grazing on coral reefs. *Science* **311**, 98–101.

Mumby, P.J. & Harborne, A.R. 2010. Marine reserves enhance the recovery of corals on Caribbean reefs. *PLoS ONE* **5**, e8657.

Mumby, P.J., Harborne, A.R., Williams, J., Kappel, C.V., Brumbaugh, D.R., Micheli, F., Holmes, K.E., Dahlgren, C.P., Paris, C.B. & Blackwell, P.G. 2007. Trophic cascade facilitates coral recruitment in a marine reserve. *Proceedings of the National Academy of Sciences of the United States of America* **104**, 8362–8367.

Mumby, P.J. & Steneck, R.S. 2008. Coral reef management and conservation in the light of rapidly-evolving ecological paradigms. *Trends in Ecology and Evolution* **23**, 555–563.

Mumby, P.J. & Wabnitz, C.C. 2002. Spatial patterns of aggression, territory size, and harem size in five sympatric Caribbean parrotfish species. *Environmental Biology of Fishes* **63**, 265–279.

Myers, R.F. 1989. *Micronesian Reef Fishes: A Practical Guide to the Identification of the Coral Reef Fishes of the Tropical Central and Western Pacific*. Guam: Coral Graphics.

Nash, K.L., Graham, N.A.J. & Bellwood, D.R. 2013. Fish foraging patterns, vulnerability to fishing and implications for the management of function across scales. *Ecological Applications* **23**, 1632–1644.

Nash, K.L., Graham, N.A.J., Januchowski-Hartley, F.A. & Bellwood, D.R. 2012. Influence of habitat condition and competition on foraging behavior of parrotfish. *Marine Ecology Progress Series* **457**, 113–124.

Norström, A.V., Nyström, M., Lokrantz, J. & Folke, C. 2009. Alternative states on coral reefs: beyond coral-macroalgal phase shifts. *Marine Ecology Progress Series* **376**, 295–306.

Nyström, M., Folke, C. & Moberg, F. 2000. Coral reef disturbance and resilience in a human-dominated environment. *Trends in Ecology and Evolution* **15**, 413–417.

Nyström, M., Graham, N.A.J., Lokrantz, J. & Norström, A.V. 2008. Capturing the cornerstones of coral reef resilience: linking theory to practice. *Coral Reefs* **27**, 795–809.

O'Connor, S., Ono, R. & Clakson, C. 2011. Pelagic fishing at 42,000 years before the present and the maritime skills of modern humans. *Science* **334**, 1117–1121.

Odum, E.P. 1968. Energy flow and ecological ecosystems: a historical review. *American Zoologist* **8**, 11–18.

O'Gorman, E.J., Yearsley, J.M., Crowe, T.P., Emmerson, M.C., Jacob, U. & Petchey, O.L. 2011. Loss of functionally unique species may gradually undermine ecosystems. *Proceedings of the Royal Society B* **278**, 1886–1893.

Ohta, I. & Tachihara, K. 2004. Larval development and food habits of the marbled parrotfish, *Leptoscarus vaigiensis*, associated with drifting algae. *Ichthyological Research* **51**, 63–69.
Ong, L. & Holland, K.N. 2010. Bioerosion of coral reefs by two Hawaiian parrotfishes: species, size, differences and fishery implications. *Marine Biology* **157**, 1313–1323.
Paddack, M.J., Cowen, R.K. & Sponaugle, S. 2006. Grazing pressure of herbivorous coral reef fishes on low coral-cover reefs. *Coral Reefs* **25**, 461–472.
Paddack, M.J., Reynolds, J.D., Aguilar, C., Appeldoorn, R.S., Beets, J., Burkett, E.W., Chittaro, P.M., Clarke, K., Esteves, R., Fonseca, A.C., Forrester, G.E., Friedlander, A.M., García-Sais, J., González-Sansón, G., Jordan, L.K.B., McClellan, D.B., Miller, M.W., Molloy, P.P., Mumby, P.J., Nagelkerken, I., Nemeth, M., Navas-Camacho, R., Pitt, J., Polunin, N.V.C., Catalina Reyes-Nivia, M., Robertson, D.R., Rodríguez Ramírez, A., Salas, E., Smith, S.R., Spieler, R.E., Steele, M.A., Williams, I.D., Wormald, C., Watkinson, A.R. & Côté, I.M. 2009. Recent region-wide declines in Caribbean reef fish abundance. *Current Biology* **19**, 590–595.
Pandolfi, J.M., Bradbury, R.H., Sala, E., Hughes, T.P., Bjorndal, K.A., Cooke, R.G., McArdle, D., McClenachan, L., Newman, M.J.H., Paredes, G., Warner, R.R. & Jackson, J.B.B. 2003. Global trajectories of the long-term decline of coral reef ecosystems. *Science* **15**, 955–958.
Parenti, P. & Randall, J.E. 2000. An annotated checklist of the species of the labroid fish families Labridae and Scaridae. *Ichthyological Bulletin* **68**, 1–97.
Parenti, P. & Randall, J.E. 2011. Checklist of the species of the families Labridae and Scaridae: an update. *Smithiana Bulletin* **13**, 29–44.
Penin, L., Michonneau, F., Baird, A.H., Connolly, S.R., Pratchett, M.S., Kayal, M. & Adjeroud, M. 2010. Early post-settlement mortality and the structure of coral assemblages. *Marine Ecology Progress Series* **408**, 55–64.
Penin, L., Michonneau, F., Carroll, A. & Adjeroud, M. 2011. Effects of predators and grazers exclusion on early post-settlement coral mortality. *Hydrobiologia* **663**, 259–264.
Perry, T.P., Murphy, G.N., Kench, P.S., Smithers, S.G., Edinger, E.N., Steneck, R.S. & Mumby, P.J. 2013. Caribbean-wide decline in carbonate production threatens coral growth. *Nature Communications* **4**, article 1402.
Petchey, O.L. & Gaston, K.J. 2006. Functional diversity: back to basics and looking forward. *Ecology Letters* **6**, 741–758.
Peyrot-Clausade, M., Chabanet, P., Conand, C., Fontaine, M.F., Letourneur, Y. & Harmelin-Vivien, M. 2000. Sea urchin and fish bioerosion on La Réunion and Moorea reefs. *Bulletin of Marine Science* **66**, 477–485.
Polunin, N.V.C. & Klumpp, D.W. 1989. Ecological correlates of foraging periodicity in herbivorous reef fishes of the Coral Sea. *Journal of Experimental Marine Biology and Ecology* **126**, 1–20.
Polunin, N.V.C. & Klumpp, D.W. 1992. A trophodynamic model of fish production on a windward reef tract. In *Plant-Animal Interactions in the Marine Benthos*, D.M. John et al. (eds). Systematics Association Special Volume 46. Oxford, UK: Oxford University Press, 213–233.
Pratchett, M.S. 2005. Dietary overlap among coral-feeding butterflyfishes (Chaetodontidae) at Lizard Island, northern Great Barrier Reef. *Marine Biology* **148**, 373–382.
Pratchett, M.S., Hoey, A.S., Feary, D.A., Bauman, A.G., Burt, J.A. & Riegl, B.M. 2013. Functional composition of *Chaetodon* butterflyfishes at a peripheral and extreme coral reef location, the Persian Gulf. *Marine Pollution Bulletin* **72**, 333–341.
Pratchett, M.S., Hoey, A.S., Wilson, S.K., Messmer, V. & Graham, N.A.J. 2011. Changes in biodiversity and functioning of reef fish assemblages following coral bleaching and coral loss. *Diversity* **3**, 424–452.
Pratchett, M.S., Munday, P.L., Wilson, S.K., Graham, N.A.J., Cinner, J.E., Bellwood, D.R., Jones, G.P., Polunin, N.V.C. & McClanahan, T.R. 2008. Effects of climate-induced coral bleaching on coral-reef fishes: ecological and economic consequences. *Oceanography and Marine Biology: An Annual Review* **46**, 251–296.
Price, S.A., Wainwright, P.C., Bellwood, D.R., Kazancioglu, E., Collar, D.C. & Near, T.J. 2010. Functional innovations and morphological diversification in parrotfish. *Evolution* **64**, 3057–3068.
Purcell, S.W. 2000. Association of epilithic algae with sediment distribution on a windward reef in the northern Great Barrier Reef, Australia. *Bulletin of Marine Science* **66**, 199–214.
Purcell, S.W. & Bellwood, D.R. 2001. Spatial patterns of epilithic algal detrital resources on a windward coral reef. *Coral Reefs* **20**, 117–125.

Raimondi, P.T. & Morse, A.N.C. 2000. The consequences of complex larval behavior in a coral. *Ecology* **81**, 3193–3211.

Randall, J.E. 1965. Grazing effects on seagrasses by herbivorous reef fishes in the West Indies. *Ecology* **46**, 255–260.

Randall, J.E. 1967. Food habits of reef fishes of the West Indies. *Studies in Tropical Oceanography (Miami)* **5**, 665–847.

Randall, J.E. 1974. The effect of fishes on coral reefs. In *Proceedings of the 2nd International Coral Reef Symposium* **1**, A.M. Cameron et al. (eds). Brisbane, Australia: Great Barrier Reef Committee, 159–166.

Rasher, D.B., Hoey, A.S. & Hay, M.E. 2013. Consumer diversity interacts with prey defenses to drive ecosystem function. *Ecology* **94**, 1347–1358.

Reinthal, P.N. & Macintyre, I.G. 1994. Spatial and temporal variations in grazing pressure by herbivorous fishes: Tobacco Reef, Belize. *Atoll Research Bulletin* **425**, 1–11.

Rhodes, K.L., Tupper, M.H. & Wichilmel, C.B. 2008. Characterization and management of the commercial sector of the Pohnpei coral reef fishery, Micronesia. *Coral Reefs* **27**, 443–454.

Risk, M.J., Sammarco, P.W. & Edinger, E.N. 1995. Bioerosion in *Acropora* across the continental shelf of the Great Barrier Reef. *Coral Reefs* **14**, 79–86.

Robertson, D.R. 1982. Fish feces as fish food on a Pacific coral reef. *Marine Ecology Progress Series* **7**, 253–265.

Robertson, D.R. & Gaines, S. 1986. Interference competition structures habitat use in local assemblage of coral reef surgeonfishes. *Ecology* **67**, 1372–1383.

Robertson, D.R., Karg, F., Moura, R.L., Victor, B. & Bernardi, G. 2006. Mechanisms of speciation and faunal enrichment in Atlantic parrotfishes. *Molecular Phylogenetics and Evolution* **40**, 795–807.

Rosenblatt, R.H. & Hobson, E.S. 1969. Parrotfishes (Scaridae) of the eastern Pacific, with a generic rearrangement of the Scarinae. *Copeia* **1969**, 175–176.

Rotjan, R.D. & Dimond, J.L. 2010. Discriminating causes from consequences of persistent parrotfish corallivory. *Journal of Experimental Marine Biology and Ecology* **390**, 188–195.

Rotjan, R.D., Dimond, J.L., Thornhill, D.J., Leichter, J.J., Helmuth, B., Kemp, D.W. & Lewis, S.M. 2006. Chronic parrotfish grazing impedes coral recovery after bleaching. *Coral Reefs* **25**, 361–368.

Rotjan, R.D. & Lewis, S.M. 2005. Selective predation by parrotfishes on the reef coral *Porites astreoides*. *Marine Ecology Progress Series* **305**, 193–201.

Rotjan, R.D. & Lewis, S.M. 2006. Parrotfish abundance and selective corallivory on a Belizean coral reef. *Journal of Experimental Marine Biology and Ecology* **335**, 292–301.

Rotjan, R.D. & Lewis, S.M. 2008. The impact of coral predators on tropical reefs. *Marine Ecology Progress Series* **367**, 73–91.

Rotjan, R.D. & Lewis, S.M. 2009. Predators selectively graze reproductive structures in a clonal marine organism. *Marine Biology* **156**, 569–577.

Russ, G. 1984. Distribution and abundance of herbivorous grazing fishes in the central Great Barrier Reef. II. Patterns of zonation of mid-shelf and outershelf reefs. *Marine Ecology Progress Series* **20**, 35–44.

Russ, G.R. 2003. Grazer biomass correlates more strongly with production than with biomass of algal turfs on a coral reef. *Coral Reefs* **22**, 63–67.

Sammarco, P.W. 1980. *Diadema* and its relationship to coral spat mortality: grazing, competition, and biological disturbance. *Journal of Experimental Marine Biology and Ecology* **45**, 245–272.

Sánchez, J.A., Gil, M.F., Chasqui, L.H. & Alvarado, E.M. 2004. Grazing dynamics on a Caribbean reef-building coral. *Coral Reefs* **23**, 578–583.

Sazima, C., Bonaldo, R.M., Krajewski, J.P. & Sazima, I. 2005. The Noronha wrasse: a 'jack-of-all-trades' follower. *Aqua, Journal of Ichthyology and Aquatic Biology* **9**, 97–108.

Scheffer, M., Bascompte, J., Brock, W.A., Brovkin, V., Carpenter, S.R., Dakos, V., Held, H., van Nes, E.H., Rietkerk, M. & Sugihara, G. 2009. Early-warning signals for critical transitions. *Nature* **461**, 53–59.

Scoffin, T.P., Stearn, C.W., Boucher, D., Frydl, P., Hawkins, C.M., Hunter, I.G. & MacGeachy, J.K. 1980. Calcium carbonate budget of a fringing reef on the west coast of Barbados. Part 2. Erosion, sediments and internal structure. *Bulletin of Marine Science* **30**, 475–508.

Smith, L.L., Fessler, J.L., Alfaro, M.E., Streelman, J.T. & Westneat, M.W. 2008. Phylogenetic relationships and the evolution of regulatory gene sequences in the parrotfishes. *Molecular Phylogenetics and Evolution* **49**, 136–152.

Smith, M.D. & Knapp, A.K. 2003. Dominant species maintain ecosystem function with non-random species loss. *Ecology Letters* **6**, 509–517.

Smith, S.V., Kimmener, W.J., Laws, E.A., Brock, R.E. & Walsh, T.W. 1981. Kaneohe Bay sewage diversion experiment: perspectives on ecosystem responses to nutritional perturbation. *Pacific Science* **35**, 279–395.

Steneck, R.S., Graham, M.H., Bourque, B.J., Corbett, D., Erlandson, J.M., Estes, J.A. & Tegner, M.J. 2002. Kelp forest ecosystems: biodiversity, stability, resilience and future. *Environmental Conservation* **29**, 436–459.

Stephenson, W. & Searles, R.B. 1960. Experimental studies on the ecology of intertidal environments at Heron Island. I. Exclusion of fish from beach rock. *Marine and Freshwater Research* **11**, 241–268.

Streelman, J.T., Alfaro, M., Westneat, M.W., Bellwood, D.R. & Karl, S.A. 2002. Evolutionary history of the parrotfishes: biogeography, ecomorphology and comparative diversity. *Evolution* **56**, 961–971.

Targett, T.E. & Targett, N.M. 1990. Energetics of food selection by the herbivorous parrotfish *Sparisoma radians*: roles of assimilation efficiency, gut evacuation rate, and algal secondary metabolites. *Marine Ecology Progress Series* **66**, 13–21.

Terborgh, J., Lopez, L., Nuñez, P., Rao, M., Shahabuddin, G., Orihuela, G., Riveros, M., Ascanio, R., Adler, G.H., Lambert, T.D. & Balbas, L. 2001. Ecological meltdown in predator-free forest fragments. *Science* **294**, 1923–1926.

Tilman, D. 1999. The ecological consequences of changes in biodiversity: a search for general principles. *Ecology* **80**, 1455–1474.

Tilman, D., Knops, J., Wedin, D., Reich, P., Ritchie, M. & Siemann, E. 1997. The influence of functional diversity and composition on ecosystem processes. *Science* **277**, 1300–1302.

Trapon, M.L., Pratchett, M.S., Adjeroud, M., Hoey, A.S. & Baird, A.H. 2013a. Post-settlement growth and mortality rates of juvenile scleractinian coral in Moorea, French Polynesia versus Trunk Reef, Australia. *Marine Ecology Progress Series* **488**, 157–170.

Trapon, M.L., Pratchett, M.S. & Hoey, A.S. 2013b. Spatial variation in abundance, size and orientation of juvenile corals related to the biomass of parrotfishes on the Great Barrier Reef, Australia. *PLoS ONE* **8**, e57788.

Trapon, M.L., Pratchett, M.S., Hoey, A.S., Baird, A.H. 2013c. Influence of fish grazing and sedimentation on the early post-settlement survival of the tabular coral *Acropora cytherea*. *Coral Reefs* **32**, 1051–1059.

Tuya, F., Sanchez-Jerez, P. & Haroun, R.J. 2006. Populations of inshore serranids across the Canarian Archipelago: relationships with human pressure and implications for conservation. *Biological Conservation* **128**, 13–24.

van Rooij, J.M., de Jong, E., Vaandrager, F. & Videler, J.J. 1996a. Resource and habitat shared by the spotlight parrotfish, *Sparisoma viride*, a Caribbean reef herbivore. *Environmental Biology of Fishes* **47**, 81–91.

van Rooij, J.M., Kroon, F.J. & Videler, J.J. 1996b. The social and mating system of the herbivorous reef fish *Sparisoma viride*: one-male versus multi-male groups. *Environmental Biology of Fishes* **47**, 353–378.

Vergés, A., Bennett, S. & Bellwood, D.R. 2012. Diversity among macroalgae-consuming fishes on coral reefs: a transcontinental comparison. *PLoS ONE* **7**, e45543.

Veron, J.E.N. 2000. *Corals of the World*. Townsville, Australia: Australian Institute of Marine Science.

Wainwright, P.C., Bellwood, D.R. & Westneat, M.W. 2002. Ecomorphology of locomotion in labrid fishes. *Environmental Biology of Fishes* **65**, 47–62.

Wainwright, P.C., Leo Smith, W., Price, S.A., Tang, K.L., Sparks, J.S., Ferry, L.A., Kuhn, K.L., Eytan, R.I. & Near, T.J. 2012. The evolution of pharyngognathy: a phylogenetic and functional appraisal of the pharyngeal jaw key innovation in labroid fishes and beyond. *Systematic Biology* **61**, 1001–1027.

Warner, R.R. 1984. Deferred reproduction as a response to sexual selection in a coral reef fish: a test of the life historical consequences. *Evolution* **38**, 148–162.

Webb, P.W. 2004. Response latencies to postural disturbances in three species of teleostean fishes. *The Journal of Experimental Biology* **207**, 955–961.

Welsh, J.Q. & Bellwood, D.R. 2012a. Spatial ecology of the steephead parrotfish (*Chlorurus microrhinos*): an evaluation using acoustic telemetry. *Coral Reefs* **31**, 55–65.

Welsh, J.Q. & Bellwood, D.R. 2012b. How far do schools of roving herbivores rove? A case study using *Scarus rivulatus*. *Coral Reefs* **31**, 991–1003.

Welsh, J.Q. & Bellwood, D.R. 2012c. Regional endothermy in a coral reef fish. *PLoS ONE* **7**, e33187.

Westneat, M.W. & Alfaro, M.E. 2005. Phylogenetic relationships and evolutionary history of the reef fish family Labridae. *Molecular Phylogenetics and Evolution* **36**, 370–390.

Westneat, M.W., Alfaro, M.E., Wainwright, P.C., Bellwood, D.R., Grubich, J.R., Fessler, J.L., Clements, K.D. & Smith, L.L. 2005. Local phylogenetic divergence and global evolutionary convergence of skull function in reef fishes of the family Labridae. *Proceedings of the Royal Society B* **272**, 993–1000.

Williams, L.D. & Polunin, N.V.C. 2001. Large-scale associations between macroalgal cover and grazer biomass on mid-depth reefs in the Caribbean. *Coral Reefs* **19**, 358–366.

Williams, D.M., Russ, G. & Doherty, P.J. 1986. Reef fishes: large-scale distributions, trophic interactions and life cycles. *Oceanus* **29**, 76–82.

Wilson, S.K., Bellwood, D.R., Choat, J.H. & Furnas, M.J. 2003. Detritus in the epilithic algal matrix and its use by coral reef fishes. *Oceanography and Marine Biology: An Annual Review* **41**, 279–309.

Wilson, S.K., Depczynski, M., Fisher, R., Holmes, T.H., O'Leary, R.A. & Tinkler, P. 2010. Habitat associations of juvenile fish at Ningaloo Reef, Western Australia: the importance of coral and algae. *PLoS ONE* **5**, e15185.

Wilson, S.K., Graham, N.A.J., Pratchett, M.S., Jones, G.P. & Polunin, N.V.C. 2006. Multiple disturbances and the global degradation of coral reefs: are reef fishes at risk or resilient? *Global Change Biology* **12**, 2220–2234.

Wismer, S., Hoey, A.S. & Bellwood, D.R. 2009. Cross-shelf benthic community structure on the Great Barrier Reef: relationships between macroalgal cover and herbivore biomass. *Marine Ecology Progress Series* **376**, 45–54.

Worm, B., Barbier, E.B., Beaumont, N., Duffy, J.E., Folke, C., Halpern, B.S., Jackson, J.B.C., Lotze, H.K., Micheli, F., Palumbi, S., Sala, E., Selkoe, K.A., Stachowicz, J.J. & Watson, R. 2006. Impacts of biodiversity loss on ocean ecosystem services. *Science* **314**, 787–790.

Wulff, J.L. 1997. Parrotfish predation on cryptic sponges of Caribbean coral reefs. *Marine Biology* **129**, 41–52.

Zemke-White, W.L., Choat, J.H. & Clements, K.D. 2002. A re-evaluation of the diel feeding hypothesis for marine herbivorous fishes. *Marine Biology* **141**, 571–579.

Oceanography and Marine Biology: An Annual Review, 2014, **52**, 133-200

Taylor & Francis

LIMITS TO UNDERSTANDING AND MANAGING OUTBREAKS OF CROWN-OF-THORNS STARFISH (*ACANTHASTER* SPP.)

MORGAN S. PRATCHETT[1], CIEMON F. CABALLES[1], JAIRO A. RIVERA-POSADA[1,2] & HUGH P.A. SWEATMAN[2]

[1]*ARC Centre of Excellence for Coral Reef Studies, James Cook University, Townsville, QLD 4811, Australia*
E-mail: morgan.pratchett@jcu.edu.au
[2]*Australian Institute of Marine Science, PMB 3, Townsville MC, Queensland 4810, Australia*

Outbreaks of crown-of-thorns starfish (*Acanthaster* spp.) remain a major cause of coral mortality in the Indo-Pacific, contributing to widespread and accelerating degradation of coral reef environments. This review examines the evidence for and against the principal hypotheses put forward to explain spatial and temporal patterns of outbreaks and also explores whether it is possible or feasible to intervene and limit ongoing degradation caused by crown-of-thorns starfish. The inherent biological characteristics of *Acanthaster* spp., such as exceptional fecundity, early maturation, and extreme flexibility in resource use, clearly contribute to extreme fluctuations in their abundance. Of the many hypotheses put forward to explain the occurrence of outbreaks, none has universal or unequivocal support. Clearly, however, the high incidence and severity of outbreaks at many reef locations cannot be sustained because anthropogenic changes to marine environments either have caused fundamental shifts in the population dynamics of *Acanthaster* spp. or have undermined the capacity of reef ecosystems to withstand these periodic disturbances. Reducing the incidence or severity of outbreaks of *Acanthaster* spp. is critical for reversing widespread and protracted declines in coral cover throughout the Indo-Pacific. Improved efficiency of direct controls provides opportunities to limit the progression and spread of outbreaks if detected early, but effective management of *Acanthaster* spp. really depends on definitive knowledge and appropriate action to address the ultimate causes of outbreaks. There are considerable, but not insurmountable, challenges to addressing important and persistent knowledge gaps relating to the biology of *Acanthaster* spp. This research is fundamental to ensure the persistence of coral reef ecosystems, especially given other emerging threats associated with global climate change.

Introduction

Coral reefs are increasingly regarded as one of the world's most threatened ecosystems. Not only have reef ecosystems suffered a long history of degradation (e.g., Pandolfi et al. 2005), but also climate change is expected to have a greater effect on coral reefs than almost any other ecosystem (Walther et al. 2002, Hoegh-Guldberg & Bruno 2010). Anthropogenic degradation of coral reef ecosystems began centuries ago with extensive exploitation and harvesting of large vertebrate species (Pandolfi et al. 2005). More recently, there have been sustained declines in the abundance of corals (e.g., Gardner et al. 2003, Bellwood et al. 2004, De'ath et al. 2012) and associated shifts in the biological and physical structure of benthic habitats (Hughes et al. 2010). Across the Caribbean, average coral cover declined from approximately 50% in 1977 to less than 10% in 2001, representing an average annual loss of 1.67% (Gardner et al. 2003). In the Indo-Pacific, average annual coral loss was 1.05% between 1982 and 2003 and is accelerating (Bruno & Selig 2007). Approximately 19% of the world's coral reefs have been effectively destroyed, meaning that more than 90% of coral has been lost, and there is little prospect of recovery (Wilkinson 2008). Moreover, 35% of reefs face a similar fate within the next 10–40 years unless there is effective management action to halt or reverse ongoing coral loss (Wilkinson 2008).

Scleractinian corals are fundamental to the geomorphology, biodiversity, and productivity of coral reef ecosystems (e.g., Hoegh-Guldberg 2004, Wild et al. 2004, Pratchett et al. 2008, Stella et al. 2011). Most important, corals contribute to both biological and physical habitat structure (Pratchett et al. 2008), providing essential resources (food and shelter) for many reef organisms (Jones et al. 2004, Cole et al. 2008, Stella et al. 2011); high diversity of distinct microhabitats (e.g., Messmer et al. 2011); and habitat structure that mediates important biological interactions, such as competition (e.g., Munday 2001, Holbrook & Schmitt 2002) and predation (e.g., Caley & St John 1996, Beukers & Jones 1998, Coker et al. 2014). There is considerable correlative and experimental evidence showing that reef locations with high cover and diversity of scleractinian corals support greater abundance and diversity of coral reef organisms, especially fishes (e.g., Carpenter et al. 1982, Jones 1988, Holbrook et al. 2000, Munday 2000, Jones et al. 2004, Messmer et al. 2011). Accordingly, extensive loss of live corals leads to marked declines in the abundance and diversity of coral reef fishes (e.g., Wilson et al. 2006, Pratchett et al. 2008, 2011). Effects of coral loss are even more pronounced and affect a greater diversity of reef organisms when combined with loss of structural complexity, either due to direct physical disturbances (e.g., severe tropical storms) that damage coral skeletons or the gradual decomposition and erosion of corals killed by biological disturbances (Pratchett et al. 2008).

Major causes of coral loss vary geographically. Most notably, the current status of coral reefs throughout the world is strongly reflective of the timing and extent of human colonisation (e.g., Pandolfi et al. 2005). The most degraded reef environments (East Africa, South-East Asia, and the Caribbean) are in areas with large human populations (Wilkinson 2004), which reflects the overarching effects of chronic (press) disturbances, such as overfishing, pollution, sedimentation, and eutrophication. There are also a range of acute (pulse) disturbances that strongly influence the structure and dynamics of coral reef assemblages (e.g., Gilmour et al. 2013). At the global scale, the most pronounced acute disturbances are episodes of mass bleaching, linked to increasing ocean temperatures (Hoegh-Guldberg 1999, Hughes et al. 2003); severe tropical storms (Gardner et al. 2003); or rapid and pronounced increases (variously called 'plagues', Vine 1973; 'outbreaks', Weber & Woodhead 1970; or 'infestations', Endean 1977) in the abundance of crown-of-thorns starfish and other coral predators.

In the Caribbean, high levels of coral mortality since 1977 are variously attributed to severe tropical storms (e.g., Gardner et al. 2003), increasing incidence of coral disease (e.g., Aronson & Precht 2001), or recent episodes of mass bleaching (Williams & Bunkley-Williams 2000). In reality, it is the combination of different disturbances that is responsible for sustained and ongoing declines in the cover of scleractinian corals across much of the Caribbean and associated shifts in the biological and physical structure of reef habitats. For example, phase shifts from coral- to macroalgae-dominated systems in Jamaica can be traced back to overfishing of herbivorous fishes, which had already occurred by the 1960s (Hughes 1994). However, marked changes in the biological and physical structure of reef habitats were not apparent until severe storms (e.g., Hurricane Allen in 1981, Hurricane Gilbert in 1988) and the mass mortality of the sea urchin *Diadema antillarum* in 1983. Each of these disturbances had an important and independent contribution to the resulting degradation of coral reef systems (Hughes 1994).

In the Indo-Pacific, outbreaks of crown-of-thorns starfish (*Acanthaster* spp.) have long been considered one of the major causes of coral loss (e.g., Pearson 1981, Bruno and Selig 2007, De'ath et al. 2012). The first well-documented outbreaks occurred in southern Japan in the late 1950s (Yamazato 1969 in Yamaguchi 1986) and on Australia's Great Barrier Reef (GBR) in the early 1960s (Pearson & Endean 1969). However, there are several earlier reports (as far back as the 1930s) of high densities of *Acanthaster* spp. that probably represented outbreaks (Dana 1970, Vine 1973). The severity and extent of coral loss caused by outbreaks of *Acanthaster* spp. in the 1960s and 1970s generated considerable concern about the fate of coral reefs (e.g., Cornell & Surowiecki 1972). In reviewing recovery of coral communities from different disturbances, Pearson (1981) reported that "damage caused by *Acanthaster* infestations during the last 10 to 15 years has been more extensive and dramatic than that caused by any other natural or man-made disturbance" (p. 110). Since that time, emerging threats associated with climate change, such as coral bleaching (Hoegh-Guldberg 1999) and disease (Bruno et al. 2007), have become the major focus of coral reef science and management (Hughes et al. 2003). However, outbreaks of crown-of-thorns starfish continue to occur throughout the Indo-Pacific (e.g., Pratchett et al. 2011, De'ath et al. 2012, Baird et al. 2013), and at many locations the effects of severe outbreaks have been far greater than combined effects of all other major disturbances, including climate-induced coral bleaching (e.g., Trapon et al. 2011, De'ath et al. 2012). Importantly, it is the combined effect of outbreaks of *Acanthaster planci* and other diverse disturbances (e.g., sedimentation, cyclones, and bleaching) that have caused sustained and accelerating degradation of coral reef ecosystems throughout the Indo-Pacific (e.g., Jones et al. 2004, Pratchett et al. 2006).

The purpose of this review is to synthesise established knowledge and insights on the causes and consequences of outbreaks of crown-of-thorns starfish, focusing on the increased understanding and research undertaken in the last two decades (mostly since 1990), given that there were several substantive reviews of the research on the biology and ecology of *Acanthaster* spp. between 1960 and 1990 (e.g., Potts 1981, Moran 1986, Birkeland & Lucas 1990, Endean & Cameron 1990). Since 1990, there have not been any major comprehensive reviews of the biology and population dynamics of *Acanthaster* spp., but there have been numerous commentaries on specific issues (e.g., Brodie 1992, Brodie et al. 2005) related to causes or consequences of population outbreaks. In the last few years, there has also been renewed interest in outbreaks of *Acanthaster* spp., largely attributable to fresh outbreaks of crown-of-thorns starfish at many locations throughout the Indo-Pacific (e.g., Kayal et al. 2012, Baird et al. 2013). Moreover, there is an increasing realisation that urgent action is needed to reverse sustained and ongoing declines in live coral cover that are occurring throughout the world (e.g., Gardner et al. 2003, Bellwood et al. 2004, De'ath et al. 2012), and of all the factors that are contributing to degradation of coral reef ecosystems in the Indo-Pacific, outbreaks of *Acanthaster* spp. are considered to be the most amenable to direct and immediate intervention (cf. climate-induced coral bleaching, increasing prevalence of coral disease, increasing severity of tropical storms). Controlling outbreak populations of *Acanthaster* spp. is considered one of the most promising strategies to halt or reverse widespread declines in live coral cover (e.g., De'ath et al. 2012) and thereby improve the capacity of reef systems to cope with inevitable threats caused by sustained and ongoing climate change as well as other more direct anthropogenic disturbances. A second purpose of this review is to highlight gaps in our knowledge of the biology of *Acanthaster* spp. that persist, despite five decades of research, and constrain both understanding of the causes and effective management of outbreaks.

Biology of *Acanthaster* spp.

The crown-of-thorns starfish, specifically *Acanthaster planci* (Linnaeus 1758), was first described based on the original description by Plancus and Gualtieri (Vine 1973). Crown-of-thorns starfish have since been reported on coral reefs throughout the tropical Indo-Pacific from the Red Sea (e.g., Goreau 1964) to Panama (e.g., Glynn 1973) but have never been recorded in the Caribbean or Atlantic Ocean. Crown-of-thorns starfish are also found in a wide range of latitudes, from 34°N on subtropical reefs in the Ryukyu Islands, Japan (Yamaguchi 1986), to 32°S at Lord Howe Island (DeVantier & Deacon 1990). Marked geographic differences in appearance and allelic frequencies within the broad geographic range suggest that there might be at least two species of crown-of-thorns starfish (Benzie 1999), distributed within the Indian and Pacific oceans, respectively. Recent molecular sampling (632 bp from the COI region) of crown-of-thorns starfish (nominally, *Acanthaster planci*) from throughout the Indo-Pacific revealed that there are four strongly differentiated clades from distinct regions of the Indo-Pacific that probably represent distinct species: (1) Red Sea, (2) southern Indian Ocean, (3) northern Indian Ocean, and (4) Pacific (Vogler et al. 2008). However, Vogler et al. (2013) found no genetic differentiation between the crown-of-thorns starfish in the far eastern Pacific (sometimes considered to be a distinct species, *Acanthaster ellisii*; e.g., Barham et al. 1973, but see Glynn 1974) from the remainder of the Pacific. For this reason, *Acanthaster* spp. is used when referring to the entire species complex, whereas *A. planci* is reserved for use when referring explicitly to crown-of-thorns starfish from the Pacific. Another well-described species, *A. brevispinus* Fisher 1917, is known from deep-water habitats in the western Pacific, but this species is rarely found in coral reef habitats (Birkeland & Lucas 1990) and is not considered in this review.

Moran (1986) described crown-of-thorns starfish as "one of the most well-known animals in coral reef ecosystems" (p. 1) and went on to say that the biology of this animal has been particularly well studied. It is true that there was considerable research on the basic biology of *Acanthaster* spp. in the 1970s and 1980s, including studies on reproductive biology (e.g., Lucas 1973); diet (e.g., Brauer et al. 1970, Branham et al. 1971, Ormond et al. 1973, Glynn 1974); and behaviour (e.g., Barnes et al. 1970, Moran et al. 1985). However, crown-of-thorns starfish remain something of an enigma, with relatively little known about their demography and population dynamics. Moore (1990) suggested that intermittent outbreaks might be attributable to fundamental switches in the inherent life-history characteristics between endemic and epidemic characters, but this has never been explicitly tested.

Most of what is known about the reproductive biology and life cycle of crown-of-thorns starfish (Figure 1) comes from detailed studies of *Acanthaster planci* in the western Pacific (e.g., Lucas 1973, Yamaguchi 1973, Nishihira & Yamazato 1974, Conand 1984, Babcock & Mundy 1992a,b). While the reproductive biology and life history are likely to be broadly similar for other *Acanthaster* spp. from the Indian Ocean, this needs to be verified as geographical (and taxonomic) differences in their biology might account for marked geographical differences in the incidence and severity of outbreaks (as discussed further in this review). Comprehensive and detailed information about well-studied aspects of the biology of *A. planci* was provided by Moran (1986). Rather than repeat this information, this review limits discussion to advances in the biological knowledge since 1986 and also considers key aspects of the biology that are fundamental in understanding the proximal causes of outbreaks, including inherent constraints of reproductive success, and the structure (age and size) of normal versus outbreak populations. There is widespread recognition that the biology (especially the reproduction and early life history) of crown-of-thorns starfish is key to understanding when and why outbreaks occur (e.g., Birkeland 1982, 1989a).

Fecundity

One of the most important biological traits of *Acanthaster* spp., which is particularly relevant to major population fluctuations, is their enormous reproductive potential (Endean 1982, Conand 1984). Large female starfish can produce up to 65 million eggs per season (Conand 1984, Kettle & Lucas 1987). It has long been recognised that *Acanthaster* spp. release millions of eggs each time they spawn (e.g., Pearson & Endean 1969), but Conand (1983, 1984) provided the first quantitative analysis of size-based fecundity for *Acanthaster planci*, following detailed studies in Noumea. These data correspond closely with similar research undertaken by Kettle and Lucas (1987), who also found disproportionate increases in fecundity with increasing size, ranging from 0.5–2.5 million

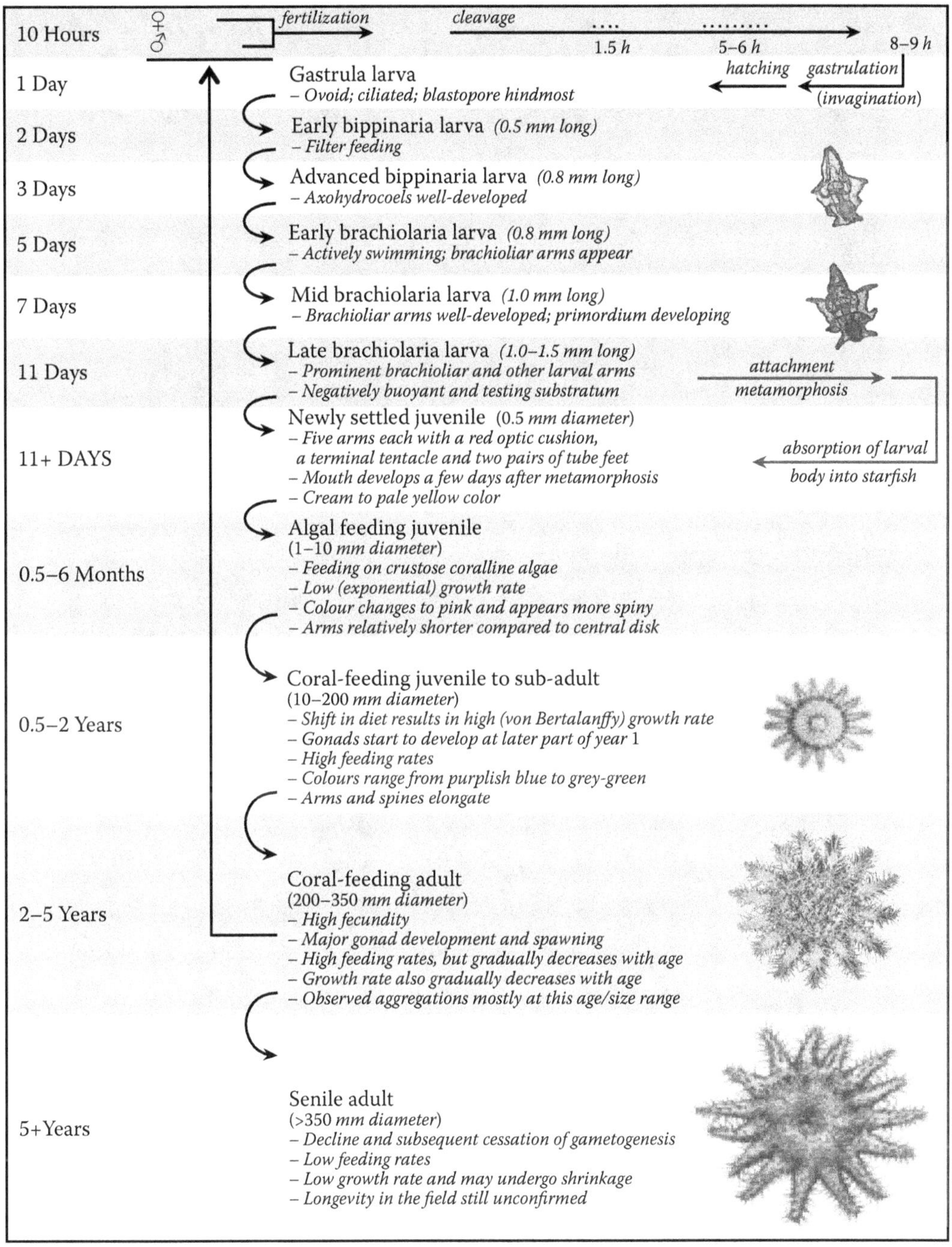

Figure 1 Complete life cycle of *Acanthaster planci* adapted from laboratory rearing studies by Yamaguchi (1973) and Lucas (1984) and compilations by Moran (1986) and Birkeland & Lucas (1990).

eggs per year for individuals less than 30 cm to 46–65 million eggs per year for starfish that have a 40-cm diameter (Birkeland & Lucas 1990).

Initiation of gametogenesis is clearly related to both age and size of *Acanthaster* spp. Gonad development in laboratory-reared starfish (Yamaguchi 1973, Lucas 1984) and in a field population in Fiji (Zann et al. 1987) started before the animals reached the age of 2 years, and the largest

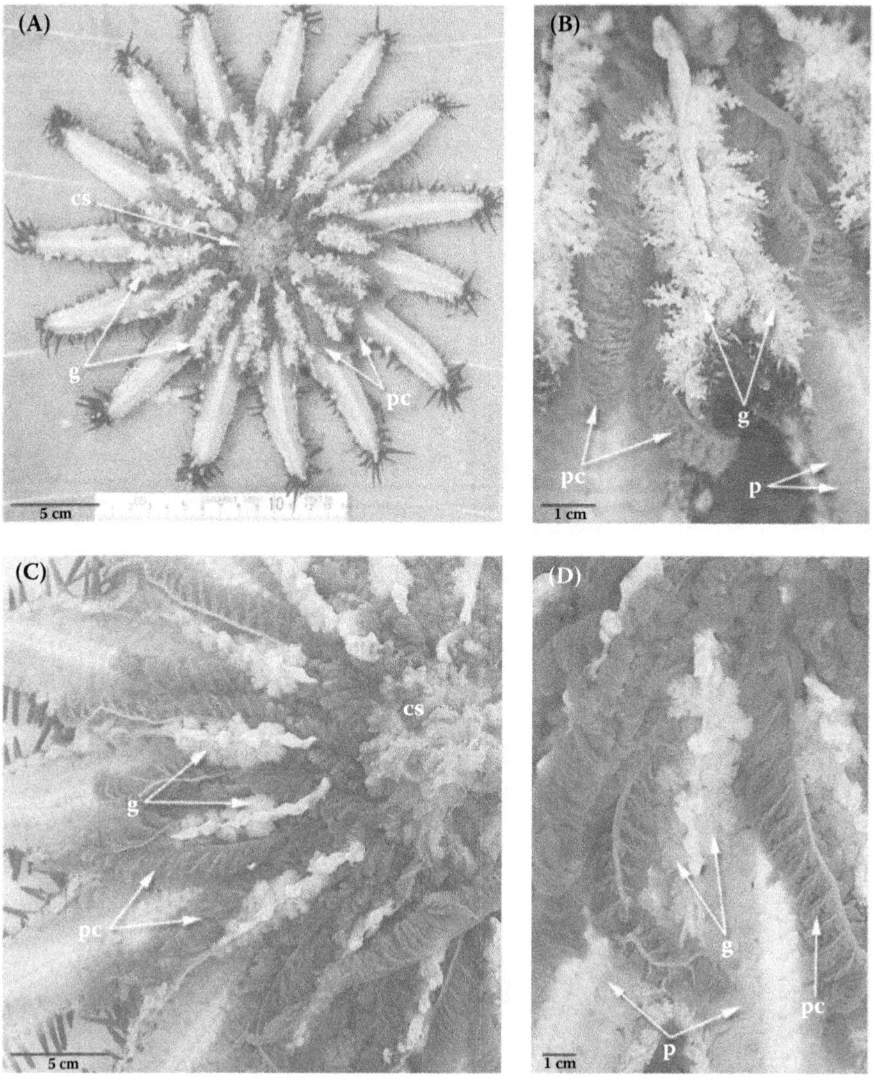

Figure 2 (See also colour figure in the insert) Aboral view of *Acanthaster planci* showing internal arrangement of gonads (g), pyloric caeca (pc), podia (p), and cardiac stomach (cs) in (A) and (B) male and (C) and (D) female specimens. (Photographs taken by C.F. Caballes.)

individuals in any given cohort were also the first to exhibit gametogenesis (Lucas 1984, Zann et al. 1987). Gonads appear as aciniform rows along each side of the inner wall of the proximal part of each arm (Figure 2). When arms are dissected to expose internal digestive and reproductive organs, male and female starfish are readily distinguishable as testes are cream or pale yellow in colour and have smaller, more numerous lobes (Figure 2A, 2B) compared to ovaries, which appear as larger, spherical, yellow (sometimes almost orange) lobes (Figure 2C, 2D). It is also apparent that the size (weight or volume) and maturation stage of gonads are consistent among arms, such that gonad somatic indices are generally based on subsampling of only one to three arms (Lucas 1973, Conand 1984, Ogura et al. 1985, but see Yokochi & Ogura 1987). Changes in the size of gonads parallels changes in gonad index and size of oocytes (Yamazato & Kiyan 1973) and swelling in the proximal region of arms.

Larger starfish increasingly partition energy towards reproduction (ova production) at the expense of the body wall and pyloric caeca (Kettle & Lucas 1987). Gonad indices in female *Acanthaster planci* are usually higher compared to males, and this disparity becomes more pronounced at the peak of the breeding season (Cheney 1974, Conand 1984, Yokochi & Ogura 1987, Babcock & Mundy 1992a). Spermatogenic development in testes is mainly reflected in the thickness of the germinal layer (Yamazato & Kiyan 1973). In females, the stage of gametogenic cycle can be assessed by looking at the size of oocytes, the presence or absence of layers of connective tissue that bind oocytes together, and ovulation (Babcock & Mundy 1992a).

While the gametogenic cycle of asteroids may be regulated by endogenous (intrinsic) factors such as age, size, and nutritional status or by exogenous (extrinsic) factors such as temperature, photoperiod, and food availability (reviewed in Mercier & Hamel 2009), there are few studies of the role of nutritional status on gametogenesis and fecundity in *Acanthaster* spp. Cheney (1974) found that starving *Acanthaster planci* (by placing them in a cage for 1 month without food) resulted in reabsorption of gonads, atrophy of the pyloric caeca, and a decrease in the overall size (diameter and weight) of individual starfish. Other environmental factors (e.g., extreme temperatures, reduced salinity, and limited food availability) may also exert exogenous control on gametogenesis and fecundity, but this has not been investigated. For the large part, scientists are still coming to terms with spatial and temporal variation in occurrence of spawning, let alone understanding variation in the fecundity of individual starfish and relating this to local environmental conditions.

Spawning

Like most asteroids, *Acanthaster planci* is a gonochoristic species, whereby male and female individuals must be in close proximity and spawn simultaneously to effectively reproduce (Babcock et al. 1994). This is important because reproductive success may be greatly constrained when there is a highly biased sex ratio (e.g., Stump 1994) or if densities of starfish are low and the distance between individuals is large (Vine 1973). Initial studies on the sex ratio of *A. planci*, based on sampling of outbreak populations (e.g., Pearson & Endean 1969, Nishihira & Yamazato 1974), suggested that there are generally an equal number of males and females. However, strongly male-biased sex ratios have been recorded in several populations (e.g., Stump 1994, C. Caballes et al. unpublished data, 2013). In September 2011, 93 large starfish (>30-cm diameter) were sampled in Guam (Hospital Point), of which only 12 were female (C. Caballes et al. unpublished data, 2013). Similarly, at Lizard Island (northern GBR) in March 2013, there were 115 starfish collected across a range of different locations (with starfish ranging in size from 15 to 47 cm), of which only 2 individuals were female (C. Caballes et al. unpublished data, 2013). Of the remaining individuals, 76 were male, and 37 were immature or non-reproductive. Both these populations were sampled in the aftermath of peak outbreak densities, and the strong male bias may reflect generally lower survival of females, which invest much more energy in reproduction (Stump 1994). However, if there is strong male sex bias in low-density populations generally, then this has the potential to greatly limit reproductive success.

Broadcast spawners (e.g., fishes, corals, and many marine invertebrates) release copious quantities of gametes during spawning but typically achieve low fertilization rates unless (1) individuals are highly aggregated, (2) spawning is synchronised, and (3) spawning occurs in low-to-moderate flow conditions (e.g., Mercier & Hamel 2009). In Okinawa, Okaji (1991) found that dispersed populations spawned later and over a much longer period compared with aggregated populations. Okaji (1991) suggested spawning is also more synchronised within aggregated populations, which results in much higher reproductive success. Moreover, Cheney (1974) found that aggregated individuals had consistently higher gonado-somatic indices (indicative of fecundity and reproductive potential) compared with those from dispersed populations. More important, fertilization rates for *Acanthaster* spp. and other free-spawning marine invertebrates decline precipitously with increasing distance

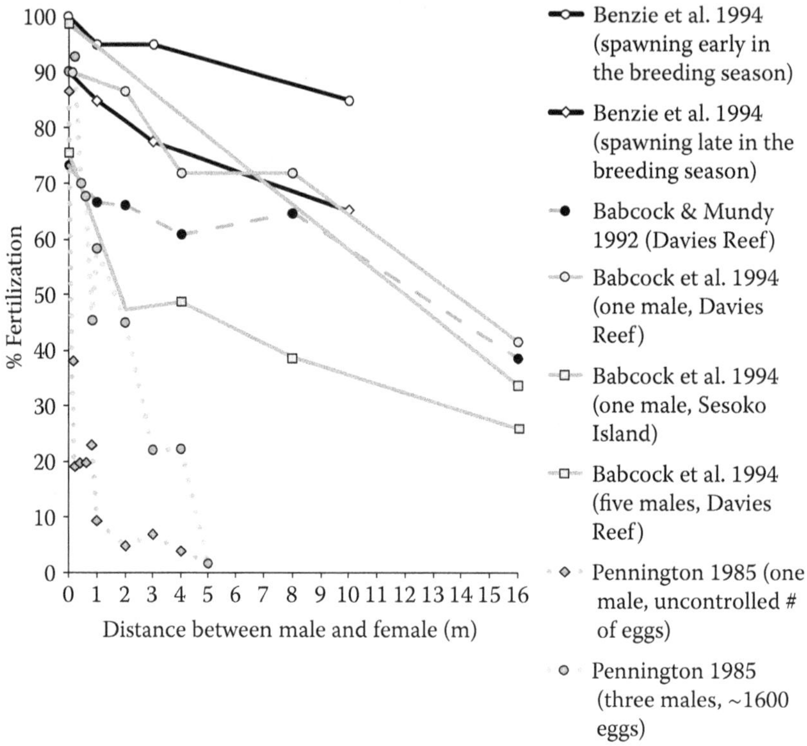

Figure 3 Fertilization rates for female *Acanthaster planci* (solid and dashed lines) versus *Strongylocentrotus droebachiensis* (dotted lines) at varying distances downstream from spawning males.

between male and female individuals (Levitan et al. 1992). For *Acanthaster* spp., fertilization success is close to 100% when male starfish spawn adjacent to spawning females (Benzie et al. 1994) but declines with increasing distance between spawning individuals (Figure 3). However, fertilization rates recorded when males and females are separated by relatively large distances (>20 m) are significantly greater than those of many other marine invertebrates (Yund 1990, Grosberg 1991, Levitan et al. 1991; Figure 3). This disparity could be caused by the greater size and fecundity of *Acanthaster* spp. and the large quantity of gametes released during spawning (Babcock & Mundy 1992a, Babcock et al. 1994) or greater capacity for fertilization at low sperm concentrations (Benzie & Dixon 1994). Even so, minimising the distance between spawning individuals by aggregating will lead to a marked increase in reproductive success, suggesting that chance aggregation of adults on reefs with very low overall densities of *Acanthaster* spp. may be sufficient to precipitate an outbreak (e.g., Vine 1973).

During spawning, gametes are shed from aboral rows of gonopores along the sides of each arm. Exudates from spawning females appear as translucent spherical grains, while males exude milky clouds of sperm. Despite the conspicuousness of spawning starfish, there have been relatively few observations of natural spawning in the field (Table 1). It is still unclear whether *Acanthaster* spp. spawns just once each year or whether there are multiple spawning events concentrated within a particular spawning period. At Lodestone Reef in the GBR, Lucas (1973) showed that periods of most active gametogenesis corresponded to periods of increasing temperature and marked changes in the photoperiod. As for many other marine invertebrates (e.g., corals; Baird et al. 2009), temperature appears to be the most important cue for seasonality in spawning. In warmer locations, there is a tendency for starfish to reach maximum gonad maturity and begin spawning whenever

Table 1 Observations of spontaneous natural spawning in the field

Region	Location	Date	Time of day	Depth	Extent of spawning	Reference
Japan	Kii Peninsula	17 Jul 1973	Evening	—	—	Hayashi 1975
	Kii Peninsula	15 Aug 1973	Evening	—	—	Hayashi 1975
	Iriomote Island	9 Jun 1984	Afternoon	4–8 m	3 males and 1 female spawned	Yokochi 1985
	Okinawa	9 Jun 1988	Daytime	15 m	Several individuals spawned	Kishimoto, 1989
	Okinawa	4–5 Jul 1988	1500	—	—	Cited in Babcock & Mundy 1992a
	Okinawa	13 Jun 1990	1500	—	3 male individuals spawned	Cited in Babcock & Mundy 1992a
	Okinawa	17 Jun 1990	1425–1645	—	6 males and 1 female spawned	Cited in Babcock & Mundy 1992a
	Okinawa	19 Jun 1990	Afternoon	—	Single male	Cited in Babcock & Mundy 1992a
Red Sea	N Gulf of Aqaba	24 Jul 2004	1823, 1859	10 m, 21 m	Two individuals spawned (water temperature 24.5°C)	D. Zakai personal observation 2004
Hawaii	Kalohi Channel	23–24 Apr 1970	—	—	Several starfish spawned	Branham et al. 1971
Guam	Tanguisson Reef	21 Apr 2006	1500	5 m	Single female (~35-cm diameter) in a massive aggregation spawned for 30 min	C. Caballes personal observation 2006
Maldives	North Malé Atoll	Apr 1991	—	—	—	Ciarapica & Passeri 1993
Papua New Guinea	Yule Island	4 Jul 1989	1630		50–100 individuals spawned	Cited in Babcock & Mundy 1992a
Central GBR, Australia	Arlington Reef	10 Jan 1968	1330	3 m	Dense aggregation; spawning lasted 30 min and involved several males and only one female spawned	Pearson and Endean 1969
	John Brewer Reef	20 Jan 1983	1500	—	Single female	Cited in Babcock & Mundy 1992a
	Lizard Island	Feb 1983	0900	—	20–30 males; no females	Gladstone 1987
	Rib Reef	21 Jan 1984	1615	—	2–3 males spawned on top of boulder	Cited in Babcock & Mundy 1992a
	Rib Reef	13 Dec 1984	1600	4 m	More than 50 clustered individuals	Cited in Birkeland & Lucas 1990
	Wheeler Reef	5 Jan 1987	1520–1800	3–8 m	40–100 starfish spread over a 2400-m^2 area	Cited in Birkeland & Lucas 1990

Table 1 (Continued) Observations of spontaneous natural spawning in the field

Region	Location	Date	Time of day	Depth	Extent of spawning	Reference
	Wheeler Reef	5 Jan 1987	1600	—	Single female (adjacent starfish did not spawn)	Cited in Babcock & Mundy 1992a
	Bowden Reef	17 Jan 1988	—	—	3–4 starfish	Cited in Babcock & Mundy 1992a
	Bowden Reef	18 Jan 1988	1430	—	1 starfish	Cited in Babcock & Mundy 1992a
	Hayman Island	5 Dec 1990	1534–1706	2.5 m	10 males; single female was last to begin spawning and spawned for the shortest time	Gladstone 1992
	Davies Reef	7 Dec 1990	2145	≤7 m	68% of 129 starfish (38 females and 50 males) spawned over a 2-h period	Babcock & Mundy 1992a
	Davies Reef	17 Dec 1990	1700	≤7 m	3 male individuals spawned	Babcock & Mundy 1992a
	Davies Reef	11 Dec 1991	1630	1–4 m	Single female initially spawned, followed by > 50 starfish (mostly males)	Babcock & Mundy 1992b
	Davies Reef	12 Dec 1991	1930	1–4 m	8 males and 1 female spawned	Babcock & Mundy 1992b
	Davies Reef	13 Dec 1991	2030	1–4 m	Single male spawned	Babcock & Mundy 1992b
	Davies Reef	23 Jan 1992	2030	1–4 m	2 males released gametes through gonopores from only a few arms	Babcock & Mundy 1992b
	South/ Palfrey Island	18 Dec 2009	1630	1 m	12 males released copious amounts of sperm; most were exposed, and individuals were well spaced from each other (3–10 m apart)	L. Vail & A. Hoggett personal observation 2009
Fiji	Muaivuso	1 Feb 1970	1400–1600	0.5–4 m	Dense clusters; many individuals spawned	Owens 1971

sea-surface temperatures exceed 27°C (Figure 4). At higher latitudes, where temperatures never reach this threshold (e.g., Hawaii, New Caledonia), gametogenesis and spawning are often concentrated in months when the temperature first starts to rise (Figure 4). Breeding and spawning seasons of *Acanthaster* spp. at higher latitudes are mostly shorter, well-defined events compared with more protracted gametogenesis and spawning at lower latitudes (e.g., Guam, Palau) (Table 2). Mature gonads have been reported year round in *A. planci* from Guam (e.g., Cheney 1972) but tend to be found in only a few months of each year at most other locations (Table 2).

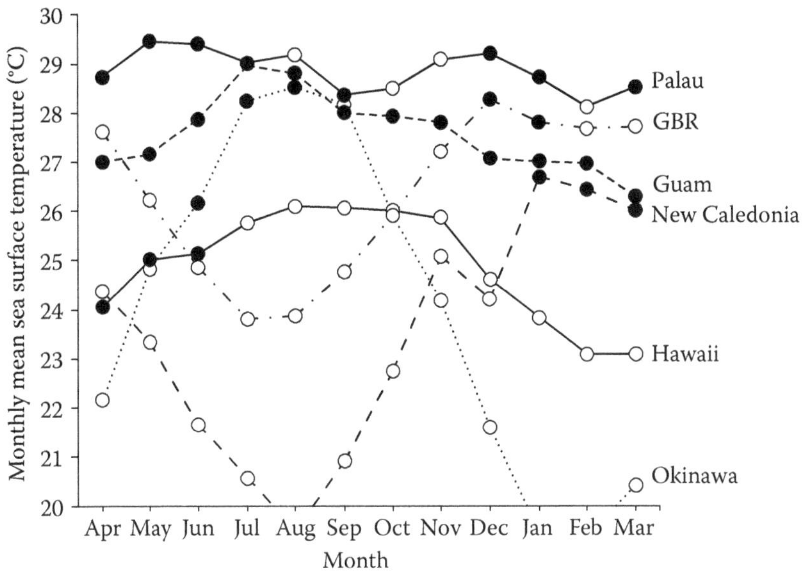

Figure 4 Seasonal variation in sea-surface temperatures and spawning times for *Acanthaster planci* at different locations throughout the Pacific. Spawning (filled circles) tends to occur only when temperatures are above 27°C, although at locations where temperatures never reach this level (e.g., Hawaii), spawning is restricted to months when temperatures first start to rise. Data sources: Palau, Idip 2003; Lizard Island, GBR, L. Vail unpublished data; Guam, Cheney 1974; New Caledonia, Conand 1984; Hawaii, Cheney 1974; and Okinawa, Japan, Yamazato & Kiyan 1973.

Compilation of the limited observations of natural spawning of *Acanthaster* spp. (Table 1) shows that spawning occurs mostly in the late afternoon, although it has occasionally been observed in the early morning (Gladstone 1987, Kishimoto 1989) and after dark (Babcock & Mundy 1992b). There is no apparent link between the timing of spawning and lunar phases or tidal cycles (Babcock & Mundy 1992a), suggesting that spawning is not linked to environmental cues. It is known that spawning by one individual will often instigate spawning by other individuals in the local proximity (e.g., Babcock & Mundy 1992a) or within aquaria. However, this does not explain the synchronous behavioural changes observed in many aggregations of *Acanthaster* spp., whereby individuals become particularly active, move to shallow promontories, and adopt the characteristic arched posture prior to the release of gametes (Babcock & Munday 1992a). Laboratory experiments by Beach et al. (1975) revealed that pheromones extracted from *Acanthaster planci* ovaries and testes synchronise spawning among neighbouring animals and also induce movement towards spawning individuals. Beach et al. (1975) argued that gametogenic cycles are influenced by local environmental conditions (mainly temperature), but chemically mediated communication between ripe individuals is necessary to coordinate gamete release. Consistent with this hypothesis, spawning by *Acanthaster* spp. is often synchronised at localised scales, within populations, but not among populations (Babcock & Mundy 1992a, Yasuda et al. 2010). Babcock & Mundy (1992a) saw *A. planci* spawning in one area of Davies Reef on the GBR, while *A. planci* in other parts of the same reef were not.

Spawning by *Acanthaster* spp. is concentrated in summer months at most locations, but the evidence is contradictory regarding whether individual starfish spawn once annually (e.g., Babcock & Mundy 1992a,b) or spawn multiple times (e.g., batch spawn) each year (e.g., Conand 1983). In Noumea, Conand (1983) found that all female starfish examined contained oocytes at vastly different stages of maturity, suggesting that individual *Acanthaster planci* spawn sequentially over several months. However, in the central GBR Babcock and Mundy (1992a) suggested that *A. planci*

Table 2 Peak seasons in annual reproductive cycle of *Acanthaster planci* from different locations based on gonad development and spawning

Latitude	Location	Month (April to March) 04	05	06	07	08	09	10	11	12	01	02	03	Reference
33°N	SW Honshu[a]				●	●								Hayashi 1975
28°N	Amami-Ohshima[b]		○	○	●	○								Yasuda et al. 2010
26.5°N	Okinawa[c,d,e]	○	○	○	●	●	○	▢		○		○	○	Yamazato & Kiyan 1973
	Okinawa[c,d]	○	○	●	●	○	○	○	○	○	○	○	○	Okaji 1989
	Okinawa[b]		○	○	●	○	○							Yasuda et al. 2010
26°N	Kerama Islands[b]		○	●	●		○							Yasuda et al. 2010
25°N	Gulf of California[d]	●												Dana & Wolfson 1970
24.5°N	Miyako Island[b]		○	●	○			○						Yasuda et al. 2010
24°N	Iriomote[c]	○	○	●	○	○	○	○	○	○			○	Yokochi & Ogura 1987
	Iriomote[c]	○	●	●	○	○	○	○	○	○			○	Habe et al. 1989
24°N	Sekisei Lagoon[b]		●	○	○	○								Yasuda et al. 2010
20–21°N	Red Sea[c]					●	●							Crump 1971
	Red Sea[c]					●	●							Moore 1985
21°N	Hawaii[a,b]	●	◌	●		○			○		○			Branham et al. 1971
13°N	Guam[f]								●	●				Chesher 1969
	Guam[b]	▢	▢	▢	▢	▢	●	●	●	▢	▢	▢	▢	Cheney 1974
9°N	Panama[f]										●			Glynn 1974

7°N	Palau[c]	●	●	●	●	○	●	○	○	◍	◍	○	●	Idip 2003
7°N	Philippines[c]	●	●		○		○		○		○		○	Bos et al. 2013
5–15°N	Micronesia[d]					●	●							Eldredge 1970
	Micronesia[c]	◍	◍	◍	◍	◍	◍	◍	◍	◍	◍	◍	◍	Cheney 1972
4°N	Maldives[a]	●											●	Ciarapica & Passeri 1993
9°S	Solomon Islands[d]												●	Eldredge 1970
14°S	W Samoa[d]									●	●			Garlovsky & Bergquist 1970
17–19°S	Central GBR[a,c]					○	○	○	○	●	●	○	○	Pearson & Endean 1969
	Central GBR[c,e]	○		○		○	○	○	○	●	●			Lucas 1973
	Central GBR [a,c,d]							○	○	●	●	●		Babcock & Mundy 1992a
	Central GBR[a,c]					○	○	○	●	●	●	○		Babcock & Mundy 1992b
	Central GBR[a]									●				Gladstone 1992
18°S	Fiji[a]											●		Owens 1971
19–23°S	SE Polynesia[d]									●	◍	◌	◍	Devaney & Randall 1973
20°S	NW Australia[b]		○					○	●	●	●	●		Wilson & Marsh 1974
22°S	New Caledonia [b,c]	○	○		○			○	●	●	●	●	○	Conand 1984
27°S	South Africa[d]											●	●	Schleyer 1998
32°S	Lord Howe Island[d]												●	DeVantier & Andrews 1987

Note: ○, months surveyed/sampled; ●, spawning/breeding season; ◍, mature gonads but only partial or no spawning; ◌, not surveyed/sampled but spawning assumed.

Method used to describe annual reproductive cycle of *Acanthaster planci*: [a]field observation of spawning; [b]changes in proportion of ripe and spent gonads; [c]changes in gonad index; [d]gonad condition or histology of ovaries and testes; [e]changes in oocyte size and frequency; [f]no method given.

spawn just once, mainly in early December. Babcock and Mundy (1992a,b) recorded changes in the size and density of oocytes in *A. planci* at Davies Reef over much of the year, but concentrated sampling over the austral summer (October to February). They saw a marked drop in the proportion of females that were gravid, as well as marked reductions in mean size of gonads, in early December. Accordingly, spawning was observed in the evening of 7 December 1990 at 2145, with 88 (of 129) starfish seen spawning along a single, shallow (1- to 4-m) transect. Elsewhere on the GBR, spawning has been reported from December and January to February (Table 2), but it is unclear whether this reflects geographic variation in the timing of annual spawning or whether *A. planci* on the GBR can spawn multiple times in a year. It is clear that starfish at some locations (e.g., Guam) are gravid year round (Cheney 1974), and reproductive behaviour is likely to vary geographically in response to environmental regimes and food availability. More detailed studies of the gametogenic cycle, maturation, and spawning of *Acanthaster* spp. are needed, combined with systematic sampling of individually tagged starfish to establish spawning behaviour of individuals as well as populations. This information is critical for potentially linking the precise timing of spawning behaviour and associated reproductive success to spatial and temporal anomalies in local conditions, such as nutrient dynamics (e.g., Fabricius et al. 2010).

Larval development

The life cycle of *Acanthaster* spp. is typical of most asteroids, with larval development divided into two distinct bipinnaria stages (pelagic feeding larva characterised by bilateral arrangement of the pre- and post-oral ciliated swimming and feeding bands) and three brachiolaria stages (feeding larva characterised by the presence of brachiolar arms and attachment disk on the preoral lobe) prior to metamorphosis and settlement (Figure 1). Fertilized embryos develop to the blastula stage after 8–9 h, then hatch after approximately 1 day as free-swimming gastrula larvae (Lucas 1982). After 2–4 days, larvae have a completely formed alimentary canal and start filter-feeding on unicellular algae and other suspended particulate matter (Yamaguchi 1973). Larvae then proceed to the brachiolaria stage, developing brachiolar arms, which will eventually be used to locate favourable substrate prior to settlement (Henderson & Lucas 1971). The typical planktonic larval duration (PLD) for *Acanthaster planci* is 11 days (Figure 1), although, like many marine invertebrates (Richmond 1987, Graham et al. 2008), these starfish can delay settlement and significantly extend their PLD (Yamaguchi 1973). The rate of development also varies with temperature (Henderson & Lucas 1971) and food availability (Lucas 1982), such that the PLD can range from 9 to 42 days.

Factors that limit the rate and success of larval development of *Acanthaster* spp. (mainly, *Acanthaster planci*) have received a great deal of attention in the past years, as this was considered key to understanding the initiation of population outbreaks (e.g., Lucas 1982, Olson & Olson 1989). Much of this research has concentrated on larval nutrition and the extent to which larval survivorship is constrained by abundance of phytoplankton, which in turn depend on the levels of nutrients. Phytoplankton are generally considered to be the main food source for larval *Acanthaster* spp., as both natural phytoplankton and cultured unicellular algae (single or mixed species) have been successfully used to rear larvae under laboratory conditions (Henderson & Lucas 1971, Lucas 1975, 1982, Uchida & Nomura 1987, Keesing et al. 1996, Okaji 1996, Fabricius et al. 2010). However, Lucas (1982) suggested that the amount of phytoplankton required to maintain cultured larvae was much higher than generally occurs in near-reef waters (e.g., within the GBR lagoon), leading to suggestions that larvae are severely food limited except during major phytoplankton blooms ('larval starvation hypothesis'; Lucas 1982). Similarly, Fabricius et al. (2010) reported minimal survival of larval *Acanthaster planci* at chlorophyll concentrations below 0.25 $\mu g.l^{-1}$, whereas larval survival increased approximately 8-fold with each doubling of chlorophyll concentrations up to 3.0 $\mu g.l^{-1}$. These experiments were designed to answer specific questions about the larval nutrition of *A. planci* and were limited in their representation of model ecosystems. For instance, any potential predators on larval crown-of-thorns starfish (≥100 μm length) were explicitly excluded from the experimental system by filtering (25-μm filters) incoming seawater. Even so, these data are used to argue that outbreaks of *A. planci* may arise from pulses of recruits produced by temporary increases in productivity linked to high rainfall events and associated floods (Fabricius et al. 2010). It is not clear whether predators or competitors would alter the survivorship of larval crown-of-thorns starfish sufficiently to affect the relationship with nutrient concentrations, but these experiments need to be repeated and extended before any conclusions about limitations to larval survivorship can be made. It is also controversial whether *Acanthaster* spp. are food limited, as several studies have successfully reared larvae at normal (low) chlorophyll concentrations (Olson 1987, Johnson et al. 1991) though it is increasingly clear that food limitation may constrain rates of development, if not survival.

Olson & Olson (1989) suggested that larval *Acanthaster planci* are capable of exploiting a diversity of different sources of nutrition. Under normal conditions of low plankton abundance, high levels of survivorship were facilitated by the ability of *A. planci* to utilise dissolved organic matter (DOM) and bacteria (Olson & Olson 1989). Importantly, larvae reared in *in situ* culture chambers showed no sign of food limitation and were able to develop at near-maximal rates despite low levels of plankton (Olson 1987). Moreover, nutrient enrichment did not result in increased survivorship but did result in a slight increase in the rate of development (Olson 1987). Accordingly, Hoegh-Guldberg (1994) showed that dissolved free amino acids (DFAAs) can supply significant amounts of energy for developing larvae. Microscopic analyses by Ayukai (1994) showed that *A. planci* larvae generally consume only large phytoplankton, as there was limited evidence of ultraplankton (<5 μm) or bacteria within stomach contents. However, *A. planci* may rely on alternative sources of prey (including ultraplankton and bacteria) when preferred prey (large phytoplankton) are in limited abundance. Given increased knowledge of patterns of resource use, food limitation experiments with developing larvae should be repeated to test explicitly whether high concentrations of DFAA, free-living bacteria, or ultraplankton may compensate for the limited abundance of phytoplankton under field conditions. Experimental work needs to be corroborated with field assessment of the nutritional condition of crown-of-thorns larvae before, during, and after major flood events to specifically relate this to local phytoplankton concentrations. In the past, it was difficult to distinguish the larvae of *Acanthaster* spp. from those of other echinoderms, but Roper (1997) developed a technique to stain crown-of-thorns larvae with fluorescently labelled monoclonal antibodies, which readily distinguish *Acanthaster* spp. within plankton samples.

As for many marine organisms, predation on larval *Acanthaster* spp. is expected to be high, especially during the late brachiolaria stage, when larvae come within the vicinity of reefs and attempt to settle. However, logistical challenges to sampling *Acanthaster* spp. larvae in the field make it difficult to quantify natural rates of larval mortality and rates of predation. Unlike coral eggs, which are heavily preyed on by planktivorous fishes (e.g., Pratchett et al. 2001), *Acanthaster* spp. gametes and larvae are often avoided by planktivorous fishes and invertebrates (Yamaguchi 1974, 1975). Chemical analyses by Lucas et al. (1979) showed that eggs and larvae of *Acanthaster* spp. contain saponins, which presumably make them less palatable. During a spawning event at Blue Pearl Bay (Hayman Island, GBR), Gladstone (1992) observed that planktivorous fish feeding nearby ignored gametes released by spawning *Acanthaster planci*. Reef fishes within the vicinity of a spawning *A. planci* at Arlington Reef, central GBR, also ignored gametes, except for one species of damselfish, *Abudefduf curacao*, which was observed feeding on eggs shed by the spawning female (Pearson & Endean 1969). Butterflyfishes *Chaetodon auripes* and *Chaetodon falcula* have also been observed to feed on *A. planci* gametes in Okinawa and the Maldives, respectively (Keesing & Halford 1992, Ciarapica & Passeri 1993).

Keesing and Halford (1992) suggested that spawning usually occurs late in the afternoon or at night, when the impact of visual predators would be minimised. However, little is known about when or where developed larvae actually settle or whether natural rates of predation on larvae that are trying to settle are high or low. Chesher (1969) suggested that predation by filter-feeders, such as corals, would inflict significant mortality on settling larvae. Yamaguchi (1973) documented that *Pocillopora damicornis* will feed on the larvae of *A. planci* and other coral reef asteroids. However, predation by corals is likely to have limited influence on overall survivorship for two reasons. Firstly, Ormond & Campbell (1974) found that *A. planci* larvae could detect and readily avoid live corals. Secondly, there are many areas of coral reef substrata that have relatively low cover of live coral where starfish larvae could settle without being consumed by coral polyps (Reichelt et al. 1990b). Chesher (1969) suggested that larvae are attracted to aggregations of adult *A. planci* because they have already removed corals that would otherwise prey on the larvae. More likely, however, is that the feeding activities of adult *A. planci* increase the availability of microhabitats (dead but intact coral colonies) that are conducive to settlement.

Abiotic factors play an important role in development and survivorship of larval *Acanthaster* spp. (e.g., Henderson & Lucas 1971, Lucas 1973). Laboratory rearing experiments suggested that *Acanthaster planci* have a narrow temperature tolerance. Optimal temperatures for larval development appear to be between 26°C and 31°C (e.g., Lucas 1973, Lamare et al 2014). Temperatures of 32°C or greater appear lethal, while larvae simply did not complete development at temperatures below 25°C (Henderson & Lucas 1971, Lucas 1973). In Guam, larvae were successfully reared at 27–29°C, while larvae reared at temperatures below 25°C did not advance to brachiolaria stage and showed regression to earlier stages, even though they were observed to feed vigorously (Yamaguchi 1973). Similar temperature ranges were used to rear larvae through to settlement in the Red Sea (28°C and 29°C; Ormond & Campbell 1974) and southern Japan (fluctuating between 25°C and 30.3°C; Uchida & Nomura 1987). Importantly, temperature tolerance of *A. planci* varies with developmental stages (Johnson & Babcock 1994). The late brachiolaria stage appears to be the most temperature sensitive (Habe et al. 1989), and this may constrain settlement to areas with relatively warm temperatures (26–31°C). However, greater temperature tolerance in early larval stages allows larvae to withstand exposure to cooler waters and slowly continue normal development during oceanic transport. Hatched gastrula larvae can tolerate temperatures between 13°C and 34°C, and bipinnaria larvae can tolerate temperatures of 14.5–32°C (Habe et al. 1989). However, the rate of development is greatly accelerated at higher temperatures: Habe et al. (1989) found that embryonic development is completed in 31 h at 20°C but only 11 h at 32°C.

Although echinoderms are generally sensitive to changes in salinity (Diehl 1986), the early larval stages of *Acanthaster* spp. appear to tolerate wide ranges in salinity. Gastrula larvae can tolerate a salinity range of 21–45‰ and bipinnaria larvae can tolerate 21–50‰ salinity (Habe et al. 1989). Henderson (1969) also found that bipinnaria larvae can tolerate abrupt salinity changes from 36‰ down to 21‰, and they developed more rapidly at lower salinities. Despite the robust early larval stages, late brachiolaria and metamorphosing stages are less tolerant to salinity and rupture with 2‰ changes in salinity (Henderson & Lucas 1971). Larval development and metamorphosis was completed at 26‰, but not at 22‰ (Lucas 1973). On the GBR, salinity in nearshore water is influenced by river discharge, and during times of flooding and heavy rainfall, salinity levels of 30‰ or less are often recorded (Brodie et al. 2005). Lucas (1973) showed that larval survival was 3-fold higher at 30‰ compared to ambient conditions, suggesting that temporary declines in salinity bought about by periods of heavy rainfall and run-off may actually enhance larval survivorship (Birkeland 1982). The tolerance of unfertilised gametes, especially sperm, to lowered salinity is unknown. Sperm of *Acanthaster planci* have been shown to be sensitive to changes in temperature and seawater chemistry (Uthicke et al. 2013) and may therefore be sensitive to changes in salinity (but see Greenwood & Bennett 1981).

Settlement

Towards the end of the brachiolaria stage, the brachiolar arms of *Acanthaster* spp. larvae elongate to improve locomotion while supporting the weight of the starfish primordium (Olson et al. 1988).

At this time, larvae start to drift downward and flex the anterior body dorsally to orient the brachiolar arms against the substratum to test its suitability for settlement (Yamaguchi 1973). Based on laboratory experiments (Henderson & Lucas 1971, Ormond et al. 1973, Yamaguchi 1973, Ormond & Campbell 1974, Lucas 1975, Johnson et al. 1991, Keesing & Halford 1992, Johnson & Sutton 1994) and field observations (Yokochi & Ogura 1987, Zann et al. 1987), larval *Acanthaster planci* are particular about where they settle. However, Lucas (1975) suggested that these strong settlement preferences do not necessarily limit settlement success as long as larvae are transported over coral reef habitats while they are still competent to settle. Larvae appear to settle preferentially in habitats with fine-scale topographic complexity, so that the larvae are completely hidden within carbonate matrix, or among coral rubble, prior to metamorphosis (Lucas 1975). This may be an adaptation to minimise larval mortality at settlement and during metamorphosis, which is suggested to be in excess of 85% (Keesing & Halford 1992). Natural rates of post-settlement mortality are extremely difficult to measure, but as for many coral reef organisms (e.g., Trapon et al. 2013), there is increasing realisation that biological interactions (competition and predation) are probably important in determining patterns of settlement (Keesing & Halford 1992). Ormond and Campbell (1974) showed that skeletons of dead colonies of *Acropora hyacinthus* were favoured over the other coral skeletons they tested (*Acropora diversa*, *Pocillopora verrucosa*, and *Stylophora pistillata*), probably because the branch spaces more closely match the size of the larvae. However, other work has suggested that settlement cues for *A. planci* are more strongly influenced by biofilms (e.g., encrusting algae and associated bacteria and detritus), rather than the microhabitat complexity. For example, *A. planci* will not generally settle on glass or ceramic tiles (Henderson & Lucas 1971, Ormond & Campbell 1974, Johnson et al. 1991) unless these substrates are first conditioned in natural environments to generate fine growth of microalgae (Henderson & Lucas 1971).

Observations of recently settled *Acanthaster planci* in Suva Reef, Fiji (Zann et al. 1987), and Ryukyu Islands, Japan (Yokochi & Ogura 1987), revealed a strong association with coralline algae (e.g., *Porolithon onkodes*), which is expected given that newly settled starfish feed almost exclusively on coralline algae (Yamaguchi 1973, Lucas 1984, Zann et al. 1987). Yamaguchi (1973) observed *A. planci* larvae settling directly on dead coral encrusted with coralline algae (*Porolithon* sp.) but found no settlement on bleached coralline algae or on pieces of beach rock covered with filamentous algae (but see Henderson & Lucas 1971). Johnson et al. (1991) also found high rates of settlement on coral rubble with high coverage of the coralline algae, *Lithothamnium pseudosorum*, but reported significantly lower settlement on tiles colonised by non-calcareous crustose red algae (*Peyssonellia* sp.) and other species of coralline algae (*Porolithon onkodes*, *Neogoniolithon foslei*). Techniques developed for large-scale culture of *A. planci* larvae have achieved high rates of settlement on thalli of *L. pseudosorum* (Ayukai et al. 1996, Keesing et al. 1996). Treatment of highly inductive shards of *L. pseudosorum* with antibiotics reduced settlement to low levels, signifying that induction of settlement and metamorphosis of *A. planci* may be mediated by chemical cues produced by epiphytic bacteria (Johnson et al. 1991). Settlement and metamorphosis were inhibited in the absence of bacteria; larvae always settled on sections of thallus with high densities of bacteria, but not in areas where epiphytic bacteria were sparse (Johnson et al. 1991, Johnson & Sutton 1994). However, surface bacteria did not induce settlement when isolated from soluble algal compounds, suggesting that bacteria require the algal substratum to produce inductive compounds or that compounds from both the bacteria and coralline algae are required to induce settlement and metamorphosis (Johnson & Sutton 1994).

Juvenile ecology

Following settlement, metamorphosis of *Acanthaster* spp. brachiolaria larvae occurs with the absorption of the anterior part of the larval body into the starfish primordium (Yamaguchi 1973), which emerges 2 days later as a five-arm juvenile starfish 0.3 to 0.7 mm in diameter with two pairs of tube feet, a terminal tentacle, and a red optic cushion on each arm (Henderson & Lucas 1971, Yamaguchi 1973, Lucas 1975). After 3 weeks, these juvenile starfish start adding arms at 2-week intervals, and the body turns pink, which camouflages the juvenile starfish against the coralline algae on which it is feeding (Yamaguchi 1973, Lucas 1975, Birkeland & Lucas 1990). During this phase, juvenile starfish do not feed on coral tissue, possibly to avoid damage caused by mesenteric filaments when coming in contact with coral polyps (Yamaguchi 1973). While feeding on coralline algae, growth rates of *A. planci* are slow (1.5–2.6 mm/month) but increase rapidly when starfish switch to feeding on scleractinian corals at around 6 months (Yamaguchi 1974, Zann et al. 1987). Maximal growth rates (16.7–25.0 mm/month) occur between 5 and 18 months after *A. planci* switch to feeding on live corals but slow substantially after this period, presumably because starfish begin diverting energy from somatic growth as they become sexually mature (Birkeland & Lucas 1990).

Although juvenile *Acanthaster* spp. contain saponins, they are still likely to be extremely vulnerable to predation (Keesing & Halford 1992). Small juveniles are cryptic and are mostly active at night (Zann et al. 1987), presumably to avoid visual predators like reef fishes. This highly cryptic and nocturnal behaviour continues until starfish reach at least 15-cm diameter, at an age of approximately 20 months (Zann et al. 1987), after which starfish are much more active during daylight hours. As a consequence, relatively few juvenile *Acanthaster* spp. have been found, despite extensive field sampling (Doherty & Davidson 1988, Johnson et al. 1992). Estimates of recruitment (e.g., Doherty & Davidson 1988, Zann et al. 1990) are often based on the emergence of relatively old (e.g., 1- to 2-year-old) individuals. The abundance and distribution of these older juveniles are likely to differ greatly from patterns at settlement because of high rates of post-settlement mortality (e.g., Keesing & Halford 1992) and likely movement of larger juvenile starfish as they switch to eating scleractinian corals (Endean & Cameron 1990).

Adult growth and longevity

The demography of *Acanthaster* spp. is extremely plastic in that adult growth and longevity seem to be strongly dependent on local environmental conditions, such as food availability, temperature, and wave exposure (e.g., Ormond & Campbell 1971, Kenchington 1977, Lucas 1984). For this reason, there has been considerable controversy surrounding even the most basic demographic questions, such as whether growth of *Acanthaster* spp. is determinate (or more precisely, asymptotic; e.g., Yamaguchi 1974, Lucas 1984) or indeterminate (Kenchington 1977). This issue was discussed at length by Moran (1986) but never resolved. Lucas (1984) argued that *Acanthaster planci* from the GBR reached a maximum size of approximately 340-mm diameter at approximately 3 years of age (Figure 5), after which they entered a period of senescence and had a maximum longevity of 4–5 years. These assertions are clearly at odds with extensive and increasing records of *A. planci* that are up to 750 mm in diameter and 8 years of age (Endean & Cameron 1990, Stump 1996). Based on their regenerative ability, as well as physical and chemical defences, Endean (1982) suggested that adult crown-of-thorns starfish would have low mortality and should live for decades (see also Ebert 1973). Large *A. planci* have been recorded, up to 750 mm in diameter (Stump 1996), but mostly on the GBR and often outside active outbreaks. At high densities, *Acanthaster* spp. may have highly constrained, finite growth and survivorship, which is possibly linked to strong intraspecific competition and rapid depletion of prey resources during major outbreaks (Kettle 1990, Mills 2012). However, recent studies (e.g., Pan et al. 2010) clearly showed that starfish in outbreak populations can grow well beyond 350 mm and can live for more than 8 years.

An important question for demographic studies (and improved understanding of the timing and therefore possible causes of outbreaks) of *Acanthaster* spp. is whether variation in the sizes of starfish within a given population reflects distinct cohorts and therefore the range in ages of individuals (e.g., Pratchett 2005). The alternative is that substantial variation in growth rates among individuals from a single cohort will obscure any relationship between size and age and that this

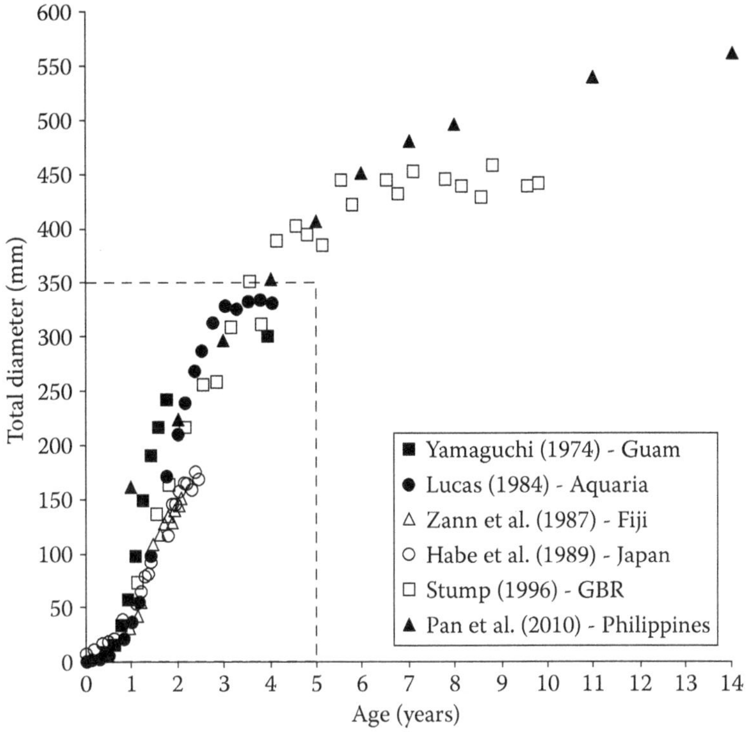

Figure 5 Relationship between total diameter (millimetres) and age (years) for *Acanthaster planci* based on data from throughout the western Pacific. There is a strong and consistent size-age relationship, despite initial suggestions that marked plasticity in initial growth, combined with rapid attainment of asymptotic size, would obscure any such relationship. The growth curve in the formative part of the life history (incorporating the initially slow growth in the first 6 months) is best explained by a sigmoidal function. After 1 year, however, the generalised growth curve is largely consistent with von Bertalanffy growth equations. The dashed line indicates the presumed maximum size and age of *A. planci* prior to 1990, based on the apparent asymptote at 340-mm diameter for laboratory-reared starfish (Lucas 1984). Field-based studies, however, suggest that there is a much larger asymptotic size.

alone accounts for the range of sizes within outbreaking populations (Stump & Lucas 1990). To test this, size-independent proxies of individual age have been explored, including spine length, age pigments, and pigment bands (equivalent to growth rings) on spines (Birkeland & Lucas 1990). The most reliable and widely adopted technique, developed by Stump and Lucas (1990), involves estimating individual ages based on pigment banding on spines. Mark-recapture and tetracycline staining of *Acanthaster planci* in the field (Stump & Lucas 1990, Stump 1994) confirmed that growth bands on the longest spines taken from the aboral surface of upper arms were laid down seasonally; thus, banding couplets are reflective of age in years. However, growth bands only become apparent after sexual maturity, at around 2 years of age (Stump 1996). Moreover, caution must be taken in selecting the most appropriate spines (Stump & Lucas 1999) as marked differences in putative ages (ranging from 1 to 17 years) may be obtained from spines taken randomly across the surface of individual starfish (Souter et al. 1997).

Initial studies of post-settlement growth of *Acanthaster planci* revealed marked changes in growth rates at different life stages (e.g., Lucas 1984). The growth pattern is sigmoid (e.g., Lucas 1984, Stump 1996), with slow growth both when starfish first settle and feed on calcareous algae and when starfish attain sexual maturity at approximately 2+ years of age. However, when considering only the individuals 1+ years of age, growth can be described effectively using von Bertalanffy growth functions (Figure 5), where the L_∞ (asymptotic size) is between 474 mm based on data from Stump (1996) and 580 mm based on data from Pan et al. (2010). By combining size-at-age data across all previous studies, including laboratory-based measurements of Lucas (1984), it is apparent that there are distinct differences in the reported size (diameter) of starfish in consecutive age classes; 30–100 mm for starfish with an estimated age between 1 and 2 years, 100–250 mm for 2- to 3-year-old starfish, 250–300 mm for those 3–4 years, 300–400 mm for 4–5 years, and greater than 380 mm for 5+ years (Figure 5). Increased research and further validation of growth relationships in different locations (especially outside the Pacific) may be needed to provide increased resolution of ages necessary to establish the precise timing for the initiation of different outbreaks. However, it is generally accepted that wide ranges in the size of individuals, especially within single populations, reflect marked differences in ages rather than extreme variation in growth of starfish from the same cohort (Stump 1996).

Lack of an effective method of tagging individuals has greatly impeded field studies of the demography of *Acanthaster* spp. (Glynn 1982). Early attempts to tag starfish involved embedded tags, but starfish quickly ejected tags (within weeks) or dropped the arm to which tags were attached. Some tags also lead to high rates of infection, or greatly modified individual behaviour, undermining efforts to document 'natural' rates of growth and survivorship (Moran 1986). Later tagging efforts avoided damaging the dermal tissues, instead attaching harnesses around the oral disk or coloured bands to individual spines (e.g., Keesing & Lucas 1992), but again tags had a limited lifespan. Stump (1996) marked *Acanthaster planci* using tetracycline injections to stain their spines. In combination with counts of arms and records of the position of madreporites, this could be used to confirm the identity of individual starfish. However, this technique did not allow easy recognition of individuals in the field. In 2012, J. Rivera-Posada (unpublished data) used loops of relatively inert nylon monofilament that were passed through predrilled holes in the skeletal elements of the aboral disk and between the ambulacral ridge of the arms. Holes were drilled with Kirschner wires at low speed to avoid thermal osteonecrosis or fracture that could lead to rapid ejection of tags. Needles and cardiocatheters were then used to pass the nylon through the arm and skeletal elements, securing the two ends with small brass connector sleeves. These tags last at least 4 weeks in aquaria but are yet to be tested in the field. By threading coloured plastic beads on to the nylon loop, it will be possible to identify a large number of tagged *Acanthaster* spp., with wide application in studies of demography and individual behaviour (e.g., movement).

Feeding behaviour

The crown-of-thorns starfish is just one of many different coral reef organisms that feed on scleractinian corals (Glynn 1988, Cole et al. 2008, Rotjan & Lewis 2008). However, the capacity of *Acanthaster* spp. to deplete local cover of scleractinian corals is far greater than for any other corallivorous species (Glynn 1988, Birkeland 1996a, Carpenter 1996). Most corallivores are limited in their rate of feeding because scleractinian corals have only a thin veneer of living tissue over the surface of an indigestible calcareous skeleton (Keesing 1990). As a consequence, corallivores must selectively pick live tissues from the surface of corals (which tends to limit the rate of feeding) or else ingest large quantities of calcium carbonate, which is energetically costly (Motta 1988). In contrast, *Acanthaster* spp. (and other corallivorous starfish) are extremely well adapted to feed on scleractinian corals as they can digest tissue from a large area of coral surface at once. These asteroids feed by everting their stomach through their oral opening and spreading it over the surface of live corals or any other benthic prey (Jangoux 1982). Enzymes are then secreted through the gastric tissues that digest coral tissues within 3–5 h (Goreau 1964, Brauer et al. 1970). *Acanthaster planci* has a much larger stomach for its size than other corallivorous asteroids (e.g., *Culcita novaeguineae*), enabling it to consume scleractinian corals two to five times faster than other starfish of equivalent size (Birkeland 1989a). Moreover, Benson (1975) suggested that *Acanthaster* spp. possess the most specialised enzyme system for digestion of wax esters, which are a major component of coral tissues. Even so, crown-of-thorns starfish consume a maximum of 150–250 cm^2 of live coral per day, depending on their body size (Chesher 1969, Glynn 1973).

A key facet of the feeding behaviour of *Acanthaster* spp., which has a major bearing on their ecological impact, is their feeding preferences (Moran 1986). Moran (1986) identified many factors that influence feeding preferences of *Acanthaster* spp., including (1) the nutritional content and growth form of corals; (2) coral defences (e.g., mesenterial filaments, nematocysts, and secondary metabolites); (3) coral defence by commensal infauna (mostly trapezid crabs); (4) the distribution of corals; (5) local environmental conditions; and (6) prior conditioning of individual starfish (see also Birkeland & Lucas 1990). This combination of different factors was expected to lead to complex patterns of feeding preferences, which vary with geographical variation in the composition of coral assemblages (Birkeland & Lucas 1990) and with size and abundance of the starfish (Moran 1986). To test this, we compiled published data on the proportional consumption of different coral genera by *Acanthaster* spp. relative to their availability at different locations throughout the Pacific (Figure 6). These data do not represent feeding preferences *per se* because they do not strictly assess the selection of one prey type over another (e.g., Keesing 1990, De'ath & Moran 1998, Pratchett 2007). Moreover, the relative consumption of different coral genera is generally inferred from

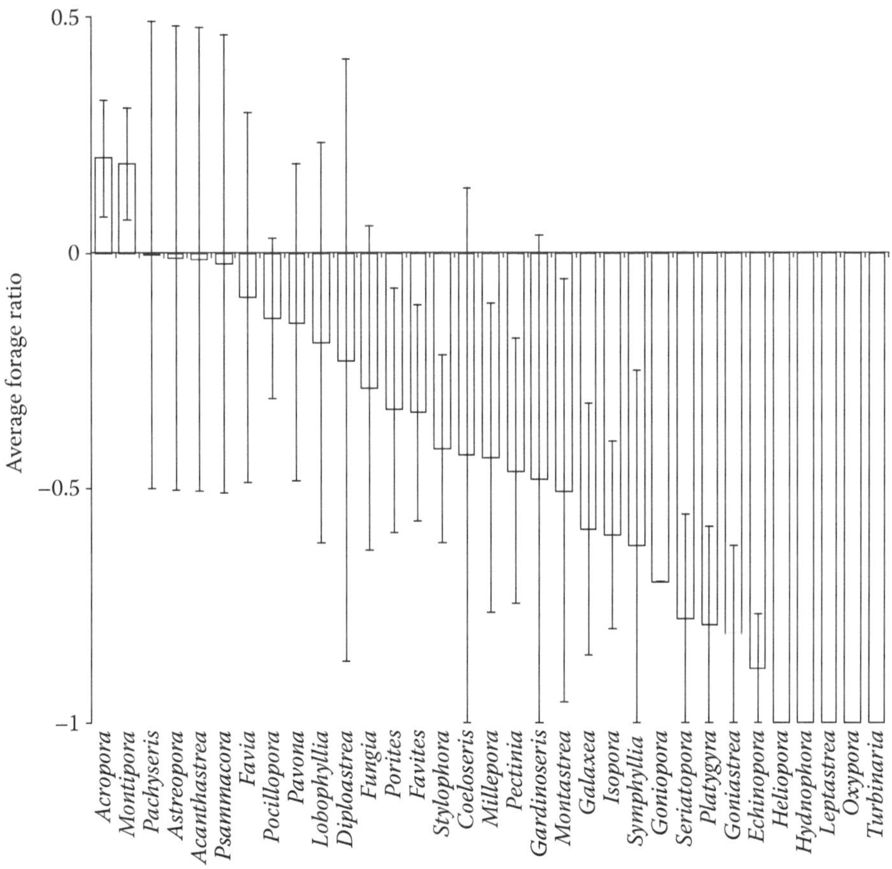

Figure 6 Variation in forage ratios (averaged across seven studies in the Pacific) for different coral genera, indicative of general feeding preferences. Forage ratios compare proportional consumption of coral genera to their relative abundance in the local area, following the work of Tokeshi & Daud (2011). Data sources: Glynn 1974, Bouchon 1985, Keesing 1992, Chess et al. 1997, Pratchett et al. 2009a, Pratchett 2010, Baird et al. 2013.

changes in the abundance or mortality of different coral genera during outbreaks, although many factors (e.g., routine coral mortality, coral disease, predation by other corallivores, or bleaching) may have contributed to coral loss (e.g., Pratchett et al. 2009a). However, studies that directly compared mortality rates between sites with and without outbreak densities of *A. planci* (e.g., Pratchett et al. 2009a, Pratchett 2010) found that coral mortality was negligible in sites where *A. planci* were rare or had only recently been reported.

Variation in forage ratios, averaged over seven studies (Figure 6) and six different geographical locations across the Pacific (Uva Island, eastern Pacific, Glynn 1974; Hawaii, Chess et al. 1997; French Polynesia, Bouchon 1985; Papua New Guinea, Pratchett et al. 2009a; GBR, Keesing 1992, Pratchett 2010; Indonesia, Baird et al. 2013), shows that *Acropora* and *Montipora* were the most preferred coral genera. These two coral genera were consistently consumed in greater proportions than would be expected from their availability. At the other extreme, there were four genera of scleractinian corals (*Hydonophora*, *Leptastrea*, *Oxypora*, and *Turbinaria*) as well as *Heliopora*, that did not decline in abundance during outbreaks of *A. planci*, presumably because they are non-preferred coral prey. Average forage ratios for virtually all other corals (all except *Acropora* and *Montipora*) were negative, although there was variation in apparent preference for individual coral genera among studies and locations (Figure 6). Reported forage ratios for *Pachyseris, Astreopora, Acanthastrea, Psammacora*, *Lobophyllia*, and *Diploastrea* ranged from −1 to close to 1, indicating that abundance did not change during outbreaks of *Acanthaster planci* at some locations, while there was extensive depletion at other locations. For example, there was a 63% decline in abundance of *Acanthastrea* during an outbreak of *A. planci* at Moorea, French Polynesia, in 1980 (Bouchon 1985), but there was no apparent change in abundance of *Acanthastrea* during a recent outbreak of *A. planci* in Sumatra, Indonesia (Baird et al. 2013).

Despite marked variation in forage ratios for some coral genera, the rank order of coral genera revealed by comparing average forage ratios are in accord with more detailed studies of feeding preferences conducted under laboratory conditions (e.g., Brauer et al. 1970, Collins 1975, Ormond et al. 1976, Pratchett 2007) and field-based studies of specific feeding behaviour (De'ath & Moran 1998). Prior studies of the feeding preferences of *Acanthaster* spp. have mostly compared among *Acropora*, *Pocillopora*, and *Porites*, showing that *Acropora* is most preferred and *Porites* is least preferred (reviewed by Moran 1986). Explanations for these overarching feeding preferences (e.g., nutritional content, chemical deterrents, or colony defence) are lacking, and they are likely to vary in their importance depending on the coral taxa (Pratchett 2007). Comparing *Acropora* spp. with *Pocillopora* spp. (and other Pocilloporidae), Pratchett (2001) showed that differential feeding preferences were largely attributable to differences in the effectiveness with which commensal infauna (and especially trapezid crabs) defend their coral hosts (see also Glynn 1987). Notably, *A. planci* did not distinguish between *Acropora* spp. and pocilloporid corals when commensals were removed from all corals (Pratchett 2001). *Porites* spp. also contain symbiotic organisms (*Pedum spondyloideum* and *Spirobranchus giganteus*); rather than preventing *A. planci* from eating their host colony, these organisms enhance the survivorship of a few adjacent coral polyps, which may enable subsequent regeneration of the colony (e.g., DeVantier & Endean 1988). The avoidance of *Porites* spp. by *Acanthaster* spp. is generally ascribed to a combination of low nutritional value, chemical deterrents to feeding (De'ath & Moran 1998), and the inability of starfish to attach to large, smooth colonies, leading to high probability of dislodgement except in calm conditions.

The relative preference of *Acanthaster* spp. for certain corals (e.g., *Acropora* spp.) could result in shifts in the structure of coral assemblages during outbreaks. However, even the least-preferred corals are consumed during extremely severe outbreaks or when coral prey is scarce (e.g., Chesher 1969, Pearson & Endean 1969). This is particularly apparent when comparing effects of outbreaks of *A. planci* at Moorea in 1980–1981 (Bouchon 1985) with those of subsequent outbreaks in 2007–2012 (Kayal et al. 2012). In 1980–1981, Bouchon (1985) recorded proportional declines in the abundance of the major coral genera *Acropora*, *Pocillopora,* and *Porites* of 45.5%, 0%, and 0%, respectively. In contrast, Kayal et al. (2012) reported comprehensive extirpation of *Acropora* spp. and *Pocillopora* spp. and greater than 95% decline in local cover of *Porites* spp. Kayal et al. (2012) did report that there was sequential depletion of these major genera, with *Acropora* removed first, then *Pocillopora*, and then *Porites*. Clearly, selective effects of *A. planci* are most apparent during outbreaks that cause only moderate loss of corals (Birkeland & Lucas 1990). The relative consumption of different coral genera also appears to be different for low-density (non-outbreak) populations of *Acanthaster* spp. (e.g., Tokeshi & Daud 2011), with feeding preferences appearing to be largely dictated by the proximity of corals to appropriate shelter.

Outbreaks of crown-of-thorns starfish

Extreme variability in adult abundance is common among marine organisms, particularly those with planktonic larvae (e.g., Roughgarden et al. 1988). However, few marine organisms show changes in abundance of the magnitude, or rate, shown by crown-of-thorns starfish. The abundance of *Acanthaster* spp. can increase by as much as six orders of magnitude within 1 to 2 years (reviewed by Birkeland & Lucas 1990). At Tutuila Island, American Samoa, the overall abundance of *A. planci* increased from 1–2 starfish in 1976 to more than 200,000 starfish in late 1977 (Birkeland & Randall 1979). Similarly, at Tanguisson Reef, Guam, densities of *A. planci* increased from less than 0.1 starfish.ha^{-1} to more than 1,000 starfish.ha^{-1} during the course of 1967 (Chesher 1969). More recently, Kayal et al. (2012) reported maximum densities of *A. planci* of 151,650 starfish/km^2 along the northern barrier reef of Moorea, French Polynesia, with densities increasing more than 10-fold over the course of just 1 year. The combined feeding activities of high densities of crown-of-thorns starfish also cause extensive coral depletion (e.g., Chesher 1969, Kayal et al. 2012), leading to widespread concern that outbreaks of *Acanthaster* spp. are becoming more frequent and more prevalent across the Indo-Pacific (e.g., Brodie et al. 2005).

Despite extensive and increasing reports of 'outbreaks' of *Acanthaster* spp. across the Indo-Pacific, there are major inconsistencies and deficiencies in published reports that constrain rigorous and comprehensive analyses of the geographical extent and recurrence of outbreaks. These problems are partly caused by inherent complexities in defining and comparing the extent and severity of outbreaks among different locations. On the GBR, for example, outbreaks are reported at the scale of the entire reef system (Sweatman et al. 2008), recognising inherent coupling of outbreaks among the well-connected reefs that make up this vast reef system (Reichelt et al. 1990a, Moran et al. 1992). Elsewhere, however, outbreaks are often reported from individual locations, let alone individual reefs (e.g., Koonjul et al. 2003). The incidence of outbreaks of crown-of-thorns starfish on small, isolated, or unpopulated reefs is important for understanding the potential causes of (and anthropogenic contribution to) outbreaks of *Acanthaster* spp., but there are likely to be many different factors that influence the initiation and impacts of outbreaks, which are rarely considered in published reports of localised outbreaks. In the Maldives, for example, outbreaks of *Acanthaster planci* were reported at many different atolls in the early 1990s (Ciarapica & Passeri 1993), which runs counter to Birkeland's (1982) assertions that outbreaks really only occur on high islands (discussed in the section on geographical incidence of recent outbreaks). However, no attempt was made to quantify temporal and spatial patterns in starfish densities, or to assess the population structure of localised outbreaks, to understand how these outbreaks were initiated or spread among nearby atolls.

Defining outbreaks of crown-of-thorns starfish

Although there are conspicuous differences in the densities of *Acanthaster* spp. between outbreaking and non-outbreaking populations, rigorous definitions of outbreaks are elusive. Potts (1981)

Table 3 Operational criteria used to distinguish outbreak densities of *Acanthaster* spp. and non-outbreaking (normal) densities

Minimum threshold	Reference
30–40 starfish per km^2	Clark & Weitzman 2006
14 starfish per 1000 m^2	Endean & Stablum 1975
40 starfish per 20-min swim	Pearson & Endean 1969
100 starfish per 20-min swim or manta tow	Chesher 1969
10 starfish per 1-min spot check	Pearson & Garrett 1978
30–50 starfish in 20 min	Faurea 1989
1 starfish per 24 m^2	Pearson & Endean 1969
100 starfish per hectare	Dana et al.1972
260 starfish per hectare	Glynn 1973
150 feeding scars per 250 m^2	Lison de Loma et al. 2006

defined outbreaks of crown-of-thorns starfish as "any large aggregation of many hundreds or thousands of individuals which persist at high densities for months or years and causes extensive mortality among coral over large areas of reef" (Potts 1981 p. 65). This definition encompasses aspects of both the biological (pronounced and unexplained increases in the abundance of a species) and ecological (rapid increases in the abundance of a species beyond that which can be sustained by local resources) definitions of population outbreaks, but these qualitative definitions are not particularly useful when attempting to rigorously account for the incidence and occurrence of outbreaks. Normal background densities of *Acanthaster planci* may be spatially variable (e.g., Glynn 1990), so it is important to distinguish between periodic outbreaks versus highly localised and chance aggregations of starfish from populations with persistent moderate densities (Moore 1990). Various quantitative definitions for outbreaks have thus been proposed (Table 3).

Moran & De'ath (1992) proposed an operational definition of 1,500 starfish.km^{-2} for outbreaks of *Acanthaster planci* on Australia's GBR by estimating the actual densities of starfish (after accounting for sampling bias) that exceed sustainable limits. Substantial coral mortality was only observed at reefs with more than 1,500 starfish.km^{-2} (equivalent to 15 starfish.ha^{-1} or 0.22 starfish per 2-min manta tow), suggesting that this is the maximum density of starfish that can be sustained by reefs with average coral cover (Moran & De'ath 1992). Similar estimates of the maximum sustainable density of *A. planci* (1,000 starfish.km^{-2}) were obtained by relating feeding rates of starfish to the average annual turnover in well-established coral assemblages (Keesing 1990). It was recognised, however, that the density of *Acanthaster* spp. that can be sustained on any given reef will vary enormously in time and space, depending on the abundance and composition of scleractinian corals, as well as the size and distribution of starfish. Keesing and Lucas (1992) estimated that densities of 10–15 starfish per hectare could be sustained in areas with more than 20% coral cover. However, sustained declines in coral cover on reefs across the Pacific (Bruno & Selig 2007) mean that outbreaks of *Acanthaster* spp. that could once be sustained would now contribute to accelerating coral loss. On Australia's GBR, for example, average coral cover has declined from 28.0% in 1982 to 13.8% in 2012, caused in large part by outbreaks of *Acanthaster planci* (De'ath et al. 2012). Sustained declines in mean coral cover could also increase vulnerability to current and future outbreaks.

Whereas the density of *Acanthaster* spp. provides the most relevant measure to compare the extent and severity of outbreaks (Moran 1986), it is not easy to obtain precise estimates of starfish abundance. Notably, density estimates for *Acanthaster* spp. depend greatly on the temporal and spatial scales of sampling (Endean & Cameron 1990). Even during outbreaks, the density of starfish varies within and among reefs (e.g., and among habitats) and may also vary on short timescales within any given location. Changes in the behaviour of starfish after they attain adult size (switching from generally cryptic and nocturnal to diurnally active) also significantly affect the ease with which starfish can be surveyed and resulting estimates of starfish densities (Endean & Cameron 1990). Outbreak populations of *Acanthaster* spp. tend to be highly aggregated, such that maximum recorded densities may be large, whereas there are few or no starfish at nearby locations (e.g., Pratchett 2005). It is possible, therefore, that densities of starfish on just one or a few transects may exceed threshold densities indicative of an outbreak (e.g., 1,500 starfish/km^2; Moran & De'ath 1992), whereas densities averaged across the entire reef or reef system remain well below these levels. For this reason, some monitoring studies distinguish between 'spot' versus 'reef-wide' outbreaks (Engelhardt 1999) or 'incipient' versus 'active' outbreaks (e.g., Sweatman et al. 2008). Moran (1986) called for standardised methods for quantifying the abundance of *Acanthaster* spp., recommending repeated and intensive surveys of starfish at individual reefs using area-based surveys to quantify densities, combined with measurements of maximum diameter, to clearly establish the fine-scale temporal and spatial dynamics of outbreak populations. The few studies that have conducted intensive sampling at a single reef throughout the course of an outbreak (e.g., Pratchett 2005, Kayal et al. 2012) have provided significant new insights into the causes and consequences of outbreaks of *Acanthaster* spp.

Primary versus secondary outbreaks

Outbreaks of *Acanthaster* spp. are thought to arise in two different ways (e.g., Potts 1981, Johnson 1992), which will be reflected in marked differences in population structure (Figure 7). In many cases, outbreak populations comprise individuals of similar size (<150 mm variation between the smallest vs. largest individuals) with a unimodal size structure (Figure 7B) suggestive of a single massive influx of new recruits (Birkeland & Lucas 1990). These outbreak populations generally consist of only a single cohort or year class (e.g., Dana et al. 1972, Glynn 1973, Zann et al. 1987) and appear as dramatic increases in the abundance of starfish within weeks to months (e.g., Chesher 1969, Branham et al. 1971). However, there are many cases for which outbreak populations include individuals of a great range of sizes with a multimodal size structure (Figure 7A). In some instances, outbreak populations have been shown to include starfish from at least five or six different cohorts (e.g., Stump 1996, Engelhardt 1999, Pratchett 2010), and there has been a gradual increase in the density of *Acanthaster* spp. for many years, indicating the progressive accumulation of individuals over several successive recruitment events (e.g., Zann et al. 1987, 1990, Stump 1996, Pratchett 2005).

Factors that contribute to instantaneous increases in the recruitment of *Acanthaster* spp. (e.g., increased survivorship of starfish larvae; Lucas 1973, Birkeland 1982), leading to rapid and dramatic increases in starfish densities, are likely to be different from those that cause slow, progressive increases in starfish densities (Johnson 1992). Importantly, the sustained and gradual accumulation of crown-of-thorns starfish from multiple successive recruitment events may represent a mechanism by which outbreaks are initiated ('primary outbreaks'; Endean 1974, Johnson 1992, Stump 1996), which then give rise to massive numbers offspring causing subsequent outbreaks on nearby and downstream reefs ('secondary outbreaks'; Endean 1974) and cause massive devastation over large areas (Reichelt et al. 1990a). Birkeland and Lucas (1990) suggested that gradual increases in the densities of *Acanthaster* spp. over several years (regardless of final densities) do not really fit with the definition of an outbreak, arguing that 'outbreaks' must arise suddenly. The distinction between primary versus secondary outbreaks, however, might be fundamental to understanding the initiation of outbreaks of *Acanthaster* spp. (e.g., Pratchett 2005, Fabricius et al. 2010) as distinct from waves of outbreaks that are a predictable consequence of large, established breeding populations.

Endean's (1974) distinction between primary and secondary outbreaks, although logical, is not easy to apply because ecologists rarely have sufficient information to identify the sources of outbreaks with certainty, so the identification of primary outbreaks is largely conjectural. Intensive monitoring of *Acanthaster planci* at Lizard Island, in the northern GBR, during the 1990s provided

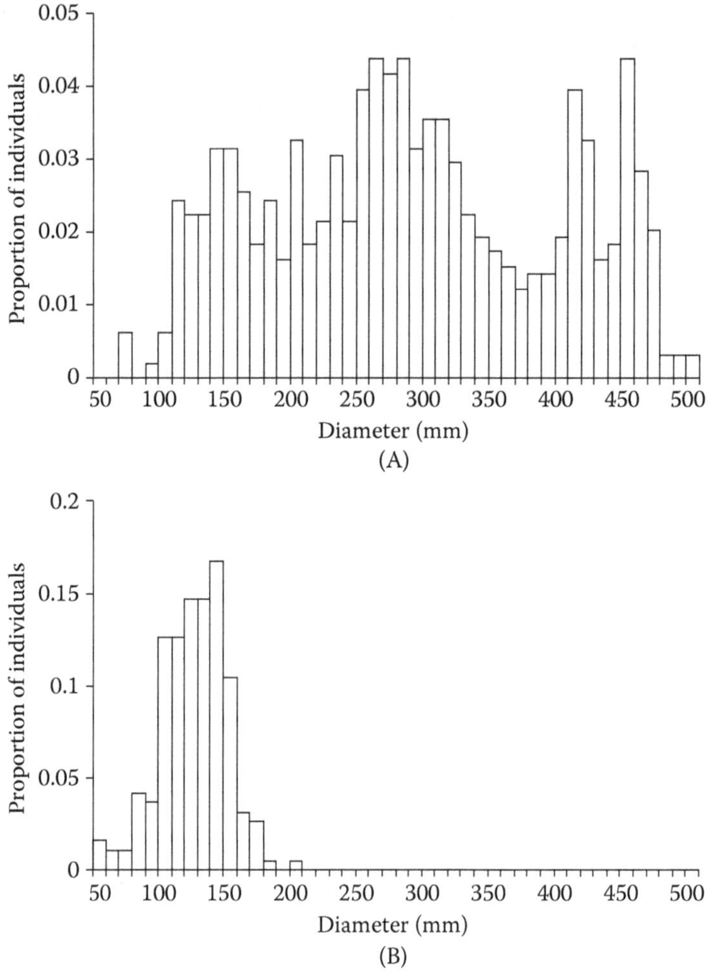

Figure 7 Contrasting size structure of outbreak populations of *Acanthaster planci* in (A) July 1995 at Lizard Island on Australia's Great Barrier Reef (Stump 1996) and (B) September 1985 at Suva, Fiji (Zann et al. 1987). At Lizard Island, the population comprised individuals ranging in size by more than 430 mm (n = 317) and multiple age classes, suggestive of gradual accumulation of starfish over several years (i.e., 'primary outbreak'; Endean 1977). In Fiji, the entire population comprised starfish ranging in size by less than 150 mm (n = 158), dominated by a single year class and indicative of a single mass recruitment (i.e., 'secondary outbreak'; Endean 1977).

a good example of a primary outbreak: The number of starfish increased gradually until it reached outbreak densities in 1996 (Sweatman et al. 1998, Pratchett 2005). Moreover, at peak densities of crown-of-thorns starfish, the local population comprised individuals ranging in size from 110 to 620 mm, and there was evidence of repeated annual recruitment from 1992 to 1997 (Pratchett 2005). The source of recruits settling at Lizard Island is not known, but they may represent the progeny of the adult starfish that were already present on reefs around Lizard Island (albeit in low densities) prior to the outbreak. With a high level of self-recruitment, these outbreaks may have resulted from incremental increases in the reproductive output of the initial reproductive population and all of their subsequent progeny, leading to exponential growth in population size. Detailed studies of this outbreak (e.g., Stump 1996, Pratchett 2005, 2010) did not provide any real clues regarding the cause(s). However, given the gradual accumulation of starfish over more than 5 years, any factor responsible for the initial onset of outbreaks is likely to be subtle and difficult to detect.

Secondary outbreaks of *Acanthaster* spp. are known mainly from the GBR and southern Japan (Potts 1981), where waves of outbreaks spread among well-connected reef systems after the initiation of one or more primary outbreaks (e.g., Kenchington 1977, Yasuda et al. 2009). However, outbreak populations with highly constrained size structure, characteristic of secondary outbreaks, have also been reported at many small and isolated reef systems throughout the Pacific (e.g., Birkeland 1982), suggesting that the extensive dispersal of larvae from primary outbreaks may initiate successive outbreaks in this area. To evaluate this, Timmers et al. (2012) explored genetic structure and connectivity among outbreak populations of *Acanthaster planci* at 23 sites across the Pacific Ocean, including the north-western (e.g., Guam), north-central (e.g., Hawaii), and south-central Pacific (French Polynesia) regions. Strong regional, archipelagic, and interreef genetic structuring of *A. planci* populations indicate that larval dispersal is highly constrained, and that high densities of larvae do not spread across open ocean expanses to initiate secondary outbreaks at distant reefs (see also Vogler et al. 2013). These data suggest either that primary outbreaks have gone undetected within many small archipelagos or that, in some circumstances, primary outbreaks may result from a single massive influx of larvae (Birkeland 1982).

Recent advances in genetic research, including a now-extensive library of microsatellite markers specific for *Acanthaster planci* (Yasuda et al. 2006, 2007, Wainwright et al. 2012), provide greatly improved opportunities to study the initiation (e.g., source populations) and subsequent spread of outbreaks. As such, future studies of outbreak populations not only should record the density of *Acanthaster* spp. but also should measure maximum diameter and obtain genetic samples (6–10 tube feet placed in 95% ethanol) for a representative sample of individual starfish.

Geographical incidence of recent outbreaks (1990–2013)

Since 1990, at least 246 outbreaks of *Acanthaster* spp. have been reported from across the Indo-West Pacific (Figure 8), which is three times the number of outbreaks reported (82) prior to 1990 (e.g., Moran 1986, Birkeland & Lucas 1990). Although mostly qualitative, these recent reports represent outbreaks or infestations that are distinct in space and time, at either a particular group of reefs (mostly at the scale of small island nations in the Pacific) or discrete episodes of outbreaks within a given geographic location (e.g., outbreaks of *Acanthaster planci* are reported in the Society Islands, in French Polynesia in 1969–1970, 1979–1982, and 2006–2009). This apparent increase in the number of reported outbreaks may or may not indicate a real increase in the spatial or temporal incidence of outbreaks; it is surely attributable in part to an increase in surveys and monitoring of reef environments, combined with increased awareness of *Acanthaster* spp. in this more recent period (see also Baird et al. 2013).

One of the key arguments put forward as evidence of the role of humans in causing or exacerbating outbreaks of crown-of-thorns starfish is that the incidence of devastating outbreaks has increased in recent times (e.g., Brodie et al. 2005). However, in locations where there have been recurrent outbreaks since the 1960s (e.g., GBR, Japan, and Society Islands), there is no evidence of increased frequency of outbreaks (or more specifically, a decrease in the period between successive outbreaks). On the GBR, it was 17 years between the start of the first recorded outbreak (1969) and the emergence of the second major outbreak (1979), and it was the same period (17 years) between the start of the third (1993) and start of the most recent, fourth, major outbreak (2010). In French Polynesia, it was only 10 years between the first two reported outbreaks (1969–1970 vs. 1979–1982), but then 27 years until the devastating outbreak that occurred in 2006–2009 (Kayal et al. 2012). Similarly, in Japan it was almost 20 years between the most recent wave of outbreaks and previous high-frequency, almost-chronic outbreaks that occurred throughout the 1960s and

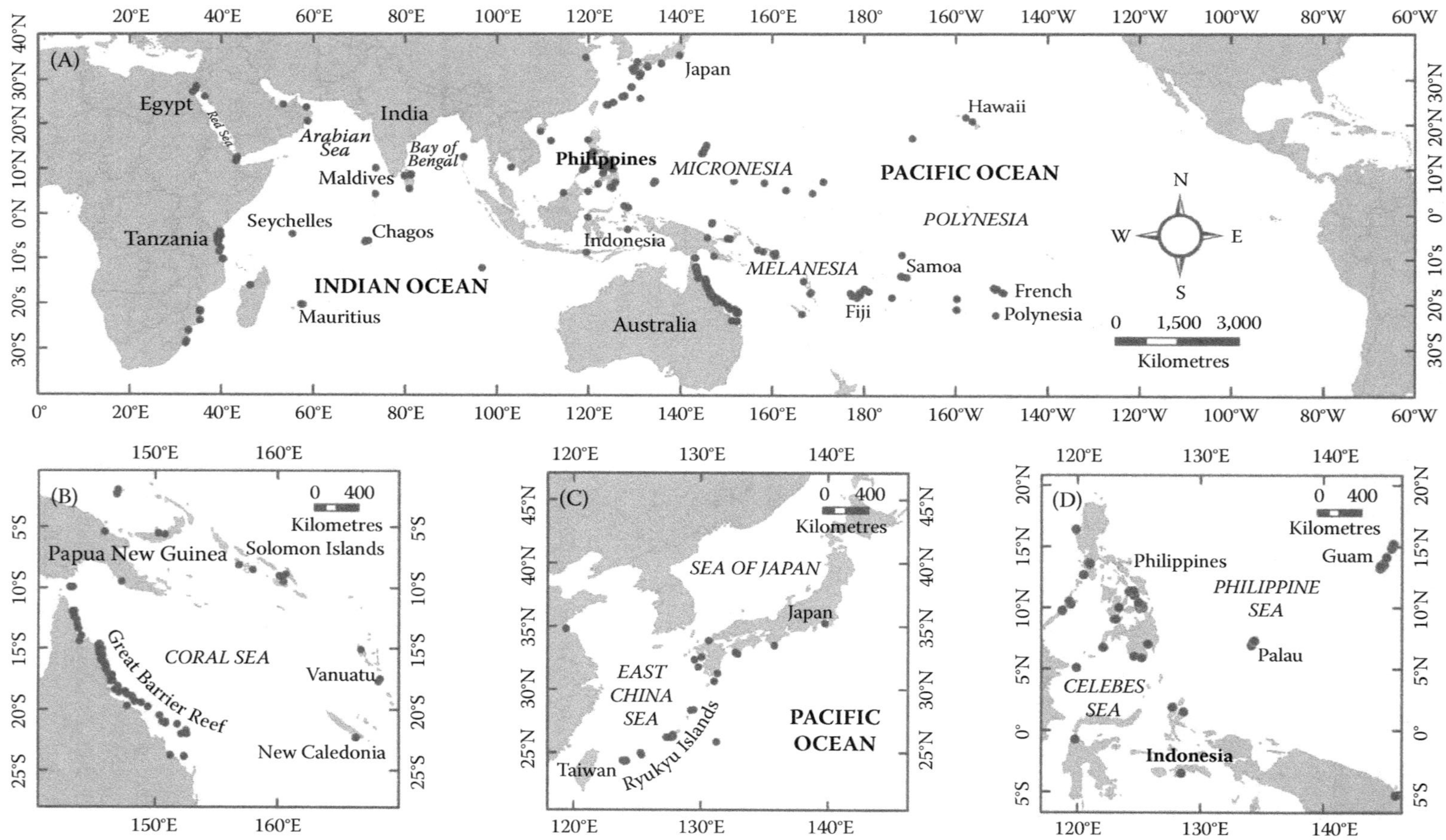

Figure 8 Geographical spread of reported outbreaks of *Acanthaster* spp. Each dot indicates a specific reef where outbreaks of crown-of-thorns starfish have been reported since 1990. No account is made of the extent, severity, or recurrence of outbreaks at each location.

early 1970s (Yasuda et al. 2009). If anything, the period between successive outbreaks is becoming longer, as might be predicted if the initiation of renewed outbreaks requires substantial recovery of coral prey (e.g., Bradbury et al. 1985, Bradbury & Antonelli 1990, Fabricius et al. 2010).

Global degradation of coral reef ecosystems, especially sustained declines in the abundance or cover of scleractinian corals, is widely attributed to the increasing frequency of acute disturbances (e.g., Seymour & Bradbury 1999, Fabricius et al. 2010), such that major disturbances occur too often to allow effective (although not necessarily complete) recovery of coral assemblages between successive disturbances. Moreover, the time required for coral assemblages to recover from periodic disturbances might be increasing (Seymour & Bradbury 1999). The minimum interval between outbreaks of *Acanthaster* spp. at individual reefs will be set by local recovery of coral assemblages (Bradbury et al. 1985, Fabricius et al. 2010) because low levels of coral cover would prevent outbreaks from becoming established. This implies that the overall incidence of outbreaks of *Acanthaster* spp. should decline with sustained and increasing degradation of coral reef ecosystems. In particular, there should be longer periods between successive outbreaks on individual reefs. Alternatively, limited abundance of coral prey might constrain the overall abundance or persistence of *Acanthaster* spp. while outbreaks continue to occur at equivalent or increasing frequency (Fabricius et al. 2010) or might even become chronic (e.g., Zann et al. 1990, Mendonça et al. 2010), thereby causing accelerated degradation of reef ecosystems. The current wave outbreaks on the GBR provide a good opportunity to test whether outbreaks of *Acanthaster planci* can become established on reefs with low coral cover, as many reefs in the central GBR were impacted by a severe tropical storm (category 5 cyclone Yasi) that tracked across the GBR in early 2011 (e.g., Lukoschek et al. 2013). Coral cover remains at or below 5% on many reefs exposed to this disturbance, and preferred coral prey of *Acanthaster* spp. (e.g., *Acropora*) are particularly scarce (e.g., Lukoschek et al. 2013).

Despite increases in the overall number of reported outbreaks, the geographical distribution of outbreaks reported in 1990–2012 is fairly consistent with the record prior to 1990 (e.g., Moran 1986, Birkeland & Lucas 1990). Most notably, the vast majority (167/246) of outbreaks of *Acanthaster* spp. are reported from the Pacific Ocean (Figure 9), mainly from the western Pacific, including Australia's GBR, southern Japan, Micronesia, and Melanesia. There has been a slight increase in the proportion of outbreaks reported from the Indian Ocean, increasing from 11.0% in 1956–1989 to

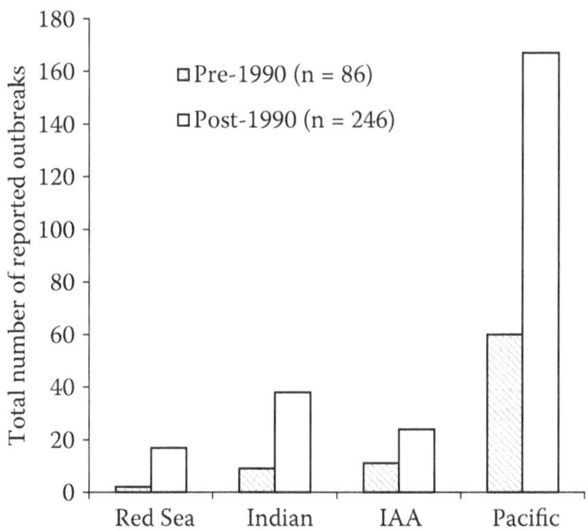

Figure 9 Number of reported outbreaks in different regions of Indo-Pacific (Red Sea, Indian Ocean, Indo-Australia Archipelago, and Pacific Ocean) pre-1990 versus post-1990. No account is made of the extent, severity, or recurrence of outbreaks at each location.

15.4% in 1990–2012, with unprecedented outbreaks of *Acanthaster* spp. recorded in some isolated locations, such as Chagos (R. Roche unpublished data 2013). In most locations, outbreaks have now been reported at least twice, and up to four times, since the 1960s within the same area, if not the same reef, showing that outbreaks of *Acanthaster* spp. are not rare, singular events. The greatest concentrations of outbreaks have occurred within extensive reef systems of the GBR and southern Japan and within the Philippines (Figure 8). However, there have also been recent outbreaks on many small, isolated, and relatively unpopulated reef systems.

Analyses of all reported outbreaks since 1990 showed that the majority of outbreaks (139 of 246) have occurred on arrays of platform reefs along continental shelves, such as the GBR. An additional 72 (56%) outbreaks have been reported on high islands, such as Moorea in French Polynesia. The remaining outbreaks (35 of 246) have been reported from low-lying islands, atolls, or completely submerged reef platforms. It is likely that the causes, especially anthropogenic contributions to the initiation of outbreaks, are likely to vary among these different reef systems (Birkeland 1982). Notably, continental and high island reefs will be much more subject to terrestrial run-off (Birkeland 1982), which is one possible cause of outbreaks (see the section on the nutrient enrichment hypothesis). The flip side to this is that human populations on low-lying islands and atolls have reduced capacity for agriculture and are much more reliant on coral reef fisheries (McClanahan et al. 2002), and excessive exploitation of predatory fishes has also been linked to increased severity of outbreaks of *A. planci* (Dulvy et al. 2004). To assess whether outbreaks occur more or less than expected on different types of reefs, the relative frequency of outbreaks needs to be compared to the proportional area of continental and high island reefs versus low-lying atolls and submerged reefs, but such data are not readily available.

Fine-scale patterns of outbreaks

Australia's Great Barrier Reef, the GBR, extending 2,000 km along Australia's eastern coast, is the world's largest reef system and has the dubious distinction of supporting both the most extensive and the greatest frequency of outbreaks of *Acanthaster* spp. The first documented outbreak was detected in 1962 at Green Island (Pearson & Endean 1969), although high densities of starfish were reported on other GBR reefs in the 1950s and earlier (Vine 1973, Ganter 1987). Since 1962, there have been three additional outbreak episodes on the GBR, commencing in 1979, 1993, and 2010. The initiation and progression of the first two outbreaks, based on extensive but largely uncoordinated sampling across the GBR, was discussed at length by Moran (1986). The publication of Moran's review in 1986 roughly coincided with the advent of systematic monitoring on the GBR to document patterns of outbreaks of *Acanthaster planci* in time and space (e.g., Sweatman et al. 2008). When monitoring began in 1985, there were active outbreaks on many reefs near Townsville (~18.5°S) and in the Swain reefs well off shore from Gladstone (~22°S) (Figure 10). Subsequent monitoring documented the progressive spread in the southward extent of active outbreaks (Figure 10), consistent with patterns reported in previous outbreaks (Reichelt et al. 1990a). However, outbreaks in the Swain region (southern GBR) appear to occur independently of the wave of outbreaks that affect reefs from Cooktown to Mackay (http://www.aims.gov.au/docs/research/biodiversity-ecology/threats/cots-animation.html).

The initiation and spread of outbreaks of *Acanthaster planci* on the GBR have been fairly consistent in all four recorded outbreaks, including the current outbreak that started in 2010. Each of the outbreaks appears to have been initiated on midshelf reefs in the north-central region of the GBR (the 'initiation box'; Figure 11) between Lizard Island (14.6°S) and Cairns (17°S) and mainly in the northern portion of this area. During the initial waves of outbreaks (in 1962 and 1979), high densities of *A. planci* were first detected (or at least reported) on reefs close to Cairns (e.g., Green Island), although increases in densities of *A. planci* may have occurred even earlier on reefs to the north, as

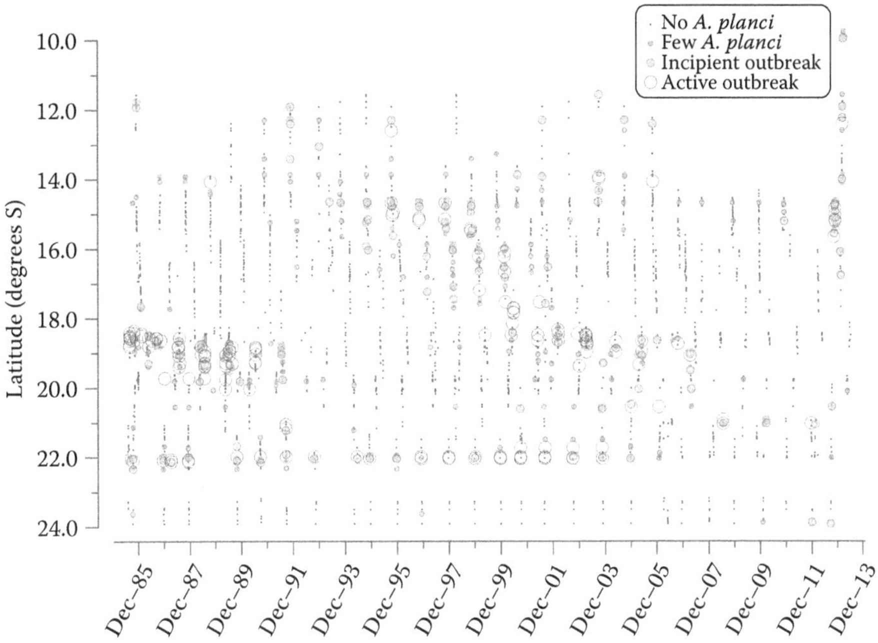

Figure 10 Occurrence of *Acanthaster planci* on reefs of the Great Barrier Reef (arranged by latitude) between 1985 and 2013 based on manta tow surveys of reef perimeters. Note that the same reef may be represented in several years. Reefs are divided into four categories: dots indicate reefs with low densities of *A. planci* (≤3 per reef); smaller grey symbols indicate reefs with more than 3 starfish but less than incipient outbreak densities (mean ≥ 0.22 *A. planci* per 2-min tow); larger grey symbols indicate reefs with incipient outbreak densities of *A. planci* (mean ≥ 0.22 *A. planci* per tow); large unfilled symbols indicate reefs with active outbreak densities (mean ≥ 1.0 *A. planci* per tow).

recorded during the start of outbreaks in 1993 and 2010. Limited temporal and spatial resolution of monitoring of GBR reefs (e.g., Sweatman et al. 1998) mean that it is still unclear whether outbreaks start from a single reef within this area or arise simultaneously on a suite of closely positioned reefs (Pratchett 2005). In 1994, outbreaks of *Acanthaster* spp. were first recorded at Lizard Island but also were reported soon after (within the same year) at several other reefs within the immediate area, including Linnet, North Direction, and Rocky Islet (Sweatman et al. 1998). The exact timing and location of primary outbreaks are critical for understanding and identifying potential causes or triggers of outbreaks on the GBR, which suggests that this area should be monitored intensively whenever initiation of a new wave of outbreaks seems possible. Alternatively, the pattern of initiation and spread of outbreaks may be reconstructed using the size-frequency distribution of high-density populations on disparate reefs (*sensu* Kenchington 1977), combined with newly developed genetic markers (Yasuda et al. 2006, Wainwright et al. 2012) to potentially map the directionality of dispersal among reefs within the initiation box.

The recurring pattern on the GBR is the southward progression of outbreaks from 17–20°S, at a rate of 1° of latitude every 3 years (Kenchington 1977, Reichelt et al. 1990a). This is attributed to progressive colonisation of reefs by larvae spawned at the outbreak front, which disperse only 1–2° of latitude (Kenchington 1977). Outbreaks of crown-of-thorns starfish also appear to spread northward from the initiation area, but the northward progression is slower and less consistent (Moran 1992), although reefs in the northern region have not been surveyed as frequently as those in other regions of the GBR. The progression of outbreaks northward and southward on the GBR has been attributed to movement of adult starfish between reefs (e.g., Talbot & Talbot 1971), but differences

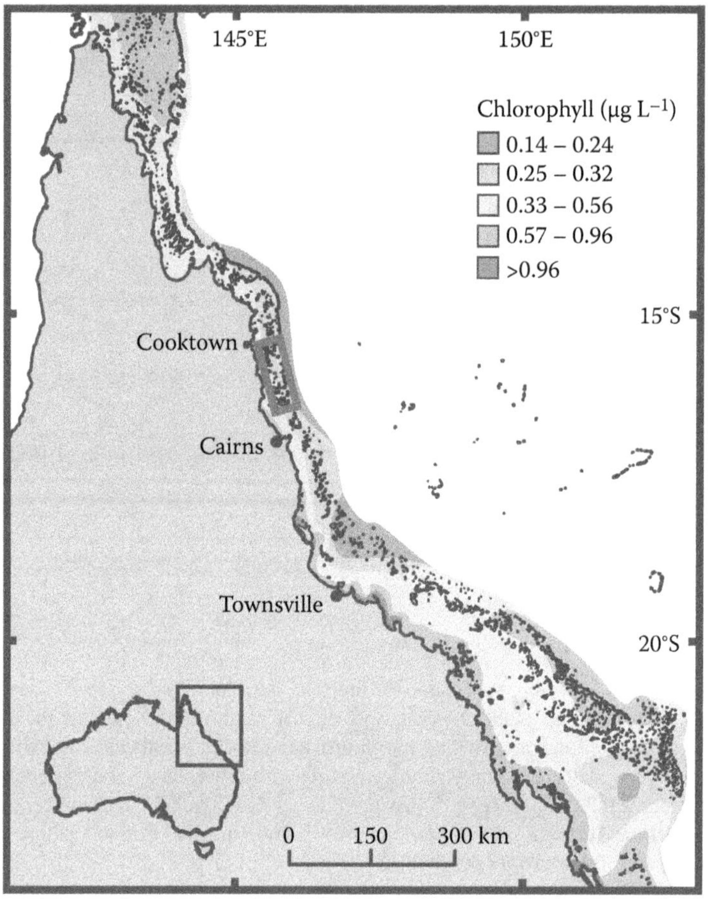

Figure 11 (See also colour figure in the insert) Map of Australia's Great Barrier Reef showing the variation in average ('normal') patterns of productivity (chlorophyll a concentrations). Data source: E-atlas (http://maps.e-atlas.org.au/). Also shown is the small area north of Cairns (the 'initiation box') where primary outbreaks of *Acanthaster planci* are thought to be initiated and then spread (mostly southwards), causing progressive waves of secondary outbreaks (e.g., Fabricius et al. 2010).

in the size-frequency distribution of disparate populations seem more likely to be the result of larval dispersal (Kenchington 1977), with prevailing currents (Black & Moran 1991).

Disturbances caused by outbreaks of crown-of-thorns starfish

Acanthaster spp. have gained considerable notoriety, not only because of their tendency to undergo rapid and dramatic increases in populations size but also because outbreak populations can cause extensive and widespread depletion of scleractinian corals (e.g., Chesher 1969, Pearson & Endean 1969, Randall 1973, Colgan 1987). At low densities (<10 starfish.ha^{-1}), *Acanthaster* spp. have a negligible impact on scleractinian coral cover (e.g., Glynn 1973, Zann et al. 1990). However, the combined feeding activities of high densities of crown-of-thorns starfish (up to 151,650 starfish/km^2; Kayal et al. 2012) can cause rapid and pronounced coral loss. In Guam, for example, high densities (>1000 starfish per hectare) of *Acanthaster planci* persisted for 30 months in 1967–1969 and killed virtually all scleractinian corals (>90% coral mortality) from 1 to 68 m depth along a 38-km stretch of coastline (Chesher 1969). Similarly, in French Polynesia, large aggregations of *A. planci* systematically moved around the entire circumference of Moorea and removed more than 96% of the coral (Kayal et al. 2012). Whereas climate-related disturbances (e.g., bleaching and disease) are a major concern for reef managers around the world, especially since 1998 (Knowlton 2001), outbreaks of *Acanthaster* spp. continue to occur throughout the Indo-Pacific region (e.g., Pratchett 2005, Baine 2006) and have so far been responsible for far more coral mortality than has been attributed to climate-related disturbances (Bruno & Selig 2007, Osborne et al. 2011, Trapon et al. 2011, De'ath et al. 2012).

Although outbreaks of *Acanthaster* spp. can cause massive and widespread coral depletion (e.g., Chesher 1969, Pearson & Endean 1969, Randall 1973, Colgan, 1987), this is not always the case. Outbreaks of *Acanthaster* spp. vary greatly, not only in the number, size, and density of starfish, but also in their effects (Moran 1986). For example, high densities of *Acanthaster planci* persisted for more than 18 months (1969–70) at Molokai Island, Hawaii, but had negligible impacts on the abundance of scleractinian corals (Branham et al. 1971). Similarly, in Panama, high-density populations of *A. planci* caused minimal reductions in the live cover of scleractinian corals (Glynn 1974, 1976). Large aggregations of many hundreds to thousands of crown-of-thorns starfish have been reported on reefs throughout the Indian and Pacific oceans, including Panama (Glynn 1974, 1976); Samoa (Birkeland & Randall 1979); Micronesia (Chesher 1969, Colgan 1987); southern Japan (Nishihira & Yamazato 1974, Keesing 1992); the GBR (e.g., Moran et al. 1988, Reichelt et al. 1990a); Cocos-Keeling Islands (Colin 1977); and the Red Sea (Ormond & Campbell 1974). However, incidences of large-scale destruction of scleractinian corals by outbreaks of *Acanthaster* spp. have occurred primarily within the southern and western Pacific (Moran 1986, Birkeland & Lucas 1990). More specifically, devastating outbreaks of *Acanthaster* spp. have been most apparent in southern Japan (Nishihira & Yamazato 1974, Keesing 1992), on Australia's GBR (Pearson & Endean 1969), in Micronesia (Chesher 1969, Colgan 1987), and in French Polynesia (Trapon et al. 2011, Kayal et al. 2012).

Variation in the effects of outbreaks of *Acanthaster* spp. in different regions of the Pacific might be explained by the relative dominance of *Acropora* in local coral assemblages (Birkeland & Lucas 1990); *Acropora* tends to dominate in the southern and western Pacific and is consistently among the corals that are first and worst affected during outbreaks (e.g., Pratchett et al. 2009a, Pratchett 2010). In the northern and eastern Pacific (e.g., the main Hawaiian islands and Panama), *Acropora* is relatively scarce and coral assemblages tend to be dominated by *Pocillopora*, which is much less susceptible to crown-of-thorns attack (e.g., Glynn 1974, 1976), although it can be damaged extensively by extremely severe outbreaks (e.g., Adam et al. 2011, Kayal et al. 2012). Geographic variation in the effects of *Acanthaster* spp. may also result from differences in the population dynamics and behaviour among the four nominal sister species distributed in different parts of the Indo-Pacific (Vogler et al. 2008). *Acanthaster* spp. from throughout the Indo-Pacific ostensibly look and behave the same way, but devastating impacts of crown-of-thorns starfish appear to be confined to the Pacific, which is the geographical range of *Acanthaster planci*. This warrants explicit comparisons of demographic rates, feeding rates, and feeding preferences among *Acanthaster* spp. from each of the four distinct subpopulations identified by Vogler et al. (2008), extending the studies in the Pacific to the Red Sea and both southern and northern Indian Ocean regions.

Effects of *Acanthaster* spp. can also vary at much smaller scales, within and among adjacent coral reefs (e.g., Pratchett 2010). Highest densities of *Acanthaster* spp., and hence the greatest depletion of live coral cover, tend to occur on the leeward side of reefs (Laxton 1974, Pratchett 2005), which is caused by either (1) high abundance of preferred coral prey (e.g., monospecific stands of staghorn *Acropora*) in these habitats or (2) reduced water flow and turbulence in back reef environments, which can potentially dislodge starfish while feeding in more exposed locations (Endean & Stablum 1973). However, crown-of-thorns starfish are rarely found within shallow, semi-enclosed lagoons, which often support many suitable coral prey (e.g., Pratchett 2005, Pratchett et al. 2011), suggesting that other environmental variables also influence their fine-scale distribution. *Acanthaster* spp. have rarely been observed on mesophotic (>30 m deep) reefs of the GBR (T. Bridge, unpublished data 2013), even though these habitats are extensive (Harris et al. 2013) and can support a high cover of scleractinian corals (Bridge et al. 2013). These habitats could provide refuges for coral species to reseed shallow water reef habitats in the aftermath of devastating outbreaks of *Acanthaster* spp. (Bridge et al. 2013).

Directional shifts in coral composition

Disturbances influence the structure of ecological communities either through selective mortality of particular species or by random, localised mass mortality across a wide range of different species (often termed 'catastrophic mortality') that clears space for recolonisation (Petraitis et al. 1989). Small-scale or relatively discrete disturbance events (e.g., predation events) usually have a disproportionate impact on certain individuals or species (Petraitis et al. 1989) and so exert a strong structuring influence on populations and communities. Such events may, for example, increase diversity and promote coexistence of species by reducing the abundance of competitively dominant species and allowing inferior competitors to persist (e.g., Porter 1972, 1974). However, disturbances that differentially affect species may also reduce diversity by disproportionately affecting rare species, thereby increasing the dominance of already-abundant species (e.g., Glynn 1974, 1976). Catastrophic disturbances, meanwhile, eliminate most (if not all) of the species in an area and may contribute to increased species diversity by preventing dominant species from monopolising all available resources (Petraitis et al. 1989). Although effects of *Acanthaster* spp. on coral assemblages are essentially a predatory interaction, resulting loss of scleractinian corals can be catastrophic (e.g., Chesher 1969), and the specific effects will depend on the frequency and severity of major outbreaks.

Endean and Cameron (1985) suggested that the severity of any given disturbance should be judged not by how much coral is killed but by the type of coral killed. Notably, the removal of long-lived and slow-growing species (e.g., *Porites* spp.) is likely to have longer-term impacts on community structure compared with selective removal of short-lived, fast-growing species (Endean & Cameron 1985). Like most disturbances (e.g., freshwater plumes, Jokiel et al. 1993; coral bleaching, Marshall & Baird 2000), outbreaks of *Acanthaster* spp. tend to have a disproportionate impact on fast-growing coral species (e.g., *Acropora*), which also recruit abundantly and often recover rapidly in the aftermath of major disturbances (e.g., Linares et al. 2011). However, *Acanthaster planci* do sometimes feed on large and old colonies of massive *Porites* (Done 1985). The specific effects of crown-of-thorns outbreaks (and other recurrent disturbances) on the composition of coral assemblages will depend on the frequency and severity of these disturbances. Severe but infrequent disturbances are likely to have a disproportionate effect on slow-growing species that are incapable of recovering between successive disturbances (Done 1985). However, frequent moderate outbreaks (especially in isolated locations) are likely to cause the localised extirpation of corals that are favoured by *Acanthaster* spp. (e.g., Berumen & Pratchett 2006, Pratchett et al. 2011).

A striking example of persistent shifts in the structure of coral assemblages has been documented in Moorea, French Polynesia, where a major outbreak of *Acanthaster planci* in 1980–1981 led to the disproportionate loss of *Acropora* corals (Bouchon 1985). Ongoing disturbances (including cyclones, coral bleaching, and a further outbreak of *A. planci*) have since prevented recovery of *Acropora* spp., such that the dominant coral genera in 2009 were *Pocillopora* and *Porites* (Berumen & Pratchett 2006, Adjeroud et al. 2009, Pratchett et al. 2011, 2013), which were relatively unaffected by outbreaks of *A. planci*. Given sufficient time between disturbances, *Acropora* might be expected to recover eventually and regain its former dominance in Moorea. However, Pratchett et al. (2011) measured recruitment rates for *Acropora* and other dominant coral genera (*Porites* and *Pocillopora*) in Moorea and showed that the relative abundance of new recruits strongly reflected the current patterns of adult abundance. This is evidence of a positive-feedback mechanism, which is likely to reinforce and sustain the altered community structure (Knowlton 1992, Nyström et al. 2008).

Selective depletion of certain coral species, linked to marked feeding preferences of *Acanthaster* spp. (e.g., Brauer et al. 1970, Collins 1975, Ormond et al. 1976, Colgan 1987, Keesing 1990, De'ath & Moran 1998), has the capacity to greatly alter coral diversity (e.g., Porter 1972, 1974, Glynn 1974, Colgan 1987) and therefore habitat heterogeneity. In the eastern Pacific, *Acanthaster planci* tend to avoid the most abundant coral, *Pocillopora* (Glynn 1974, 1976, 1980). By feeding mostly on rare coral species, this further increases the dominance of *Pocillopora,* leading to declines in coral diversity (see also Branham et al. 1971). In the Indo-West Pacific, however, *A. planci* feeds predominantly on *Acropora* spp. and *Montipora* spp. (e.g., Ormond et al. 1976, Colgan 1987, Keesing 1990, De'ath & Moran 1998, Pratchett 2010), which are relatively abundant and often competitively dominant corals. This presumably increases the prevalence of other less-abundant coral species, which Porter (1972, 1974) suggested might lead to increases of coral diversity. Pratchett (2010) explicitly tested for changes in coral diversity during a moderate outbreak of *Acanthaster planci* at Lizard Island, northern GBR, and showed that despite disproportionate effects on preferred corals, coral diversity declined (rather than increased) with declines in coral cover. Contrary to Porter's (1972) suggestion, it seems unlikely that outbreaks of *A. planci* (even relatively moderate outbreaks) would ever cause increases in coral diversity because crown-of-thorns starfish are not sufficiently averse to rare corals. In the absence of preferred corals, *Acanthaster* spp. will certainly feed on other less-preferred corals (Moran 1986, Keesing 1990, Endean & Cameron 1990), so further outbreaks on highly degraded reefs are likely to lead to even more extensive depletion of corals, with collateral effects on the local diversity of other reef-associated organisms.

Indirect effects of outbreaks of crown-of-thorns starfish

Extensive coral depletion caused by outbreaks of *Acanthaster* spp., as well as directional shifts in the composition of coral assemblages, have broad impacts on a wide variety of coral reef organisms. Outbreaks of crown-of-thorns starfish have been linked to increased abundance of soft corals (e.g., Endean 1971, Chou & Yamazato 1990), algae (Larkum 1988), urchins (Belk & Belk 1975), and herbivorous fish species (Endean & Stablum 1973, Wass 1987), while causing declines in abundance of coral-dependent fishes (e.g., Sano et al. 1984, 1987, Williams 1986, Munday et al. 1997) and motile invertebrates (Garlovsky & Bergquist 1970). Changes in the abundances of these reef organisms are the indirect result of massive reductions in the abundance of scleractinian corals. For example, increases in the abundance of urchins (specifically, *Echinometra mathaei* and *Diadema* spp.) following outbreaks of *Acanthaster planci* have been related to increased food availability, as algae colonise skeletons of dead, but intact, corals (e.g., Belk & Belk 1975, Larkum 1988). Given the potential severity of starfish outbreaks, it is not surprising that many coral reef organisms are indirectly affected.

Most studies that have considered secondary effects of outbreaks of *Acanthaster* spp. have measured changes in the abundance of coral reef fishes associated with localised coral loss (e.g., Williams 1986, Sano et al. 1987, Hart et al. 1996, Munday et al. 1997, Adam et al. 2011, Pratchett et al. 2012). Aside from herbivorous fish species (e.g., Endean & Stablum 1973, Wass 1987, Adam et al. 2011), most fishes tend to decline in abundance in the aftermath of major outbreaks of *Acanthaster* spp. that cause extensive coral loss (e.g., Bouchon-Navaro et al. 1985, Williams 1986, Sano et al. 1987, Munday et al. 1997). These effects are most pronounced for specialist fishes that depend on corals for food (e.g., butterflyfishes; Williams 1986) or habitat (e.g., coral-dwelling gobies, Munday et al. 1997; coral-dwelling damselfishes, Pratchett et al. 2012). At least 133 species (and 11 different families) of coral reef fishes feed on scleractinian corals (Cole et al. 2008), the majority (69 species) of which are butterflyfishes (family Chaetodontidae). Also, 320 species (from 39 different families) have been shown to use live corals as habitat (Coker et al. 2014). These data suggest that 8–10%

of coral reef fishes will be directly and adversely affected by extensive coral depletion (see also Munday et al. 2008). Localised depletion of preferred cover may ultimately lead to local or global extinction of fishes that are directly reliant on corals (e.g., Kokita & Nakazono 2001, Munday 2004). This is particularly so for highly specialised fishes that rely on only a limited suite of coral species, although it is important to look at other biological traits that may offset extinction risk in these species (Lawton et al. 2011).

The effects of severe outbreaks of *Acanthaster planci* also extend well beyond the few fishes that are directly dependent on live corals (e.g., Sano et al. 1987), especially where coral depletion is associated with loss of habitat structure. Extensive coral depletion caused by large or persistent outbreaks of *A. planci* (e.g., Chesher 1969, Pearson and Endean 1969, Colgan 1987, Sano et al. 1987) almost invariably leads to marked declines in habitat and topographical complexity, which are critical for sustaining a high diversity of reef fishes and other reef-associated organisms (Wilson et al. 2006, Pratchett et al. 2009b). Once dead, the exposed skeletons of scleractinian corals are susceptible to biological and physical erosion (Hutchings 2011). Over time, skeletons of erect branching corals (e.g., *Acropora* and *Pocillopora*) break down into coral rubble (Sheppard et al. 2002), whereas more robust skeletons of massive corals (e.g., *Porites*) may become dislodged or gradually eroded *in situ* (Sheppard et al. 2002). The structural collapse of dead coral skeletons takes 4–7 years (Pratchett et al. 2008), and if there is no substantial recovery of corals in the meantime, then the corresponding loss of habitat structure and topographic complexity can have far-reaching effects on the abundance and diversity of fishes (Sano et al. 1984, 1987, Pratchett et al. 2009b). In southern Japan, for example, Sano et al. (1987) reported more than 65% fewer individuals and species of fishes at a reef that had been devastated by localised outbreaks of *Acanthaster planci*, compared with nearby reefs with extensive growth of staghorn *Acropora*. On the reef that had been devastated by *A. planci*, extensive stands of *Acropora* corals were rapidly eroded, converting once highly complex 3-dimensional habitats into flat, homogeneous rubble fields (Sano et al. 1987).

Causes of *Acanthaster* outbreaks

Unifying theories for population outbreaks were proposed by scientists working in terrestrial environments long before outbreaks of the crown-of-thorns starfish were even known to occur (e.g., MacArthur 1955, Elton 1958). Both MacArthur (1955) and Elton (1958) argued that population outbreaks are manifestations of inherent instability within certain systems, attributed to either (1) particular life-history characteristics (e.g., high fecundity, short generation times, high mortality during their early life history, and generalised patterns of prey and habitat use) that predispose an organism to major fluctuations in population size or (2) major changes in the physical or biological environment that release the outbreaking population from usual regulating factors (e.g., Andrewartha & Birch 1984, Berryman 1987). Numerous hypotheses have been put forward to explain the occurrence of population outbreaks of *Acanthaster* spp. (reviewed by Moran 1986, Birkeland & Lucas 1990). These hypotheses generally fall into two groups that place importance either on factors affecting recruitment rates (i.e., 'natural causes hypothesis', Vine 1973; 'larval recruitment hypothesis', Lucas 1973; 'terrestrial run-off hypothesis', Birkeland 1982) or on changes in the behaviour or survivorship of post-settlement individuals (i.e., 'predator removal hypothesis', Endean 1969; 'adult aggregation hypothesis', Dana et al. 1972; 'prey-threshold hypothesis', Antonelli & Kazarinoff 1984). Whereas several of these hypotheses have been considered biologically improbable (e.g., Potts 1981, Birkeland & Lucas 1990), no single hypothesis has universal support. Sudden and dramatic increases in the abundance of starfish must involve successful recruitment (Birkeland & Lucas 1990), but both pre- and post-recruitment processes are likely to contribute to the dynamic nature of *Acanthaster* populations (Bradbury & Antonelli 1990), as has also been shown for many other marine organisms (e.g., Jones 1987, 1991, Hughes 1990). Many biologists and theoretical ecologists concur that single-factor hypotheses that seek to explain the occurrence of crown-of-thorns outbreaks in all locations and at all times are likely to oversimplify the population dynamics for this organism (reviewed by Birkeland & Lucas 1990, Bradbury & Antonelli 1990). It is also important to recognise that *Acanthaster* spp., probably more so than any other coral reef organism, are predisposed to major fluctuations in population abundance (Birkeland 1989b).

Given the life-history characteristics of *Acanthaster* spp., it is almost harder to explain the persistence of low-density populations than it is to explain outbreaks (Endean & Cameron 1990). On any given reef, it is likely that outbreaks will occur periodically through the effects of random environmental variation on reproductive success or larval survival. As discussed previously, *Acanthaster* spp. have extremely high fecundity (e.g., Conand 1983, 1984), and fertilisation rates, as well as developmental rates and survivorship of larvae, are highly subject to the vagaries of local environmental conditions. Moore (1990) examined the characteristics of key locations where *Acanthaster* spp. occur, but never in outbreak densities, and suggested that a combination of (1) low and fragmented coral cover (causing individuals to be dispersed), (2) hydrodynamic conditions that cause larvae to be retained rather than exported, and (3) relatively high populations of predators (reducing starfish numbers and causing individuals to disperse) that prevent outbreaks from occurring.

Understanding the causes of crown-of-thorns outbreaks has been greatly hindered by a lack of data on fine-scale temporal and spatial changes in the population structure and dynamics of *Acanthaster* spp. (Moran 1986). In particular, there are few data on changes in the distribution, density, and spawning behaviour of *Acanthaster* spp. in the period immediately preceding an outbreak. This is because most studies of outbreak populations (e.g., Chesher 1969, Pearson & Endean 1969, Branham et al. 1971, Sakai 1985) are initiated after starfish densities have already increased to outbreak levels. Also, few studies have continually monitored changes in the structure and dynamics of *Acanthaster* populations at regular intervals over an extended period, encompassing an entire outbreak cycle (Moran 1986). On the GBR, for example, long-term and extensive monitoring of the distribution and abundance of *Acanthaster planci* is undertaken by the Australian Institute of Marine Science (Sweatman et al. 2011), but the methods developed to sample over vast reef areas prohibits the collection of detailed information on population structure and reproductive condition. Inherent trade-offs in the collection of fine-scale biological information versus broad-scale surveys to detect changes in the abundance of *Acanthaster* spp. across extensive reef areas represent one of the greatest challenges to understanding the processes that contribute to the initiation of new and distinct outbreaks. On the GBR, it may be possible to focus detailed surveys in the area where primary outbreaks are known to occur. Elsewhere, however, the distribution of primary versus secondary outbreaks is largely unknown.

Natural versus anthropogenic drivers

Reviews of the effects of disturbances on coral reefs (e.g., Pearson 1981) invariably distinguish between natural (e.g., storms and other weather events) versus anthropogenic (e.g., overfishing and pollution) disturbances. The connotations of this are obvious, in that it is the anthropogenic disturbances that are considered responsible for the recent (anthropocene) degradation of coral reef ecosystems (e.g., Hughes et al. 2003) and need to be managed. However, the distinction between natural and anthropogenic disturbances is not always clear (e.g., Potts 1981). Severe tropical storms (cyclones, hurricanes, and typhoons), for example, are recurrent disturbances that have impacted coral reefs throughout their evolution and development. However, Webster et al. (2005) reported that anthropogenic climate change is increasing the severity, if not the frequency, of severe tropical storms (but see Klotzbach 2006, Landsea et al. 2006), with obvious ramifications for coral reef ecosystems. The role of anthropogenic activities in causing or exacerbating outbreaks of crown-of-thorns starfish is also highly controversial (e.g., Potts 1981).

When extensive outbreaks of *Acanthaster* spp. were documented in the late 1960s (Chesher 1969, Pearson & Endean 1969), it was immediately assumed that these were new and unprecedented phenomena linked to human activity, such as coastal development (Chesher 1969), pesticides, and pollutants (Randall 1972), or excessive harvesting or predatory organisms (Endean 1977). In support of this, Endean (1982) pointed out that the first affected reefs (e.g., Green Island) on the GBR were those that had greatest human visitations. Several scientists (e.g., Dana 1970, Vine 1970, 1973, Weber & Woodhead 1970) put forward the contrary view that outbreaks of *Acanthaster* spp. were a natural occurrence that had simply gone unnoticed prior to the 1960s. Rapid increases in the number of reports of 'infestations' and 'plagues' of *Acanthaster* spp. from throughout the Indo-Pacific following initial awareness of the issue were taken as evidence for this (Vine 1973). Some argued that outbreaks had occurred in both the recent (e.g., Vine 1973) and distant past (e.g., Walbran et al. 1989a,b).

There are many anecdotal accounts of crown-of-thorns starfish occurring in high densities at locations across the Indo-Pacific well before outbreaks were reported by scientists (Vine 1970, 1973). Fishers in Santa Ysabel in the Solomon Islands recalled a time in the 1930s when night fishing was hazardous because of the abundance of *Acanthaster planci* (Vine 1970). Former trochus and pearl shell divers on the GBR recalled seeing large numbers of the starfish on individual reefs going back to the 1930s (Ganter 1987). In 1913, H.L. Clark collected three *A. planci* from the reef flat of Mer Island in the eastern Torres Strait (far northern GBR) without diving (Clark 1921), suggesting that starfish must have been common. These anecdotal records suggest that localised outbreaks have occurred in the past (Vine 1973). However, the previous occurrence of waves of outbreaks as seen on the GBR in recent decades cannot be confirmed in the absence of systematic, broad-scale monitoring.

Further evidence of past outbreaks (over geological time) has been sought in the form of mesodermal skeletal elements of *Acanthaster* spp. in the sediment record. Skeletal elements from *A. planci* have been found in numerous sediment cores from GBR reefs (Frankel 1977, 1978, Walbran et al.1989a,b), but reconstructing a history of *A. planci* numbers and drawing conclusions about the existence of past outbreaks is not simple because of disturbance by burrowing organisms, varying sedimentation rates, and differential compaction of the sediments (Fabricius & Fabricius 1992, Keesing et al. 1992, Pandolfi 1992). Only a few cores from a limited number of reefs have been studied intensively, and the resulting reconstructions are relatively variable. Even if outbreaks did occur prior to the 1950s (which appears likely), the key question is whether outbreaks are occurring more frequently in recent times, and if this reflects increasing anthropogenic changes to marine and coastal environments (Brodie 1992).

While debate continues about the role of anthropogenic activities in causing or exacerbating outbreaks of *Acanthaster* spp., it is clear that the current regime of disturbances to which most coral reefs are subject cannot be sustained (Gardner et al. 2003, Bruno & Selig 2007, De'ath et al. 2012, Fabricius 2013). On the GBR, for example, De'ath et al. (2012) reported a 50.7% decline in mean coral cover across 214 reefs that have been repeatedly and comprehensively surveyed (using reef-wide manta tows) from 1985 to 2012. The timing and rates of coral loss varied spatially (see also Sweatman et al. 2011), but these data are evidence of significant reef-wide habitat degradation, largely attributable to recurrent outbreaks of *Acanthaster planci* that compound other large-scale and persistent disturbances (Osborne et al. 2011, De'ath et al. 2012). Over this 27-year period, outbreaks of *A. planci* affected 49% of reefs, and hind casting showed that coral cover would have increased at 0.89% per year (as opposed to annual declines of 0.53%) were it not for impacts of crown-of-thorns starfish (De'ath et al. 2012). Similarly, in the central Pacific, the reefs surrounding Moorea Island in French Polynesia have been subject to a high frequency of different disturbances, including seven distinct episodes of mass coral bleaching, two major cyclones, and two outbreaks of *A. planci* since 1979 (Trapon et al. 2011). Despite this frequency and diversity of disturbances in Moorea, significant long-term coral loss and degradation of reef environments are clearly attributable to the devastating effects of *A. planci* outbreaks in 1980–1981 and 2007–2011 (Adam et al. 2011, Trapon et al. 2011, Kayal et al. 2012). It has been suggested that if disturbances of this frequency and magnitude had occurred throughout the geological period (Holocene) during which contemporary coral assemblages evolved and coral reefs developed, then the biological and physical structure would be vastly different (Randall 1972). Most notably, large colonies of slow-growing massive corals (mostly *Porites*) could not withstand extremely severe outbreaks of *Acanthaster* spp. that cause high rates of mortality across all preferred and non-preferred corals (e.g., Endean & Cameron 1985, Done et al. 1988, Done 1992).

A significant and increasing effect of *Acanthaster* spp. on coral assemblages and reef ecosystems is not in itself evidence that outbreaks are unnatural (Birkeland & Lucas 1990). Rather, other anthropogenic disturbances (e.g., fishing and harvesting, sedimentation, eutrophication, and pollutants) might have undermined the capacity of reef ecosystems to withstand these periodic disturbances, eroding their resilience and leading changes in the ecosystem responses to persistent and ongoing disturbances. Even more likely is that the pervasive effects of humans on coastal ecosystems have fundamentally altered the structure and function of both crown-of-thorns populations and reef ecosystems, forever altering any semblance of a natural system. Managing ongoing effects of *Acanthaster* spp. is conditional on identifying the specific factor(s) that cause or exacerbate contemporary outbreaks, and much of the current discussion is centred around one of two alternative hypotheses: (1) nutrient enrichment and (2) predatory release (e.g., Birkeland & Lucas 1990, McClanahan et al. 2002, Brodie et al. 2005, Mendonça et al. 2010, Fabricius 2013).

Nutrient enrichment hypothesis

The notion that outbreaks of *Acanthaster* spp. may arise due to enhancement of larval survivorship through nutrient enrichment has been proposed several times (e.g., Pearson & Endean 1969, Lucas 1973, Nishihira & Yamazato 1974, Birkeland 1982, Brodie 1992, Brodie et al. 2005, Fabricius et al. 2010). Birkeland (1982) suggested that outbreaks of *Acanthaster planci* at several locations in Micronesia and Polynesia tended to occur 3 years after extremely heavy rainfall events, often preceded by extended droughts. Birkeland (1982) argued that such events provide a pulse of nutrients that stimulate phytoplankton blooms, which supplement otherwise-limited food for crown-of-thorns larvae (Lucas 1973). However, enhanced survival of larval crown-of-thorns starfish may also be related to specific environmental conditions (low salinity and high temperatures) at times and in locations affected by river run-off (Henderson 1969, Lucas 1973). Fundamental to this terrestrial run-off hypothesis is the notion that outbreaks occur suddenly, and 3 years following heavy rainfall periods (Birkeland 1982), which accounts for the time required for larval starfish to settle on the reef, metamorphose into the adult form, begin feeding on corals, and attain sufficient size (200–300 mm) to emerge from the reef matrix and become readily apparent (Figure 1).

However, some of Birkeland's (1982) findings have since been questioned, based on inconsistencies in either the initiation of outbreaks or the timing of severe tropical storms and peak rainfall events (Endean & Cameron 1990). In Guam, for example, major outbreaks were first reported in 1967 (Chesher 1969), meaning that the larval recruitment event that led to this outbreak would have preceded the drought-breaking rains in July 1965 by 18 months (Endean & Cameron 1990). Similarly for the GBR, Fabricius et al. (2010) argued that each of the major episodes of outbreaks was initiated by a major flooding event (in 1958, 1974, 1991, and 2008) that contributed to increased survival of larvae (see also Day 2000). However, the time between flooding events and subsequent outbreaks of *A. planci* ranges from 2 to 5 years (Table 4). The purported years of major flooding events given in Fabricius et al. (2010) also do not correspond to the documented incidence of major drought-breaking floods, based on barium/calcium ratios in long-lived colonies of *Porites* from Havanah and Pandora reefs (McCulloch et al. 2003). Some of the biggest floods (1968 and 1981) recorded using this method (McCulloch et al. 2003) did not appear to initiate outbreaks (Table 4), although Fabricius et al. (2010) did stress that floods must occur in November to January to benefit crown-of-thorns larvae. Also, floods may not cause outbreaks of *Acanthaster* spp. if they occur too

Table 4 Timing of major flood events on Australia's Great Barrier Reef (specifically, peak flow events from the Burdekin River, located 19.6°S) relative to the agreed start of each of the four major waves of outbreaks

Peak flood events		Start of corresponding outbreak	Time between flood and outbreak	Location
McCulloch et al. 2003	Fabricius et al. 2010			
1958	1958	1962	4 years	Green Island (16.7°S)
1968	—	—	—	
1970	—	—	—	
—	1974	1979	5 years	Green Island (16.7°S)
1981	—	—	—	
1988	—	1993	5 years	Lizard Island (14.6°S)
1991	1991	1993	2 years	Lizard Island (14.6°S)
1998	—	—	—	
NA	2008	2010	2 years	Lizard Island (14.6°S)

soon after the preceding outbreak so that the coral cover has not had time to recover sufficiently to sustain a new outbreak (Fabricius 2013).

Extended delays (>3 years) between flood events and reported outbreaks on the GBR in 1962 and 1979 (Table 4) may be attributed to limitations in the detection of outbreaks prior to the implementation of reef-wide monitoring in 1985. For more recent outbreaks, however, high densities of *A. planci* would have already recruited on reefs around Lizard Island when flooding events occurred in 1991 and 2008. Moreover, these recent outbreaks almost certainly developed through several consecutive years of high recruitment from 1994 to 1998 (e.g., Pratchett 2005), making links to individual flooding events somewhat tenuous. In addressing these observations, Fabricius et al. (2010) reported that "floods have reached or crossed this part of the shelf in 1991, 1994, 1995 and 1996" (p. 603), inferring that this was an unusual period with a high frequency of flooding events. It is possible that initial outbreaks that occurred on the GBR in 1958, 1971, and 2008 also comprised multiple cohorts and consecutive years of high recruitment, but there are no data on population structure to assess this explicitly.

Further complexities associated with linking outbreaks of *Acanthaster planci* on the GBR to periodic major flooding events relate to the spatial patterns of outbreaks (Brodie et al. 2005). Notably, outbreaks of *A. planci* tend to occur predominantly on midshelf reefs (Moran 1986), rather than inshore reefs where the influence of terrestrial run-off is greatest (Brodie et al. 2005), or on offshore reefs. Also, outbreaks are initiated north of Cairns (close to either Cairns at 17.0°S or Lizard Island at 14.5°S), rather than on reefs in immediate proximity to major river systems (Brodie 1992). Fabricius et al. (2010) argued that the confluence of high nutrients and midshelf reefs is limited to northern latitudes, between 14.5°S and 17.0°S, and it is only here that nutrient concentrations exceed minimal thresholds (>0.25–0.5 $\mu g.l^{-1}$) necessary for survivorship of crown-of-thorns larvae. They suggested that flood plumes from major rivers (particularly the Burdekin River) travel northward, staying close to the coast, but are deflected off shore by Cape Grafton, a promontory just south of Cairns. There are, however, large areas of the GBR (e.g., in the far northern section and Swains) that have long-term average chlorophyll concentrations greater than 0.5 $\mu g.l^{-1}$ (Figure 11). Also, chlorophyll concentrations in summer months (November–May), which is when *A. planci* spawn, are generally greater than 0.5 $\mu g.l^{-1}$ throughout the Wet Tropics (16°S to 19°S), independent of any major flooding events (Brodie et al. 2005).

If the productivity of waters on the GBR is consistently below levels (0.25–0.5 $\mu g.l^{-1}$) needed to sustain larval growth and survivorship (e.g., Fabricius et al. 2010), it makes it hard to explain the southward propagation of waves of outbreaks that cause widespread devastation. The conventional wisdom (Brodie 1992, Brodie et al. 2005, Fabricius et al. 2010) is that once primary outbreaks have become established, the immense numbers of larvae produced by high densities of well-fed starfish will generate enough late-stage larvae to settle successfully on downstream reefs (to the south) regardless of low rates of larval survival. It is certainly true that three waves of outbreaks have propagated effectively through the midshelf reefs (>25 km off shore) north and south of Townsville (~19°S), where chlorophyll concentrations rarely exceed 0.25 $\mu g.l^{-1}$. However, it is unclear whether this is because of the sheer volume of larvae spawned on reefs to the north (of which only a small proportion actually survive) or evidence of the capacity of larvae to successfully develop and settle despite relatively oligotrophic conditions (e.g., Olson 1987). It does seem illogical that low nutrient levels would prevent the formation of primary outbreaks on reefs south of 16°S and yet allow for the extensive formation of devastating secondary outbreaks. Moreover, simulation models do not reproduce the southward propagation of waves of outbreaks when using high levels of larval mortality expected to occur when chlorophyll concentrations are less than 0.25 $\mu g.l^{-1}$ (Fabricius et al. 2010). A more parsimonious explanation might be that primary outbreaks are established over several years and independent of any flood events, but the subsequent spread of outbreaks might be conditional on years of high larval survivorship, which is facilitated by major flood events that enhance food availability.

A logical extension of the terrestrial run-off hypothesis is that outbreaks would be expected to occur more frequently on high islands compared to atolls (the 'high island hypothesis'; Birkeland 1982) because of localised elevated nutrient concentrations due to runoff. Tsuda (1971) and Pearson (1975) noted that outbreaks occurred predominantly on reefs near high islands or along continental shelves. Birkeland (1982) showed that 'large populations' of *Acanthaster planci* were reported on 19 (of 23) high islands compared to 2 (of 22) atolls across Micronesia and Polynesia based on data presented by Marsh & Tsuda (1973). Analyses of all reported outbreaks since 1990 showed that outbreaks have occurred on at least 35 low islands or atolls across the Indo-Pacific (e.g., Eniwetok Atoll and Majuro Atoll in the Marshall Islands), although the majority of outbreaks (29% and 56%, respectively) were reported from high islands (e.g., Moorea, French Polynesia) and continental shelves (e.g., the GBR). Outbreaks on low islands and atolls cannot be readily linked to terrestrial run-off. However, there might be other sources of nutrients that cause plankton blooms and thereby enhance larval survival away from high islands or major rivers, including (1) upwellings (e.g., Sweatman et al. 2008, Mendonça et al. 2010), (2) bioturbation and resuspension of sediments by severe tropical storms, and (3) oceanographic features that create high-productivity fronts (Houk et al. 2007). On the GBR, it is evident that here have been outbreaks on reefs in the Swains region that occur almost independently and asynchronously with waves of outbreaks propagating from north of Cairns (Sweatman et al. 1998). Since the Swain Reefs are more than 100 km off shore, they are rarely if ever exposed to flood plumes, although there is some evidence that upwelling occurs in the region (Kuchler & Jupp 1988).

Houk et al. (2007) described interannual fluctuations in ocean productivity in the northern Pacific associated with the Transition Zone Chlorophyll Front (TZCF), which may explain the irregular occurrence of outbreaks across this region. However, these periods of peak productivity coincide with the 'emergence' (presumably from deeper water) of high densities of adult *Acanthaster planci* (Houk et al. 2007, Houk & Raubani 2010) rather than settlement of larvae, which must have occurred about 2 years earlier; it is unclear how this phenomenon relates to secondary outbreaks. Despite the growing enthusiasm for the nutrient enrichment hypothesis, several authors have cautioned against its broad applicability (e.g., Potts 1981, Olson & Olson 1989, Endean & Cameron 1990, Lane 2012). Lane (2012) pointed out that outbreaks of *Acanthaster* spp. have been occurring despite overarching declines in global ocean productivity. Also, there is no evidence of increased incidence of crown-of-thorns outbreaks in areas with enhanced nutrient concentrations (because of either periodic high precipitation and high erosion rates or areas with upwellings) within the Indo-Australia archipelago or 'Coral Triangle' (Lane 2012). It is also important to remember that much of the potential importance of nutrient enrichment depends on larvae of *Acanthaster* spp. being generally food limited (Lucas 1982, Okaji et al. 1997, Fabricius et al. 2010).

Olson (1985, 1987) set out to test Lucas's suggestion that larval *A. planci* rarely complete development at phytoplankton concentrations that are typical of GBR waters in the absence of flood plumes. He developed an apparatus that allowed larvae to be supplied with seawater with controlled concentrations of phytoplankton while being held in chambers underwater in the field. He found that *A. planci* larvae developed at near-maximal rates at ambient chlorophyll levels in the absence of phytoplankton blooms (Olson 1987). He therefore suggested that fluctuations in larval food resources have a limited role in explaining interannual variation in larval recruitment, which must underlie sudden increases in the abundance of *Acanthaster* spp. (Olson 1987).

Okaji (1966) tried to continue this line of research using the same apparatus, as well as a modified form with inline filters, to further manipulate phytoplankton densities. In spite of changing the chambers every 2 days, he found that chlorophyll concentrations within the chambers increased over the course of his experiment to well above ambient levels, presumably through contamination and retention of phytoplankton. He concluded that the apparatus was unreliable and abandoned the approach in favour of the laboratory experiments described later by Fabricius et al. (2010).

Olson (1987) did not measure chlorophyll concentrations in the chambers in the course of his main experiments, but he did specifically test for such an effect in a pilot experiment (Olson 1985) and found that chlorophyll concentrations in the experimental chambers remained at or below ambient levels after 2 days in shallow water and bright sunlight. Based on this observation, the larval chambers were changed every 2 days during the main experiment (Olson 1987) specifically to prevent any accumulation of phytoplankton. These divergent results reinforce the importance of repeating and extending the limited studies relating growth and survival of *Acanthaster* larvae to phytoplankton concentration; they also emphasise the need to confirm the findings in the field.

Despite discrepancies and inconsistencies in the spatial and temporal occurrence of outbreaks (e.g., Table 4), high rainfall, terrestrial run-off, and elevated nutrients are likely to increase the likelihood that outbreaks of *Acanthaster* spp. will actually occur, but they do not necessarily account for all recorded outbreaks. Importantly, the nutrient enrichment hypothesis provides one of the only plausible mechanisms by which anthropogenic activities may have exacerbated outbreaks of *Acanthaster* spp. (increasing their severity or frequency) over recent decades (Brodie 1992, Brodie et al. 2005). When proposing the terrestrial run-off and high island hypotheses, Birkeland (1982) maintained that outbreaks were essentially a natural phenomenon triggered by irregular rainfall events. He did acknowledge that land clearing may increase nutrient concentrations in coastal environments following terrestrial run-off (Birkeland 1982). If so, there might also be a signal whereby outbreaks of *Acanthaster* spp. are greatest in areas closest to heavily populated coastlines (but see Lane 2012) or coinciding with periods of extensive settlement and clearing of coastal land. However, outbreaks have been reported on isolated and unpopulated reefs (e.g., Chagos), but they might be less frequent or less severe than on reefs subjected to increased anthropogenic influences. Unfortunately, inconsistencies in monitoring and reporting of nominal outbreaks make it virtually impossible to test these ideas explicitly. This should be a priority for future monitoring studies.

Predator removal hypothesis

One of the earliest hypotheses to account for outbreaks of *Acanthaster* spp. was the predator removal hypothesis (Endean 1969), which assumed that populations of crown-of-thorns starfish are normally regulated by high levels of post-settlement or adult predation (Endean 1969, McCallum 1987, 1990). This hypothesis was given increased credibility by two recent studies (Dulvy et al. 2004, Sweatman 2008) that reported increased incidence or severity of outbreaks of crown-of-thorns starfish in areas subject to fisheries exploitation (see also Ormond et al. 1990). Sweatman (2008) compared the rates of occurrence of starfish outbreaks on GBR reefs that were open to fishing and on reefs where fishing was prohibited. Because outbreaks come in waves, only those reefs that were close to known outbreaks were considered; 75% of reefs that were open to fishing suffered outbreaks, compared to 20% for reefs that had been closed to fishing for a minimum of 5 years (Sweatman 2008). These studies (Ormond et al. 1990, Dulvy et al. 2004, Sweatman 2008) did not reveal the mechanistic basis for the observed results but suggested that harvesting of coral reef fishes may increase the likelihood that outbreaks of *Acanthaster* spp. will occur by removing one of the key regulatory mechanisms that prevent extreme population fluctuations.

When the predator removal hypothesis was initially proposed, the principal known predator of adult *Acanthaster* spp. was the giant triton, *Charonia tritonis* (Pearson & Endean 1969). Harvesting of tritons for sale as curios was suggested to have reduced predator numbers in the decades leading up to the first recorded outbreak in the 1960s (Endean 1969). However, the little that is known of the ecology of *C. tritonis* suggests that even at preharvest densities they would not be effective in controlling outbreak populations of crown-of-thorns starfish. For example, when large *C. tritonis* were placed in a cage and supplied with abundant adult *A. planci*, average consumption was only 0.7 starfish per triton per week (Pearson & Endean 1969).

More recently, attention has focused on predation by fishes, particularly emperors (family Lethrinidae) because these are relatively large, generalist benthic carnivores that feed on and around areas of coral (e.g., Sweatman 1997, Mendonça et al. 2010). They are also widely fished, so there is the possibility that their numbers have been reduced through overexploitation, which could be related to the apparent increase in frequency of outbreaks on the GBR. Other known predators of *Acanthaster* spp. are triggerfishes and pufferfishes (Campbell & Ormond 1970, Owens 1971), although numbers of these fishes are unlikely to have changed through exploitation, at least on the GBR. Direct evidence that any fisheries target species are major predators of *Acanthaster* spp. is meagre (Sweatman 1997). Remains of *Acanthaster* spp. have been found in the gut contents of some fisheries target species (Randall et al. 1978, Birdsey 1988; Table 5), but only rarely. Ormond et al. (1990) pointed out that prey-switching behaviour, which is a critical component of the population models, means that a lack of observations of predation or the absence of *Acanthaster* spp. in predators' gut contents when starfish densities are low does not necessarily mean that predation by fishes is not important in regulating starfish populations.

Few studies have sampled the gut contents of potential predators in locations where *Acanthaster* spp. are known to be present. However, Sweatman (1997) examined the gut contents of 98 lethrinids that were caught close to an area with outbreak densities of adult *Acanthaster planci* on the GBR and did not find any starfish remains. Similarly, Mendonça et al. (2010) examined gut contents for more than 20 large (~50 cm total length) individuals of potential starfish predators, including snappers *Lutjanus bohar* and *Lutjanus johni*, emperors *Lethrinus* spp., and *Cheilinus lunulatus*, at reefs infested with *Acanthaster* spp. but failed to detect starfish remains in guts of any of the fishes. It is possible that these fishes rarely feed on adult *Acanthaster* spp. but instead target juvenile starfish (Endean 1976). Endean (1976) recorded remains of juvenile *A. planci* in the guts of a Queensland grouper, *Epinephelus lanceolatus*, although predation rates on juvenile starfish are difficult to quantify.

The cryptic behaviour of small (<200-mm diameter) *Acanthaster* spp. is often attributed to predator avoidance (Yokochi & Ogura 1987, Zann et al. 1987, Birkeland & Lucas 1990). Moreover, many small and large starfish have missing or regenerating arms (Table 6), which is considered evidence of recent predator attacks (McCallum et al. 1989). Given the high proportion of starfish with missing arms, reflective of partial or incomplete predation, it may be that mortality rates attributable to predation are also high (McCallum et al. 1989). In the Philippines, Rivera-Posada et al. (2014) showed that more than 70% of juvenile (110- to 200-mm diameter) *Acanthaster planci* had missing or regenerating arms. Rates of injury were much lower in both smaller (probably because of highly cryptic behaviour) and larger starfishes, which have increased protection from longer, more sturdy spines (Rivera-Posada et al. 2014) compared with starfish of intermediate sizes. Moreover,

Table 5 Fishes that feed on *Acanthaster* spp.

Family	Species	Reference
Serranidae	*Epinephelus lanceolatus*[L,K]	Endean 1976
Lethrinidae	*Lethrinus miniatus*[L]	Keesing & Halford 1992, Sweatman 1995
	Lethrinus nebulosus[L]	Birdsey 1988, Keesing & Halford 1992
	Lethrinus atkinsoni[L]	Sweatman 1995
Pomacanthidae	*Euxiphipops sexstriatus*[r]	Moran 1992
	Pomacanthus semicirculatus[r]	Moran 1992
	Holacanthus passer[r]	Glynn 1984
Chaetodontidae	*Chaetodon auriga*[r]	Moran 1992
Labridae	*Thalassoma lunare*[r]	Rivera-Posada et al. 2013
	Thalassoma lucasanum[r]	Glynn 1984
	Thalassoma hardwicki[r]	Moran 1992
	Chaetodon citrinellus[r]	Glynn 1984
	Cheilinus diagrammus[r]	Moran 1992
	Cheilinus fasciatus[r]	Moran 1992
	Cheilinus undulatus[L,K]	Ormond & Campbell 1974, Randall et al. 1978, Keesing & Halford 1992
Pomacentridae	*Abudefduf curacao*[e]	Pearson & Endean 1969
	Pomacentrus moluccensis[r]	Moran 1992
	Chromis caerulea[r]	Moran 1992
	Chromis dimidiatus[l]	Lucas 1975
Gobiidae	*Cryptocentrus* sp.[r]	Moran 1992
Tetraodontidae	*Arothron hispidus*[L,K]	Ormond et al. 1973
	Arothron manilensis[r]	Rivera-Posada et al. 2013
	Arothron nigropunctatus[r]	Moran 1992
	Arothron stellatus[L,K]	Keesing & Halford 1992
Balistidae	*Balistoides viridescens*[L,K]	Ormond et al. 1973
	Pseudobalistes flavimarginatus[L,K]	Ormond & Campbell 1974, Owens 1971

Note: We clearly distinguish those animals that prey on live starfish (L) and have the capacity to kill (K) juvenile or adult starfish, as these species may be important in regulating abundance of *Acanthaster* spp. Fishes that have only been seen to feed on remains of dead or dying starfish (r), eggs (e), or larvae (l) may also feed on live juvenile and adult starfish, but this is unconfirmed.

the rates of injury were much higher within areas closed to fishing compared to nearby fished areas, providing a potential mechanism to explain why outbreaks are less prevalent in areas closed to fishing (Dulvy et al. 2004, Sweatman et al. 2008). However, observations of lethal predation on *A. planci* are rare, and the relationship between partial predation and rates of mortality is completely unknown. Field experiments have failed to find intense predation by fishes on juvenile *Acanthaster* spp. When small (25- to 79-mm diameter), laboratory-reared *A. planci* were placed in a semi-natural setting in an area of one reef on the GBR where likely fish predators were present, predation rates were only 0.13% of starfish per day (Sweatman 1995). Also, when similarly small juvenile *A. planci* were presented to emperors on a nearby reef, only a minority of juveniles were consumed, and many were rejected after mouthing and survived the encounter (Sweatman 1995). These results suggest that predation is unlikely to constrain survivorship of juvenile *A. planci*, although the results are based on only a single reef and single time.

The circumstances in which predators could regulate *Acanthaster* spp. have been investigated using population models (McCallum 1987, 1990, 1992, Ormond et al. 1990), which focus on the feasibility of fish predators controlling the numbers of crown-of-thorns starfish on an individual reef

Table 6 Geographical variation in the proportion of *Acanthaster* spp. with missing or damaged arms, which is considered to be a proxy for differing levels of predation

Location	Year	% with arm damage	Reference
Western Australia	1985	64	Simpson & Grey 1988
Hawaii	1972	60	Branham 1973
Philippines	2012	59	Rivera-Posada et al. 2014
Guam	1991	59	Lawrence 1991
Papua New Guinea	1970	50	Pyne 1970
GBR	1994	50	Stump 1996
Western Australia	1985	47	Simpson & Grey 1988
Ryukyu Islands	1984	46	Nakamura 1986
Guam	1981	43	Glynn 1982
GBR	1987	40	McCallum et al. 1989
Ryukyu Islands	1985	35	Nakamura 1986
Ryukyu Islands	1986	33	Nakamura 1986
GBR	1967–1968	33	Pearson & Endean 1969
Ryukyu Islands	1985	32	Nakamura 1986
Sudan	1984	30	Moore 1985
Sudan	1969	29	Ormond and Campbell 1971
Western Australia	1985	25	Simpson & Grey 1988
Ryukyu Islands	1984	20	Nakamura 1986
Sudan	1970	20	Ormond 1971
Panama	1980–1981	17	Glynn 1982
Fiji	1984–1985	13	Zann et al. 1987
Sudan	1970	4	Ormond 1971
Sudan	1984	2	Moore 1985

Note: Levels of damage are generally highest among smallest individuals, but few studies explicitly distinguished between size classes, so data are presented as the overall percentage of individuals with any evidence of damage. Data are arranged from highest to lowest, rather than by geographical location.

rather than considering their role in the initiation and perpetuation of waves of outbreaks. Potential predators need to be able to take a range of prey so they can persist when densities of *Acanthaster* spp. are low but then switch their attention to crown-of-thorns starfish as numbers rise. These type II and type III functional responses (Holling 1965) are generally characteristic of vertebrate predators, reinforcing the focus on large fishes (e.g., Lethrinidae). Most of the parameters in the models have never been measured in the field, but using plausible estimates for numbers of starfish that recruit and for rates of prey consumption by fishes, the models suggest that realistic densities of fishes may regulate local populations of *Acanthaster* spp. over a range of recruitment rates and might therefore be important in preventing the gradual development of primary outbreaks. However, mass recruitment of *Acanthaster* spp. would quickly swamp fish predators (McCallum 1987, 1992, Ormond et al. 1990), limiting the capacity to prevent sudden or secondary outbreaks.

Pathogenesis and outbreak cycles

Outbreaks of *Acanthaster* spp. often end with precipitous declines in starfish densities (e.g., Chesher 1969, Moran et al. 1985, Moran 1986, Pratchett 2005). In many cases (e.g., Chesher 1969, Endean 1969, Pearson & Endean 1969), these declines follow extensive depletion of scleractinian corals, prompting suggestions that starfish either die from starvation or move *en masse* to find food (Endean 1969). An alternative explanation, however, is that *Acanthaster* spp. are highly vulnerable to pathogenesis (e.g., Zann et al. 1987, Pratchett 1999), and that rapid transmission of disease(s) among high-density populations ultimately leads to population collapse. Echinoderms generally are highly susceptible to disease (Jangoux & Lawrence 1989), and disease has been implicated in mass mortalities of numerous species of urchins and starfish (e.g., Dungan et al. 1982, Lessios et al. 1984). Moreover, the operation of disease (independent of any other regulatory mechanisms) can lead to predictable fluctuations in the abundance of host animals (e.g., May & Anderson 1979), possibly accounting for cyclical outbreaks of *Acanthaster* spp., such as have been observed on the GBR.

There is extensive evidence that crown-of-thorns starfish are particularly prone to disease and are often seen to exhibit symptoms of pathogenesis (Birkeland & Lucas 1990). Burkholder (1973), for example, showed that *Acanthaster planci* had much lower resistance to bacterial infection compared with many other starfish. Importantly, *A. planci* had weak resistance to Gram-positive bacteria and almost no resistance against Gram-negative bacteria (Burkholder 1973). Accordingly, crown-of-thorns starfish often succumb to bacterial infection when kept in captivity (Lucas 1984, Sutton et al. 1988). Symptoms of disease in *Acanthaster* spp. include lesion formation, tissue degeneration, loss of turgor, and collapsed spines, which indicate bacterial infection (Sutton et al. 1988). Sutton et al. (1988) isolated several potential pathogens, including *Vibrio harveyi*, *V. tubiashi*, *V. campbellii*, *Pseudomonas*, and *Moraxella* bacteria from starfish exhibiting the aforementioned symptoms. Similar symptoms were also seen in a crown-of-thorns starfish found at Lizard Island (northern GBR) in 1999 (Pratchett 1999). While attempts to identify the pathogenic organism(s) causing these symptoms were unsuccessful, Pratchett (1999) did show that tissues removed from the sick and dying starfish could be used to infect other seemingly healthy starfish in close proximity, but without direct contact. Perhaps more important, the detection of this dying starfish in 1999 coincided with the rapid disappearance of starfish in the aftermath of a significant outbreak (Pratchett 2005) and preceded the local depletion of scleractinian corals (Pratchett 2010).

Rapid declines in abundance of crown-of-thorns starfish following almost total consumption of coral prey (e.g., Chesher 1969, Moran 1986) might also be partly attributable to pathogenesis, whereby severe limits to energetic, nutrient, or vitamin intake increases susceptibility to opportunistic pathogens that are ubiquitous in the marine environment. Mills (2012) demonstrated that *Acanthaster planci* at high densities with *ad libitum* access to food invest far greater energy into immune defence compared to individuals living at low densities, presumably as a response to the increased likelihood of infection at high densities. If *A. planci* continue to invest considerable energy in immune defence even when prey are scarce, energy reserves may be rapidly depleted, actually making starfish more prone to disease (Rivera-Posada 2012). Accordingly, Sutton et al. (1988) showed that bacteria isolated from diseased *A. planci* did not induce disease when injected into healthy (well-fed) individuals. Sutton et al. (1988) also showed that animals collected from reefs with limited coral cover were particularly prone to disease, as opposed to starfish collected prior to extensive depletion of coral resources. Similarly, significant incidence of disease was recorded in Fiji during the late 1980s when *Acanthaster* spp. settled onto reefs that had been almost completely denuded of potential prey (Zann et al. 1990).

Despite anecdotal and opportunistic observations of pathogenesis in *Acanthaster* spp., there has been little rigorous research on the vulnerability of these starfish to disease. Several potentially pathogenic organisms, including *Vibrio owensii*, *V. rotiferianus*, *V. fortis*, *V. harveyi*, *V. natriegens*, and *Photobacterium eurosenbergii* have been isolated from *Acanthaster planci* (Rivera-Posada et al. 2011a,b). Birkeland & Lucas (1990) also reported a marine bacterium described as Type A bacterium, which was found on more than 50% of individuals examined and appeared to be a specific symbiont of *Acanthaster planci*. Zann et al. (1990), meanwhile, attributed the mass mortality of juvenile *A. planci* in Fiji to a sporozoan pathogen. However, this pathogen was not isolated or identified. Ongoing research on pathogenesis in *Acanthaster* spp. may provide critical insights for understanding outbreak dynamics but may also be utilised to control outbreak populations (e.g., Rivera-Posada et al. 2012, 2013) and thereby reduce ongoing degradation of coral reef habitats in the Indo-Pacific.

Managing outbreaks of crown-of-thorns starfish

Managing coral reefs in the face of significant and increasing disturbances (e.g., climate-induced coral bleaching and coral disease) and increasing anthropogenic pressures associated with burgeoning populations in many tropical nations (e.g., Bell et al. 2009) represents a considerable challenge—more so because most major disturbances cannot be managed directly. Outbreaks of *Acanthaster* spp. are one of the few disturbances on coral reefs that may be amenable to direct intervention, by increased efficiency of direct control (Rivera-Posada et al. 2012), development of effective biological controls (Endean 1969), or ultimately addressing the anthropogenic factors (e.g., eutrophicaton or overfishing) that may have initiated or exacerbated outbreaks near heavily populated or highly modified coastal environments (Kenchington & Kelleher 1992, Brodie & Waterhouse 2012). Irrespective of their cause(s), control of outbreaks of *Acanthaster* spp. may provide the best opportunity to reverse ongoing coral loss and reef degradation throughout the Indo-Pacific (Endean & Cameron 1990), which is important for sustaining not only the ecological integrity of reef systems but also the critical goods and services that are derived from coral reef environments (Pratchett et al. 2008). Further, emerging threats posed by climate change provide a renewed imperative to try to reverse the ongoing degradation of coral reef ecosystems to maximise the adaptive capacity and resilience of coral reef organisms (Hughes et al. 2003). If outbreaks of *Acanthaster* spp. could be prevented, it is suggested that this alone could reverse sustained declines on the GBR (De'ath et al. 2012). The same is probably true for other locations in the western Pacific, where outbreaks of *Acanthaster planci* have recurring and devastating effects on coral assemblages.

Direct control

Since the first documented outbreaks of *Acanthaster* spp. in the 1950s and 1960s, the immediate response has been to try to collect or kill the burgeoning numbers of starfish (e.g., Barnes 1966, Raymond 1986). These responses reflected the pervading view that outbreaks of *Acanthaster* spp. were a direct consequence of anthropogenic changes in the marine environment and posed a real and immediate threat to the future of coral reefs. Despite subsequent controversy about the extent to which outbreaks may or may not be natural (Sapp 1999), there have continued to be extensive and widespread control efforts aimed at eradicating or excluding *Acanthaster* spp. from reef environments and thereby minimising damage caused to coral assemblages (Figure 12). The commonest methods used to protect reef areas from crown-of-thorns starfish are hand collections of individual starfish for disposal on shore (e.g., Yamaguchi 1986) or injections with toxic chemicals to kill starfish *in situ* (e.g., Johnson et al. 1990). Early eradication programmes also cut *Acanthaster* spp. in small pieces and left them on the reef, but this process is not recommended because of their phenomenal capacity to survive and regenerate after significant physical damage (Owens 1971, Sweatman & Butler 1993, Messmer et al. 2013). Several other initiatives, such as the construction of a physical barrier or fences to exclude starfish from accessing areas of live coral, have also been attempted (e.g., Bell et al. 1986) but are only feasible on small scales. In the past five decades (mostly since the 1970s), at least 17 million starfish have been killed or removed from reefs across the Indo-Pacific, mainly in southern Japan and on Australia's GBR (Figure 12), at an estimated cost of US$15–44 million. Given the extensive and widespread efforts to eradicate outbreaks of crown-of-thorns starfish, it is clear that the collective of coral reef scientists and managers have already come to a decision about whether we *should* intervene and control outbreak populations of *Acanthaster* spp., so the question now is whether we *could* intervene and control outbreak populations of *Acanthaster* spp.

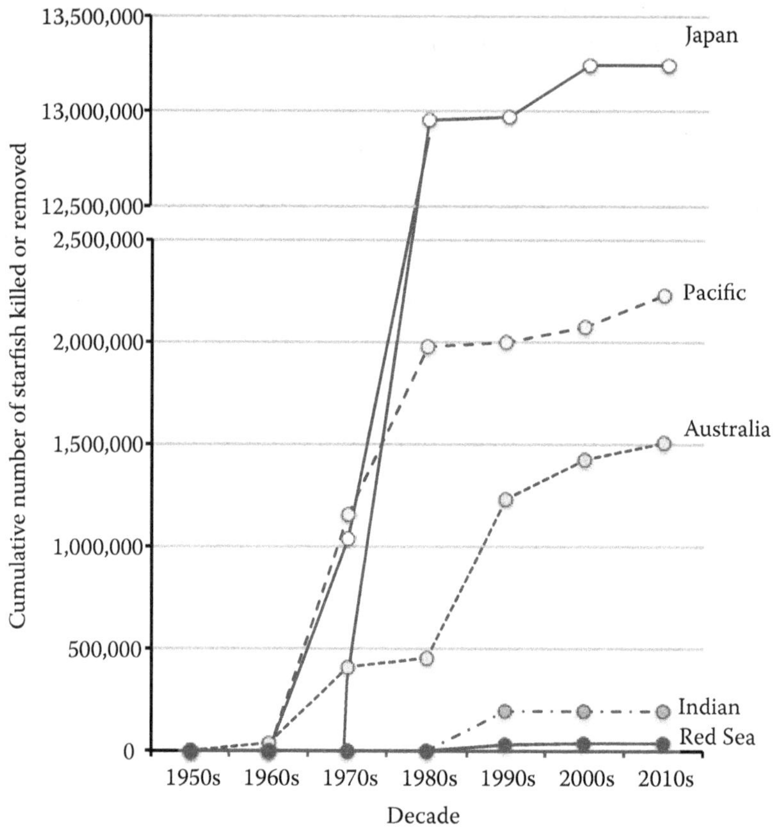

Figure 12 Cumulative number of crown-of-thorns starfish that have been killed *in situ* or removed from coral reefs across the Indo-Pacific, beginning with eradication programs on Australia's Great Barrier Reef and in Southern Japan in the 1960s.

The largest and longest-running control programme for *Acanthaster planci* was undertaken in Japan, where approximately 13 million starfish were collected between 1970 and 1983, at an estimated cost of US$6–7 million (Yamaguchi 1986). Despite this prolonged effort, chronic infestations of *Acanthaster* spp. still killed in excess of 90% of coral across vast areas of fringing reefs. The few examples where control programmes have been effective, either in eradicating crown-of-thorns starfish from areas of reefs or preventing significant declines in local coral cover, have tended to be small in spatial scale (Moran et al. 1988, Birkeland & Lucas 1990), but perhaps even more important is the early detection and rapid response to apparent increases in the density of *Acanthaster* spp. Fisk & Power (1999) suggested that effective controls are only ever going to be feasible on small areas of reef and considered starfish control to be an important tactical response to maintain coral cover at important tourism sites. However, this requires a significant long-term commitment to detect and eradicate starfish that continue to invade from outside the protected areas. Fisk & Power (1999) showed that control efforts would need to be undertaken every 1–2 weeks throughout the course of an outbreak, while effectiveness of localised control efforts is greatly increased by explicitly targeting known aggregations of starfish that occur outside the area protected. The feasibility and effectiveness of large-scale (e.g., reef-wide) control programmes has been continually questioned (e.g., Kenchington & Pearson 1982) because it not clear that measures required to effectively protect small patches of reefs can be achieved simply by scaling up effort (e.g., number of diver hours) in proportion to reef area. Even if such scaling were relevant, it would require inordinate effort to protect entire reef systems, such as the GBR (Birkeland 1989b, Endean & Cameron 1990, Bos et al. 2013). However, it may be feasible to protect individual reefs, and for the GBR, effective control of primary outbreaks on a small number of reefs may be all that is required if these localised episodes are the basis of subsequent reef-wide outbreaks (Kenchington 1977, Moran et al. 1988, Reichelt et al. 1990a, Fabricius et al. 2010).

A widely used method for eradicating *Acanthaster planci* is to inject starfish with sodium bisulphate (e.g., GBR, Hoey & Chin 2004; Papua New Guinea, Pratchett et al. 2009a), which is considered not only to have limited side effects for other reef organisms but also to be one of the most efficient methods for directly killing starfish *in situ*. Even so, each starfish has to be extracted from the reef matrix and then injected multiple times all over the oral disk to ensure complete mortality. Significant increases in efficiency could be achieved using a toxin that could be administered with a single dose and anywhere on the starfish. Rivera-Posada et al. (2012) demonstrated that single injections of low concentrations of oxbile and oxgall induced rapid death of *A. planci*, representing a novel and potentially efficient method for controlling crown-of-thorns starfish. The recommended protocol involves injecting each starfish just once, preferably in the upper portion of one of the arms, with 10 ml of 8–10 $g.l^{-1}$ of oxbile mixed in fresh water (Rivera-Posada et al. 2013). In tank-based trials, 100% of starfish injected with 10 ml of 8 $g.l^{-1}$ of oxbile died within 24 h (Rivera-Posada et al. 2013). Moreover, there was no evidence of unintended side effects for other coral reef organisms (e.g., scleractinian corals, echinoderms, and fishes). This method could increase the number of starfish that can be killed on a single dive by a factor of 10 because each starfish need only be injected once and it is not necessary to extract the animals from the coral. Field trials are currently under way; if successful, this will be a great advance in the efficiency of control programmes, which has been deemed necessary before attempting large-scale reef-wide controls of *Acanthaster* spp. (e.g., Bos et al. 2013).

This discovery that oxbile could be used to effectively kill *Acanasther planci* arose from research regarding the pathogenesis of crown-of-thorns starfish (Rivera-Posada 2012). Initially, thiosulfate-citrate-bile-sucrose agar (TCBS), which is the primary plating medium used for the isolation and growth of *Vibrio* bacteria (Kobayashi et al. 1963), was injected into starfish to explore naturally occurring bacteria on and in *Acanasther planci* (Rivera-Posada 2011a,b). However, *A. planci* that were injected with TCBS rapidly developed symptoms of bacterial infection (Rivera-Posada et al. 2011a), which under the right conditions was spread among nearby starfish. Initiation of disease following injection of TCBS into *Acanthaster planci* was attributed to increased activity of normally resident *Vibrio* bacteria. However, the proteins in TCBS (peptone, oxgall, and yeast) also caused an allergic reaction in crown-of-thorns starfish, thereby contributing to their rapid mortality (Rivera-Posada et al. 2012). Similarly, gastric mucin from pigs was shown in the 1980s to be toxic to *Acanthaster* spp. (Lassig 1991).

Current research (e.g., Rivera-Posada et al. 2013) is focused on the use of oxbile to control outbreaks of crown-of-thorns starfish, which does not invoke transmissible disease and thereby ensures that only injected starfish will actually die. Ultimately, however, it may be possible to exploit the inherent susceptibility of echinoderms to disease (Rivera-Posada 2012) to control outbreak populations of *Acanthaster* spp. The use of transmissible diseases not only would greatly increase the effectiveness of control efforts but also would introduce the risk of interspecific transmission (mainly to other species of echinoderms). Caballes et al. (2012) tested for disease transmission between *Acanthaster* spp. and another common starfish on coral reefs, *Linckia guildingi*. After exposure (and direct contact) to diseased *Acanthaster planci*, four of the five *Linckia guildingi* developed skin lesions. Moreover, *Vibrio rotiferianus*, which was previously reported as a pathogen isolated from lesions of experimentally infected crown-of-thorns starfish, was also isolated from lesions on the surface of *Linckia guildingi*. Importantly, all *L. guildingi* fully recovered after 14–46 days, but these findings highlight the need for extensive testing on the specifics of potentially pathogenic organisms before they can be used for biological control.

There are several other organisms, such as sporozoans, fungi, and ciliates, that could potentially induce pathogenesis in *Acanthaster planci*. The key is to identify potential pathogens that have selective effects on *Acanthaster* spp. while minimising the risk of secondary infections among other coral reef organisms (e.g., Byrne et al. 1997). Protozoan parasites, for example, often have high host specificity, which reduces the risk of interspecific transmission of inducible diseases. There are also several other parasitic organisms that are known to infect echinoderms, and these organisms warrant further investigation. Three genera of gregarine sporozoans that infect the digestive tracts of echinoderms are *Cystobia*, *Lithocystis*, and *Urospora*. *Cystobia* species (*C. grassei*, *C. holothuriae*, *C. irregularis*, *C. schneiden*) cause disease in the gut-associated haemal system and coelomic cavity of echinoderms (Jangoux 1987a,b), inducing the same clinical signs of disease described by Zann et al. 1990. Another potential parasite is the ciliate *Orchitophyra stellarum*, which is found in the testes of the asteroid *Asterias rubens*. The ciliate mostly parasitises male gonads, causing a progressive breakdown of germinal tissue (Byrne et al. 1997), rendering the starfish sterile. There are also protophytan algae (e.g., *Coccomyxa ophiura* and *C. astericola*) that cause lethal lesions in the body wall of asteroids and ophiuroids. It is currently unknown whether these pathogens can infect *Acanthaster* spp. or whether infection will necessarily lead to increased mortality or reduced reproductive output. However, the potential for greatly increasing the efficiency and effectiveness of existing control efforts warrants further research in this area (all the while recognising the need for careful and rigorous tests of the risks involved).

Addressing ultimate causes of crown-of-thorns outbreaks

Irrespective of increases in the efficiency of direct controls, these solutions will only ever provide temporary or short-term solutions for minimising effects of *Acanthaster* spp. on coral reef ecosystems (Birkeland & Lucas 1990). Long-term or permanent solutions, meanwhile, will require improved understanding of the ultimate cause(s) of outbreaks. When outbreaks are initiated or exacerbated by anthropogenic effects, be it elevated nutrient concentrations associated with terrestrial run-off (Fabricius et al. 2010) or overfishing (e.g., Dulvy et al. 2004), then these causes need to be addressed explicitly to reduce the likelihood that outbreaks will recur or at least reduce the frequency of future outbreaks (Fabricius et al. 2010). Changes to water quality and fisheries management are already being implemented in some locations because they represent 'no-regret' strategies: Improvements in water quality (by minimising run-off of sediments, nutrients, and pollutants), and reductions in fisheries exploitation (largely through the increasing establishment of no-take marine protected areas) are likely to increase the resilience of reef ecosystems (e.g., Fabricius 2005, Wooldridge 2009, McCook et al. 2010), even if they do not actually limit outbreaks of *Acanthaster* spp.

Declining water quality in coastal environments (caused by land clearing, urbanisation and coastal development, and increased fertilizer use in adjacent catchments) is a pervasive threat to coral reefs throughout the world (Fabricius 2005). On the GBR, changed land use in adjacent catchments is suggested to have resulted in a 6-fold increase in nitrogen and a 9-fold increase in the amount of phosphorus entering the GBR lagoon (Kroon et al. 2012). Increasing nutrients directly affect the abundance, composition, and resilience (especially reproductive capacity) of reef-building corals and increase relative abundance of macroalgae or heterotrophic filter-feeders (Fabricius 2005, Fabricius et al. 2005), leading to overall degradation of reef ecosystems. For this reason, the Australian federal government and Queensland state government have collectively invested $375 million in Reef Plan, which aims to reduce anthropogenic dissolved inorganic nitrogen loads in priority areas along the Queensland coast by 50% by 2018, along with marked reductions in pesticide loads, and thereby eliminate detrimental effects on the health and resilience of the GBR (State of Queensland 2013). Spatially explicit modelling suggests that improvements in water quality across the GBR will reduce macroalgal cover by 39% and increase diversity of hard corals by 16% on a subset of reefs (28%) that are currently exposed to water quality conditions outside recommended thresholds (De'ath & Fabricius 2010). Moreover, these benefits are additional to any reductions in the frequency of outbreaks of *A. planci* that might occur as a consequence of improved water quality (De'ath & Fabricius 2010).

The number of no-take marine protected areas on coral reefs has increased considerably in recent decades, coincident with accelerating reef degradation and increasing expectations of the benefits that no-take areas can provide (Graham et al. 2011). No-take areas are one of the foremost strategies used to reduce exploitation of coral reef fishes and have definitive benefits for abundance and biomass of fishes, especially large-bodied, predatory fishes (Graham et al. 2011). The ecological consequences of removing large predatory fishes from coral reef ecosystems are poorly understood but may include cascading effects on ecosystem structure, function, and diversity (Dulvy et al. 2004). Most notably, Dulvy et al. (2004) revealed a direct relationship between densities of predatory fishes (associated with differences in local fishing pressure) and the local severity of outbreaks of *Acanthaster planci* in Fiji. Similarly, Sweatman (2008) showed that there was a lower frequency of outbreaks of the *A. planci* in no-take areas on the GBR, compared to fished areas. Despite these observations, the mechanistic basis for an ecological link between exploited fishes and the local incidence of crown-of-thorns outbreaks remains uncertain. It may be that predatory fishes are important in regulating the abundance of *Acanthaster* spp., but only at specific and early stages in their life history (Rivera-Posada et al. 2014), but this requires further research across a range of different locations.

Whatever the underlying mechanism, Sweatman (2008) suggests that increases in the areal extent of fisheries closures might reduce the reef-wide incidence of outbreaks of *Acanthaster* spp. This effect may also be amplified if fewer reefs with starfish outbreaks mean less-effective propagation of outbreaks from reef to reef through the central GBR. More important, coral cover was also shown to be higher inside no-take areas on the GBR, compared to fished areas (McCook et al. 2010), which is an obvious consequence of reduced incidence of outbreaks. More generally, the evidence that no-take areas benefit coral assemblages is limited (Graham et al. 2011), and differences in coral cover between areas that are open versus closed to fishing are most apparent where destructive fishing methods lead to declines in coral cover on fished reefs (e.g., Baird et al. 2005). However, no-take areas are likely to have benefits extending far beyond the protection of exploited fish species and are considered fundamental to sustaining the resilience of coral reef ecosystems (Hughes et al. 2007).

Conclusions

Outbreaks of *Acanthaster* spp. are considered to be one of the most significant disturbances affecting coral reefs in the Indo-Pacific (Pearson 1981, Birkeland 1996b, Bruno & Selig 2007). In his review of especially influential species and potentially 'keystone species' on tropical coral reefs, Birkeland (1996b) devoted a considerable portion of his discussion to *Acanthaster* spp. The importance of crown-of-thorns starfish is also reflected in the plethora of studies (at least 1200 published studies) on this single species or species complex (see reviews by Potts 1981, Moran 1986, Birkeland & Lucas 1990, and references therein). In spite of this considerable research effort, many questions about the biology of *Acanthaster* spp. remain unanswered, which greatly limits the understanding and hence the potential to manage outbreaks. The most fundamental question is what regulates the normal abundance of *Acanthaster* spp., thereby preventing chronic infestations of starfish throughout the Indo-Pacific.

There is no doubt that the *Acanthaster* spp. are predisposed to major population fluctuations because of inherent properties of their life history (e.g., immense fecundity, short generation times), and it is easy to conceive how small changes in distribution or behaviour could lead to rapid and pronounced increases in the abundance of starfish (Potts 1981, Moran 1986, Birkeland & Lucas 1990, Fabricius 2013). Most notably, there is likely to be limited reproductive success (specifically fertilization success) in low-density, preoutbreak populations of *Acanthaster* spp. where individuals tend to be highly dispersed (Endean & Cameron 1990). This may be further exacerbated by strongly male-biased sex ratios observed in some low-density populations (e.g., C. Cabelles et al. unpublished data 2013). If, however, there is a chance aggregation of male and female crown-of-thorns starfish, then the number of resulting progeny will go from effectively zero to many millions virtually overnight. However, this alone cannot explain periodic outbreaks because there is likely to be strong selection for *Acanthaster* spp. to aggregate every time they spawn, whereas outbreaks tend to occur at most every 7–15 years (Potts 1981). It is more likely that slow and progressive increases in the abundance of *Acanthaster* spp. on reefs that have small persistent populations are a necessary precursor to a marked increase in reproductive success, which may then initiate outbreaks (Moore 1990). The reproductive biology of *Acanthaster* spp. is geared towards extensive but infrequent spawning, which would be expected to lead to marked interannual fluctuations in reproductive success and recruitment, yet for much of the time and on most reefs within their range, crown-of-thorns starfish occur at low (almost negligible) densities (Birkeland & Lucas 1990). This suggests that there are other regulatory factors that limit reproductive success, larval survival, settlement rates, or post-settlement growth and survivorship.

The many hypotheses used to explain the initiation of outbreaks (or relax normal regulatory processes) of *Acanthaster* spp. are not mutually exclusive, and their importance is likely to vary both spatially and temporally (Birkeland & Lucas 1990, Bradbury & Antonelli 1990). The critical question is what, if anything, can be done to reduce the frequency or severity of crown-of-thorns outbreaks and thereby minimise or even reverse sustained declines in abundance of corals in many parts of the Indo-Pacific. The threat of the ensuing reef degradation to coastal fisheries and other reef-based industries (e.g., tourism) suggests that tropical nations should invest heavily in the eradication of *Acanthaster* spp., especially if reef-wide controls are potentially feasible. The effectiveness of control programmes will be further enhanced by focusing on detecting and eradicating primary outbreaks before they can initiate widespread secondary outbreaks (Fabricius 2013). Long-term or permanent solutions depend on definitive knowledge and appropriate actions to address the ultimate cause(s) of outbreaks. However, no-regret strategies can be implemented even in the absence of complete knowledge: Improvements in water quality and reductions in fisheries exploitation should benefit the resilience of reef ecosystems and may reduce long-term effects of *Acanthaster* spp. whether or not they reduce the incidence of outbreaks.

Acknowledgements

This contribution was supported by the ARC Centre of Excellence for Coral Reef Studies and Australian Institute of Marine Science. We are grateful to colleagues and collaborators, especially past and present members of AIMS Long Term Monitoring Program (LTMP), who have contributed unpublished data, as well as D.J. Pratchett, who helped ready this manuscript for publication. This manuscript benefitted greatly from discussions with R. Brainard, P. Doherty, M. Hall, T. Hughes, P. Muir, L. Vail and all participants at the crown-of-thorns starfish international workshop, Fitzroy Island, July 5–6, 2012.

References

Adam, T.C., Schmitt, R.J., Holbrook, S.J., Brooks, A.J., Edmunds, P.J., Carpenter, R.C. & Bernardi, G. 2011. Herbivory, connectivity, and ecosystem resilience: response of a coral reef to a large-scale perturbation. *PloS ONE* **6**, e23717.

Adjeroud, M., Michonneau, F., Edmunds, P.J., Chancerelle, Y., Lison de Loma, T., Penin, L.,Thibaut, L., Vidal-Dupiol, J., Slavat, B. & Galzin, R. 2009. Recurrent disturbances, recovery trajectories, and resilience of coral assemblages on a South Central Pacific reef. *Coral Reefs* **28**, 775–780.

Andrewartha, H.G. & Birch, L.C. 1984. *The Ecological Web*. Chicago, Illinois: University of Chicago Press.

Antonelli, P.L. & Kazarinoff, N.D. 1984. Starfish predation of a growing coral reef community. *Journal of Theoretical Biology* **107**, 667–684.

Aronson, R.B. & Precht, W.F. 2001. White-band disease and the changing face of Caribbean coral reefs. *Hydrobiologia* **460**, 25–38.

Ayukai, T. 1994. Ingestion of ultraplankton by the planktonic larvae of the crown-of-thorns starfish, *Acanthaster planci*. *Biological Bulletin* **186**, 90–100.

Ayukai, T., Keesing, J.K., Cartwright, C.M., Halford, A.R., Hall, K., Hocking, D. & Okaji, K. 1996. Practical handbook for large scale rearing of the larvae and juveniles of crown-of-thorns starfish. Australian Institute of Marine Science report no. **22**, 1–29. Townsville, Australia: Australian Institute of Marine Science.

Babcock, R.C. & Mundy C.N. 1992a. Reproductive biology, spawning and field fertilization rates of *Acanthaster planci*. *Marine and Freshwater Research* **43**, 525–533.

Babcock, R.C. & Mundy C.N. 1992b. Seasonal changes in fertility and fecundity in *Acanthaster planci*. *Proceedings of the Seventh International Coral Reef Symposium* **2**, 757–761.

Babcock, R.C., Mundy, C.N. & Whitehead, D. 1994. Sperm diffusion models and *in situ* confirmation of long-distance fertilization in the free-spawning asteroid *Acanthaster planci*. *Biological Bulletin* **186**, 17–28.

Baine, M.S.P. 2006. A major outbreak of crown-of-thorns starfish in Bootless Bay, Central Province, Papua New Guinea. *Coral Reefs* **25**, 607.

Baird, A.H., Campbell, S., Anggoro, A.W., Fadli, N., Herdiana, Y., Katawijaya, T., Legawa, R., Mahyiddin, D., Mukminin, A., Pratchett, M.S., Rudi, E., Siregar, A.M. & Trilestari, S. 2005. Achenese reefs in the wake of the tsunami. *Current Biology* **15**, 1926–1930.

Baird, A.H., Guest, J.R. & Willis, B.L. 2009. Systematic and biogeographical patterns in the reproductive biology of scleractinian corals. *Annual Review of Ecology, Evolution, and Systematics* **40**, 551–571.

Baird, A.H., Pratchett, M.S., Hoey, A.S., Campbell, S. & Herdiana, Y. 2013. Outbreaks of *Acanthaster planci* in Indonesia: a major cause of coral mortality in the coral triangle. *Coral Reefs* **32**, 803–812.

Barham, E.G., Gowdy, R.W. & Wolfson, F.H. 1973. *Acanthaster* (Echinodermata, Asteroidea) in the Gulf of California. *Fisheries Bulletin* **71**, 927–942.

Barnes, D.J., Brauer, R.W. & Jordan, M.R. 1970. Locomotory response of *Acanthaster planci* to various species of coral. *Nature* **228**, 342–344.

Barnes, J.H. 1966. The crown-of-thorns starfish as a destroyer of coral. *Australian Natural History* **15**, 257–261.

Beach, D.H., Hanscomb, N.J. & Ormond, R.F.G. 1975. Spawning pheromone in crown-of-thorns starfish. *Nature* **254**, 135–136.

Belk, M.S. & Belk, D. 1975. Observation of algal colonization on *Acropora aspera* killed by *Acanthaster planci*. *Hydrobiologia* **46**, 29–32.

Bell, J.D., Kronen, M., Vunisea, A., Nash, W.J., Keeble, G., Demmke, A., Ponitifex, S. & Andréfouët, S. 2009. Planning the use of fish for food security in the Pacific. *Marine Policy* **33**, 64–76.

Bell, R., Kettle, B. & Stump, R. 1986. An evaluation of the effectiveness of artificial barriers for meso-scale crown-of-thorns management. Townsville, Australia: Great Barrier Reef Marine Park Authority.

Bellwood, D.R., Hughes, T.P., Folke, C. & Nystrom, M. 2004. Confronting the coral reef crisis. *Nature* **429**, 827–833.

Benson, A.A. 1975. Wax as a medium for energy exchange in the coral reef community food web. *Proceedings of the 13th Pacific Science Association Congress* **1**, 123.

Benzie, J.A. 1999. Major genetic differences between crown-of-thorns starfish (*Acanthaster planci*) populations in the Indian and Pacific Oceans. *Evolution* **53**, 1782–1795.

Benzie, J.A.H., Black, K.P., Moran, P.J. & Dixon, P. 1994. Small-scale dispersion of eggs and sperm of the crown-of-thorns starfish (*Acanthaster planci*) in a shallow coral reef habitat. *Biological Bulletin* **186**, 153–167.

Benzie, J.A.H. & Dixon, P. 1994. The effects of sperm concentration, sperm: egg ratio, and gamete age on fertilization success in crown-of-thorns starfish (*Acanthaster planci*) in the laboratory. *Biological Bulletin* **186**, 139–152.

Berryman, A.A. 1987. The theory and classification of outbreaks. In *Insect Outbreaks*, B. Barbosa & J.C. Shultz (eds). San Diego, California: Academic Press, 3–30.

Berumen, M.L. & Pratchett, M.S. 2006. Persistent disturbance and long- term shifts in the structure of fish and coral communities at Tiahura Reef, Moorea. *Coral Reefs* **25**, 647–653.

Beukers, J.S. & Jones, G.P. 1998. Habitat complexity modifies the impact of piscivores on a coral reef fish population. *Oecologia* **114**, 50–59.

Birdsey, R. 1988. Large reef fishes as potential predators of *Acanthaster planci*: A pilot study by alimentary tract analysis of predatory fishes from reefs subject to *Acanthaster* feeding. Townsville, Australia: Great Barrier Marine Park Authority.

Birkeland, C. 1982. Terrestrial runoff as a cause of outbreaks of *Acanthaster planci* (Echinodermat: Asteroidea). *Marine Biology* **69**, 175–185.

Birkeland, C. 1989a. The influence of echinoderms on coral-reef communities. In *Echinoderm Studies* Volume 3, M. Jangoux & J.M. Lawrence (eds). Rotterdam, the Netherlands: Balkema, 1–77.

Birkeland, C. 1989b. The faustian traits of crown-of-thorns starfish. *American Scientist* **77**, 155–163.

Birkeland, C. 1996a. Geographic differences in ecological processes on coral reefs. In *Life and Death of Coral Reefs*, C. Birkeland (ed.). New York: Chapman and Hall, 273–297.

Birkeland, C. 1996b. Why some species are especially influential on coral-reef communities and others are not. *Galaxea* **13**, 77–84.

Birkeland, C. & Lucas, J.S. 1990. *Acanthaster planci: Major Management Problem of Coral Reefs*. Boca Raton, Florida: CRC Press.

Birkeland, C. & Randall, R.H. 1979. Report on the *Acanthaster planci* (Almea) studies on Tutuila, American Samoa. Samoa: Office of Marine Resources, Government of Samoa.

Black, K.P. & Moran, P.J. 1991. Influence of hydrodynamics on the passive dispersal and initial recruitment of larvae of *Acanthaster planci* (Echinodermata: Asteroidea) on the Great Barrier Reef. *Marine Ecology Progress Series* **69**, 55–65.

Bos, A.R., Gumanao, G.S., Mueller, B. & Saceda-Cardoza, M.M.E. 2013. Management of crown-of-thorns sea star (*Acanthaster planci* L.) outbreaks: removal success depends on reef topography and timing within the reproduction cycle. *Ocean and Coastal Management* **71**, 116–122.

Bouchon, C. 1985. Quantitative study of scleractinian coral communities of Tiahura Reef (Moorea Island, French Polynesia). *Proceedings of the Fifth Coral Reef Congress* **6**, 279–284.

Bouchon-Navaro, Y., Bouchon, C. & Harmelin-Vivien, M.L. 1985. Impact of coral degradation on a Chaetodontid fish assemblage (Moorea, French Polynesia). *Proceedings of the 5th International Coral Reef Symposium* **5**, 427–432.

Bradbury, R.H. & Antonelli, P.L. 1990. What controls outbreaks? In *Acanthaster and the Coral Reef: A Theoretical Perspective*, R.H. Bradbury (ed.). Berlin: Springer, 278–290.

Bradbury, R.H., Hammond, L.S., Moran, P.J. & Reichelt, R.E. 1985. Coral reef communities and the crown-of-thorns starfish: evidence for qualitatively stable cycles. *Journal of Theoretical Biology* **113**, 69–80.

Branham, J.M. 1973. The crown of thorns on coral reefs. *BioScience* **23**, 219–226.

Branham, J.M., Reed, S.A., Bailey, J.H. & Caperon, J. 1971. Coral-eating sea stars *Acanthaster planci* in Hawaii. *Science* **172**, 1155–1157.

Brauer, R.W., Jordan, M.R. & Barnes, D.J. 1970. Triggering of the stomach eversion reflex of *Acanthaster planci* by coral extracts. *Nature* **228**, 344–346.

Bridge, T.C., Hughes, T.P., Guinotte, J.M. & Bongaerts, P. 2013. Call to protect all coral reefs. *Nature Climate Change* **3**, 528–530.

Brodie, J., Fabricius, K., De'ath, G. & Okaji, K. 2005. Are increased nutrient inputs responsible for more outbreaks of crown-of-thorns starfish? An appraisal of the evidence. *Marine Pollution Bulletin* **51**, 266–278.

Brodie, J. & Waterhouse, J. 2012. A critical review of environmental management of the 'not so Great' Barrier Reef. *Estuarine, Coastal and Shelf Science* **104**, 1–22.

Brodie, J.E. 1992. Enhancement of larval and juvenile survival and recruitment in *Acanthaster planci* from the effects of terrestrial run-off: a review. *Australian Journal of Marine and Freshwater Research* **43**, 539–554.

Bruno, J.F. & Selig, E.R. 2007. Regional decline of coral cover in the Indo-Pacific: timing, extent, and subregional comparisons. *PLoS ONE* **2**, e711.

Bruno, J.F., Selig, E.R., Casey, K.S., Page, C.A., Willis, B.L., Harvell, C.D., Sweatman, H. & Melendy, A.M. 2007. Thermal stress and coral cover as drivers of coral disease outbreaks. *PLoS Biology* **5**, e124.

Burkholder, P.R. 1973. The ecology of marine antibiotics and coral reefs. In *Biology and Geology of Coral Reefs*, Volume 2, O.A. Jones & R. Endean (eds). New York: Academic Press, 117–182.

Byrne, M., Cerra, A., Nishigaki, T. & Hoshi, M. 1997. Infestation of the testes of the Japanese sea star *Asterias amurensis* by the ciliate *Orchitophyra stellarum*: a caution against the use of this ciliate for biological control. *Diseases of Aquatic Organsims* **28**, 235–239.

Caballes, C.F., Schupp, P.J., Pratchett, M.S. & Rivera-Posada, J.A. 2012. Interspecific transmission and recovery of TCBS-induced disease between *Acanthasther planci* and *Linckia guildingi*. *Diseases of Aquatic Organisms* **100**, 263–267.
Caley, M.J. & St John, J. 1996. Refuge availability structures assemblages of tropical reef fishes. *Journal of Animal Ecology* **65**, 414–428.
Campbell, A.C. & Ormond, R.F.G. 1970. The threat of 'crown-of-thorns' starfish (*Acanthaster planci*) to coral reefs in the Indo-Pacific area: observations on a normal population in the Red Sea. *Biological Conservation* **2**, 246–251.
Carpenter, K.E., Miclat, R.I., Albaladejo, V.D. & Corpuz, V.T. 1982. The influences of substrate structure on the local abundance and diversity of Philippine reef fishes. *Proceedings of the 4th International Coral Reef Symposium* **2**, 497–502.
Carpenter, R.C. 1996. Invertebrate predators and grazers. In *Life and Death of Coral Reefs*, C. Birkeland (ed.). New York: Chapman and Hall, 198–229.
Cheney, D.P. 1972. Spawning and aggregation of *Acanthaster planci*. *University of Guam Technical Report* **3**, 11.
Cheney, D.P. 1974. Spawning and aggregation of *Acanthaster planci* in Micronesia. *Proceedings of the Second International Coral Reef Symposium* **1**, 591–594.
Chesher, R.H. 1969. Destruction of the Pacific corals by the sea star *Acanthaster planci*. *Science* **165**, 280–283.
Chess, J.R., Hobson, E.S. & Howard, D.F. 1997. Interactions between *Acanthaster planci* (Echinodermata, Asteroidea) and scleractinian corals at Kona, Hawai'i. *Pacific Science* **51**, 121–133.
Chou, L.M. & Yamazato, K. 1990. Community structure of coral reefs within the vicinity of Motobu and Sesoko, Okinawa, and the effects of human and natural influences. *Galaxea* **9**, 9–75.
Ciarapica, G. & Passeri, L. 1993. An overview of the Maldivian coral reefs in Felidu and North Male atoll (Indian Ocean): platform drowning by ecological crises. *Facies* **28**, 33–65.
Clark, C. & Weitzman, B. 2006. Population study survey of *Acanthaster planci*, the crown of thorns starfish on the northwest coast of Moorea, French Polynesia. USA: Université de Californie Santa Cruz.
Clark, H.L. 1921. The echinoderm fauna of Torres Strait: its composition and its origin. *Papers from the Department of Marine Biology Carnegie Institution Washington* **10**, 1–224.
Coker, D.J., Wilson, S.K. & Pratchett, M.S. 2014. Importance of live coral habitat for reef fishes. *Review in Fish Biology and Fisheries* **24**, 89–126.
Cole, A.J., Pratchett, M.S. & Jones, G.P. 2008. Diversity and functional importance of coral-feeding fishes on tropical coral reefs. *Fish and Fisheries* **9**, 286–307.
Colgan, M.W. 1987. Coral reef recovery on Guam (Micronesia) after catastrophic predation by *Acanthaster planci*. *Ecology* **68**, 1592–1605.
Colin, P.L. 1977. The reefs of Cocos-Keeling Atoll, Eastern Indian Ocean. *Proceedings of the 3rd International Coral Reef Symposium* **1**, 63–68.
Collins, A.R.S. 1975. Biochemical investigation of two responses involved in the feeding behaviour of *Acanthaster planci* (L.). III. Food preferences. *Journal of Experimental Biology and Ecology* **17**, 87–94.
Conand, C. 1983. *Abondance, cycle sexuel et relations biométriques de l'étoile* Acanthaster planci *en Nouvelle-Calédonie*. Nouméa, New Caledonia: Office de la recherche scientifique et téchnique outre-mer, Centre de Nouméa, Océanographie.
Conand, C. 1984. Distribution, reproductive cycle and morphometric relationships of *Acanthaster planci* (Echinodermata: Asteroidea) in New Caledonia, western tropical Pacific. *Proceedings of the Fifth International Echinoderm Conference*, 499–506.
Cornell, J. & Surowiecki, J. (eds) 1972. *The Pulse of the Planet; A State of the Earth Report from the Smithsonian Institution Center for Short-lived Phenomena*. New York: Crown.
Crump, R. 1971. Reproductive condition of *Acanthaster planci* in the Port Sudan area of the Red Sea. In *New Studies on the Crown of Thorns Starfish (*Acanthaster planci*) from Investigations in the Red Sea*, C.H. Roads & R.F.G Ormond (eds). Cambridge, UK: Coral Starfish Research Group, 1971.
Dana, T. & Wolfson, A. 1970. Eastern Pacific crown-of-thorns starfish populations in the lower Gulf of California. *Transactions of the San Diego Society of Natural History* **16**, 83–90.
Dana, T.F. 1970. *Acanthaster:* a rarity in the past? *Science* **169**, 894.
Dana, T.F., Newman, W.A. & Fager, E.W. 1972. *Acanthaster* aggregations: interpreted as primarily responses to natural phenomena. *Pacific Science* **26**, 355–372.

Day, K., 2000. Summary of investigations into a possible link between terrestrial run-off and crown-of-thorns starfish outbreaks on the Great Barrier Reef. Unpublished report by Ken Day, Department of Natural Resources, Mines and Energy, Brisbane, Australia.

De'ath, G. & Fabricius, K. 2010. Water quality as a regional driver of coral biodiversity and macroalgae on the Great Barrier Reef. *Ecological Applications* **20**, 840–850.

De'ath, G., Fabricius, K.E., Sweatman, H. & Puotinen, M. 2012. The 27–year decline of coral cover on the Great Barrier Reef and its causes. *Proceedings of the National Academy of Sciences of the United States of America* **109**, 17995–17999.

De'ath, G. & Moran, P.J. 1998. Factors affecting the behaviour of crown-of-thorns starfish (*Acanthaster planci* L.) on the Great Barrier Reef: 2: feeding preferences. *Journal of Experimental Marine Biology and Ecology* **220**, 107–126.

Devaney, D.M. & Randall, J.E. 1973. Investigations of *Acanthaster planci* in southeastern Polynesia during 1970–1971. *Atoll Research Bulletin* **169**, 1–23.

DeVantier, L.M. & Andrews, G.J. 1987. Report on surveys of the distribution, abundance and impact of *Acanthaster planci* on the fringing reefs of Lord Howe Island and the Solitary Islands, March to April 1987. Townsville, Australia: Australian Institute of Marine Science.

DeVantier, L.M. & Deacon, G. 1990. Distribution of *Acanthaster planci* at Lord Howe Island, the southern-most Indo-Pacific reef. *Coral Reefs* **9**, 145–148.

DeVantier, L.M. & Endean, R. 1988. The scallop *Pedum spondyloideum* mitigates the effects of *Acanthaster planci* predation on the coral host *Porites:* host defence facilitated by exaptation? *Marine Ecology Progress Series* **47**, 293–301.

Diehl, W.J. 1986. Osmoregulation in echinoderms. *Comparative Biochemistry and Physiology Part A: Physiology* **84**, 199–205.

Doherty, P.P. & Davidson, J.J. 1988. Monitoring the distribution and abundance of juvenile *Acanthaster planci* in the central Great Barrier Reef. *Proceedings of the 6th International Coral Reef Symposium* **2**, 131–136.

Done, T. 1992. Constancy and change in some Great Barrier Reef coral communities: 1980–1991. *American Zoologist* **32**, 655–662.

Done, T.J. 1985. Effects of two *Acanthaster* outbreaks on coral community structure: the meaning of devastation. *Proceedings of the 5th International Coral Reef Symposium* **5**, 315–320.

Done, T.J., Osborne, K. & Navin, K.F. 1988. Recovery of coral communities post-*Acanthaster:* progress and prospects. *Proceedings of the 6th International Coral Reef Symposium* **2**, 137–142.

Dulvy, N.K., Freckleton, R.P. & Polunin, N.V. 2004. Coral reef cascades and the indirect effects of predator removal by exploitation. *Ecology Letters* **7**, 410–416.

Dungan, M.L., Miller, T.E. & Thompson, D.A. 1982. Catastrophic decline of a top carnivore in the Gulf of California rocky intertidal zone. *Science* **216**, 989–991.

Ebert, T.A. 1973. Estimating growth and mortality rates from size data. *Oecologia* **11**, 281–298.

Eldredge, L.G. 1970. *Acanthaster Newsletter* **2**. Mangilao, Guam: Marine Laboratory, University of Guam.

Elton, C.S. 1958. *The Ecology and Invasions by Animals and Plants*. London: Chapman and Hall, London.

Endean, R. 1969. Report on investigations made into aspects of the current *Acanthaster planci* (crown of thorns) infestations of certain reefs of the Great Barrier Reef. Brisbane, Australia: Queensland Department of Primary Industries (Fisheries Branch).

Endean, R. 1971. The recovery of coral reefs devastated by catastrophic events with particular reference to current *Acanthaster planci* plagues in the tropical Indo-West Pacific region. *Journal of the Marine Biological Association of India* **13**, 1–13.

Endean, R. 1974. *Acanthaster planci* on the Great Barrier Reef. *Proceedings of the Second International Coral Reef Symposium* **1**, 563–576.

Endean, R. 1976. Destruction and recovery of coral reef communities. In *Biology and Geology of Coral Reefs Vol. 3. Biology 2,* O.A. Jones & R. Endean (eds). New York: Academic Press, 215–254.

Endean, R. 1977. *Acanthaster planci* infestations of reefs of the Great Barrier Reef. *Proceedings of Third International Coral Reef Symposium* **1**, 185–191.

Endean, R. 1982. Crown-of-thorns starfish on the Great Barrier Reef. *Endeavour* **6**, 10–14.

Endean, R. & Cameron, A.M. 1985. Ecocatastrophe on the Great Barrier Reef. *Proceedings of the 5th International Coral Reef Symposium* **5**, 309–314.

Endean, R. & Cameron, A.M. 1990. *Acanthaster planci* population outbreaks. In *Coral Reefs: Ecosystems of the World 25*, Z. Dubinsky (ed.). New York: Elsevier, 419–437.

Endean, R. & Stablum, W. 1973. A study of some aspects of the crown-of-thorns starfish (*Acanthaster planci*) infestations of Australia's Great Barrier Reef. *Atoll Research Bulletin.* **167**, 1–62.
Endean, R. & Stablum, W. 1975. Population explosions of *Acanthaster planci* and associated destruction of hard-coral cover of reefs of the Great Barrier. *Environmental Conservation* **2**, 247–256.
Engelhardt, U. 1999. Fine-scale surveys of crown-of-thorns starfish (*Acanthaster planci*) in the central Great Barrier Reef region. CRC Reef Research Centre technical report no. 30. Townsville, Australia: CRC Reef Research Centre.
Fabricius, K.E. 2005. Effects of terrestrial runoff on the ecology of corals and coral reefs: review and synthesis. *Marine Pollution Bulletin* **50**, 125–146.
Fabricius, K. 2013. *Acanthaster planci.* In *Starfish: Biology and Ecology of the Asteroidea,* J.M. Lawrence (ed.). Baltimore, Maryland: John Hopkins University Press, 132–141.
Fabricius, K.E. & Fabricius, F.H. 1992. Re-assessment of ossicle frequency patterns in sediment cores: rate of sedimentation related to *Acanthaster planci. Coral Reefs* **11**, 109–114.
Fabricius, K., De'ath, G., McCook, L., Turak, E. & Williams, D.M. 2005. Changes in algal, coral and fish assemblages along water quality gradients on the inshore Great Barrier Reef. *Marine Pollution Bulletin* **51**, 384–398.
Fabricius, K.E., Okaji, K. & De'ath, G. 2010. Three lines of evidence to link outbreaks of the crown-of-thorns seastar *Acanthaster planci* to the release of larval food limitation. *Coral Reefs* **29**, 593–605.
Faurea, G. 1989. Degradation of coral reefs at Moorea Island (French Polynesia) by *Acanthaster planci. Journal of Coastal Research* **5**, 295–305.
Fisk, D.A. & Power, M.C. 1999. Development of cost-effective control strategies for crown-of-thorns starfish. CRC Reef Research Centre Technical Report No. 25. Townsville, Australia: CRC Reef Research Centre.
Frankel, E. 1977. Previous *Acanthaster* aggregations in the Great Barrier Reef. *Proceedings of the 3rd International Coral Reef Symposium* **1**, 201–208.
Frankel, E. 1978. Evidence from the Great Barrier Reef of ancient *Acanthaster* aggregations. *Atoll Research Bulletin* **220**, 75–93.
Ganter, R. 1987. Oral history of human use and experience of crown of thorns starfish on the Great Barrier Reef. Report to the Great Barrier Reef Marine Park Authority. Townsville, Australia: Great Barrier Reef Marine Park Authority.
Gardner, T.A., Cote, I.M., Gill, J.A., Grant, A. & Watkinson, A.R. 2003. Long-term region-wide declines in Caribbean corals. *Science* **301**, 958–960.
Garlovsky, D.F. & Bergquist, A. 1970. Crown-of-thorns starfish in Western Samoa. *South Pacific Bulletin* **20**, 47–49.
Gilmour, J.P., Smith, L.D., Heyward, A.J., Baird, A.H. & Pratchett, M.S. 2013. Recovery of an isolated coral reef system following severe disturbance. *Science* **340**, 69–71.
Gladstone, W. 1987. A spine-bending story. *Australian Natural History* **22**, 303–305.
Gladstone, W. 1992. Observations of crown-of-thorns starfish spawning. *Marine and Freshwater Research* **43**, 535–537.
Glynn, P.W. 1973. *Acanthaster*: effect on coral reef growth in Panama. *Science* **180**, 504–506.
Glynn, P.W. 1974. The impact of *Acanthaster* on corals and coral reefs in the eastern Pacific. *Environmental Conservation* **1**, 295–304.
Glynn, P.W. 1976. Some physical and biological determinants of coral community structure in the Eastern Pacific. *Ecological Monographs* **46**, 431–456.
Glynn, P.W. 1980 Defence by symbiotic crustacea of host corals elicited by chemical cues from a predator. *Oecologia* **47**, 287–290.
Glynn, P.W. 1982. Individual recognition and phenotypic variability in *Acanthaster planci* (Echinodermata: Asteroidea). *Coral Reefs* **1**, 89–94.
Glynn, P.W. 1984. An amphinomid worm predator of the crown-of-thorns sea star and general predation on asteroids in eastern and western Pacific coral reefs. *Bulletin of Marine Science* **35**, 54–71.
Glynn, P.W. 1987. Some ecological consequences of coral-crustacean guard mutualisms in the Indian and Pacific Ocean. *Symbiosis* **4**, 301–324.
Glynn, P.W. 1988. Predation on coral reefs: some key processes concepts and research directions. *Proceedings of the 6th International Coral Reef Symposium* **1**, 51–62.
Glynn, P.W. 1990. Coral mortality and disturbances to coral reefs in the tropical eastern Pacific. *Elsevier Oceanography Series* **52**, 55–126.

Goreau, T.F. 1964. On the predation of coral by the spiny starfish *Acanthaster planci* (L.) in the southern Red Sea. *Bulletin of the Sea Fisheries Research Station of Israel* **35**, 23–26.

Graham, E.M., Baird, A.H. & Connolly, S.R. 2008. Survival dynamics of scleractinian coral larvae and implications for dispersal. *Coral Reefs* **27**, 529–539.

Graham, N.A.J., Ainsworth, T.D., Baird, A.H., Ban, N.C., Bay, L.K., Cinner, J.E., De Freitas, D.M., Diaz-Pulido, G., Dornelas, M., Dunn, S.R., Fiedlman, P.I.J., Foret, S., Good, T., Kool, J., Mallela, J., Penin, L., Pratchett, M.S. & Williamson, D.H. 2011. From microbes to people: tractable benefits of no-take area for coral reefs. *Oceanography and Marine Biology: An Annual Review* **49**, 105–136.

Greenwood, P.J. & Bennett, T. 1981. Some effects of temperature-salinity combinations on the early development of the sea urchin *Parechinus angulosus* (Leske). Fertilization. *Journal of Experimental Marine Biology and Ecology* **51**, 119–131.

Grosberg, R.K. 1991. Sperm-mediated gene flow and the genetic structure of a population of a colonial ascidian *Botryllus schlosseri*. *Evolution* **45**, 130–142.

Habe, T., Yamamoto, G., Nagai, A., Kosaka, M., Ogura, M., Sawamoto, S., Ueno, S. & Yokochi, H. 1989. Studies on the conservation and management of coral reefs and the control of *Acanthaster planci* juveniles. Japan: Ministry of Education, Science and Culture.

Harris, P.T., Bridge, T. C, Beaman, R.J., Webster, J.M., Nichol, S.L. & Brooke, B.P. 2013. Submerged banks in the Great Barrier Reef, Australia, greatly increase available coral reef habitat. *ICES Journal of Marine Science* **70**, 284–293.

Hart, A.M., Klumpp, D.W. & Russ, G.R. 1996. Response of herbivorous fishes to crown-of-thorns starfish *Acanthaster planci* outbreaks. II Density and biomass of selected species of herbivorous fish and fish-habitat correlations. *Marine Ecology Progress Series* **132**, 21–30.

Hayashi, K. 1975. Occurrences of the crown-of-thorns starfish, *Acanthaster planci* (L.), along the southern coast of Kii Peninsula, Japan. *Bulletin of the Marine Park Research Stations* **1**, 1.

Henderson, J.A. 1969. Preliminary observations on the rearing and development of *Acanthaster planci* (L.) (Asteroidea larvae). *Fisheries Notes, Queensland Department of Harbours and Marine* **3**, 69–75.

Henderson, J.A. & Lucas, J.S. 1971. Larval development and metamorphosis of *Acanthaster planci* (Asteroidea). *Nature* **232**, 655–657.

Hoegh-Guldberg, O. 1994. Uptake of dissolved organic matter by larval stage of the crown-of-thorns starfish *Acanthaster planci*. *Marine Biology* **120**, 55–63.

Hoegh-Guldberg, O. 1999. Climate change, coral bleaching and the future of the world's coral reefs. *Marine and Freshwater Research* **50**, 839–866.

Hoegh-Guldberg, O. 2004. Coral reefs in a century of rapid environmental change. *Symbiosis* **37**, 1–31.

Hoegh-Guldberg, O. & Bruno, J.F. 2010. The impact of climate change on the world's marine ecosystems. *Science* **328**, 1523–1528.

Hoey, J. & Chin, A. 2004. Crown-of-thorns starfish. In *The State of the Great Barrier Reef On-line, Great Barrier Reef Marine Park Authority, Townsville*, A. Chin (ed.). Townsville, Australia: Great Barrier Marine Park Authority, 1–13.

Holbrook, S.J., Forrester, G.E. & Schmitt, R.J. 2000. Spatial patterns in abundance of a damselfish reflect availability of suitable habitat. *Oecologia* **122**, 109–120.

Holbrook, S.J. & Schmitt, R.J. 2002. Competition for shelter space causes density-dependent predation mortality in damselfishes. *Ecology* **83**, 2855–2868.

Holling, C.S., 1965. The functional response of predators to prey density and its role in mimicry and population regulation. *Memoirs of the Entomological Society of Canada* **97**, 5–60.

Houk, P., Bograd, S. & van Woesik, R. 2007. The transition zone chlorophyll front can trigger *Acanthaster planci* outbreaks in the Pacific Ocean: Historical confirmation. *Journal of Oceanography* **63**, 149–154.

Houk, P. & Raubani, J. 2010. *Acanthaster planci* outbreaks in Vanuatu coincide with ocean productivity, furthering trends throughout the Pacific Ocean. *Journal of Oceanography,* **66**, 435–438.

Hughes, T.P. 1990. Recruitment limitation, mortality, and population regulation in open systems: a case study. *Ecology* **71**, 12–20.

Hughes, T.P. 1994. Catastrophes, phase shifts, and large-scale degradation of a Caribbean coral reef. *Science* **265**, 1547–1551.

Hughes, T.P., Baird, A.H., Bellwood, D.R., Card, M., Connolly, S.R., Folke, C., Grosberg, R., Hoegh-Guldberg, O., Jackson, J.B.C., Kleypas, J., Lough, J.M., Marshall, P., Nystrom, M., Palumbi, S.R., Pandolfi, J.M., Rosen, B. & Roughgarden, J. 2003. Climate change, human impacts and the resilience of coral reefs. *Science* **301**, 929–933.

Hughes, T.P., Bellwood, D.R., Folke, C.S., McCook, L.J. & Pandolfi, J.M. 2007. No-take areas, herbivory and coral reef resilience. *Trends in Ecology and Evolution* **22**, 1–3.

Hughes, T.P., Graham, N.A., Jackson, J.B., Mumby, P.J. & Steneck, R.S. 2010. Rising to the challenge of sustaining coral reef resilience. *Trends in Ecology & Evolution* **25**, 633–642.

Hutchings, P. 2011. Bioerosion. In *Encyclopedia of Modern Coral Reefs*, D. Hopley (ed). Nederlands; Springer, 139–156.

Idip, D., Jr. 2003. Annual reproduction cycle of *Acanthaster planci* (L.) in Palau. *Proceedings of the First Coral Reef Conference, Palau International Coral Reef Center Publication* **04-001**, 87–91.

Jangoux, M. 1982. Food and feeding mechanisms, Asteroidea. In *Echinoderm Studies*, Vol. 3, M. Jangoux & J.M. Lawrence (eds). Rotterdam, the Netherlands: Balkema, 117–159.

Jangoux, M. 1987a. Diseases of Echinodermata. I. Agents microorganisms and protistans. *Diseases of Aquatic Organisms* **2**, 147–162.

Jangoux, M. 1987b. Diseases of Echinodermata. II. Agents metazoans (Mesozoa to Bryozoa). *Diseases of Aquatic Organisms* **2**, 205–234.

Jangoux, M. & Lawrence, J.M. 1989. *Echinoderm Studies*, Vol. 3. Rotterdam, the Netherlands: Balkema.

Johnson, C. 1992. Reproduction, recruitment and hydrodynamics in the crown-of-thorns phenomenon on the Great Barrier Reef: introduction and synthesis. *Marine and Freshwater Research* **43**, 517–523.

Johnson, C.R. & Sutton, D.C. 1994. Bacteria on the surface of crustose coralline algae induce metamorphosis of the crown-of-thorns starfish *Acanthaster planci. Marine Biology* **120**, 305–310.

Johnson, C.R., Sutton, D.C., Olson, R.R. & Giddins, R. 1991. Settlement of crown-of-thorns starfish: role of bacteria on surfaces of coralline algae and a hypothesis for deepwater recruitment. *Marine Ecology Progress Series* **71**, 143–162.

Johnson, D.B., Moran, P.J. & Driml, S. 1990. Evaluation of a crown-of-thorns starfish (*Acanthaster planci*) control program at Grub Reef (central Great Barrier Reef). *Coral Reefs* **9**, 167–171.

Johnson, D.B., Moran, P.J., Baker, V.J., Christie, C.A., Miller, I.R., Miller-Smith, B.A. & Thompson, A.A. 1992. An attempt to locate high density populations of juvenile crown-of-thorns starfish (*Acanthaster planci*) on the central Great Barrier Reef. *Coral Reefs* **11**, 122 only.

Johnson, L.G. & Babcock, R.C. 1994. Temperature and the larval ecology of the crown-of-thorns starfish, *Acanthaster planci. Biological Bulletin* **187**, 304–308.

Jokiel, P.L., Hunter, C.L. & Watarai, L. 1993. Ecological impact of a fresh-water 'reef kill' in Kaneohe Bay, Oahu, Hawaii. *Coral Reefs* **12**, 177–184.

Jones, G.P. 1987. Competitive interactions among adults and juveniles in a coral reef. *Ecology* **71**, 1691–1698.

Jones, G.P. 1988. Experimental evaluation of the effects of habitat structure and competitive interactions on the juveniles of two coral reef fishes. *Journal of Experimental Marine Biology and Ecology* **123**, 115–126.

Jones, G.P. 1991. Postrecruitment processes in the ecology of coral reef fish populations: a multifactorial perspective. In *The Ecology of Fishes on Coral Reefs,* P.F. Sale (ed.). San Diego, California: Academic Press, 294–328.

Jones, G.P., McCormick, M.I., Srinivasan, M. & Eagle, J.V. 2004. Coral decline threatens fish biodiversity in marine reserves. *Proceedings of the National Academy of Sciences of the United States of America* **101**, 8251–8253.

Kayal, M., Vercelloni, J., De Loma, T.L., Bosserelle, P., Chancerelle, Y., Geoffroy, S., Stievenart, C., Michonneau, F., Penin, L., Planes, S. & Adjeroud, M. 2012. Predator crown-of-thorns starfish (*Acanthaster planci*) outbreak, mass mortality of corals, and cascading effects on reef fish and benthic communities. *PloS ONE* **7**, e47363.

Keesing, J.K. 1990. *Feeding biology of the crown-of-thorns starfish,* Acanthaster planci *(Linnaeus).* PhD thesis, James Cook University of North Queensland, Townsville, Australia.

Keesing, J.K. 1992. Influence of persistent sub-infestation density *Acanthaster planci* (L.) and high density *Echinometra mathaei* (de Blainville) populations on coral reef community structure in Okinawa, Japan. *Proceedings of the 7th International Coral Reef Symposium* **2**, 769–779.

Keesing, J.K., Bradbury, R.H., DeVantier, L.M., Riddle, M.J. & De'Ath, G. 1992. Geological evidence for recurring outbreaks of the crown-of-thorns starfish: a reassessment from an ecological perspective. *Coral Reefs* **11**, 79–85.

Keesing, J.K. & Halford, A.R. 1992. Field measurement of survival rates of juvenile *Acanthaster planci*: techniques and preliminary results. *Marine Ecology Progress Series* **85**, 107–114.

Keesing, J.K. & Lucas, J.S. 1992. Field measurement of feeding and movement rates of the crown-of-thorns starfish *Acanthaster planci* (L.). *Journal of Experimental Marine Biology and Ecology* **156**, 89–104.

Keesing, J.K., Wiedermeyer, W.L., Okaji, K., Halford, A.R., Hall, K.C. & Cartwright, C.M. 1996. Mortality rates of juvenile starfish *Acanthaster planci* and *Nardoa* spp. measured on the Great Barrier Reef, Australia and in Okinawa, Japan. *Oceanologica Acta* **19**, 441–448.

Kenchington, R.A. & Kelleher, G. 1992. Crown-of-thorns management conundrums. *Coral Reefs* **11**, 53–56.

Kenchington, R.A. & Pearson, R.G. 1982. Crown of thorns starfish on the Great Barrier Reef: a situation report. *Proceedings of the 4th International Coral Reef Symposium* **2**, 597–600.

Kenchington, R.A. 1977. Growth and recruitment of *Acanthaster planci* (L.) on the Great Barrier Reef. *Biological Conservation* **11**, 103–118.

Kettle, B.T. 1990. *Variations in biometric and physiological parameters of* Acanthaster planci *(L.) (Echinodermata; Asteroidea) during the course of a high density outbreak.* PhD thesis, James Cook University, Australia.

Kettle, B.T. & Lucas, J.S. 1987. Biometric relationships between organ indices, fecundity, oxygen consumption and body size in *Acanthaster planci* (L.) (Echinodermata; Asteroidea). *Bulletin of Marine Science* **41**, 541–551.

Kishimoto, F. 1989. *Acanthaster planci* spawning. *Marine Parks Journal* **81**, 2 only.

Klotzbach, P.J. 2006. Trends in global tropical cyclone activity over the past twenty years (1986–2005). *Geophysical Research Letters* **33**, GL025881.

Knowlton, N. 1992. Thresholds and multiple stable states in coral reef community dynamics. *American Zoologist* **32**, 674–682.

Knowlton, N. 2001. The future of coral reefs. *Proceedings of the National Academy of Science U S A* **98**, 5419–5425.

Kobayashi, T., Enomoto, S., Sakazaki, R. & Kuwahara, S. 1963. A new selective medium for pathogenic vibrios: T.C.B.S. Agar (Modified Nakanishi's Agar). *Japanese Journal of Bacteriology* **18**, 387–391.

Kokita, T. & Nakazono, A. 2001. Rapid response of an obligately corallivorous filefish *Oxymonacanthus longirostris* (Monacanthidae) to a mass coral bleaching event. *Coral Reefs* **20**, 155–158.

Koonjul, M.S., Mangar, V. & Luchmun. 2003. Eradication on crown of thorns starfish (*Acanthaster planci*) infestation in a path reef in the lagoon off Ile Aux Cerfs, Mauritius. Mauritius: Food and Agricultural Research Council.

Kroon, F.J., Kuhnert, P.M., Henderson, B.L., Wilkinson, S.N., Kinsey-Henderson, A., Abbott, B., Brodie, J.E. & Turner, R.D.R. 2012. River loads of suspended solids, nitrogen, phosphorus and herbicides delivered to the Great Barrier Reef lagoon. *Marine Pollution Bulletin* **65**, 167–181.

Kuchler, D., Jupp, D.L.B., 1988. Shuttle photograph captures massive phytoplankton bloom in the Great Barrier Reef. *International Journal of Remote Sensing* **9**, 1299–1301.

Lamare, M., Pecorino, D., Hard, N., Liddy, M., Byrne, M. & Uthicke, S. 2014. The thermal tolerance of crown-of-thorns (*Acanthaster planci*) embryos and bipinnaria larvae: implications for spatial and temporal variation in adult populations. *Coral Reefs* **33**, 207–219.

Landsea, C.W., Harper, B.A., Hoarau, K. & Knaff, J.A. 2006. Climate change. Can we detect trends in extreme tropical cyclones? *Science* **313**, 452.

Lane, D.J. 2012. *Acanthaster planci* impact on coral communities at permanent transect sites on Bruneian reefs, with a regional overview and a critique on outbreak causes. *Journal of the Marine Biological Association of the United Kingdom* **92**, 803–809.

Larkum, A.W.D. 1988. High rates of nitrogen fixation on coral skeletons after predation by the crown of thorns starfish *Acanthaster planci*. *Marine Biology* **97**, 503.

Lassig, B. 1991. Final report on the results of COTSAC management related research: 1985–1989. Townsville, Australia: Great Barrier Reef Marine Park Authority.

Lawrence, J.M., 1991. Analysis of characteristics of echinoderms associated with stress and disturbance. In *Biology of Echinodermata*, T. Yanagisawa et al. (eds). Rotterdam, the Netherlands: Balkema, 11–26.

Lawton, R.J., Messmer, V., Bay, L.K. & Pratchett, M.S. 2011. High gene flow across large geographic scales reduces extinction risk for highly specialised coral-feeding butterflyfishes. *Molecular Ecology* **17**, 3584–3598.

Laxton, J.H. 1974. Aspects of the ecology of the coral-eating starfish *Acanthaster planci*. *Biological Journal of the Linnean Society* **6**, 19–45.

Lessios, H.A., Robertson, D.R. & Cubit, J.D. 1984. Spread of *Diadema* mass mortality through the Caribbean. *Science* **226**, 335–337.

Levitan, D.R., Sewell, M.A. & Chia, F.-S. 1991. Kinetics of fertilization in the sea urchin *Strongylocentrotus franciscanus*: interaction of gamete dilution, age, and contact time. *Biological Bulletin* **181**, 371–378.

Levitan, D.R., Sewell, M.A. & Chia, F.S. 1992. How distribution and abundance influence fertilization success in the sea urchin *Strongylocentrotus franciscanus*. *Ecology* **73**, 248–254.

Linares, C., Pratchett, M.S. & Coker, M.S. 2011. Recolonisation and growth of *Acropora hyacinthus* following climate-induced coral bleaching on the Great Barrier Reef. *Marine Ecology Progress Series* **438**, 97–104.

Lison de Loma, T., Chancerelle, Y. & Lerouvreur, F. 2006. Evaluation des densités d'*Acanthaster planci* sur l'île de Moorea. French Polynesia: CRIOBE CNRS-EPHE.

Lucas, J.S. 1973. Reproductive and larval biology of *Acanthaster planci* (L.) in Great Barrier Reef waters. *Micronesica* **9**, 197–203.

Lucas, J.S. 1975. Environmental influences on the early development of *Acanthaster planci* (L.). In *Crown-of-Thorns Starfish Seminar Proceedings, Brisbane*. Canberra, Australia: Australian Government Publishing Service, 109–121.

Lucas, J.S. 1982. Quantitative studies of feeding and nutrition during larval development of the coral reef asteroid *Acanthaster planci* (L.). *Journal of Experimental Marine Biology and Ecology* **65**, 173–193.

Lucas, J.S. 1984. Growth, maturation and effects of diet in *Acanthaster planci* (L.) (Asteroidea) and hybrids reared in the laboratory. *Journal of Experimental Marine Biology and Ecology*, **79**, 129–147.

Lucas, J.S., Hart, R.J., Howden, M.E. & Salathe, R. 1979. Saponins in eggs and larvae of *Acanthaster planci* (L.) (Asteroidea) as chemical defences against planktivorous fish. *Journal of Experimental Marine Biology and Ecology* **40**, 155–165.

Lukoschek, V., Cross, P., Torda, G., Zimmerman, R. & Willis, B.L. 2013. The importance of coral larval recruitment for the recovery of reefs impacted by cyclone Yasi in the central Great Barrier Reef. *PLoS ONE* **8**, e65363.

MacArthur, R.H. 1955. Fluctuations of animal populations and a measure of community stability. *Ecology* **36**, 533–536.

Marsh, J.A. & Tsuda, R.T. 1973. Population levels of *Acanthaster planci* in the Mariana and Caroline Islands, 1969–1972. *Atoll Research Bulletin* **170**, 1–16.

Marshall, P.A. & Baird, A.H. 2000. Bleaching of corals on the Great Barrier Reef: differential susceptibilities among taxa. *Coral Reefs* **19**, 155–163.

May, R.M. & Anderson, R.M. 1979. Population biology of infectious diseases: part II. *Nature* **280**, 455–461.

McCallum, H.I. 1987. Predator regulation of *Acanthaster planci*. *Journal of Theoretical Biology* **127**, 207–220.

McCallum, H.I. 1990. Effects of predation on *Acanthaster*: age-structured metapopulation models. In *Acanthaster and the Coral Reef: A Theoretical Perspective*, R.H. Bradbury (ed.). Berlin: Springer, 208–219.

McCallum, H.I. 1992. Completing the circle: stock-recruitment relationships and *Acanthaster*. *Marine and Freshwater Research* **43**, 653–662.

McCallum, H.I., Endean, R., Cameron, A.M. 1989. Sublethal damage to *Acanthaster planci* as an index of predation pressure. *Marine Ecology Progress Series* **56**: 29–36.

McClanahan, T., Polunin, N. & Done, T. 2002. Ecological states and the resilience of coral reefs. *Conservation Ecology* **6**, 18.

McCook, L.J., Ayling, T., Cappo, M., Choat, J.H., Evans, R.D., De Freitas, D.M., Heupel, M., Hughes, T.P., Jones, G.P., Mapstone, B., Marsh, H., Mills, M., Molloy, F.J., Pitcher, C.R., Pressey, R.L., Russ, G.R., Sutton, S., Sweatman, H., Tobin, R., Wachenfeld, D.R. & Williamson, D.H. 2010. Adaptive management of the Great Barrier Reef: a globally significant demonstration of the benefits of networks of marine reserves. *Proceedings of the National Academy of Sciences of the United States of America* **107**, 18278–18285.

McCulloch, M., Fallon, S., Wyndham, T., Hendy, E., Lough, J. & Barnes, D. 2003. Coral record of increased sediment flux to the inner Great Barrier Reef since European settlement. *Nature* **421**, 727–730.

Mendonça, V.M., Al Jabri, M.M., Al Ajmi, I., Al Muharrami, M., Al Areimi, M. & Al Aghbari, H.A. 2010. Persistent and expanding population outbreaks of the corallivorous starfish *Acanthaster planci* in the northwestern Indian Ocean: are they really a consequence of unsustainable starfish predator removal through overfishing in coral reefs, or a response to a changing environment? *Zoological Studies* **49**, 108–123.

Mercier A. & Hamel J.-F. 2009. Endogenous and exogenous control of gametogenesis and spawning in echinoderms. *Advances in Marine Biology* **55**, 1–302.

Messmer, V., Jones, G.P., Munday, P.L., Holbrook, S.J., Schmitt, R.J. & Brooks, A.J. 2011. Habitat biodiversity as a determinant of fish community structure on coral reefs. *Ecology* **92**, 2285–2298.

Messmer, V., Pratchett, M.S. & Clark, T.D. 2013. Capacity for regeneration in crown-of-thorns starfish, *Acanthaster planci*. *Coral Reefs* **32**, 461.

Mills, S.C. 2012. Density-dependent prophylaxis in the coral-eating crown-of-thorns sea star, *Acanthaster planci*. *Coral Reefs* **31**, 603–612.

Moore, R.J. 1985. A study of an outbreak area of the crown-of-thorns starfish *Acanthaster planci*. Report of the Queen Mary College 1984 Red Sea Expedition to Dunganab Bay, Sudan. London: Queen Mary College.

Moore, R.J. 1990. Persistent and transient populations of the crown-of-thorns starfish, *Acanthaster planci*. *Lecture Notes in Biomathematics* **88**, 236–277.

Moran, P.J. 1986. The *Acanthaster* phenomenon. *Oceanography and Marine Biology, An Annual Review* **24**, 379–480.

Moran, P.J. 1992. Preliminary observations of the decomposition of crown-of-thorns starfish, *Acanthaster planci* (L.). *Coral Reefs* **11**, 115–118.

Moran, P.J. & De'Ath, G. 1992. Estimates of the abundance of the crown-of-thorns starfish *Acanthaster planci* in outbreaking and non-outbreaking populations on reefs within the Great Barrier Reef. *Marine Biology* **113**, 509–515.

Moran, P.J., Bradbury, R.H. & Reichelt, R.E. 1985. Mesoscale studies of the crown-of-thorns/coral interaction: a case history from the Great Barrier Reef. *Proceedings of the 5th International Coral Reef Symposium, Tahiti* **5**, 321–326.

Moran, P.J., Bradbury, R.H. & Reichelt, R.E. 1988. Distribution of recent outbreaks of the crown-of-thorns starfish (*Acanthaster planci*) along the Great Barrier Reef: 1985–1986. *Coral Reefs* **7**, 125–137.

Moran, P.J., De'ath, G., Baker, V.J., Christie, C.A., Miller, I.R., Miller-Smith, B.A. & Thompson, A.A. 1992. Pattern of outbreaks of crown-of-thorns starfish (*Acanthaster planci* L.) along the Great Barrier Reef since 1966. *Australian Journal of Marine and Freshwater Research* **43**, 555–568.

Motta, P.J. 1988. Functional morphology of the feeding apparatus of ten species of Pacific butterflyfishes (Perciformes, Chaetodontidae): an ecomorphological approach. *Environmental Biology of Fishes* **22**, 39–67.

Munday, P.L. 2000. Interactions between habitat use and patterns of abundance in coral-dwelling fishes of the genus *Gobiodon*. *Environmental Biology of Fishes* **58**, 355–369.

Munday, P.L. 2001. Fitness consequences of habitat use and competition among coral-dwelling fishes. *Oecologia* **128**, 585–593.

Munday, P.L. 2004. Habitat loss, resource specialization, and extinction on coral reefs. *Global Change Biology* **10**, 1642–1647.

Munday, P.L., Jones, G.P. & Caley, M.J. 1997. Habitat specialisation and the distribution and abundance of coral-dwelling gobies. *Marine Ecology Progress Series* **152**, 227–239.

Munday, P.L., Jones, G.P., Pratchett, M.S. & Williams, A. 2008. Climate change and the future for coral reef fishes. *Fish and Fisheries* **9**, 261–285.

Nakamura, R. 1986. A morphometric study on *Acanthaster planci* (L.) populations in the Ryukyu Islands. *Galaxea* **5**, 223–237.

Nishihira, M. & Yamazato, K. 1974. Human interference with the coral reef community and *Acanthaster* infestation of Okinawa. *Proceedings of the Second International Coral Reef Symposium* **1**, 577.

Nyström, M., Graham, N., Lokrantz, J. & Norström, A. 2008. Capturing the cornerstones of coral reef resilience: linking theory to practice. *Coral Reefs* **27**, 795–809.

Ogura, M., Yokochi, H., Ishimaru, A., Kohno, H., Kishi, K., Ninomiya, K. & Mitsuhashi, H. 1985. Spawning period and the discovery of juveniles of the crown-of-thorns starfish, *Acanthaster planci* (L.), in the northwestern part of Iriomote Island. *Bulletin of the Institute of Oceanic Research Development, Tokai University* **7**, 25.

Okaji, K. 1989. *Delayed spawning activity and prolonged reproductive period in dispersed individuals of* Acanthaster planci *(L.) in Okinawa*. MSc thesis, Department of Marine Sciences, University of the Ryukyus, Japan.

Okaji, K. 1991. Delayed spawning activity in individuals of *Acanthaster planci* in Okinawa. In *Biology of Echinodermata*, T. Yanagisawa et al. (eds). Rotterdam, the Netherlands: Balkema, 291–295.

Okaji, K. 1996. *Feeding ecology in the early life stages of the crown-of-thorns starfish,* Acanthaster planci *(L.)*. PhD thesis, James Cook University, Australia.

Okaji, K., Ayukai, T. & Lucas, J. 1997. Selective feeding by larvae of the crown-of-thorns starfish, *Acanthaster planci* (L.). *Coral Reefs* **16**, 47–50.

Olson, R.R. 1985. *In situ* culturing of larvae of the crown-of-thorns starfish *Acanthaster planci*. *Marine Ecology Progress Series* **25**, 207–210.

Olson, R.R. 1987. *In situ* culturing as a test of the larval starvation hypothesis for the crown-of-thorns starfish, *Acanthaster planci*. *Limnology and Oceanography* **32**, 895–904.

Olson, R.R. & Olson, M.H. 1989. Food limitation of planktotrophic marine invertebrate larvae: does it control recruitment success? *Annual Review of Ecology and Systematics* **20**, 225–247.

Olson, R.R., McPherson, R. & Osborne, K. 1988. *In situ* larval culture of the crown-of-thorns starfish, *Acanthaster planci* (L.): effect of chamber size and flushing on larval settlement and morphology. In *Echinoderm Biology,* Proceedings of the Sixth International Echinoderm Conference, Victoria, R.D. Burke et al. (eds). Rotterdam, the Netherlands: Balkema, 247–251.

Ormond, R.F.G. 1971. Predation and control of *Acanthaster planci*. In *New Studies on the Crown of Thorns Starfish (*Acanthaster planci*) from Investigations in the Red Sea*, C.H. Roads & R.F.G. Ormond (eds). Cambridge, UK: Coral Starfish Research Group, 98–100.

Ormond, R., Bradbury, R., Bainbridge, S., Fabricius, K., Keesing, J., de Vantier, L., Medlay, A. & Steven, A. 1990. Test of a model of regulation of crown-of-thorns starfish by fish predators. In Acanthaster *and the Coral Reef: A Theoretical Perspective*, R.H. Bradbury (ed.). Heidelberg, Germany: Springer, 189–207.

Ormond, R.F.G. & Campbell, A.C. 1971. Observations on *Acanthaster planci* and other coral reef echinoderms in the Sudanese Red Sea. *Symposia of the Zoological Society of London* **28**, 433–454.

Ormond, R.F.G. & Campbell A.C. 1974. Formation and breakdown of *Acanthaster planci* aggregations in the Red Sea. *Proceedings of the Second International Coral Reef Symposium* **1**, 595–619.

Ormond, R.F.G., Campbell, A.C., Head, S.H., Moore, R.J., Rainbow, P.R. & Saunders, A.P. 1973. Formation and breakdown of aggregations of the crown-of-thorns starfish, *Acanthaster planci* (L.). *Nature* **246**, 167–169.

Ormond, R.F.G., Hanscomb, N.J. & Beach, D.H. 1976. Food selection and learning in the crown-of-thorns starfish *Acanthaster planci* (L.). *Marine Behaviour and Physiology* **4**, 93–105.

Osborne, K., Dolman, A.M., Burgess, S.C. & Johns, K.A. 2011. Disturbance and the dynamics of coral cover on the Great Barrier Reef (1995–2009). *PLoS ONE* **6**, e17516.

Owens, D. 1971. *Acanthaster planci* starfish in Fiji: survey of incidence and biological studies. *Fiji Agricultural Journal* **33**, 15–23.

Pan, M., Hilomen, V. & Palomares, M.L.D. 2010. Size structure of *Acanthaster planci* populations in Tubbataha Reefs Natural Parks, Sulu Sea, Philippines. In *Marine Biodiversity of Southeast Asian and Adjacent Seas*, M.L.D. Palomares & D. Pauly, D. (eds). Vancouver, Canada: University of British Columbia: Fisheries Centre Research Reports **18**(3), 70–77.

Pandolfi, J., Jackson, J.B.C., Baron, N., Bradbury, R.H., Guzman, H.M., Hughes, T.P., Kappel, C.V., Micheli, F., Ogden, J.C., Possingham, H.P. & Sala, E. 2005. Are U.S. coral reefs on the slippery slope to slime? *Science* **307**, 1725–1726.

Pandolfi, J.M. 1992. A palaeobiological examination of the geological evidence for recurring outbreaks of the crown-of-thorns starfish, *Acanthaster planci* (L.). *Coral Reefs* **11**, 87–93.

Pearson, R.G. 1975. Coral reefs, unpredictable climatic factors and *Acanthaster*. In *Crown-of-Thorns Starfish Seminar Proceedings*. Canberra, Australia: Australian Government Publishing Service, 131–134.

Pearson, R.G. 1981. Recovery and recolonisation of coral reefs. *Marine Ecology Progress Series* **4**, 105–122.

Pearson, R.G. & Endean, R. 1969. A preliminary study of the coral predator *Acanthaster planci* (L.) (Asteroidea) on the Great Barrier Reef. *Fisheries Notes, Queensland Department of Harbours and Marine* **3**, 27–55.

Pearson, R.G. & Garrett, R.N. 1978. *Acanthaster planci* on the Great Barrier Reef: swain reefs and northern surveys in 1975. *Micronesica* **14**, 259–272.

Petraitis, P.S., Latham, R.E. & Niesenbaum, R.A. 1989. The maintenance of species diversity by disturbance. *Quarterly Review of Biology* **64**, 393–418.

Porter, J.W. 1972. Predation by *Acanthaster planci* and its effect on coral species diversity. *American Naturalist* **106**, 487–492.

Porter, J.W. 1974. Community structure of coral reefs on opposite sides of the Isthmus of Panama. *Science* **186**, 543.

Potts, D.C. 1981. Crown-of-thorns starfish—man-induced pest or natural phenomenon? In *The Ecology of Pests: Some Australian Case Histories*. R.E. Kitching & R.E. Jones (eds). Melbourne, Australia: CSIRO, 55–86.

Pratchett, M.S. 1999. An infectious disease in crown-of-thorns starfish on the Great Barrier Reef. *Coral Reefs* **18**, 272.

Pratchett, M.S. 2001. Influence of coral symbionts on feeding preferences of crown-of-thorns starfish. *Marine Ecology Progress Series* **214**, 111–119.

Pratchett, M.S. 2005. Dynamics of an outbreak population of *Acanthaster planci* at Lizard Island, northern Great Barrier Reef (1995–1999). *Coral Reefs* **24**, 453–462.

Pratchett, M.S. 2007. Feeding preferences of *Acanthaster planci* (L.) under controlled conditions of food availability. *Pacific Science* **61**, 113–120.

Pratchett, M.S. 2010. Changes in coral communities during an outbreak of *Acanthaster planci* at Lizard Island, northern Great Barrier Reef (1995–1999). *Coral Reefs* **29**, 717–725.

Pratchett, M.S., Coker, D.J., Jones, G.P. & Munday, P.L. 2012. Specialisation in habitat use by coral reef damselfishes and their susceptibility to habitat loss. *Ecology and Evolution* **2**, 2168–2180.

Pratchett, M.S., Gust, N., Goby, G. & Klanten, S.O. 2001. Consumption of coral propagules represents a significant trophic link between corals and reef fish. *Coral Reefs* **20**, 13–17.

Pratchett, M.S., McCowan, D., Maynard, J.A. & Heron, S.F. 2013. Changes in bleaching susceptibility among corals subject to ocean warming and recurrent bleaching in Moorea, French Polynesia. *PLoS ONE* **8**, e70443.

Pratchett, M.S., Munday, P.L., Wilson, S.K., Graham, N.A.J., Cinner, J.E., Bellwood, D.R., Jones, G.P., Polunin, N.V.C. & McClanahan, T.R. 2008. Effects of climate-induced coral bleaching on coral-reef fishes: ecological and economic consequences. *Oceanography and Marine Biology: An Annual Review* **46**, 251–296.

Pratchett, M.S., Schenk, T.J., Baine, M., Syms, C. & Baird, A.H. 2009a. Selective coral mortality attributable to an outbreak of *Acanthaster planci* L. in Bootless Bay, Central Province, Papua New Guinea. *Marine Environmental Research* **67**, 230–236.

Pratchett, M.S., Trapon, M., Berumen, M.L., Chong-Seng, K. 2011. Recent disturbances augment community shifts in coral assemblages in Moorea, French Polynesia. *Coral Reefs* **30**, 183–193.

Pratchett, M.S., Wilson, S.K. & Baird, A.H. 2006. Declines in the abundance of *Chaetodon* butterflyfishes (Chaetodontidae) following extensive coral depletion. *Journal of Fish Biology* **69**, 1269–1280.

Pratchett, M.S., Wilson, S.K., Graham, N.A.J., Munday, P.L., Jones, G.P. & Polunin, N.V.C. 2009b. Coral bleaching and consequences for motile reef organisms: past, present and uncertain future effects. In *Coral Bleaching: Patterns and Processes, Causes and Consequences*, M. van Oppen & J. Lough (eds). Heidelberg, Germany: Springer, 139–158.

Pyne, R.R 1970. Notes on the crown-of-thorns starfish: its distribution in Papua and New Guinea (Echinodermata: Asteroidea: Acanthasteridae). *Papua New Guinea Journal of Agriculture, Forestry and Fisheries* **21**, 128–138.

Randall, J.E. 1972. Chemical pollution in the sea and the crown-of-thorns starfish (*Acanthaster planci*). *Biotropica* **1972**, 132–144.

Randall, J.E., Head, S.M. & Sanders, A.P.L. 1978. Food habits of the humphead wrasse, *Chelinus undulatus* (Labridae). *Environmental Biology of Fishes* **3**, 235–238.

Randall, R.H. 1973. Distribution of corals after *Acanthaster planci* (L.) infestation at Tanguisson Point, Guam. *Micronesica* **3**, 213–222.

Raymond, R. 1986. *Starfish Wars: Coral Death and the Crown-of-Thorns*. South Yarra, Australia: Macmillan.

Reichelt, R.E., Bradbury, R.H. & Moran, P.J. 1990a. Distribution of *Acanthaster planci* outbreaks on the Great Barrier Reef between 1966 and 1989. *Coral Reefs* **9**, 97–103.

Reichelt, R.E., Greve, W., Bradbury, R.H. & Moran, P.J. 1990b. *Acanthaster planci* outbreak initiation: a starfish-coral site model. *Ecological Modelling* **49**, 153–177.

Richmond, R.H. (1987). Energetics, competency, and long-distance dispersal of planula larvae of the coral *Pocillopora damicornis. Marine Biology* **93**, 527–533.

Rivera-Posada, J., Owens, L., Caballes, C.F. & Pratchett, M.S. 2012. The role of protein extracts in the induction of disease in *Acanthaster planci. Journal of Experimental Marine Biology and Ecology* **429**, 1–6.

Rivera-Posada, J.A. 2012. *Pathogenesis of crown-of-thorns (*Acanthaster planci *L).* PhD thesis, James Cook University, Australia.

Rivera-Posada, J.A., Caballes, C.F. & Pratchett, M.S. 2013. Lethal doses of oxbile, peptones and thiosulfate-citrate-bile-sucrose agar (TCBS) for *Acanthaster planci*; exploring alternative population control options. *Marine Pollution Bulletin* **75**, 133–139.

Rivera-Posada, J.A., Caballes, C.F. & Pratchett, M.S. 2014. Size-related variation in arm damage frequency in the crown-of-thorns sea star, *Acanthaster planci. Journal of Coastal Life Medicine* **2**, 187–195.

Rivera-Posada, J.A., Pratchett, M.S., Cano-Gomez, A., Arango-Gomez, J.D. & Owens, L. 2011a. Injection of *Acanthaster planci* with thiosulfate-citrate-bile-sucrose agar (TCBS). I. Disease induction. *Diseases of Aquatic Organisms* **97**, 85–94.

Rivera-Posada, J.A., Pratchett, M.S., Cano-Gomez, A., Arango-Gomez, J.D. & Owens, L. 2011b. Refined identification of *Vibrio* bacterial flora from *Acanthasther planci* based on biochemical profiling and analysis of housekeeping genes. *Diseases of Aquatic Organisms* **96**, 113–123.

Roper, K.G. 1997. *Identification of* Acanthaster planci *larvae using species-specific monoclonal antibodies.* PhD thesis, James Cook University, Australia.

Rotjan, R.D. & Lewis, S.M. 2008. Impact of coral predators on tropical reefs. *Marine Ecology Progress Series* **367**, 73–91.

Roughgarden, J., Gaines, S. & Possingham, H. 1988. Recruitment dynamics in complex life cycles. *Science* **241**, 1460–1466.

Sakai, K. 1985. Brief observations on the population of *Acanthaster planci* (L.) and coral assemblages around Sesoko Island, Okinawa in 1983. *Galaxea* **4**, 23–31.

Sano, M., Shimizu, M. & Nose, Y. 1984. Changes in the structure of coral reef fish communities by destruction of hermatypic corals: observational and experimental views. *Pacific Science* **38**, 51–79.

Sano, M., Shimizu, M. & Nose, Y. 1987. Long-term effects of destruction of hermatypic corals by *Acanthaster planci* infestation of reef fish communities at Iriomote Island, Japan. *Marine Ecology Progress Series* **37**, 191–199.

Sapp, J. 1999. *What Is Natural? Coral Reef Crisis.* New York: Oxford University Press.

Schleyer, M. 1998. Crown of thorns starfish in the western Indian Ocean. *Reef Encounter* **23**, 25–27.

Seymour, R.M. & Bradbury, R.H. 1999. Lengthening reef recovery times from crown-of-thorns outbreaks signal systemic degradation of the Great Barrier Reef. *Marine Ecology Progress Series* **176**, 1–10.

Sheppard, C. R., Spalding, M., Bradshaw, C. & Wilson, S. 2002. Erosion vs. recovery of coral reefs after 1998 El Niño: Chagos reefs, Indian Ocean. *AMBIO: A Journal of the Human Environment* **31**, 40–48.

Simpson, C.J. & Grey, K.A. 1988. *Crown-of-thorns starfish (*Acanthaster planci*) in the Dampier Archipelago, Western Australia.* Technical Series No. 25. Perth, Australia: Environmental Protection Authority.

Souter, D.W., Cameron, A.M. & Endean, R. 1997. Implications of sublethal predation, autotomy and regeneration: pigment bands on their spines can not be used to determine the ages of adult specimens of the corallivore *Acanthaster planci. Marine and Freshwater Research* **48**, 321–328.

State of Queensland. 2013. Reef water quality protection olan. Australia: Reef Water Quality Protection Plan Secretariat. Online. http://www.reefplan.qld.gov.au/resources/assets/reef-plan-2013.pdf/ Brisbane (viewed 13 July, 2013).

Stella, J.S., Pratchett, M.S., Hutchings, P.A. & Jones, G.P. 2011. Diversity, importance and vulnerability of coral-associated invertebrates. *Oceanography and Marine Biology: An Annual Review* **49**, 43–104.

Stump, R.J. 1994. *Age determination and life-history characteristics of* Acanthaster planci *(L.) (Echinodermata: Asteroidea).* PhD thesis, James Cook University, Australia.

Stump, R.J.W. 1996. An investigation of the methods to describe the population dynamics of *Acanthaster planci* (L.) around Lizard Island, northern Cairns Section, GBR. Technical report no. 10. Townsville, Australia: CRC Reef Research and Great Barrier Reef Marine Park Authority.

Stump, R.J.W. & Lucas, J.S. 1990. Linear growth in spines from *Acanthaster planci* (L) involving growth lines and periodic pigment bands *Coral Reefs* **9**, 149–154.

Stump, R.J.W. & Lucas, J.S. 1999. Age estimation and patterns of growth in *Acanthaster planci:* a reply to Souter et al. (1997). *Marine and Freshwater Research* **50**, 71–72.

Sutton, D.C., Trott, L., Reichelt, J.L. & Lucas, J.S. 1988. Assessment of bacterial pathogenesis in crown-of-thorns starfish, *Acanthaster planci* (L.). *Proceedings of the 6th International Coral Reef Symposium* **2**, 171–176.

Sweatman, H., Bass, D.K., Cheal, A.J., Coleman, G., Miller, I.R., Ninio, R., Osborne, K., Oxley, W.G., Ryan, D.A.J., Thompson, A.A. & Tomkins, P. 1998. Long-term monitoring of the Great Barrier Reef. Status report number 3. Townsville, Australia: Australian Institute of Marine Science.

Sweatman, H., Delean, S. & Syms, C. 2011. Assessing loss of coral cover on Australia's Great Barrier Reef over two decades, with implications for longer-term trends. *Coral Reefs* **30**, 521–531.

Sweatman, H.P.A. 1995. A field study of fish predation on juvenile crown-of-thorns starfish. *Coral Reefs* **14**, 47–53.

Sweatman, H.P.A. 1997. Commercial fishes as predators of adult *Acanthaster planci. Proceedings of the 8th International Coral Reef Symposium* **1**, 617–620.

Sweatman, H.P.A. 2008. No-take reserves protect coral reefs from predatory starfish. *Current Biology* **18**, R598–R599.

Sweatman, H.P.A. & Butler, I. 1993. An experimental investigation of the ability of adult crown-of-thorns starfish to survive physical damage. In *The Possible Causes and Consequences of Outbreaks of the Crown-of-Thorns Starfish*, U. Engelhardt & B. Lassig (eds). GBRMPA Workshop Series No **18**, 71–82. Townsville, Australia: Great Barrier Reef Marine Park Authority.

Sweatman, H.P.A., Cheal, A.J., Coleman, G.J., Emslie, M.J., Johns, K., Jonker, M., Miller, I.R. & Osborne, K. 2008. Long-term monitoring of the Great Barrier Reef, status report. 8. Townsville, Australia: Australian Institute of Marine Science.

Talbot, F.H. & Talbot, M.S. 1971. The crown-of-thorns starfish (*Acanthaster*) and the Great Barrier Reef. *Endeavour* **30**, 38–42.

Timmers, M.A., Bird, C.E., Skillings, D.J., Smouse, P.E. & Toonen, R.J. 2012. There's no place like home: crown-of-thorns outbreaks in the Central Pacific are regionally derived and independent events. *PLoS ONE* **7**, e31159.

Tokeshi, M. & Daud, J.R.P. 2011. Assessing feeding electivity in *Acanthaster planci:* a null model analysis. *Coral Reefs* **30**, 227–235.

Trapon, M.L., Pratchett, M.S., Hoey, A.S. & Baird, A.H. 2013. Influence of fish grazing and sedimentation on the early post-settlement survival of the tabular coral *Acropora cytherea. Coral Reefs* **32**, 1051–1059.

Trapon, M., Pratchett, M.S. & Penin, L. 2011. Change in coral cover and composition following 30 years of disturbances at Moorea, French Polynesia. *Journal of Marine Biology* **2011**, ID807625.

Tsuda, R.T. 1971. Status of *Acanthaster planci* and coral reefs in the Mariana and Caroline Islands, June 1970 to May 1971. University of Guam Marine Laboratory Technical Report No. 2. Guam, USA: University of Guam.

Uchida, H. & Nomura, K. 1987. On the efficiency of natural plankton culture of pelagic larvae of the crown-of-thorns starfish. *Bulletin of Marine Science* **41**, 643.

Uthicke, S., Pecorino, D., Albright, R., Negri, A. P., Cantin, N., Liddy, M., Dworjanyn, S., Kamya, P., Byrne, M. & Lamare, M. 2013. Impacts of ocean acidification on early life-history stages and settlement of the coral-eating sea star *Acanthaster planci. PloS ONE* **8(12)**, e82938.

Vine, P.J. 1970. Densities of *Acanthaster planci* in the Pacific Ocean. *Nature* **228**, 341–342.

Vine, P.J. 1973. Crown of thorns (*Acanthaster planci*) plagues: the natural causes theory. *Atoll Research Bulletin* **166**, 1–10.

Vogler, C., Benzie, J., Lessios, H., Barber, P. & Worhelde, G. 2008. A threat to coral reefs multiplied? Four species of crown-of-thorns starfish. *Biology Letters* **4**, 296–699.

Vogler, C., Benzie, J.A.H., Tenggardjaja, K., Barber, P.H. & Wörheide, G. 2013. Phylogeography of the crown-of-thorns starfish: genetic structure within the Pacific species. *Coral Reefs* **32**, 515–525.

Wainwright, B.J., Arlyza, I.S. & Karl, S.A. 2012. Eighteen microsatellite loci for the crown-of-thorns starfish, *Acanthaster planci. Conservation Genetics Resources* **4**, 861–863.

Walbran, P.D., Henderson, R.A., Faithful, J.W., Polach, H.A., Sparks, R.J., Wallace, G. & Lowe, D.C. 1989a. Crown-of-thorns starfish outbreaks on the Great Barrier Reef: a geological perspective based upon the sediment record. *Coral Reefs* **8**, 67–78.

Walbran, P.D., Henderson, R.A., Jull, A.T. & Head, M.J. 1989b. Evidence from sediments of long-term *Acanthaster planci* predation on corals of the Great Barrier Reef. *Science* **245**, 847–850.

Walther, G.R., Post, E., Convey, P., Menzel, A., Parmesan, C., Beebee, T.J., Fromentin, J., Hoegh-Guldberg, O. & Bairlein, F. 2002. Ecological responses to recent climate change. *Nature* **416**, 389–395.

Wass, R.C. 1987. Influence of *Acanthaster*-induced coral kills on fish communities at Fagatele Bay and at Cape Larsen. Biological assessment of the Fagatele Bay National Marine Sanctuary. Washington, District of Columbia: US Department of Commerce.

Weber, J.N. & Woodhead, P.J.M. 1970. Ecological studies on the coral predator *Acanthaster planci* in the South Pacific. *Marine Biology* **6**, 12–17.

Webster, P.J., Holland, G.J., Curry, J.A. & Chang, H.R. 2005. Changes in tropical cyclone number, duration, and intensity in a warming environment. *Science* **309**, 1844–1846.

Wild, C., Huettel, M., Klueter, A., Kremb, S.G., Rasheed, M.Y. & Jørgensen, B.B. 2004. Coral mucus functions as an energy carrier and particle trap in the reef ecosystem. *Nature* **428**, 66–70.

Wilkinson, C. (ed.) 2004. *Status of Coral Reefs of the World: 2004 Volume 1*. Townsville, Australia: Australian Institute of Marine Science.

Wilkinson, C. (ed.) 2008. *Status of Coral Reefs of the World: 2008*. Townsville, Australia: Global Coral Reef Monitoring Network and Reef and Rainforest Research Centre.

Williams, D.M. 1986. Temporal variation in the structure of reef slope fish communities (central Great Barrier Reef): short-term effects of *Acanthaster planci* infestation. *Marine Ecology Progress Series* **28**, 157–164.

Williams, E.H. & Bunkley-Williams, L. (2000). Marine major ecological disturbances of the Caribbean. *Infectious Disease Review* **2**, 110–127.

Wilson, B.R. & Marsh, L.M. 1974. *Acanthaster* studies on a Western Australian coral reef. *Proceedings of the 2nd International Great Barrier Reef Symposium* **1**, 621–630.

Wilson, S.K., Graham, N.A.J., Pratchett, M.S., Jones, G.P. & Polunin, N.V.C. 2006. Multiple disturbances and the global degradation of coral reefs: are reef fishes at risk or resilient? *Global Change Biology* **12**, 2220–2234.

Wooldridge, S.A. 2009. Water quality and coral bleaching thresholds: formalising the linkage for the inshore reefs of the Great Barrier Reef, Australia. *Marine Pollution Bulletin* **58**, 745–751.

Yamaguchi, M. 1973. Early life histories of coral reef asteroids, with special reference to *Acanthaster planci* (L.). In *Biology and Geology of Coral Reefs (Vol. 2: Biology)*, O.A. Jones & R. Endean (eds). New York: Academic Press, 369–387.

Yamaguchi, M. 1974. Growth of juvenile *Acanthaster planci* (L.) in the laboratory. *Pacific Science* **28**: 123–138.

Yamaguchi, M. 1975. Coral-reef asteroids of Guam. *Biotropica* **7**, 12–23.

Yamaguchi, M. 1986. *Acanthaster planci* infestations of reefs and coral assemblages in Japan: a retrospective analysis of control efforts. *Coral Reefs* **5**, 23–30.

Yamazato, K. & Kiyan T. 1973. Reproduction of *Acanthaster planci* in Okinawa. *Micronesica* **9**, 185–195.

Yasuda, N., Nagai, S., Hamaguchi, M., Lian, C.L. & Nadaoka, K. 2006. Development of microsatellite markers for the crown-of-thorns starfish *Acanthaster planci*. *Molecular Ecology Notes* **6**, 141–143.

Yasuda, N., Nagai, S., Hamaguchi, M. & Nadaoka, K. 2007. Seven new microsatellite markers for crown-of-thorns starfish *Acanthaster planci*. *Plankton and Benthos Research* **2**, 103–106.

Yasuda, N., Nagai, S., Hamaguchi, M., Okaji, K., Gerard, K. & Nadaoka, K. 2009. Gene flow of *Acanthaster planci* (L.) in relation to ocean currents revealed by microsatellite analysis. *Molecular Ecology* **18**, 1574–1590.

Yasuda, N., Ogasawara, K., Kajiwara, K., Ueno, M., Oki, K., Taniguchi, H., Kakuma, S., Okaji, K. & Nadaoka, K. 2010. Latitudinal differentiation in the reproduction patterns of the crown-of-thorns starfish *Acanthaster planci* through the Ryukyu Island Archipelago. *Plankton and Benthos Research* **5**, 156–164.

Yokochi, H. 1985. Crown-of-thorns starfish spawning (in Japanese). *Iriomote-dayori* (Tokai University) **3**, 4.

Yokochi, H. & Ogura, M. 1987. Spawning period and discovery of juvenile *Acanthaster planci* (L.) (Echinodermata: Asteroidea) at northwestern Iriomote-jima, Ryukyu Islands. *Bulletin of Marine Science* **41**, 611–616.

Yund, P.O. 1990. An *in situ* measurement of sperm dispersal in a colonial marine hydroid. *Journal of Experimental Zoology* **253**, 102–106.

Zann, L., Brodie, J., Berryman, C. & Naqasima, M. 1987. Recruitment, ecology, growth and behavior of juvenile *Acanthaster planci* (L.) (Echinodermata: Asteroidea). *Bulletin of Marine Science* **41**, 561–575.

Zann, L., Brodie, J. & Vuki, V. 1990. History and dynamics of the crown-of-thorns starfish *Acanthaster planci* (L.) in the Suva area, Fiji. *Coral Reefs* **9**, 135–144.

Colour Figure 3 (Brandl & Bellwood)

Colour Figure 3 (Bonaldo, Hoey & Bellwood)

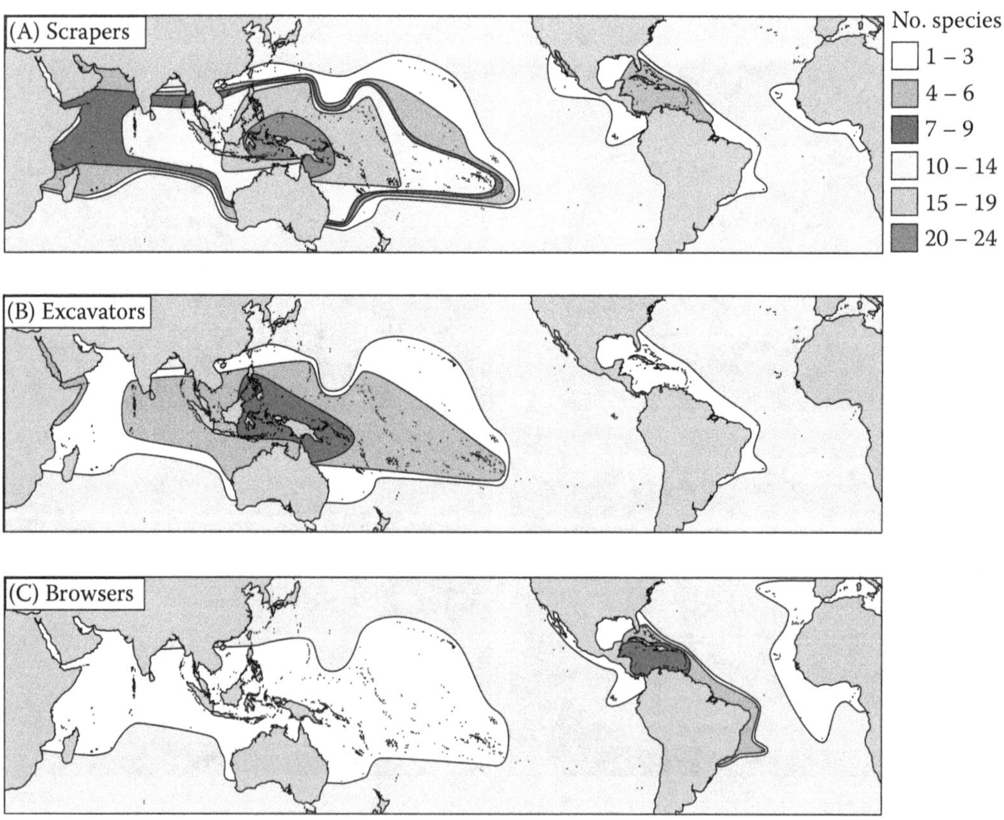

Colour Figure 4 (Bonaldo, Hoey & Bellwood)

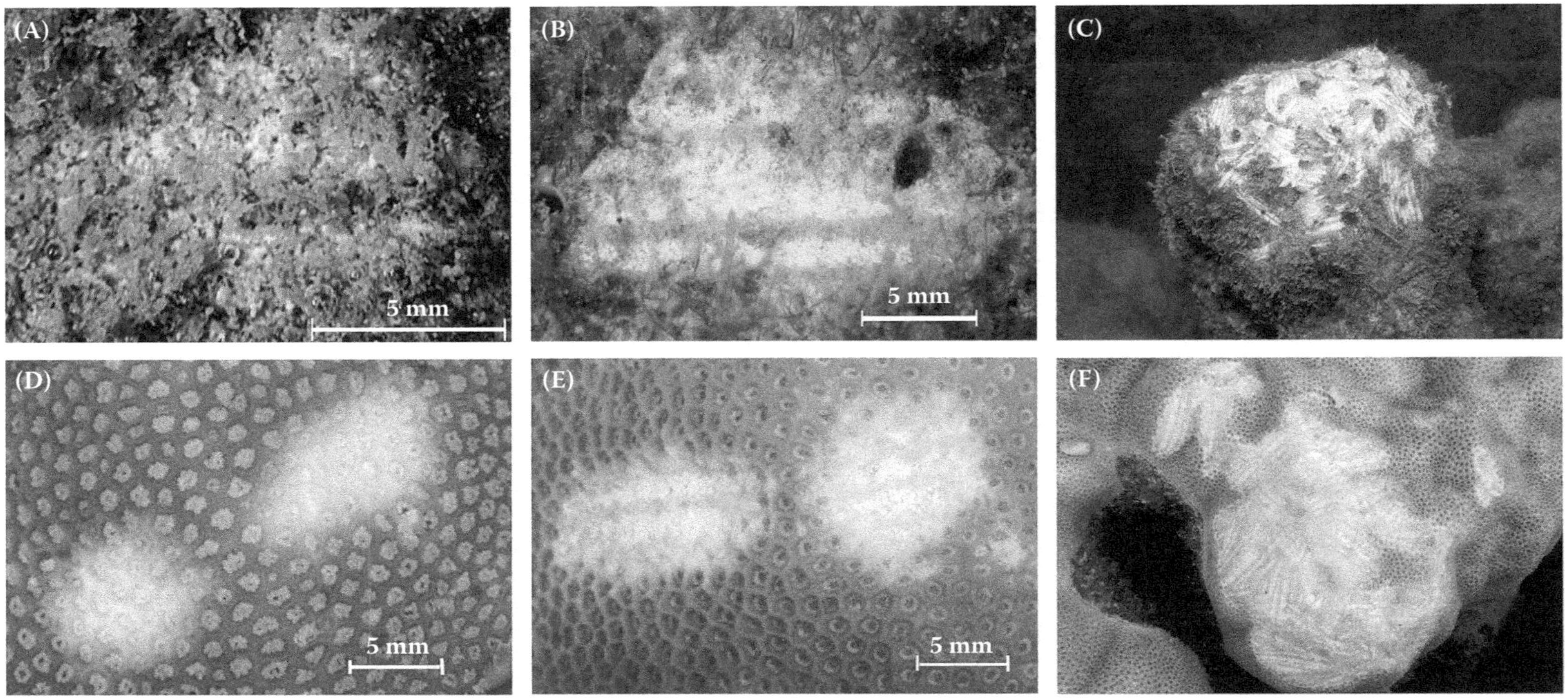

Colour Figure 5 (Bonaldo, Hoey & Bellwood)

Colour Figure 2 (Pratchett et al.)

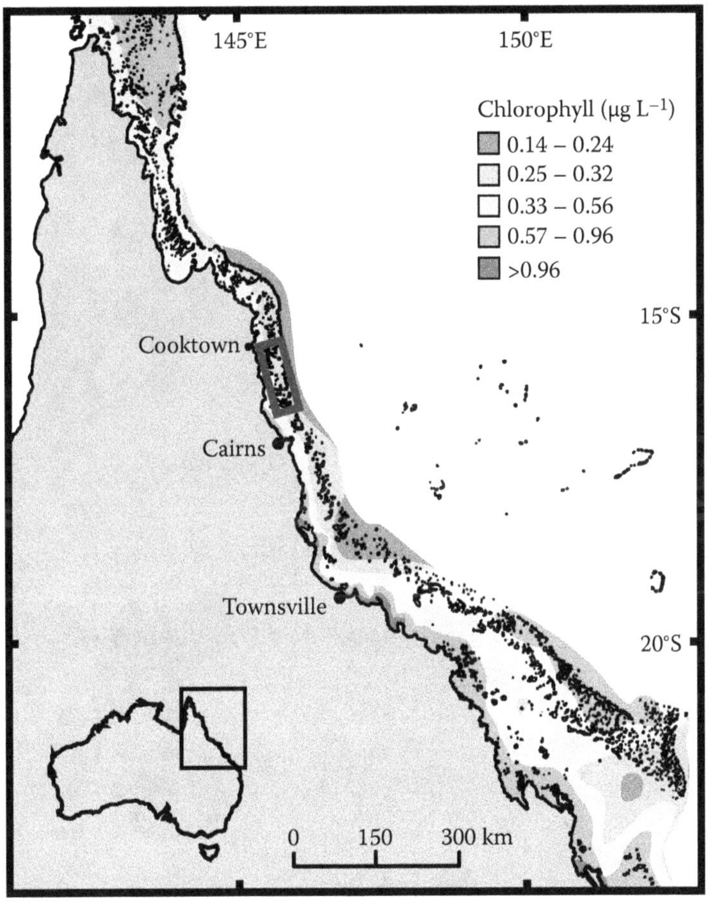

Colour Figure 11 (Pratchett et al.)

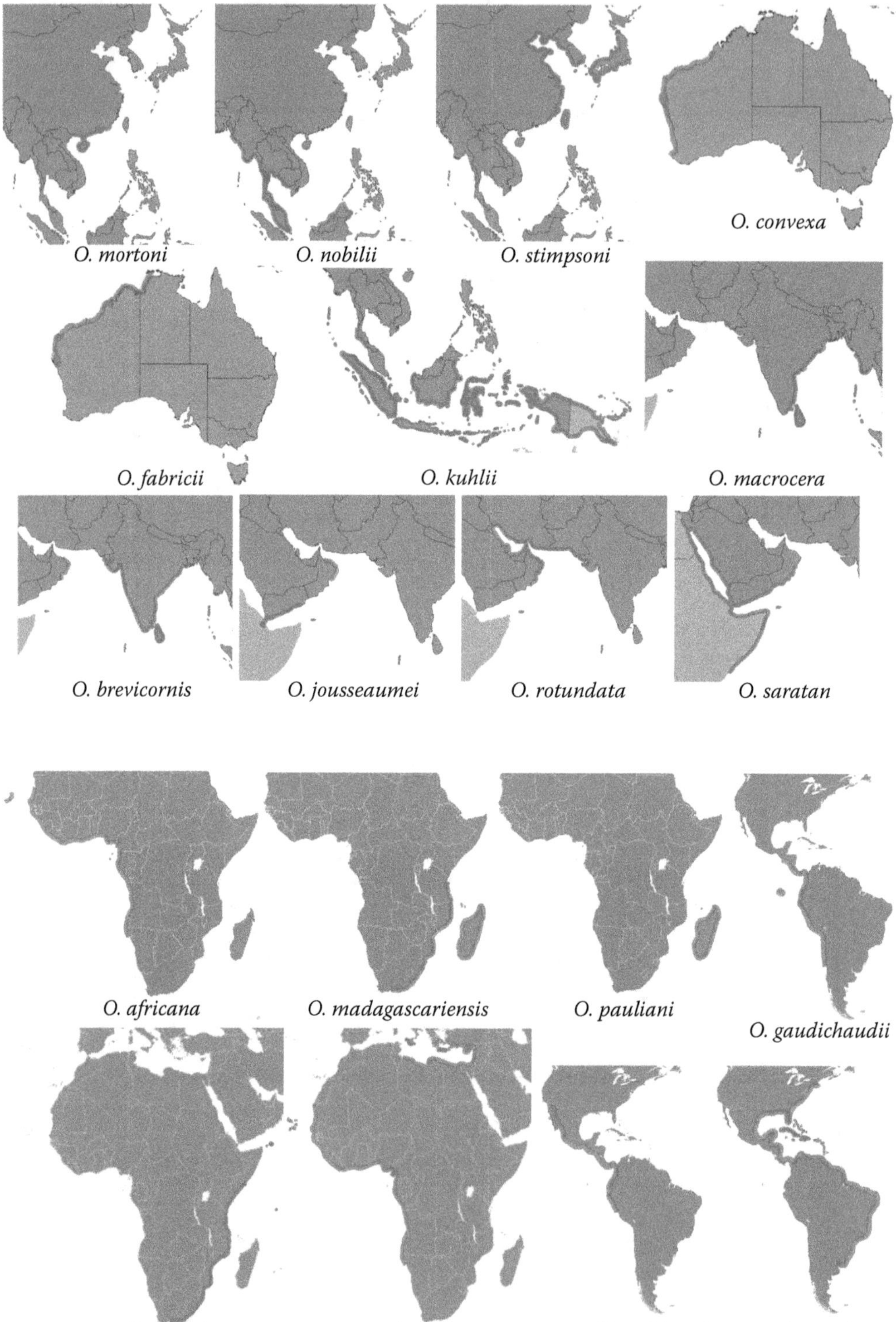

Colour Figure 1 (Lucrezi & Schlacher)

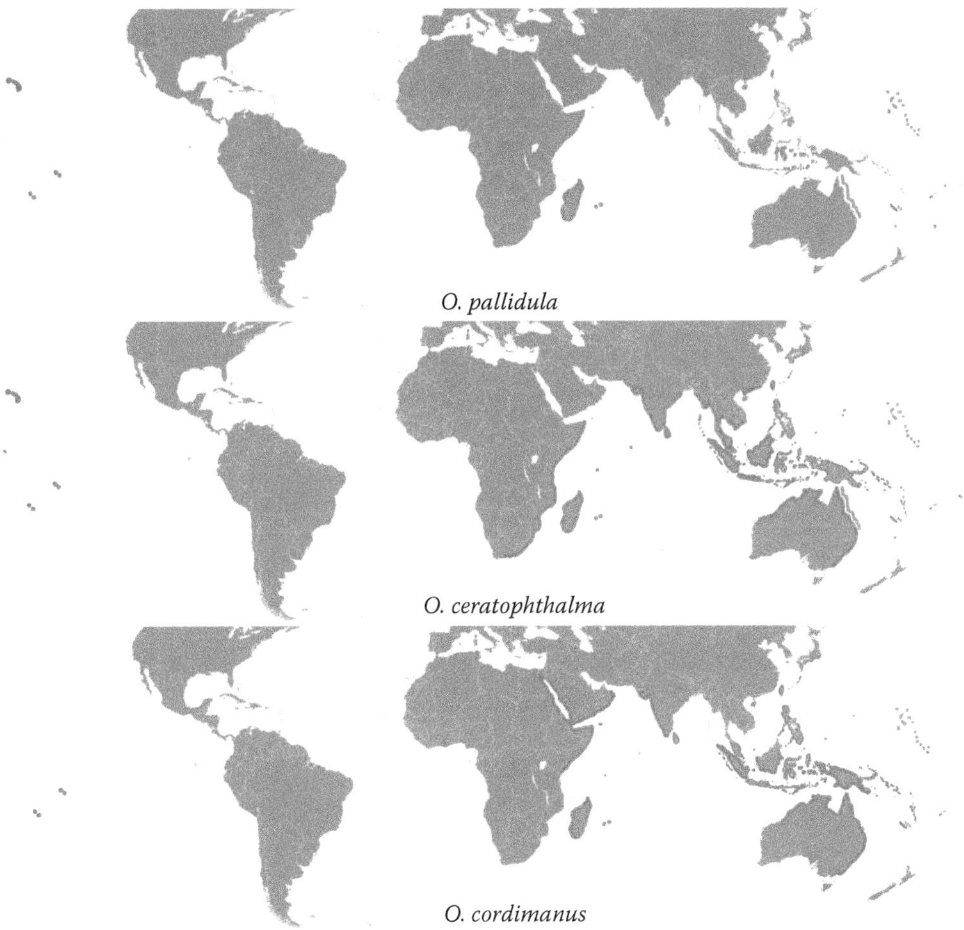

Colour Figure 1 (Lucrezi & Schlacher)

Colour Figure 2 (Lucrezi & Schlacher)

Colour Figure 7 (Lucrezi & Schlacher)

Oceanography and Marine Biology: An Annual Review, 2014, **52**, 201-256

Taylor & Francis

THE ECOLOGY OF GHOST CRABS

SERENA LUCREZI[1] & THOMAS A. SCHLACHER[2]

[1]*TREES—Tourism Research in Economic Environs and Society, North-West University, Potchefstroom, South Africa 2520*
E-mail: 23952997@nwu.ac.za
[2]*Faculty of Science, Health and Education, University of the Sunshine Coast, Maroochydore DC, Q-4558, Australia*
E-mail: tschlach@usc.edu.au

Ghost crabs are the largest invertebrates on ocean shores. They are globally abundant on many sandy beaches from the tropics to temperate latitudes. A defining ecological trait of the group is its fossorial habit: The crabs excavate deep, voluminous, and complex burrows, alternating between activity on the beach surface (e.g., foraging, mating, and dispersion) and underground microhabitats. Although not true land crabs, they have evolved a range of physiological, morphological, and behavioural adaptations to semi-terrestrial habitats and hence occupy a broad spatial gradient from coastal dunes to the swash zone. Ghost crabs are the fastest crustaceans on land and have acute senses of sight, smell, and hearing. Functionally, ghost crabs are pivotal in littoral food webs as mesopredators: They are often the apex invertebrate consumers on beaches while being predated by a diversity of reptiles, birds, and mammals that forage at the land-sea interface. As consumers, ghost crabs display extraordinary trophic plasticity, occupying several trophic levels, obtaining food through a variety of strategies, and consuming a wide diversity of prey. Their diet is broad—including predation on the eggs and young of sea turtles and shorebirds—and frequently encompasses efficient scavenging of animal carcasses.

Introduction

Ghost crabs are the largest and most conspicuous invertebrates on many ocean-exposed sandy beaches around the globe (Figure 1). They are often abundant—regularly encountered by all beach visitors (including professional ecologists)—and their large and prominent burrows are noticeable features of ocean shores from the tropics to the temperate zone (Jones 1972, Lucrezi & Schlacher 2010; Figure 2). In a functional sense, ghost crabs are the great bioturbators of beaches and form a key link in the food webs of these ecosystems, being apex invertebrate consumers, prolific scavengers of carrion, and important prey items for vertebrates (Schlacher et al. 2013a,b).

The term 'ghost crab' is thought to have its etymological origins in the pale colouration and swift movement of the crabs—a combination that, apparently, creates the fleeting image of a ghost (Cott 1929, Pearse et al. 1942). *Ocypode* means 'swift-footed', and Cott (1929) refers to the genus as the "rabbit of the crustaceans" because of its fossorial habits.

Taxonomically, ghost crabs (subfamily Ocypodinae; Rafinesque 1815) and fiddler crabs (subfamily Ucinae; Dana 1852) form the family Ocypodidae (Rafinesque 1815). Fiddler crabs tend to have more aquatic habits, occurring mainly on sheltered shores (e.g., mangroves, muddy estuaries; Davie 1994), whereas ghost crabs extend from coastal dunes to the swash zone on ocean-exposed beaches (Dahl 1953). Sakai & Türkay (2013) provide an authoritative, comprehensive, and

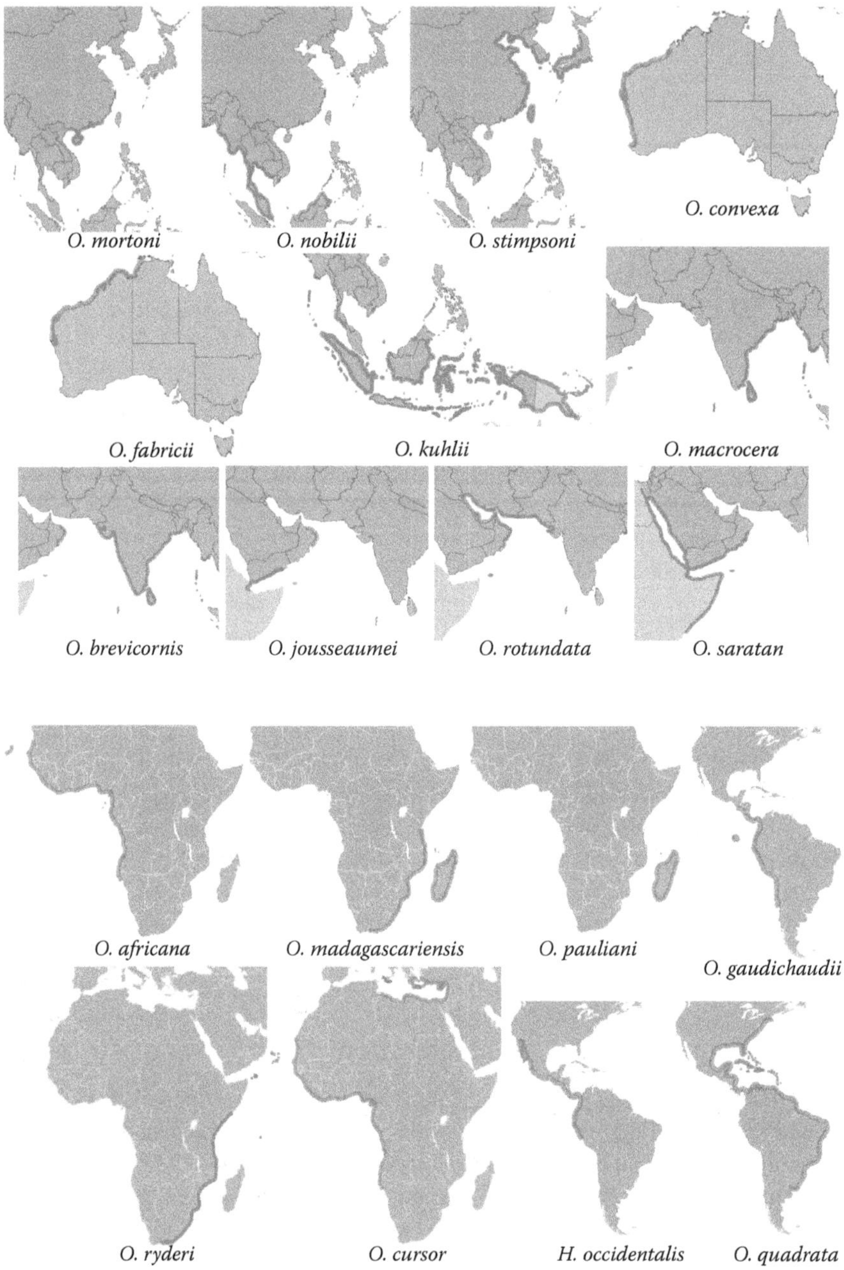

Figure 1 (See also colour figure in the insert) Distributions and ranges of individual ghost crab species around the world. For information on species ranges, see Sakai & Türkay (2013) and references therein. Additional key references for distributional ranges include Armstrong (1876), Suvatti (1950), Monod (1956), Holthuis (1961), George & Knott (1965), Day (1969), Kensley (1970), Bright & Hogue (1972), Vannini (1976), Brusca (1980), McLachlan (1980), Fischer et al. (1981), Manning & Holthuis (1981), Ziese (1985), Jones (1988), von Prahl et al. (1990), Koukouras et al. (1992), Branch et al. (1994), Hendrickx (1996), Veloso et al. (1997), De (1998), Strachan et al. (1999), Simões et al. (2001), Davie (2002), Aerts (2004), Demicheli et al. (2006), Lazarus-Agudelo & Cantera-Kintz (2007), McLay (2009), Ateş et al. (2010), Poupin (2010), Haddoud & El Gomati (2011), Coll et al. (2013). A complete list of sources used to compile the maps is available from the authors of this review. *(continued)*

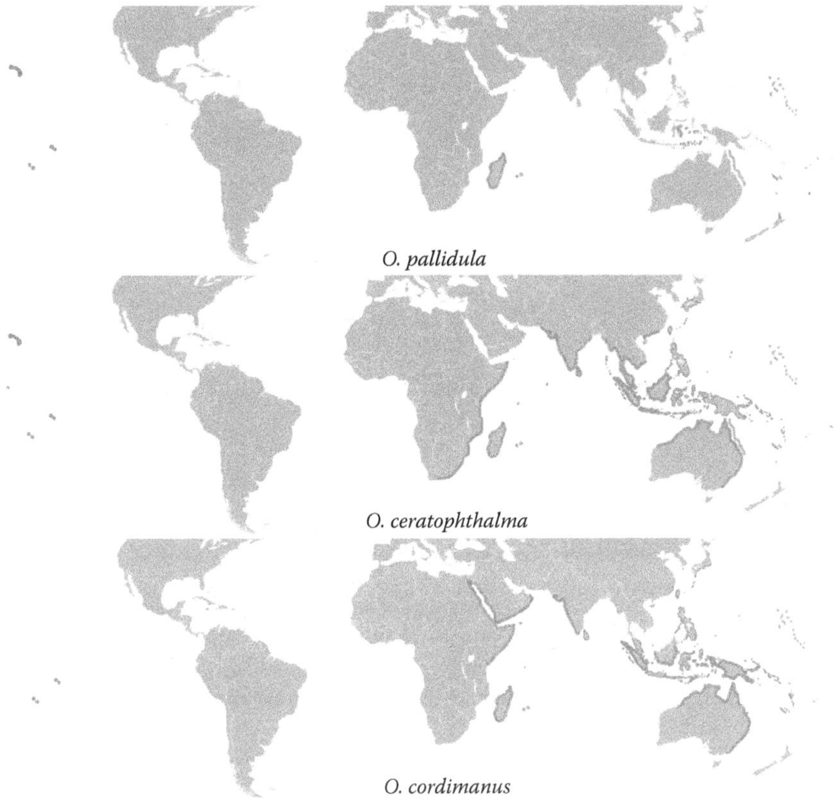

Figure 1 (continued) (See also colour figure in the insert) Distributions and ranges of individual ghost crab species around the world. For information on species ranges, see Sakai & Türkay (2013) and references therein. Additional key references for distributional ranges include Armstrong (1876), Suvatti (1950), Monod (1956), Holthuis (1961), George & Knott (1965), Day (1969), Kensley (1970), Bright & Hogue (1972), Vannini (1976), Brusca (1980), McLachlan (1980), Fischer et al. (1981), Manning & Holthuis (1981), Ziese (1985), Jones (1988), von Prahl et al. (1990), Koukouras et al. (1992), Branch et al. (1994), Hendrickx (1996), Veloso et al. (1997), De (1998), Strachan et al. (1999), Simões et al. (2001), Davie (2002), Aerts (2004), Demicheli et al. (2006), Lazarus-Agudelo & Cantera-Kintz (2007), McLay (2009), Ateş et al. (2010), Poupin (2010), Haddoud & El Gomati (2011), Coll et al. (2013). A complete list of sources used to compile the maps is available from the authors of this review.

detailed synopsis of the systematics of ghost crabs (family Ocypodidae, subfamily Ocypodinae); they recognized 22 species in two genera: *Ocypode* (21 spp.) and the monospecific *Hoplocypode* (*H. occidentalis*).

A defining taxonomic character of the Ocypodidae is the presence of setae between the bases of the walking legs; these setae are instrumental in water uptake, thought to be a primary adaptation to semi-terrestrial habits (Hartnoll 1988a). Other key adaptations associated with terrestrial affinities are large eyes with a 360° field of view (Cowles 1908) and long legs for rapid running (Hartnoll 1988a).

Ghost crabs occur worldwide, their distribution extending from the tropics to temperate latitudes (Sakai & Türkay 2013; Figure 1). Most species are greyish, but there are a number of exceptions. *Ocypode ryderi*, the 'pink ghost crab', has a pinkish colour and distinctive red joints on the chelipeds and legs (Barnard 1950, Lawrence 2003; Figure 3F); *O. madagascariensis* is more orange (Berry 1976). *Ocypode pallidula* is usually bright pink when it emerges from its burrow, but within 2 min the crab turns from pink to greyish (Hayashi & Cooke 1978). This colour change is caused by

Figure 2 (See also colour figure in the insert) Typical habitats in which ghost crabs occur. (A) Warm temperate exposed beach, Margate, KwaZulu-Natal (KZN), South Africa; (B) warm temperate exposed beach, Fraser Island, Queensland, Australia; (C) tropical beach, Cape Tribulation, Queensland, Australia; (D) warm temperate coastal dunes, Margate, KZN, South Africa; (E) subtropical exposed beach, Sodwana Bay, KZN, South Africa. (Photos by S. Lucrezi and T.A. Schlacher.)

the aggregation of pigment granules to the centre of cuticle cells (Brown 1940). *Ocypode pallidula* displays prominent bright-red patches on the ambulatory legs (Poupin et al. 2012), and *O. ceratophthalma* is slightly greenish in colour (Berry 1976, Lawrence 2003; Figure 3B). *Ocypode ceratophthalma* also displays a daily rhythm of chromatophore changes, with maximum concentrations of dark pigment during the night and maximum dispersion of dark pigment during the day (Little 1967, Stevens et al. 2013; Figure 3). This pattern, typical of young individuals rather than adults, allows ghost crabs to change colour according to the time of the day, and it is believed to provide camouflage against predators (Stevens et al. 2013; Figure 3C, 3D). *Ocypode cordimanus* has a greyish to pale-buff colour, with yellow to reddish outer edges to the chelae (Grubb 1971; Figure 3H). *Ocypode gaudichaudii*, the 'painted ghost crab' of Central America, is crimson red (Trott 1987a,b, 1998).

Ghost crabs generally have large chelae that are asymmetrical in size (Rathbun 1900, Mariappan et al. 2000, Schenk & Wainwright 2001; Figure 3). The larger chela can be on either side of the body, with even ratios between left and right claw reported for several species, including *Ocypode ceratophthalma* (Barrass 1963), *O. saratan* (Linsenmair 1967), *O. quadrata* (Haley 1969), and *O. cursor* (Strachan et al. 1999). *Ocypode ryderi* is, however, mainly 'right handed' (Vannini 1980a), and *O. gaudichaudii* is mainly 'left handed' (Trott 1987b).

Senses and physiology

Eyes and vision

The eyes of ghost crabs are amongst the most highly evolved in the crustaceans. The most conspicuous external structure is a large cornea, covering most of the eyestalk's ventral surface (Cowles 1908). The large cornea functions to make the eyes highly sensitive to light. Light sensitivity is further enhanced by the ability to respond to changes in light intensity by altering the diameter of the receptor under the cornea (Rosenberg & Langer 2001).

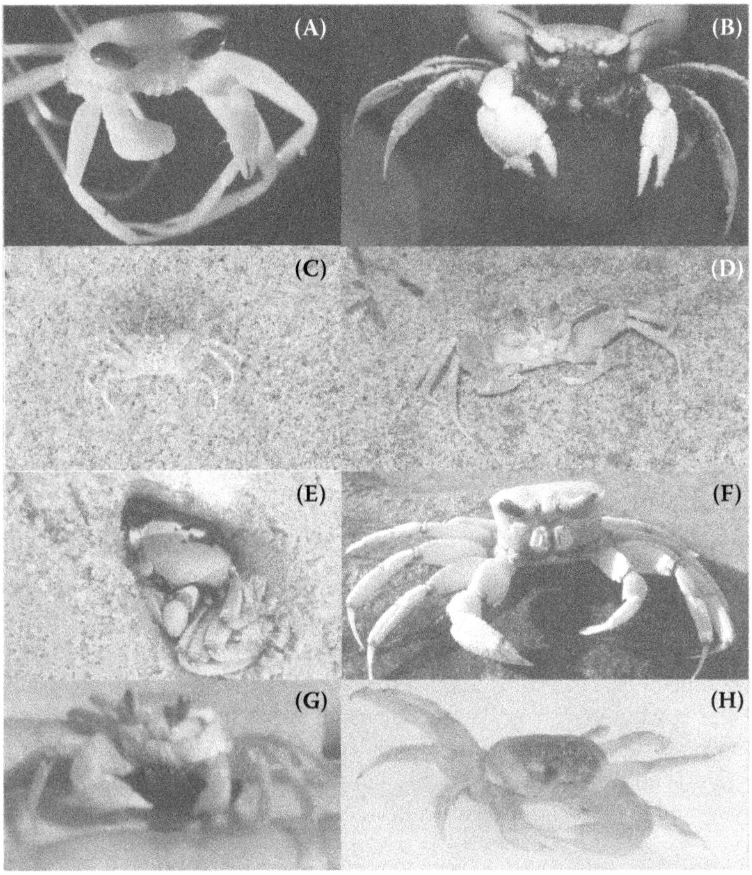

Figure 3 Typical morphological characteristics of ghost crab species. (A) Ghost crab, juvenile; (B) *Ocypode ceratophthalma*, adult; (C) *O. ceratophthalma*, juvenile, camouflaged with sand (day); (D) *O. ceratophthalma*, juvenile, camouflaged with sand (night); (E) *O. ceratophthalma*, adult, at burrow entrance; (F) *O. ryderi*, adult female; (G) *O. cordimanus*, adult, ovigerous female; (H) *O. cordimanus*, adult. (Photos by S. Lucrezi and T.A. Schlacher.)

Ghost crabs can visually detect large objects, such as a human, from a distance of at least 45 m, and the mobility of the eyestalks creates a 360° field of view (Cowles 1908). Because the cornea covers a smaller area of the dorsal part of the eyestalks, ghost crabs are less able to detect objects directly above them (Cowles 1908). The cornea grows allometrically and is proportionally larger in the juveniles of several species, for example, *Ocypode ceratophthalma* (Cott 1929, Haley 1973), *O. saratan* (Sandon 1937), *O. gaudichaudii*, and *Hoplocypode occidentalis* (Crane 1941).

Eye morphology can be a diagnostic character. For example, in adults of *Ocypode ceratophthalma*, the eyestalk extends beyond the cornea, forming a 'horn' (Figure 3), whereas in *O. cursor*, it is extended into a short conical process bearing a brush of hairs (Barnard 1950, Haley 1973). The longer eyestalks in adult males of *O. ceratophthalma* are possibly used in intraspecific recognition (Hartnoll 1988b).

Taste and smell

Ghost crabs have acute senses of smell and taste (Cowles 1908, Wellins et al. 1989). They locate food using cues from volatile chemicals detected with chemosensory hairs located on the dactyls

(Wellins et al. 1989). These hairs are the primary organs for smell, aided by receptors on the antennules (Wellins et al. 1989). The chelae and mouthparts also contain large numbers of chemoreceptors (Wolcott & Wolcott 1999). Ghost crabs will repeatedly pinch samples of potential food items and place them in the mouth using their pincers, a behaviour called *cheliped flexion* (Trott & Robertson 1984, Wolcott & Wolcott 1999).

Detection of sound and vibration

Sound is detected by Barth's myochortonal organ, a receptor located in the meral segment of the walking legs (Horch 1971, 1974, Salmon et al. 1977). The receptor responds to air and substratum-borne energies, and ghost crabs are therefore sensitive to both airborne sound and substratum vibration (Popper et al. 2001). Salmon et al. (1977) demonstrated that ghost crabs can respond to sound and vibration at distances of up to 10 m, and moist sediment enhances detection of vibrations transmitted through the substratum. Acoustic and vibratory signals are important behavioural components displayed in defending territories and courtship (Imafuku et al. 2001, Clayton 2005, 2008).

Respiration and circulation

Although ghost crabs are air breathers (*sensu lato*), they need water to create and maintain a moist environment for their respiratory organs. The water needed to moisten the areas of oxygen uptake in the branchial chambers is taken up from the sediment and the swash zone and during brief immersions in the surf zone of beaches (Borradaile 1903, Flemister 1958, Bliss 1968, Wolcott & Wolcott 1985, 2001). Water is acquired by capillary action using hydrophilic setae situated at the base of the walking legs (between the second and the third pair) and subsequently transferred to the branchial chambers, where it is absorbed across the gill membranes (Rao 1968, Wolcott 1976, 1984, 1992). Excess water is moved to the mouthparts, swallowed into the foregut, and later gradually absorbed (Wolcott 1984).

The branchial chamber contains both lungs and gills as sites of gas exchange. The lung, located on the dorsal (or posterior) part of the branchial chamber, is the main site for oxygen uptake (from air) and CO_2 excretion (in air) (Díaz & Rodríguez 1977, Whiteley et al. 1990, Farrelly & Greenaway 1994). The lung is formed by an epibranchial membrane that has numerous small folds, thus creating a large surface area (Flemister 1959, Díaz & Rodríguez 1977, Greenaway & Farrelly 1984, Al-Wassia et al. 1989, Henry 1994, Tsai & Lin 2012). In addition, the membranes forming the lung have a much shorter gas/blood diffusion distance compared with the gills, making the lung more efficient in oxygen uptake and CO_2 excretion in air (Storch & Welsch 1975).

The gills (five to seven pairs), located on the ventral (or anterior) part of the branchial chambers, are the main site of CO_2 excretion (Pearse 1929, Greenaway & Farrelly 1984, Tsai & Lin 2012). They are smaller than in truly aquatic crabs but larger than in crabs with a stronger terrestrial affinity (e.g., *Cardisoma* and *Birgus* spp.; Cameron 1981). When excreted via the gills, CO_2 is expelled into the branchial water surrounding the gills (Greenaway & Farrelly 1984, Al-Wassia et al. 1989, Eshky et al. 1990, Farrelly & Greenaway 1994, Henry 1994). This bimodal (lung/gill) form of respiration is a key adaptation to semi-terrestrial habitats (Vernberg 1956, Henry 1994, Takeda et al. 1996).

A number of biochemical adaptations aid ghost crabs to respire in dry environments: (1) high water content of the gill tissue; (2) greater activity in the gill tissue of the enzyme cytochrome *c* oxidase (the terminal enzyme in the respiratory chain); (3) a high affinity of the hemocyanin (blood pigment) for oxygen loading; (4) structural and functional properties of the hemocyanin (e.g., a dominance of hexamers) typical of terrestrial species; (5) a large oxygen reserve formed by the venous system; and (6) a high hemolymph pO_2 (Vernberg 1956, Vernberg & Vernberg 1968, Burnett 1979, Morris & Bridges 1985, 1994, Johnson 1987, A.C. Taylor personal communication in McMahon & Burggren 1988).

Water balance and osmoregulation

Water loss by evaporation is higher in ghost crabs than in other land crabs (Wolcott 1984). Greater water loss occurs due to the permeability of the exoskeleton and the high levels of activity outside the burrows (Greenaway 1988). Ghost crabs can, however, efficiently extract water from the sediment and the swash or during brief immersions in the surf zone (Borradaile 1903, Flemister 1958, Bliss 1968, Rao 1968, Wolcott 1976, 1984, 1992, Wolcott & Wolcott 1985, 2001). Species that use flotation feeding (e.g., *Ocypode quadrata*) can take up water from the buccal cavity (Powers & Bliss 1983).

Ghost crabs are strong osmoregulators (Flemister & Flemister 1951, Gross 1955, Greenaway 1988, Santos et al. 1989), using three principal physiological mechanisms: (1) ion absorption via the gills (Flemister & Flemister 1951, Flemister 1959); (2) ion excretion via the kidneys and possibly also the branchial epithelium (Flemister & Flemister 1951, Flemister 1959); and (3) ion reclamation in the antennal glands (Weihrauch et al. 2004).

When water is readily available, ghost crabs have abundant urine flow (Wolcott 1984). To avoid excessive salt loss under these conditions, ghost crabs reprocess urine by passing it to the branchial chamber, where salts and water are reabsorbed through the gills (Storch & Welsch 1975, Wolcott & Wolcott 1985). Conversely, excess salt can be eliminated by increasing the uptake of water from the sediment and flushing the surplus salt with the urine (Greenaway 1988). Where fresh water is available, ghost crabs can take this up to void excess salts in isosmotic urine with low energetic costs (Wolcott 1984).

There is considerable variation in reported salinity values tolerated by ghost crabs, but a salinity range of 10–40 psu appears to be a reasonable generalization under field conditions (Table 1). Data from laboratory experiments measuring the time or dose at which 50% of the test individuals die (LT_{50} or LD_{50}) when exposed to different levels of salinity suggest a somewhat broader range of salinity tolerances compared with field data (Table 1). Even so, ghost crabs may not survive prolonged exposures to either seawater or freshwater. In a laboratory experiment, exposure to low-salinity values resulted in animals dying within 2 days at 0.5 psu and within a week when held in water at 35 psu (Ewa-Oboho 1993).

Table 1 Summary of reported salinity tolerances of ghost crabs (*Ocypode* spp.)

Species	Salinity range/tolerance	Notes	Locality	References
O. cursor	10 psu (minimum) 10–35 psu[a]	100% mortality at 5 psu after 7 days; 5% mortality at 35 psu after 7 days; 100% mortality at 0.5 psu after 2 days (LT_{50} = 26 h)[a]	Nigeria	Ewa-Oboho (1993)
O. macrocera	10–34 psu[a]	Not reported whether death occurs at either extreme[a]	Eastern coast of India	Rao (1966, 1969), Allayie et al. (2011)
O. quadrata	32–35 psu	0–48[a] (exposure for up to 12 h; not reported whether death occurs at either extreme)	Brazil	Santos et al. (1989), Burggren et al. (1993), Santos & Moreira (1999)
O. quadrata	19–28 psu		Brazil	Vinagre et al. (2007), Branco et al. (2010)
O. quadrata	14–32 psu	Estuarine habitat	Brazil	Da Rosa & Borzone (2008)
O. saratan	38.8–40.1 psu		Oman	Kazemiyan (2008)
O. stimpsoni	Not recorded	Death after 12-h exposure to freshwater[a]	Taiwan	Tsai & Lin (2012)

[a] Laboratory conditions/experiments; all others are based on field observations of crab surface activity

Thermal biology

Temperature plays a central role in the biology and ecology of ghost crabs. It governs most physiological and metabolic processes (Vernberg & Vernberg 1968, Dittrich 1990, Weinstein & Full 1994, Weinstein et al. 1994); is critical for survival (Eshky et al. 1988, Wolcott 1988); sets distributional limits (Wolcott 1988); determines activity levels (Haley 1972, Ahn et al. 1991); influences growth and reproduction (Rao 1966, Negreiros-Fransozo et al. 2002); and affects rates of locomotion and activity (Florey & Hoyle 1976, Ahn et al. 1991, Weinstein 1998).

In the field, ghost crabs have been reported to be active at air temperatures between 12°C and 50°C (Table 2). Below 12°C, ghost crabs generally become inactive and remain inside their burrows (Leber 1977). The lower lethal limit for *Ocypode quadrata* on the beaches of North Carolina (USA) is thought to be 6–8°C (T.G. Wolcott unpublished data in Wolcott 1988). Upper temperature limits for activity have been reported in the field as 45.5–50°C for *Ocypode fabricii* and *O. ryderi* (George & Knott 1965, Dittrich 1990, Buchholz & Vetter 1993). Complementing field observations, laboratory experiments indicated that the range of temperature tolerances approximates 10–40°C in *Ocypode saratan* (Eshky et al. 1988) and 9–37°C in *O. ceratophthalma* (Florey & Hoyle 1976).

Table 2 Summary of reported temperature tolerances of ghost crabs (*Ocypode* spp.)

Species	Temperature range/tolerance	Notes	Locality	References
O. ceratophthalma	12.4°C (minimum)		Inhaca Island, Mozambique	Hughes (1966)
O. ceratophthalma	9–37°C[a]	>35°C erratic; < 15°C slow; 9°C no movement and some deaths; body temperature within 0.5°C of air temperature at 26–28°C	Hawaii	Florey & Hoyle (1976)
O. ceratophthalma	15–34°C	Burrow temperature at 25 cm depth is 13°C lower than 45°C surface temperature	Hong Kong	Chan et al. (2006)
O. cursor	22.8–35.5°C	Burrow temperature changes by < 3°C in 24 h, and at 40 cm depth can be up to 20°C lower than 51°C surface temperature	Northern Cyprus	Strachan et al. (1999)
O. cursor	18–27.5°C	Constant burrow temperature varying by < 4°C in 24 h	Turkey	Türeli et al. (2009)
O. fabricii	45.5°C (maximum)		Shark Bay, Western Australia	George & Knott (1965)
O. macrocera	15–34°C[a]	70% mortality and 30% no growth at 15°C and no moulting at 34°C	Eastern coast of India	Rao (1966, 1969), Allayie et al. (2011)
O. quadrata	16°C (minimum)	Crabs inactive in burrow in winter	Texas, USA	Haley (1972)
O. quadrata	6–8°C (minimum lethal) 12–15°C (minimum activity)		North Carolina, USA	Leber (1977), T.G. Wolcott unpublished data in Wolcott (1988)

Table 2 (Continued) Summary of reported temperature tolerances of ghost crabs (*Ocypode* spp.)

Species	Temperature range/tolerance	Notes	Locality	References
O. quadrata	30–34°C (maximum)		Georgia, USA	Robertson & Pfeiffer (1982)
O. quadrata	28–33°C (maximum)		São Paulo, Brazil	Santos et al. (1989), Burggren et al. (1993), Santos & Moreira (1999)
O. quadrata	15°C (minimum)[a]	No field activity < 15°C; burrow temperature ~ 2°C higher than beach surface at 20°C; at 30–35°C, crabs lower body temperature by up to 6°C	North Carolina, USA	Weinstein & Full (1994), Weinstein et al. (1994), Weinstein (1995)
O. quadrata	28–34°C (maximum)	Crabs inactive in burrow	Veracruz, Mexico	Valero-Pacheco et al. (2007)
O. quadrata	13.5–29.9°C		Santa Catarina, Brazil	Vinagre et al. (2007), Branco et al. (2010)
O. quadrata	15–27.9°C		Paraná, Brazil	Da Rosa & Borzone (2008)
O. quadrata	16°C (minimum)		Rio Grande do Sul, Brazil	Antunes et al. (2010)
O. ryderi	50°C (maximum)	Inferred from temperature dependence of enzyme activity	Kenya	Dittrich (1990), Buchholz & Vetter (1993)
O. saratan	10–40°C (near-lethal limits)[a]	Few large crabs survive these extremes	Jeddah, Saudi Arabia	Eshky (1985), Eshky et al. (1988)
O. saratan	20–31.3°C		Oman	Kazemiyan (2008)

[a] Laboratory conditions/experiments; all others are based on field observations of crab emergence/surface activity and/or the presence of active burrow openings.

Three principal mechanisms are used by crabs to regulate body temperature: (1) cooling through evaporative water loss that can maintain body temperature 5–6°C below ambient temperature; (2) limiting surface activity by remaining inside burrows that provide a thermal refuge against temperature extremes, with crabs at times actively plugging burrow openings; and (3) acclimatising metabolically, for example, by increasing the activity of respiratory enzymes or oxygen consumption (Vernberg & Vernberg 1968, Berry 1976, Eshky et al. 1988, Weinstein et al. 1994, Strachan et al. 1999).

Although temperature extremes can, to some degree, be buffered by crabs digging deep burrows, this mechanism is not always effective. At the limits of distribution, occasional extreme cold temperatures are known to be lethal, as shown for *Ocypode quadrata* on the East Coast of the United States (Wolcott 1988). *Ocypode quadrata* remains inside the burrow during the coldest part of the year, when temperatures below 15°C appear to trigger this 'overwintering' behaviour (Vernberg & Vernberg 1968, Grant 1983, T.G. Wolcott unpublished data in Wolcott 1988). Dormancy can reportedly last from 3 (Haley 1972) to 6 months (Leber 1982). This behaviour has also been labelled 'hibernation' (Williams 1965, Leber 1982, Manning & Lindquist 2003, Hobbs et al. 2008), although the use of the term 'hibernation' (*sensu stricto*) appears inappropriate for these ectotherms (T.G. Wolcott personal communication 6 October 2013). It has not been established to what extent mortality may occur during the overwintering phase or how metabolic processes are modified to enable the crabs to survive extended periods without feeding.

Locomotion

Ghost crabs are the fastest crustaceans on land, reaching speeds of 4 m s^{-1}, hence their colloquial name 'racing crabs' (Cott 1929, Hafemann & Hubbard 1969). Remarkably, they can employ bipedal motion during sprints and are amongst the few crabs to regularly run forwards in addition to the typical sideways motion of brachyurans (Daumer et al. 1963, Hughes 1966, Wolcott 1978, Blickhan & Full 1987, Trott 1988, Full & Weinstein 1992, Weinstein & Full 1992).

High rates of oxygen uptake and long terminal segments of the limbs, combined with hook-like dactyls (Figure 3), are important traits that make ghost crabs remarkably fleet (Cott 1929, Ayres 1938, Hafemann & Hubbard 1969, Perry et al. 2009). Growth of the walking legs can be either isometric, that is, growing proportionately to the rest of the body as in *Ocypode quadrata* (Haley 1969), or allometric, that is, growing longer with respect to the rest of the body as in *O. ceratophthalma*, *O. gaudichaudii*, and *Hoplocypode occidentalis* (Cott 1929, Crane 1941, Haley 1973).

Sideways locomotion is brief and intermittent, biomechanically characterized by up-and-down movement of the body and sideways movement of the legs in the sagittal plane, describing an irregular, asynchronous, and alternating tetrapod gait (Blickhan & Full 1987, Full & Weinstein 1992, Weinstein & Full 1992, Weinstein 1995, 2001, Spagna et al. 2007). As the crab's centre of mass rises and falls, the energy is transferred from kinetic to potential, with a saving of up to 55% (Blickhan & Full 1987). Individuals of *Ocypode ceratophthalma* use fewer and fewer legs as they run faster during sideways locomotion, first raising the fourth pair of walking legs off the ground, then raising the third pair, and finally performing a bipedal gait with the second pair of legs (Burrows & Hoyle 1973). This gait represents a true 'run', and the transition from slow to fast running has been likened to the change from a trot to a gallop in horses (Full & Tu 1991, Blickhan et al. 1993).

Running speed in ghost crabs has been reported to be a function of carapace width: Running speeds in *Ocypode ceratophthalma* increase up to 20-mm carapace width, after which they start decreasing (Burrows & Hoyle 1973). Under laboratory conditions, large individuals of *Ocypode quadrata* (70–90 g) do not exceed 0.5 m s^{-1}, whereas crabs of intermediate mass (30–50 g) move at greater speed (up to 1.6 m s^{-1}; Blickhan & Full 1987). The main reasons why larger crabs run more slowly than smaller crabs are thought to be that (1) larger crabs carry a comparatively heavier body weight supported by shorter legs (i.e., legs grow linearly, but body weight increases exponentially; Burrows & Hoyle 1973) and (2) larger crabs have lower physiological endurance, reaching maximum oxygen consumption at lower speeds than crabs of intermediate size (Blickhan & Full 1987, Full 1987, Full & Weinstein 1992).

Reproductive biology and life histories

Seasonal and lunar patterns

Temporal facets of reproduction in marine invertebrates are governed by a large number of interacting factors, including *inter alia* temperature, photoperiod, salinity, oxygen, food availability, variations in currents, and the dynamics of predators (Olive 1992, 1995, Ramirez-Llodra 2002, Al-Solamy & Hussein 2012; Table 3). Local selection can also influence the timing of breeding, particularly in species that have broad geographic ranges (Olive 1995; Table 3). A seasonal pattern of reproduction is not uncommon in marine invertebrates (Table 3), and several hypotheses have been advanced to explain its evolutionary significance and selective advantage (Olive 1995). For example, Olive (1992) put forward the functional hypotheses that seasonal reproduction enhances survival of the brood and maximizes brood fitness.

Ghost crabs show both seasonal and non-seasonal patterns of reproduction (Table 3; Figure 4). A warm climate supports optimal gonad development in ghost crabs, whereas low temperatures may delay egg development and larval release (Haley 1973, Hughes 1973, Brooke 1981, Al-Solamy & Hussein 2012). It is therefore expected that in tropical or subtropical environments ghost crabs

Table 3 Life-history characteristics of ghost crab (*Ocypode*) species

Species	Locality	Mating season	Time of larval release	Main recruitment period	Notes	Reference
O. brevicornis	Oman	Spring (May)				Clayton (2008)
O. ceratophthalma	Hawaii	All year; peak of vitellogenesis in spring/summer (April–August)			Male reproductive maturity: > 33 mm carapace width (CW); female reproductive maturity: 33 mm CW	Haley (1973)
O. ceratophthalma	Seychelles	All year; greatest activity 1 week before new moon	All year (peak at full moon); peak in winter/early spring (July–September)		Time required for eggs to mature: 2–3 weeks	Brooke (1981)
O. ceratophthalma	Iriomote-jima, Japan		Late spring/summer (June–July); only during high water and during last and first quarters of the moon			Saigusa et al. (2003)
O. ceratophthalma	Karwar, eastern coast of India				Time of megalopa stage (immediately following zoeae): 13–14 days; time between hatching and first crab stage (zoeae plus megalopa): 47–56 days[a]	Kakati (2005)
O. ceratophthalma; *O. madagascariensis*; *O. ryderi*	Eastern Cape, South Africa			Late summer (February–March)		McLachlan (1980)
O. ceratophthalma; *O. madagascariensis*; *O. ryderi*	KwaZulu-Natal, South Africa	Seasonal, peaking in spring/summer (November–March)		Summer/early autumn (February–April)	CW: megalopae at recruitment: 7.2–7.4 mm; 1 yr after recruitment: 30 mm; 2 yr after recruitment: 36 mm; maximum after 2 yr post-recruitment: 45 mm; larval duration estimated at 2–3 months	Jackson et al. (1991)
O. cursor	Northern Cyprus			Summer (August)	Egg mass contains 3000 eggs; all hatched in aquarium	Strachan et al. (1999)

Continued

Table 3 Life-history characteristics of ghost crab (*Ocypode*) species

Species	Locality	Mating season	Time of larval release	Main recruitment period	Notes	Reference
O. gaudichaudii	Panama		Late winter/early spring (March)			Crane (1939, 1940)
O. quadrata	Carolinas, USA		Spring/summer (April–July)			Coues (1871), Williams (1965)
O. quadrata	Long Island; Fire Island, New York, USA			Summer/early autumn (August–September)		Smith (1873)
O. quadrata	Loggerhead Key, Florida, USA	Spring/early summer (March–June)		Possibly late winter/ early spring (March)		Cowles (1908)
O. quadrata	Mustang Island, Texas, USA	All year; peak in summer (June–September)	Proportion of ovigerous females up to 100% in summer (August)	Summer/early autumn (July–October)	CW: gonad maturation: 24 mm in males and 26 mm females; 10–12 months between first crab stage (immediately following the megalopa stage) and maturity; lifespan estimated at ~3 yr	Haley (1969, 1972)
O. quadrata	North Carolina, USA	Summer (June–September)	New moon and full moon, after dark within 40 min of high tide	Summer/early autumn (July–September)	Larval phase duration: 34 days (at 25°C and 35 psu)[a]	Díaz & Costlow (1972), Leber (1982), Wellins et al. (1989), Peterson et al. (2006)
O. quadrata	Assateague Island, Maryland-Virginia, USA	Summer/early autumn (June–September)		Summer/autumn (June–November)	Reproduction close to new or full moon; close to the water's edge; maturity reached 13 months after recruitment	Christoffers (1986)

O. quadrata	São Paulo, Brazil	Spring/summer/ autumn (October–May)	Late summer/autumn (February–May); spring (October–November)	All year; peak in summer/early autumn (January–March)	CW: gonad maturation: 20.0 mm in males and 23.0 mm in females; ovigerous females: 34.5–39.0 mm; maximum lifespan estimate: 3 yr	Alberto & Fontoura (1999), Negreiros-Fransozo et al. (2002)
O. quadrata	Santa Catarina, Brazil	All year; peak in spring/summer (October–March)	Late summer (February–March)		No juveniles caught and no juvenile active burrows found in late spring/early summer (December–January)	Vinagre et al. (2007), Branco et al. (2010)
O. quadrata	Rio Grande do Sul, Brazil	All year; peak in the summer (December–March)				Antunes et al. (2010)
O. saratan	Chabahar Bay, Oman	All year	Spring (March–June); autumn (September–December)			Kazemiyan (2008), Kazemiyan & Hossein (2008)
O. saratan	Jeddah, Saudi Arabia	All year	All year; peak in summer (June–September) and decline in winter (December–March)	All year; peak in spring/summer (April–August) and low in winter (December–March)		Al-Solamy & Hussein (2012)
O. stimpsoni	Wakayama, Japan	Summer (July)				Imafuku et al. (2001)
Hoplocypode occidentalis	Costa Rica	Summer (August)				Hughes (1973)

[a] Laboratory conditions/experiments.

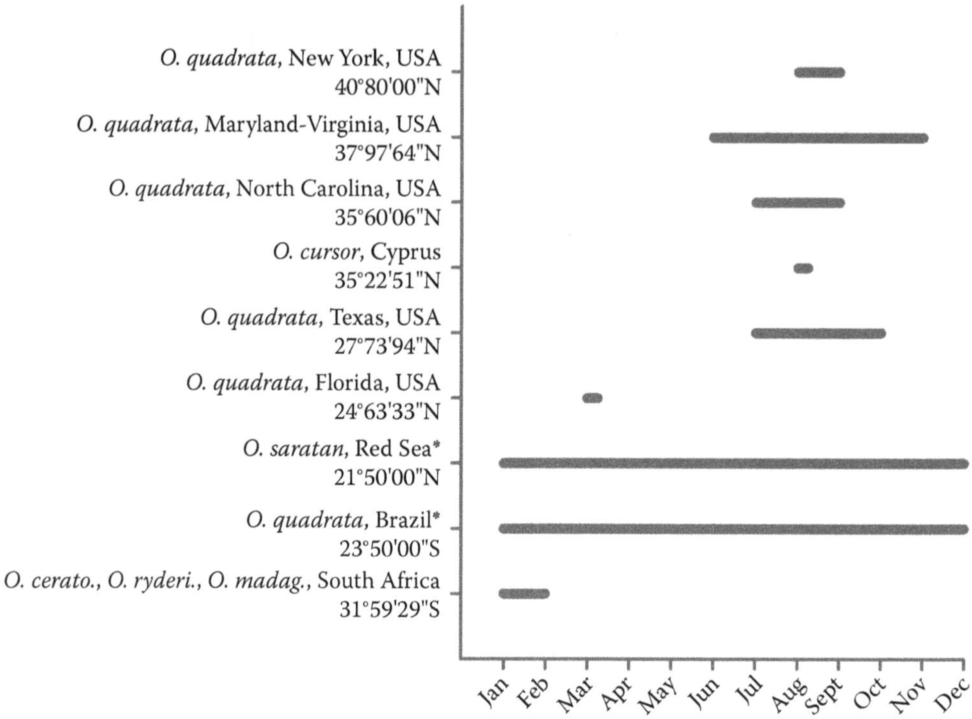

Figure 4 Recruitment period of *Ocypode* species from various localities. **Ocypode saratan* from the Red Sea has a recruitment peak between April and August and a recruitment low between December and March; *O. quadrata* from Brazil has a recruitment peak between January and March. *O. cerato.*, *Ocypode ceratophthalma*; *O. madag.*, *O. madagascariensis.*

will breed throughout the year, but some tropical species show seasonality in their reproduction (Table 3; Figure 4). This suggests that evolutionary history could be important in determining temporal attributes of reproduction (Al-Solamy & Hussein 2012).

Ghost crabs can show lunar rhythmicity in reproduction, synchronizing reproductive maturity and activities (e.g., construction and defence of copulation burrows and acoustic calling by males, maturation of ovaries, release of eggs) with the lunar cycle (Brooke 1981, Salmon 1983, Christoffers 1986, Christy 1987, Wellins et al. 1989, Saigusa et al. 2003; Table 3).

Mating

Copulation usually occurs on the beach surface. However, several species (e.g., *Ocypode ceratophthalma*, *O. saratan*, *O. gaudichaudii*, and *O. jousseaumei*) build burrows specifically for mating, termed 'copulation burrows' (Linsenmair 1967, Hughes 1973). Copulation burrows have a characteristic spiral shape, with the spiral turning clockwise or anticlockwise, depending on which side the male bears the major chela (Hartnoll 1969). The spiral often ends in a chamber (Linsenmair 1967, Brooke 1981). Copulation burrows are also flanked by a sand mound, a key visual cue thought to signal to receptive females the presence of a male ready to mate (Linsenmair 1965, 1967, Hartnoll 1969). When a female is close to the male and its burrow, the male will also stridulate (Hartnoll 1969). Mating is usually diurnal, with the sand mound the main feature used to attract females. When mating occurs at night, females are mostly attracted by sound produced by males (Hartnoll 1969). Linsenmair (1965) suggested that the sand mounds constructed by males have a second function, which is to stimulate other males to construct mounds, thus possibly synchronizing mating within a population. This group selectionist view of reproductive strategies might have sounded plausible at the time of Linsenmair's (1965) observations. However, the modern understanding of reproductive strategies, which focus on individual selection, would tend to cast doubt on this idea.

The act of mating in *Ocypode* has been described by Dunham & Gilchrist (1988). The male first climbs on the female's carapace from the rear and uses his minor cheliped to tug at, and stroke, the anterior part of the female's carapace. The male then uses the ambulatory legs to perform a series of tapping and stroking movements, followed by turning the female upside down and holding her in place with the walking legs. Finally, the tips of the gonopods are inserted into the female's gonopores, with the crabs remaining in this position for up to 1 h.

Egg bearing and release

Gravid females can be easily recognized by their abdomen, which hangs lower, held away from the ventral cephalothorax by the dark-coloured or bright-orange mass of eggs attached to the swimmerets (Crane 1939, Milne & Milne 1946; Figure 3G). Ovigerous females are usually recorded in relatively small numbers on the beach surface (Brooke 1981, Alberto & Fontoura 1999, Negreiros-Fransozo et al. 2002). It is thought that females remain inside their burrows during most of the egg development phase, leaving the burrows only to release their eggs (Fellows 1973, Trott 1998, Negreiros-Fransozo et al. 2002). Gravid females usually shift sand to obliterate the opening of their burrows; they also dig deeper burrows, probably to create a more stable thermal environment (Christy 1987, Al-Solamy & Hussein 2012). Egg-bearing females have, however, been observed to enter the surf zone to irrigate the egg mass (Milne & Milne 1946). In *Ocypode gaudichaudii*, gravid females are known to emerge at dusk to enter the water to drink, suggesting that water uptake by capillary action through the hydrophilic setae at the base of the walking legs is more difficult when the abdomen is distended by an egg mass (Koepcke & Koepcke 1953).

The period between ovulation and hatching may be up to 43 days (Haley 1972). Egg membranes usually rupture prior to release, which generally occurs at night when females emerge from their burrows and move to the swash zone (D. Rittschof unpublished data in Wolcott 1988, Saigusa et al. 2003). On contact with the water, the egg mass is ejected, and the larvae are released from the ruptured egg membranes, first sinking but then swimming within 15–30 min after release (D. Rittschof unpublished data in Wolcott 1988). Egg release has a lunar periodicity, with peaks at both full and new moon (Brooke 1981, Wellins et al. 1989, Saigusa et al. 2003; Table 3).

Larval stages and recruitment

The larval phase has five or six zoea stages that disperse in the plankton, followed by a megalopa stage that recruits to the beach (Crane 1940, Raja Bai Naidu 1951, Díaz & Costlow 1972, Kakati 2005, McDermott 2009). The duration of the larval period (a measure of dispersal potential; Hines 1986) is thought to be 2–3 months in the field (Jackson et al. 1991). However, under suitable rearing conditions in the laboratory, larval stages can be shorter: The larval period in *Ocypode quadrata* and *O. ceratophthalma* reared in the lab at 23–26°C and 30–35 psu lasts 34–42 days (Díaz and Costlow 1972, Kakati 2005; Table 3).

The larvae are free swimming in the plankton during their pelagic phase (Smith 1873, Vogt 2013), and this is the main dispersal stage. There are, however, no published data (to the best of our knowledge) on the most fundamental aspects of larval ecology and biology during this dispersal phase, including information on depths at which larvae occur in the pelagic zone, larval behaviour, and total dispersion distance.

The megalopa stage marks the passage from the sea to the beach (Smith 1873). Megalopae recruit from the plankton to the beach by becoming stranded on the sand during receding tides (Smith 1873, Crane 1940). Although the temporal patterns of recruitment have not been established, megalopae tend to be abundant on the beach at night and during high tides (Crane 1940, Raja Bai Naidu 1954). On contact with the sand, megalopae immediately excavate holes as deep as 15 cm (Raja Bai Naidu 1954). Although these holes are not true burrows, they are the sites where moulting into the first crab stage has been observed to take place (Smith 1880, Crane 1940).

Recruitment occurs at different times of the year depending on the species (Table 3; Figure 4). Recruitment in some common species (e.g., *Ocypode ceratophthalma*, *O. madagascariensis*, and *O. ryderi*) occurs during the warmer months (i.e., spring to early autumn; Table 3; Figure 4). Others (e.g., *Ocypode quadrata* and *O. saratan*) recruit throughout the year, peaking in summer (Table 3; Figure 4). Two broods per year are possible in *Ocypode quadrata* (Haley 1972, Wolcott 1978). Lunar cycles have been reported in the abundance of zoeas (most abundant during the last and first quarter; Saigusa et al. 2003), and megalopae have been reported to be abundant around the new moon (Mense et al. 1995).

Moulting and growth

Seven moults, at intervals of about 35 days, usually take place between the megalopa stage and sexual maturity (Rao 1968, Haley 1972). Sexual maturity is reached at a carapace width ranging from 19 to 26 mm depending on the species and sex (Haley 1969, Fransozo et al. 2002); males tend to reach maturity earlier than females (Haley 1969, Wolcott 1988).

Moulting occurs mainly inside the burrow, and crabs plug the burrow's opening 7–14 days prior to moulting (Fellows 1973, Henning & Klassen 1973, Greenaway 1993). Following moulting, crabs eat the exuviae to retain essential calcium and magnesium and emerge from the burrow after a week (Greenaway 1993). There is no evidence for a terminal ecdysis (i.e., the last moult) in ghost crabs, suggesting that moulting and growth continue beyond sexual maturity (Hartnoll 1988b). Males of some species may keep growing after maturation up to twice their size at maturity (Haley 1969, 1973, Al-Solamy & Hussein 2012). After reaching sexual maturity, allometric growth of the chelae is faster in males, whereas females grow a larger abdomen (Hartnoll 1974).

An adult ghost crab with a 45-mm carapace width weighs approximately 60 g (Burrows & Hoyle 1973). Haley (1972) estimated growth rates to be about 0.117 mm per day during the larval phase, about 0.093 mm per day during maturation, and about 0.025 mm per day from post-maturation to death. Data on lifespans are limited to a single species, *Ocypode quadrata*, estimated to be 3 years (Haley 1972, Alberto & Fontoura 1999).

Behaviour

Surface activity

Whereas a defining characteristic of ghost crabs is their fossorial mode of life, they are regularly active on the beach surface, mainly for feeding, burrow construction, and courtship (Hughes 1973, Clayton 2005, Türeli et al. 2009). There is a widespread perception that ghost crabs are predominantly nocturnal (Clayton 2005, Valero-Pacheco et al. 2007). For example, Palmer (1971) investigated the influence of light intensity on surface activity and experimentally demonstrated that crabs shift their nocturnal activity rhythms when exposed to an inverted light cycle. Diurnal activity on the beach surface is, however, common in the field, especially in habitats largely free from human activity and disturbance. Juveniles are also surface active during the day on beaches where humans are present (T.A. Schlacher, personal observation).

Individuals that have burrowed in the intertidal zone tend to remain in their plugged burrows while being covered by the tides (Flemister & Flemister 1951). Thus, tidal periodicity in surface activity is possible, but rarely supported by robust evidence. One-third of papers we reviewed

Table 4 Variation in emergence times from burrows and periods of activity on the beach surface in ghost crabs (*Ocypode* spp.)

Emergence and activity	Species–locality	References
Mainly nocturnal	*O. africana*–Ghana *O. brevicornis* (formerly *O. platytarsis*)–Oman *O. ceratophthalma*–French Frigate Shoals, Hawaii; Cousin, Seychelles; Marshall Islands, western Pacific Ocean; KwaZulu-Natal, South Africa *O. cordimanus*–Marshall Islands, western Pacific Ocean *O. madagascariensis*–KwaZulu-Natal, South Africa *O. macrocera*–Eastern coast of India *O. quadrata*–New Jersey, USA; North Carolina, USA *O. ryderi*–KwaZulu-Natal, South Africa *O. saratan*–Jeddah, Saudi Arabia	Gauld & Buchanan (1956), Williams (1965), Rao (1969), Lighter (1974), Horch (1975), Berry (1976), Wolcott (1978), Brooke (1981), Leber (1982), Eshky (1985), Clayton (2008), Marschhauser (2010), Al-Solamy & Hussein (2012)
Nocturnal/some diurnal activity	*O. africana*–Sierra Leone *O. ceratophthalma*–Inhaca Island, Mozambique *O. cursor*–Sierra Leone–northern Cyprus *O. quadrata*–Loggerhead Key, Florida, USA; Assateague Island, Maryland-Virginia, USA; Bermuda; Gulf of Mexico *O. ryderi* (misidentified as *O. kuhlii*)–Watamu, Kenya	Cowles (1908), Longhurst (1958), Barrass (1963), Palmer (1971), Hughes (1973), Evans et al. (1976), Powers (1977), Christoffers (1986), Strachan et al. (1999), Valero-Pacheco et al. (2007)
Mainly diurnal	*O. ceratophthalma*–Maui, Hawaii *O. fabricii*–Shark Bay, Western Australia *O. gaudichaudii*–Culebra Island, Pacific coast of Panama; Pacific coast of Costa Rica *O. pallidula*–Maui, Hawaii *O. quadrata*–Sapelo Island, Georgia, USA *O. stimpsoni*–Wakayama Prefecture, Japan	George & Knott (1965), Robertson & Pfeiffer (1982), Schober & Christy (1993), Trott (1998), Imafuku et al. (2001), Norcross-Nu'u & Brown (2006)
No clear diel pattern	*O. jousseaumei*–Oman	Clayton (2005)

reported greater activity during low tides, and the remainder made no distinction between levels of surface activity during different tidal phases (Table 4). Surface activity can also be related to the lunar phases, with a few studies reporting higher activity around the new moon (e.g., Horch 1975, Brooke 1981, Clayton 2005, Moss & McPhee 2006). This pattern could be associated with cyclic egg releases if nocturnal high tides occur predictably during a particular lunar phase on an individual beach (Saigusa et al. 2003).

Signalling

Ghost crabs use visual, auditory, and vibratory signals during territorial interactions and courtship (Hartnoll 1969, Imafuku et al. 2001, Clayton 2005, 2008). All species except *Ocypode cordimanus* have a stridulating ridge on the palm of the larger chela. This ridge is scraped against another ridge on the third joint to produce sound (Rathbun 1899, Montgomery 1931, Barnard 1950, George 1982, Cooper 1997, Khan & Ravichandran 2007, Sakai & Türkay 2013). This type of sound production is called stridulation. Ghost crabs also produce sound by knocking the major cheliped onto the sand (Hartnoll 1969, Horch & Salmon 1969, Horch 1971, 1975). Male ghost crabs emit sounds from their burrow as part of the courtship ritual (Hartnoll 1969). To avoid overlap and mixing of acoustic signals, males in the same area stagger their calls (Guinot-Dumortier & Dumortier 1960, Horch & Salmon 1969). For instance, on beaches of the Seychelles, where *Ocypode cordimanus* and *O. ceratophthalma* co-occur, the former produces a calling sound that consists mainly of a trisyllabic 'err-er-er' (with a rhythm of 'dash-dot-dot'), whereas its sympatric congener has a bird-like rasping call that consists of long-drawn-out notes, each rising in pitch (Grubb 1971).

Ghost crabs 'wave' by rhythmically raising and lowering their chelipeds (Griffin 1968). Species that are active by day use waving displays extensively, particularly during courtship and agonistic encounters (Griffin 1968, Wright 1968, Vannini 1976). Another visual signal is the use of sand mounds, which ghost crabs construct next to their burrow entrances (Linsenmair 1965, 1967). These mounds are thought to have two principal functions: (1) as a spacing signal or mechanism (Dunham & Gilchrist 1988) and (2) as an attractant for females (Linsenmair 1967, Hartnoll 1969). As noted previously, a proposed role in synchronizing mating in the population (Linsenmair 1965) is now considered unlikely.

Agonistic behaviour

Several species show complex agonistic behaviours, thought to occur in the context of competition for food and burrows and during courtship. Competitive interactions have been reported on East African beaches between *Ocypode cordimanus* and *O. ryderi* and other scavengers, such as hermit crabs (Vannini 1975, 1976), and have been inferred by Koepcke & Koepcke (1953) on Peruvian shores to occur between *Ocypode gaudichaudii* and scavenging vertebrates such as foxes and rats. Although the actual competitive interaction between these vertebrates and ghost crabs are not described, foxes and rats are known predators of ghost crabs on these beaches, suggesting a predator-prey interaction mediated by the presence of stranded carrion (see also Schlacher et al. 2013a,b).

Two forms of agonistic display generally occur: (1) lateral merus display, in which the chelae are extended laterally and held almost horizontally; and (2) chelae display, with the chelae held almost vertically in front of the crab with the tips pointing downwards (Schöne 1968, Evans et al. 1976, Clayton 2008). The former display usually occurs either as a burrow defence mechanism or as a way to discourage a conspecific from excavating close to an existing burrow (Evans et al. 1976, Lighter 1976). The latter display is often observed in encounters over food (Evans et al. 1976, Vannini 1980a). Displays may end with one opponent surrendering or may be followed by 'pouncing', by which the two crabs raise themselves, lift the first walking legs, and push one another by touching the fronts of the chelae (Schöne 1968, Evans et al. 1976, Lighter 1976). Most agonistic encounters concerning burrows end with the expulsion and chasing of the intruding crab from the burrow or territory (Lighter 1976, Clayton 2005).

Agonistic behaviour (e.g., threat display) is considered to be an important spacing mechanism amongst individuals burrowing on the beach (Evans et al. 1976, Lighter 1976, Fisher & Tevesz 1979, Vannini 1980a). Examples of this behaviour in relation to the burrow include burrow plugging, burrow filling, burrow displacement, and mound building with waving displays to obscure a neighbouring burrow (Dunham & Gilchrist 1988, Wolcott 1988). Although behaviours related to burrow spacing are often interpreted as a form of territoriality (Vannini 1980a), this is not entirely accurate because ghost crabs show little burrow fidelity (Wolcott 1988).

Orientation

The positions of the sun, polarized light, and other visual cues are important for orientation in ghost crabs (Dunham & Gilchrist 1988). *Ocypode ceratophthalma* and *O. quadrata* can detect the landward-seaward direction, using a sun compass (Schöne & Schöne 1961, Daumer et al. 1963). *Ocypode ceratophthalma* is reported to be capable of returning to its burrow (i.e., homing behaviour) from 100 to 200 m distance (Balss 1955–1956), although Hughes (1966) considered that 30 m is more likely to be the maximum distance at which crabs can relocate a burrow.

Homing is possibly aided by visual cues, such as sediment mounds constructed near the burrow (Hughes 1966). *Ocypode saratan* has been observed building sand mounds in the proximity of burrows and wandering around them in a regular manner, possibly to memorize any landmarks before setting out on longer excursions (Linsenmair 1967). Females of this species use the mounds built by males both as landmarks and for sexual attraction in courtship (Linsenmair 1967, Hartnoll 1969). It is possible that species constructing similar mounds near the burrow (e.g., *Ocypode rotundata*; Cooper 1997) use them as visual cues in orientation. Ghost crabs also employ menotactic (i.e., moving at a constant angle) relocation (i.e., when orientation of the long body axis is at a fixed angle to the stimulus). Under laboratory conditions (within a 50 cm wide circular arena), three of five individuals of *O. ceratophthalma* were able to orient menotactically to the pattern of polarized light when returning to their artificial shelter hole after having inspected a piece of bait (Daumer et al. 1963).

Across-shore distributions

General patterns of dune-to-surf distributions

Ghost crabs occupy a wide band across the dune-beach-surf gradient, extending from the lower intertidal up to 400 m inland (Williams 1965, Grubb 1971, Jones 1972, Fellows 1975, Haig 1984). Although many species have this broad across-shore distribution, there exists considerable spatial variability between species (Figure 5). Some species can occupy the full dune-beach profile (e.g., *Ocypode quadrata* in Central America and Texas; Rodriguez 1963, Powers 1977, Maccarone & Mathews 2007); others inhabit the unvegetated beach only (e.g., *O. fabricii*; Davie 2002); several others occur mainly in the supratidal (e.g., *O. cordimanus*, *O. kuhlii*, *O. ryderi*, and *O. saratan*; Macnae & Kalk 1962, McLachlan et al. 1981, 1998, Vannini & Valmori 1981, Wooldridge et al. 1981, Türkay et al. 1996); and a few species are strictly intertidal (e.g., *O. gaudichaudii* and *O. brevicornis*; Rao 1968, Trott 1998). Spatial variability in the distribution of individuals of the same species from different geographic locations is also evident (Figure 5). Examples of this are *Ocypode ceratophthalma*, *O. rotundata*, and *O. pallidula*, which can be either supratidal or intertidal, depending on locality (Alexander 1976, Lighter 1977, Titgen 1982, Jones 1988, Apel & Türkay 1999, Haragi et al. 2010, Trivedi et al. 2012).

Ghost crabs need to access water regularly (Wolcott 1984). Habitat selection can therefore be driven by the moisture content of the sediment, to the extent that some species change position according to tidal amplitudes (Barrass 1963, Hughes 1966, Fimpel 1975, Vannini 1976, Shuchman & Warburg 1978). Densely vegetated dunes may not be favourable habitats for some species, possibly because escape from predators by running is hindered and because vegetation may impede burrow construction (Henning & Klaassen 1973, Fimpel 1975).

Zonation by species, sex, and size

When several species of ghost crabs occur sympatrically, there is generally some degree of spatial separation in their distribution across the shore. For example, in East Africa, where three species co-occur, *Ocypode ceratophthalma* burrows from below low-water neaps to just above high-water springs, with peak abundance just above the mean tide level and all activity taking place in the intertidal area (Grubb 1971, Jones 1972, Hartnoll 1975, Berry 1976, Haig 1984). *Ocypode madagascariensis* occupies the midbeach region, where it feeds on stranded material (Berry 1976). *Ocypode ryderi* lives higher on the shore from high-water neaps upwards, with greatest abundances around high-water springs (Jones 1972, Hartnoll 1975, Berry 1976, Dye et al. 1981). In the Seychelles, *Ocypode cordimanus* mainly occupies the uppermost 3 m of the beach, whereas *O. ceratophthalma* lives below this zone (Grubb 1971, Haig 1984).

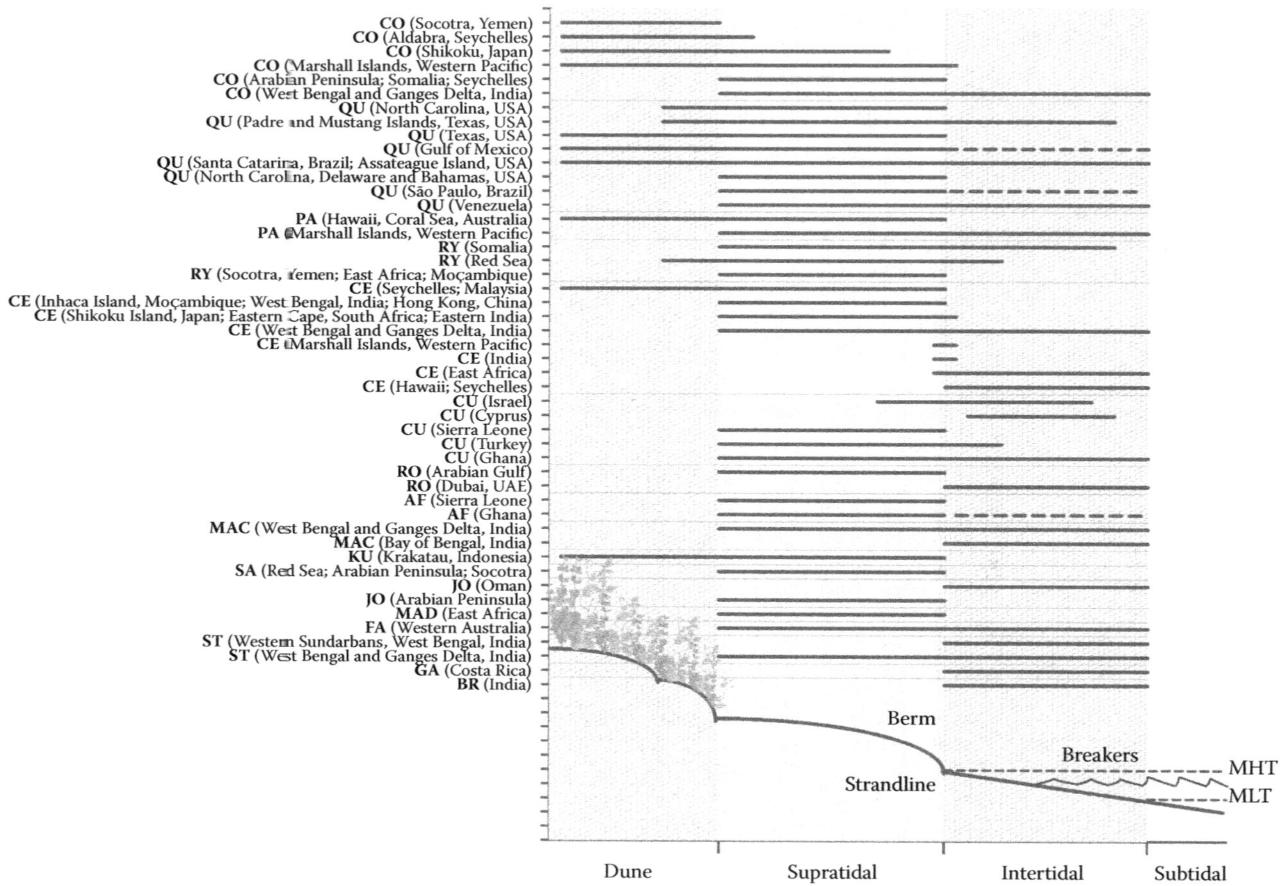

Figure 5 Distribution patterns of ghost crab species across the dune-beach-swash axis. If different distributions are reported for juveniles and adults, juveniles are denoted by a broken line and adults by a solid line. MHT, medium high tide; MLT, medium low tide; CO, *Ocypode cordimanus*; CE, *O. ceratophthalma*; QU, *O. quadrata*; RO, *O. rotundata*; RY, *O. ryderi*; AF, *O. africana*; CU, *O. cursor*; SA, *O. saratan*; MAC, *O. macrocera*; MAD, O. *madagascariensis*; JO, *O. jousseaumei*; ST, *O. stimpsoni*; PA, *O. pallidula*; KU, *O. kuhlii*; BR, *O. brevicornis*; FA, *O. fabricii*; GA, *O. gaudichaudii*.

Differences in distribution between the sexes have been reported for *Ocypode gaudichaudii* in Costa Rica, where males are more abundant on the upper shore, whereas females numerically dominate the lower intertidal, presumably because gravid females require access to the swash while the territorial males defend burrows on the upper beach (Milne & Milne 1946, Wright 1968, Trott 1998).

Juveniles are often found closer to the swash (Gauld 1960, Fisher & Tevesz 1979, Chakrabarti 1981, Simões et al. 2001, Valero-Pacheco et al. 2007, de Araújo et al. 2008, Türeli et al. 2009, Branco et al. 2010). This pattern might reflect territoriality of adults, lower tolerance of smaller individuals to dry conditions prevailing higher on the shore, or possibly cannibalism (Fisher & Tevesz 1979, Chakrabarti 1981, Ewa-Oboho 1993, Barros 2001). There are, however, exceptions (e.g., *Ocypode quadrata*) where juvenile crabs have a broad distribution across the shore, extending into the upper intertidal (Alberto & Fontoura 1999, Clum 2005, Menezes et al. 2007).

Burrows and burrowing

Burrow functions

Fossorial habits—centred on the construction of large and relatively complex burrows—are one of the defining ecological traits of ghost crabs. Burrows have multiple functions, amongst which the provision of refuges from predators and shelter from weather extremes are the most frequently cited (Christoffers 1986, Kristensen et al. 2012). Another function of the burrows is to provide crabs with a moist environment that enables them to take up oxygen and to avoid desiccation during hot conditions. Consequently, the depth of burrows is often governed by sediment moisture (Hayasaka 1935, Williams 1983, Antia 1989, Menezes et al. 2007). Burrows are also important for moulting (Christoffers 1986); sex-specific signalling (i.e., sand mounds are usually built near burrow entrances to attract females during the reproductive season; Linsenmair 1965, 1967, Hartnoll 1969); and egg development (Haley 1973). Limited or no ghost crab activity is usually observed during the winter, leading to the assumption that the aforementioned overwintering behaviour occurs inside burrows (Pearse & Wharton 1938, Williams 1965, Haley 1972, Wolcott 1978, Jackson et al. 1991, T.G. Wolcott unpublished data in Wolcott & Wolcott 1999, Manning & Lindquist 2003).

Although it is recognized that burrows can protect ghost crabs against some types of predators (e.g., birds), it is equally plausible that they present visual cues (e.g., sand mounds) for other predators, such as reptiles and mammals. The provision of such cues would, hypothetically, increase predation risk. Several mammals (baboons, foxes, coyotes, and mongooses) and reptiles (monitor lizards and iguanas) are known to dig ghost crabs from their burrows (Phillips 1940, Messeri 1978, McLachlan 1980, Iwamoto 1986, Polis & Hurd 1996, Rose & Polis 1998, Arndt 1999). Snakes (e.g., yellow rat snake *Coelognathus* sp.) have also been observed entering ghost crab burrows (M.G. Frick personal communication 4 October 2013).

Ghost crab burrows are known to be geologically important: Sites with abundant (i.e., 55–80 individuals m^{-2}) burrows form bioturbated sediment layers up to 2.4 m deep (Specht 1985, De 1998) that are more prone to erosion (De 1998), even resulting in instability of archaeological sites (Specht 1985).

Burrow types

Burrows generally have a funnel-shaped opening, leading to a tunnel that is usually circular or oval in cross section (Duncan 1986, Strachan et al. 1999). The end of the tunnel is either cigar shaped (Braithwaite & Talbot 1972) or inflated to form a chamber (Evans et al. 1976). Ghost crabs construct a wide variety of burrow shapes (Table 5; Figures 6, 7). In order of frequency, burrow shapes resemble the letters Y, J, I, and U (Table 5; Figure 6). Burrow types associated with a specific function are the spiral-shaped (S-shaped) copulation burrows (Vannini 1980b; Figure 6). Copulation burrows

Table 5 Summary of burrow attributes in ghost crabs (*Ocypode* spp. and *Hoplocypode occidentalis*), including burrow distribution across the dune-surf gradient, the relative frequency of different burrow types, and their dimensions

Species	Locality	Across-shore position	M shape	U shape	J shape	I shape	Y shape	Spiral shape	L shape	Notes	Reference
Ocypode (fossil)	Korytnica Basin, Poland	Shallow intertidal		~50%	~50%					Diameter: 4–6 cm	Radwanski (1977)
O. africana	Congo						100%				Rathbun (1921)
O. africana	Nigeria				100%						Bruce-Chwatt & Fitz-John (1951)
O. ceratophthalma	Hawaii							Yes			Fellows (1966)
O. ceratophthalma	Seychelles	Middle and lower beach/below high-tide mark			~30%	~30%	~30%	~10%		Diameter: 1–6 cm; depth: 10–50 cm; sand mound height: 10–15 cm; burrow lifespan: 8–20 h	Braithwaite & Talbot (1972), Brooke (1981)
O. ceratophthalma	Western Sundarbans, India	Upper beach/above strandline		~30%	~40%		~30%			Inclination: 70–80°; sand mound present;	Bakshi et al. (1980)
O. ceratophthalma	Gopalpur, India	Middle and upper beach/below the strandline to the base of the dunes	15%	15%	15%	15%	40%			Depth: up to 100 cm; sloping shoreward	Chakrabarti (1981)
O. ceratophthalma	Hong Kong	Middle beach/above high-water mark			23%	19%	39%	19%		Diameter: 1.5–4.6 cm; inclination: 73–85°; depth: 16–41 cm	Chan et al. (2006)

O. ceratophthalma	Sai Kaew, Thailand			21%	13%		66%				Lim et al. (2011), Yong et al. (2011)
O. ceratophthalma; *O. cordimanus*	Shikoku Island, Japan	Entire beach/below the strandline to dunes	~10%			~45%	~45%			Diameter: 1–4 cm; depth: 10–90 cm; length: up to 140 cm	Seike & Nara (2008)
O. ceratophthalma; *O. cordimanus*	SE Queensland, Australia	Middle and upper beach	3%	2%	11%	32%	50%	1%	1%		Lucrezi & Schlacher (2010)
O. ceratophthalma; *O. cordimanus*; *O. macrocera*; *O. stimpsoni*	Ganges Delta, India	Entire beach	~20%	~20%	~20%	~20%	~20%			Diameter: 3.6–5.1 cm; length: 6–87 cm; inclination: 70–90°; secondary arms in landward direction	De (2005)
O. cordimanus	Seychelles	Middle beach and upper beach/ around and above high-water springs/beach crest								Diameter: 1–6 cm	Braithwaite & Talbot (1972)
O. cordimanus	Sar Uanle, Somalia					100%					Vannini (1980b)
O. cursor	Israel	Lower and middle beach		~20% (closer to sea)			~70%		10%		Shuchman & Warburg (1978)
O. cursor	Northern Cyprus	Lower and middle beach			100%				Yes		Strachan et al. (1999)

Continued

Table 5 (Continued) Summary of burrow attributes in ghost crabs (*Ocypode* spp. and *Hoplocypode occidentalis*), including burrow distribution across the dune-surf gradient, the relative frequency of different burrow types, and their dimensions

Species	Locality	Across-shore position	M shape	U shape	J shape	I shape	Y shape	Spiral shape	L shape	Notes	Reference
O. cursor	Turkey	Supralittoral and upper eulittoral			50%				50%		Türeli et al. (2009)
O. gaudichaudii	Culebra Island, Panama			38% males; 56.7% females	27.8% males; 10% females	16.2% males; 30% females		18% males; 3.3% females		Diameter: 1.5–5.5 cm	Schober & Christy (1993)
O. jousseaumei	Oman	Middle beach/intertidal						Yes			Clayton (2005)
H. occidentalis	Pacific coast of Central America			~35%	~30%		~35%				Peters (1955)
O. pallidula	Hawaii	Middle and upper beach			~28%	~15%	~57%			Seaward orientation decreases with distance from water	Fellows (1966)
O. quadrata	Padre Island and Mustang Island, Texas, USA	Entire beach and foredunes		~25%	~25%	~25%	~25%			Diameter: 1–2.5 cm; length: 10–70°	Hill & Hunter (1973)

O. quadrata	North Carolina, USA	Upper beach and foredunes	~1%	~50%			~49%		Diameter: 2–7.5 cm	Allen & Curran (1974)
O. quadrata	Assateague Island, Maryland-Virginia, USA	Entire beach/swash to dunes		~30%	~30%		~40%			Christoffers (1986)
O. rotundata	Bandar Abbas, Iran				100%				Depth: up to 200 cm	Pretzmann (1975)
O. ryderi	Sar Uanle Beach, Somalia	Entire beach			~10%	~40%	~50%		Depth: up to 50 cm; length: up to 100 cm; openings pointing seawards	Vannini (1980b)
O. saratan	Red Sea, Egypt							100%		Linsenmair (1967)
O. stimpsoni	Japan				31%	23%	46%			Utashiro & Horii (1965)
O. stimpsoni	Western Sundarbans, India	Lower beach/ intertidal zone		~50%		~50%			Inclination: 75–85°; radial grazing marks outside entrance	Bakshi et al. (1980)

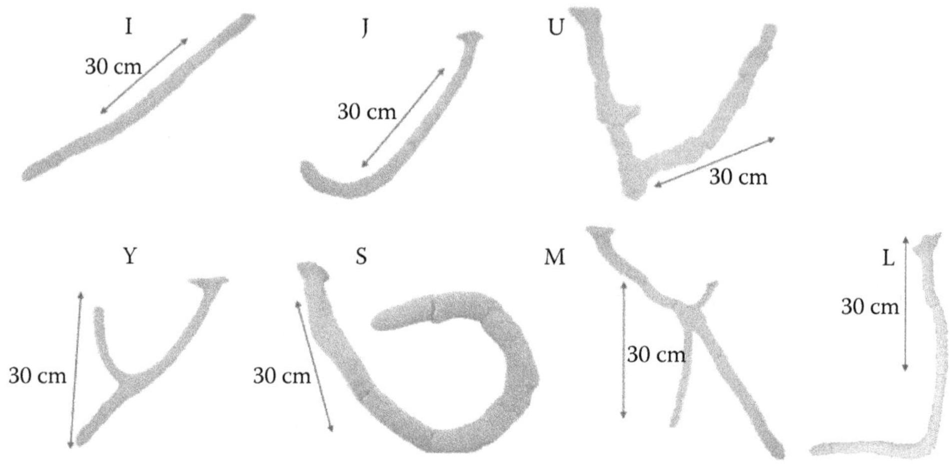

Figure 6 Examples of burrow shapes from casts taken in south-eastern Queensland, Australia (including Fraser Island, the Sunshine Coast, and North Stradbroke Island). The burrows from which the casts were taken were excavated by either *Ocypode ceratophthalma* or *O. cordimanus*, the two species occurring in the region, with exception of the spiral-shaped (S-shaped) copulation burrow, which is typical of *O. ceratophthalma* and not *O. cordimanus*. (Photos by S. Lucrezi.)

Figure 7 (See also colour figure in the insert) Burrows and burrowing in ghost crabs. (A) Burrowing ghost crab (*Ocypode ceratophthalma*), dumping fresh sand outside burrow opening, Sodwana Bay, KwaZulu-Natal (KZN), South Africa; (B) ghost crab burrow opening, surrounded by ghost crab trails, Margate, KZN, South Africa; (C) T.A. Schlacher sitting near plaster cast of a J-shaped burrow kept at its original angle of inclination with respect to the surface sand, Queensland, Australia; (D) plaster cast of a multibranched ghost crab burrow, Fraser Island, Queensland, Australia. (Photos by S. Lucrezi and T.A. Schlacher.)

are excavated exclusively by males of some species during the reproductive season (Hartnoll 1969, Hughes 1973). Rare burrow forms include multibranched (or M-shaped) burrows and L-shaped burrows (Table 5; Figures 6, 7). The former are common in *Ocypode ceratophthalma* (Takahashi 1932) and are usually the product of an abandoned burrow being re-excavated by a new occupant (Chakrabarti 1981). L-shaped burrows are typically constructed by *Ocypode cursor* (Strachan et al. 1999, Türeli et al. 2009). Burrows including Y-shaped and M-shaped types have secondary branches (Figures 6, 7), interpreted to function as escape routes or hiding places from predators (Yong et al. 2011).

Burrows can be wide (opening diameter up to 75 mm) and can reach considerable depths into the sediment (up to 200 cm; Figure 7). While burrow diameter is generally correlated with ghost crab size (Turra et al. 2005, Lucrezi et al. 2009b, Türeli et al. 2009), burrow depth and complexity are not (Chakrabarti 1980, Antia 1989, Clum 2005, Menezes et al. 2007, Seike & Nara 2008). Burrow depth is determined mainly by moisture levels, with burrows becoming deeper in drier conditions (Hayasaka 1935, Williams 1983, Antia 1989, Menezes et al. 2007).

The position of burrows on the shore also influences their complexity and lifespan. Burrows dug close to the swash zone will be less complex and have a short lifespan because of inundation by the tides. Conversely, burrows excavated in the supralittoral zone and in the dunes out of the tides' reach have a longer lifespan. These burrows also tend to be more complex (Duncan 1986, De 2005). Recent work has demonstrated that human disturbance influences burrow complexity and depth, with burrows becoming simpler in form, yet deeper and longer, on beaches subject to vehicle traffic (Lucrezi & Schlacher 2010, Schlacher & Lucrezi 2010a).

Burrow orientation is often seaward near the strandline, with the main shaft sloping away from the shore and the branches extending in a landward direction from the main shaft (Frey & Mayou 1971, Chakrabarti 1981, 1993, Hill 1982). Below the high-tide mark, burrows are short and mainly vertical, whereas burrows in the dunes are aligned normal to the dune face (Frey & Mayou 1971, Hill & Hunter 1973).

Burrow construction

Burrowing takes place mainly at night (Trevallion et al. 1970, Strachan et al. 1999, Valero-Pacheco et al. 2007, Brook et al. 2009). Ghost crabs burrow with the side of the body bearing the smaller chela facing downwards; the larger chela is the first to appear as the crab comes out of a hole after digging (Hughes 1973). Burrowing begins by scraping the sediment with the dactyls of the first to third walking legs on one side. The legs on the opposite side anchor the crab to the ground, and the fourth pair of walking legs supports the body (Savazzi 1985). As it is excavated, the sand is pushed away by the chelipeds past the front of the crab towards the entrance of the burrow (Savazzi 1985). The excavated sand is then thrown out of the burrow without any further action, carried and dumped close to the burrow to form a mound, or dumped and trampled to obliterate digging traces (Crane 1941, Barrass 1963, George & Knott 1965, Grubb 1971; Figure 7A). An individual crab can scoop up to 40 cm^3 of sand per trip (Linsenmair 1967). Burrow walls are flattened with the back of the carapace and the larger chela (Savazzi 1985).

Burrow longevity and homing

Burrow longevity depends mainly on habitat stability driven by tides: Burrows regularly inundated by tides last for considerably shorter periods (1–7 days) than those located above the high-tide mark (>10 days; Hughes 1966, Evans et al. 1976). Whether individuals return to the same burrow, or construct a new one after a bout of activity on the beach surface, depends largely on the location of a burrow. Evans et al. (1976) calculated that over 50% of burrows located in the intertidal zone are abandoned after 24 hours; this suggests that while the ghost crabs are foraging, rising tides destroy the burrows, and crabs will construct a new burrow after the activity. Further up the shore, above the reach of the tides, Hughes (1966) found that all crabs returned to the same burrow over at least 10 days, provided that densities were low. On beaches with higher densities, none of the crabs monitored returned to their burrows following activity (Hughes 1966).

Palaeontology of burrows

Fossilized specimens and burrows can be of palaeo-environmental significance and have been reported from many localities worldwide (Stephenson 1965, Frey & Mayou 1971, Radwanski 1977, Frey et al. 1984, Myint 2001, Portell 2002, Portell et al. 2003, Seike & Nara 2007, Allington-Jones et al. 2010, Garassino et al. 2010, Neto de Carvalho & Baucon 2010). Because ghost crabs construct burrows in semi-consolidated beach sediments, the presence of fossilized burrows is an important tool to trace the position and characteristics of ancient shorelines (Frey 1970, Duncan 1986, Chakrabarti 1993, Nara 1998, Nesbitt & Campbell 2006, Allington-Jones et al. 2010). Species that are known to be deposit-feeders (e.g., *Ocypode ceratophthalma*) also produce sediment pellets or balls that form characteristic radiating patterns near the burrow entrance (De 2000). These pellets are preserved in ancient sediments, thus contributing to the interpretation of palaeo-microenvironments (De 2000).

Trophic ecology

Ghost crabs display remarkable trophic plasticity, occupying a range of trophic levels, obtaining food through a variety of strategies, and consuming a wide diversity of prey items. They are often the apex invertebrate predators on sandy shores while also being preyed on by higher-level vertebrate consumers (e.g., birds, reptiles, mammals; Tables 6–8; Figure 8).

Ghost crabs consume an extraordinarily broad range of organic matter (Table 6). Diet items reported in the literature range from small interstitial diatoms (obtained via sand sifting) to the young of shore-nesting birds. Their diet encompasses a wide range of live animal prey and dead carcasses, which can include both animal tissues and plant material (Table 6; Figure 8).

Feeding modes and diet plasticity

Five modes of feeding are conventionally distinguished in ghost crabs: (1) deposit-feeding, during which crabs separate small organic particles (e.g., diatoms and meiofauna) contained within the sediments; (2) consuming larger, macroscopic plant detritus and litter (e.g., stranded algae, seagrass, seeds); (3) scavenging on animal carcasses cast ashore (Figure 8A, 8C); (4) actively predating by catching, subduing, and killing a wide range of invertebrate and vertebrate prey species (Figure 8B); and (5) cannibalism.

Deposit-feeding generally entails four steps: (1) the crabs take up small amounts of sand from the beach surface; (2) the sand is transferred to the chamber formed by the mouthparts (buccal cavity); (3) small organic particles mixed with the sand are extracted and retained; and (4) the cleaned sand is discarded from the buccal cavity in the form of feeding pellets (Crane 1941, Robertson & Pfeiffer 1982, Trott 1987a). Different species appear to employ deposit-feeding to varying degrees, most likely depending on morphological traits, the abundance and quality of sediment-bound resources, and the availability of alternative prey (e.g., carrion, wrack, invertebrates). In some instances, it can be an important adjunct to predatory behaviour, with crabs extracting up to 70% of benthic microalgae from the sand (Robertson & Pfeiffer 1982). Trott (1987a) showed that *Ocypode gaudichaudii* responded positively, in terms of feeding activity, to experimentally added chemical stimuli in the sand and suggested that contact chemoreception may be used in deposit-feeding to optimize selection of foraging patches.

Table 6 Reported diet items of ghost crabs, illustrating their remarkable trophic plasticity

Feeding mode, broad food item classification, and diet item	Reference
Deposit and detritus feeding	
Benthic microalgae (e.g., diatoms) and meiofauna (e.g., copepods, Foraminifera)	Crane (1939, 1941), Tweedie (1950), Koepcke & Koepcke (1953), Jones (1972), Robertson & Pfeiffer (1982), Trott (1988), Chartosia et al. (2010)
Detritus such as seagrass, algae, fruits, seeds, bird faeces, stranded horse manure, food discards, refuse	Cowles (1908), Koepcke & Koepcke (1953), Williams (1965), Jones (1972), Evans et al. (1976), Vannini (1976), Wolcott (1978), Steiner & Leatherman (1981), Iwamoto (1986), Trott (1988), Wellins et al. (1989), Strachan et al. (1999), Chartosia et al. (2010)
Scavenging on animal carrion	
Carcasses of stranded jellyfish, cows, rats, whales, giant clams, octopus, holothurians, horseshoe crabs, stalked barnacles on drift material, brachyurans, insects, fish, turtles, birds	Cowles (1908), Fowler (1912), Koepcke & Koepcke (1953), Peterson & Fisher (1956), Hirth (1963), Williams (1965), Hughes (1966), Jones (1972), Evans et al. (1976), Fales (1976), Wolcott (1978), McLachlan (1980), Felder (1982), Trott (1988), Wellins et al. (1989), Megyesi & Griffin (1996), Pilcher (1999), Strachan et al. (1999), Burger (2005), Branco et al. (2010)
Active predation	
Invertebrate prey	
Numerous invertebrate taxa, such as wedge clams (*Donax* spp.), gastropods, annelid worms, amphipods, isopods, shrimps, hermit crabs, mole crabs (*Emerita* spp.), box crabs (*Calappa*), fiddler crabs (*Uca* spp.), ghost crabs (cannibalism), hippid crabs, portunid crabs, sentinel crabs (*Macrophthalmus* spp.), pagurids, and various insects (e.g., dipterans, beetles, ants)	Fowler (1912), Koepcke & Koepcke (1953), Hughes (1966), Jones (1972), Smith (1975), Fales (1976), Vannini (1976), Wolcott (1978), Alexander (1979), Lomer (1985), Iwamoto (1986), Wellins et al. (1989), Strachan et al. (1999), Manning & Lindquist (2003), Branco et al. (2010), Chartosia et al. (2010)
Vertebrate prey: turtles and birds	
Hatchlings or eggs of eight turtle species (diamondback terrapin, flatback sea turtle, green sea turtle, hawksbill turtle, Kemp's ridley sea turtle, leatherback turtle, loggerhead sea turtle, olive ridley turtle)	Gibson-Hill (1950), Frazier (1971), Hill & Green (1971), Pritchard and Marquez (1973), Diamond (1976), Fowler (1979), Stancyk (1982), Dodd (1988), Arndt (1991), Conant (1991), Arndt (1994), Eckrich & Owens (1995), Smith et al. (1996), Schmidt & Burney (1998), Bouchard & Bjorndal (2000), Mendoça et al. (2001), Ali and Ibrahim (2002), Simms et al. (2002), Butler et al. (2004), Hitchins et al. (2004), Glen et al. (2005), Caut et al. (2006), Limpus (2007), Barton & Roth (2008), Marschhauser (2010), Mendoça et al. (2010), Tomillo et al. (2010), Rebelo et al. (2012), Peterson et al. (2013)
Eggs or chicks of beach- or dune-nesting birds such as terns, plovers, oystercatchers, tropicbirds, storm petrels, shearwaters	Murphy (1924), Bent (1942), Sprunt (1948), Peterson & Fisher (1956), Blus & Prouty (1979), Jackson & Jackson (1985), Phillips (1987), Loegering et al. (1995), Watts and Bradshaw (1995), Megyesi & Griffin (1996), Rose (2001), Ramos et al. (2005), Sabine et al. (2005, 2006), Feare et al. (2007), Durkin (2012), Schulte (2012)

Table 7 Rates of predation of sea turtle nests, eggs, and hatchlings by ghost crabs (*Ocypode* spp.)

Ghost crab species	Turtle species	Nests disturbed by crabs	Egg/hatchling loss	Locality	References
O. ceratophthalma; *O. kuhlii*	Green turtle (*Chelonia mydas*)	23–50%	0.4–2.7% (eggs)	Malaysia	Ali & Ibrahim (2002)
O. cordimanus	Hawksbill (*Eretmochelys imbricata*)		7.2–38.3% (eggs) 0.5–1.6% (hatchlings)	Seychelles	Hitchins et al. (2004)
O. cursor	Green turtle (*Chelonia mydas*)	1.7%	"Low" (eggs)	Northern Cyprus	Smith et al. (1996)
O. quadrata	Green turtle (*Chelonia mydas*)		11.8% (eggs)	Surinam	Hill & Green (1971)
O. quadrata	Loggerhead (*Caretta caretta*)		8–39% (hatchlings)	North Carolina, USA	Conant (1991)
O. quadrata	Loggerhead (*Caretta caretta*)	2.1%	"Low" (eggs)	Florida, USA	Schmidt & Burney (1998)
O. quadrata	Loggerhead (*Caretta caretta*)	48%	2.5% (eggs)	Florida, USA	Bouchard & Bjorndal (2000)
O. quadrata	Leatherback (*Dermochelys coriacea*)	49%	4.9% (eggs)	French Guiana	Caut et al. (2006)
O. quadrata	Loggerhead (*Caretta caretta*)	50–71%	1–18% (eggs)	Florida, USA	Barton & Roth (2008)
O. quadrata	Loggerhead (*Caretta caretta*)		24% (hatchlings)	North Carolina, USA	Peterson et al. (2013)

Since ghost crabs are omnivores, characterized by a highly flexible and diverse diet spectrum, they also frequently consume non-living plant material deposited on ocean beaches and coastal dunes (Jones 1972; Table 6). Detritus consumed by ghost crabs spans a great variety of types, including, but not limited to, seagrass, algae, terrestrial plant remains, seeds, pods, and fruits (Table 6). Ghost crabs will also take food and other edible refuse discarded by humans (Strachan et al. 1999) and have even been observed feeding on bird faeces and horse manure (Koepcke & Koepcke 1953, Wellins et al. 1989).

A recurring theme in the literature on ghost crabs is that they are efficient, frequent, and common scavengers of animal carcasses on sandy beaches and dunes worldwide (Trott 1988, Schlacher et al. 2013a,b; Table 6; Figure 8A, 8C). Scavenging, like detritus feeding, may occur at any level across the dune-surf gradient but is often concentrated around the strandline near the dunes—a band of flotsam and jetsam where most carcasses that have been cast ashore are deposited (Colombini & Chelazzi 2003, Sikes & Slowik 2010).

The great diversity of carrion consumed by ghost crabs essentially reflects the diversity of this food source on ocean shores and the generalist feeding strategy of the scavengers exploiting it (Table 6). Carrion items that ghost crabs have been observed to forage on include gelatinous zooplankton, clams, cephalopods, sea cucumbers, horseshoe crabs, stalked barnacles on driftwood, several species of brachyurans, insects, stranded fish, sea turtles and their hatchlings, and bird and marine mammal carcasses (Table 6; Figure 8A, 8C).

Ghost crabs can respond rapidly to carrion inputs to the shore and aggregate around carcass falls (Schlacher et al. 2013a,b). Vertebrates (e.g., birds and foxes) can depress this aggregative response of ghost crabs around carcasses. This functional modification of the scavenger response may occur as a result of competition for food, higher predation risk for crabs in the presence of

Table 8 List of taxa reported to consume ghost crabs (*Ocypode* spp.)

Higher taxon	Predator species	Prey species	Locality	References
Crustacea Decapoda[a]	Grapsid crab (*Geograpsus crinipes*)	*O. ceratophthalma*	Aldabra, Seychelles	Alexander (1979)
Osteichthyes	Fishes[b]	*O. ceratophthalma*; *O. ryderi*; *O.madagascariensis*	South Africa	Berry (1976)
	Polla drum (*Umbrina xanti*)	*O. gaudichaudii*	Peru	Koepcke & Koepcke (1953)
	Threadfin (*Polynemus approximans*)	*O. gaudichaudii*	Peru	Koepcke & Koepcke (1953)
Chondrichthyes	Smooth-hound shark (*Mustelus* spp.)	*O. gaudichaudii*	Peru	Koepcke & Koepcke (1953)
Reptilia	Black iguana (*Ctenosaura similis*)	*O. gaudichaudii*	Pacific coast of Costa Rica	Arndt (1999)
	Tropidurid lizard (*Tropidurus peruvianus*)	*O. gaudichaudii*	Peru	Koepcke & Koepcke (1953)
	Coastal goanna (*Varanus panoptes*)	*Ocypode* spp.	Northern Territory, Australia	Blamires (2004)
	Water monitor (*Varanus salvator*)	*O. kuhlii*	Krakatau Islands, Indonesia	Iwamoto (1986)
Aves				
Charadriiformes	Kentish plover (*Charadrius alexandrinus occidentalis*)	*O. gaudichaudii*	Peru	Koepcke & Koepcke (1953)
	Crab plover (*Dromas ardeola*)	*O. ceratophthalma*	Seychelles	Alexander (1979)
	"Shorebirds and seagulls"	*O. quadrata*	Brazil	Blankensteyn (2006)
	"Birds" (gulls)	*O. ceratophthalma*; *O. ryderi*; *O. madagascariensis*	South Africa	Berry (1976)
	Franklin's gull (*Larus pipixcan*)	*O. gaudichaudii*	Peru	Koepcke & Koepcke (1953)
	Great black-backed gull (*Larus marinus*) or kelp gull (*L. dominicanus*)	*O. gaudichaudii*	Peru	Koepcke & Koepcke (1953)
	Belcher's gull (*Larus belcheri*)	*O. gaudichaudii*	Peru	Koepcke & Koepcke (1953)
	Bristle-thighed curlew (*Numensis tahitiensis*)	*O. laevis*; *O. ceratophthalma*	Hawaii	Marks & Hall (1992)
	Sanderling (*Calidris alba*)	*O. gaudichaudii*	Peru	Koepcke & Koepcke (1953)
	Spotted sandpiper (*Actitis macularius*)	*O. gaudichaudii*	Peru	Koepcke & Koepcke (1953)
	Turnstone (*Arenaria interpres*)	*O. ceratophthalma*	Seychelles	Alexander (1979)
	Willet (*Tringa semipalmata*)	*O. gaudichaudii*	Peru	Koepcke & Koepcke (1953)

Continued

Table 8 (Continued) List of taxa reported to consume ghost crabs (*Ocypode* spp.)

Higher taxon	Predator species	Prey species	Locality	References
Falconiformes	Black hawk (*Buteogallus anthracinus*)	*O. gaudichaudii*	Costa Rica	Sherfy (1986)
	Brahminy kite (*Haliastur indus*)	*O. macrocera*	India	Alcock (1902)
	Brahminy kite (*Haliastur indus*)	*Ocypode* spp.	Eastern Australia	Smith (1992)
	"Hawks"	*O. quadrata*	Brazil	Blankensteyn (2006)
Passeriformes	Pied crow (*Corvus alba*)	*Ocypode* spp.	Seychelles	Alexander (1979), Frith (1979)
	Pied wagtail (*Motacilla aguimp*)	*O. ceratophthalma*; *O. ryderi*	Kenya	Vader (1982)
	Common myna (*Acridotheres tristis*)	*O. ceratophthalma*	Singapore	Stevens et al. (2013)
Ciconiiformes	Yellow-crowned night heron (*Nycticorax violacea*)	*O. quadrata*	Bermuda	Wingate (1982)
Strigiformes	Burrowing owl (*Speotyto cunicularia*)	*O. quadrata*	Brazil	Branco et al. (2010)
Mammalia				
Artiodactyla	Domestic pig (*Sus scrofa*)	*O. gaudichaudii*	Peru	Koepcke & Koepcke (1953)
Carnivora				
Canidae	Coyote (*Canis latrans*)	*Ocypode* spp.	Gulf of California, Mexico	Polis & Hurd (1996), Rose & Polis (1998)
	Dingo (*Canis lupus dingo*)	*Ocypode* spp.	NE Australia	Allen et al. (2012a,b)
	Domestic dog (*Canis lupus familiaris*)	*Ocypode* spp.	Fiji	Carlton & Hodder (2003)
	Golden jackal (*Canis aureus*)	*O. macrocera*	India	Alcock (1902)
	Crab-eating fox (*Cerdocyon thous*)	*O. quadrata*	Brazil	Gatti et al. (2006)
	Sechuran fox (*Lycalopex sechurae*)	*O. gaudichaudii*	Peru	Koepcke & Koepcke (1953), Huey (1969), Cossíos (2010)
	Gray fox (*Urocyon cineoargenteus*)	*Ocypode* spp.	Gulf of California, Mexico	Rose & Polis (1998)
	Blanford's fox (*Vulpes cana*)	*O. cordimanus*; *O. jousseaumei*; *O. rotundata*; *O. saratan*	Oman	Mendoça et al. (2010)
	Rüppell's fox (*Vulpes rueppellii sabaea*)	*O. cordimanus*; *O. jousseaumei*; *O. rotundata*; *O. saratan*	Oman	Mendoça et al. (2010)
	Arabian red fox (*Vulpes vulpes arabica*)	O. *cordimanus*; *O. jousseaumei*; *O. rotundata*; *O. saratan*	Oman	Mendoça et al. (2010)
	Red fox (*Vulpes vulpes*)	*Ocypode* spp.	Gulf of California, Mexico	Rose & Polis (1998)

Table 8 (Continued) List of taxa reported to consume ghost crabs (*Ocypode* spp.)

Higher taxon	Predator species	Prey species	Locality	References
Herpestidae	Cape grey mongoose (*Myonax pulverulentius*)	*O. ryderi*	South Africa	McLachlan (1980)
	Small Indian mongoose (*Herpestes auropunctatus auropunctatus*)	*O. ceratophthalma*	Hawaii	La Rivers (1948)
	Yellow mongoose (*Cyncitis pencillata*)	*O. ryderi*	South Africa	McLachlan (1980)
Mephitidae	Hog-nosed skunk (*Conepatus* spp.)	*O. gaudichaudii*	Peru	Koepcke & Koepcke, (1953)
	Striped skunk *(Mephitis mephitis*)	*O. quadrata*	Atlantic coast, USA	Amos (1966)
Procyonidae	Pygmy raccoon (*Procyon pygmaeus*)	*O. quadrata*	Cozumel Island, Mexico	McFadden et al. (2006)
	Crab-eating raccoon (*Procyon cancrivorus*)	*O. quadrata*	Brazil	Gatti et al. (2006)
	Racoon (*Procyon lotor*)	"Crabs"	California, USA	Warrick & Wilcox (1981)
	Racoon (*Procyon lotor*)	*O. quadrata*	Florida, USA	Ratnaswamy (1995), Barton & Roth (2008)
Viverridae	Genets (*Genetta* spp.)	*O. ceratophthalma*; *O. ryderi*; *O. madagascariensis*	South Africa	Berry (1976)
Rodentia	Brown rat (*Rattus norvegicus*)	*O. quadrata*	Atlantic coast of USA	Amos (1966)
	"Rats"	*O. gaudichaudii*	Peru	Koepcke & Koepcke (1953)
Primates	Yellow baboon (*Papio cynocephalus*)	*Ocypode* spp.	East Africa	Messeri (1978)

[a] Excludes cannibalism that occurs in several species of ghost crabs.

[b] Not further specified; predation by fishes in the swash and surf zone is not widely reported, but plausible.

vertebrate consumers, and actual crab predation by birds and mammals that are facultative scavengers (Schlacher et al. 2013a,b).

Ghost crabs use chemosensory hairs on the dactyls to detect volatile odours that signal food, reacting to olfactory cues from animal, plant, and faecal material (Wellins et al. 1989, Rittschof 1992). A broad range of chemical attractants has been identified, including amino acids from live prey, sugars from fresh and decomposing fruits, and amines (cadaverine, putrescine) produced during anaerobic decomposition (Trott & Robertson 1982, 1984, Dunham & Gilchrist 1988, Trott 1999). However, once carcasses have reached an advanced state of decay, they are less attractive to scavengers (Brooke 1981, Trott 1999).

Ghost crabs habitually prey on a great diversity of both invertebrate and vertebrate species on sandy shores and dunes, with more than 40 prey taxa reported in the literature (Table 6). Invertebrates recorded as prey items of ghost crabs include wedge clams (*Donax faba, D. incarnatus, D. variabilis*); gastropods; polychaetes; amphiods (including talitrids); isopods; shrimp; hermit crabs; mole crabs (*Emerita talpoida*); several genera of brachyurans (e.g., *Calappa, Uca, Lupa, Macrophthalmus, Callinectes*); and a variety of insects (Table 6; Figure 8B). Consumption of larger

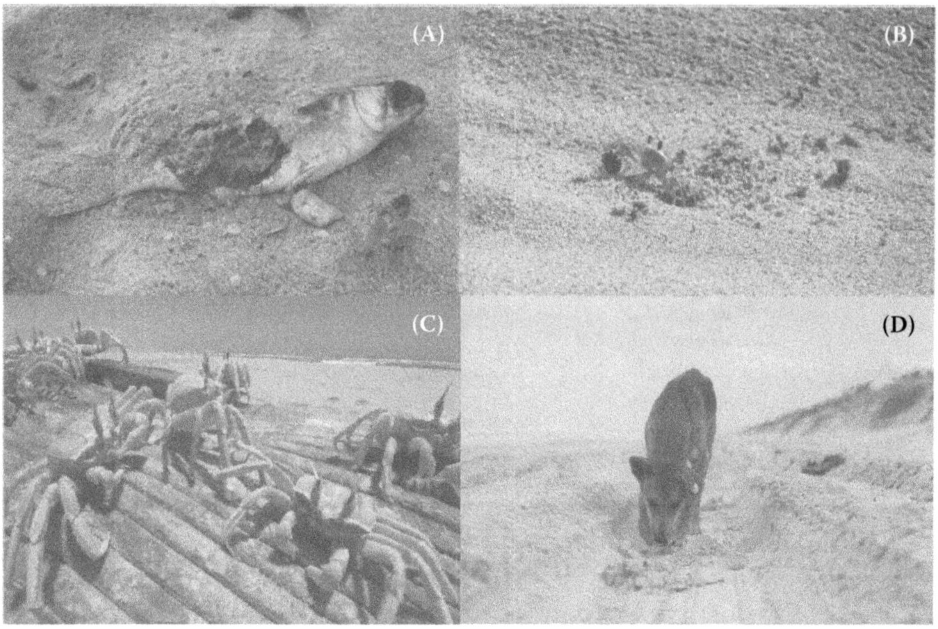

Figure 8 Ghost crabs as consumers and their predators. (A) Fish carcass surrounded by ghost crab feeding trails, Queensland, Australia; (B) ghost crab (*Ocypode ceratophthalma*) holding captured live insect and about to enter a burrow, Margate, KwaZulu-Natal, South Africa; (C) ghost crabs (*O. cursor*) feeding on a whale carcass, Cape Frio, Namibia; (D) dingo (predator of ghost crabs) walking on sandy beach, Fraser Island, Queensland, Australia. (Photos by S. Lucrezi, T.A. Schlacher, and A. Bailey.)

macrobenthos by ghost crabs can be a pivotal energy pathway in beach food webs. Wolcott (1978) calculated that on North Carolina beaches, *Ocypode quadrata* could consume between 58% and 89% of the annual production of the beach clam *Donax variabilis* and crop the majority of *Emerita talpoida* (mole crab) production. In this system, ghost crabs are therefore the major biological vector for the transfer of marine energy into beach and dune food webs.

Cannibalism has been observed in several species of ghost crabs (T.A. Schlacher personal observation). For example, in *Ocypode quadrata* conspecifics can comprise up to 1.5% of the total number of prey items taken (Wolcott 1978). Cannibalism is thought to occur when other, more commonly consumed, food items are rare, and Wolcott (1988) speculated that cannibalism of juveniles may be a mechanism to reduce intraspecific competition.

Predation on vertebrates: sea turtles and birds

Whereas predation on invertebrates is widespread, frequent, and likely to be energetically important (Wolcott 1988), the most publicized and widely cited examples of ghost crab predation concern sea turtles and birds (Fowler 1979, Sabine et al. 2005). This emphasis arises from the charismatic nature of these vertebrates and concerns about their conservation status (Maslo et al. 2011, Witherington et al. 2011). Ghost crabs predate the eggs, hatchlings, or both of all seven extant species of sea turtles and are reported to kill chicks of several bird species that nest on beaches and dunes (Tables 6, 7). Despite a sizeable number of reports about ghost crabs predating eggs and the young of sea turtles and birds (Tables 6, 7), and some support by models and assertions about the possible consequences of ghost crab predation for these vertebrates (e.g., Simms et al. 2002, Yasué & Dearden 2006), there are surprisingly few quantitative estimates of actual consumption rates, the frequency of occurrence, and the likely population consequences (Table 7).

Encounters between ghost crabs and the nests and hatchlings of sea turtles are a common occurrence on turtle-nesting beaches worldwide, and crab predation of hatchlings is unquestionably a source of mortality in turtle populations (Bouchard & Bjorndal 2000, Peterson et al. 2013). Turtle hatchlings are dug up from nests, chased down while crossing the non-vegetated part of the beach, or ambushed at the water's edge (Alexander 1979, Hitchins et al. 2004, Glen et al. 2005, Mendoça et al. 2010, Rebelo et al. 2012). There are, however, surprisingly few quantitative data on the magnitude of egg and hatchling losses that can be unequivocally attributed to ghost crabs, and the published estimates differ considerably (Table 7). In many instances, a sizeable proportion of turtle nests may be detected and exploited by crabs (e.g., Caut et al. 2006), but this does not always translate into large egg losses that are directly attributable to ghost crab predation (Table 7). Data on hatchling loss rates are sparse and span a broad range (0.5–39%; Table 7), so that generalizations about the significance of ghost crab predation are weakly supported by empirical evidence and remain speculative (Table 7).

While there are many records of predation by ghost crabs on birds' eggs and young (Table 6), it is not clear whether these observations represent isolated incidents or whether ghost crabs commonly consume eggs and chicks (Phillips 1987, Loegering et al. 1995). The few studies that recorded quantitative estimates of consumption suggest both infrequent and relatively sizeable predation rates.

In American oystercatchers (*Haematopus palliates*), nesting at Cumberland Island in Georgia, Sabine et al. (2005) recorded a nest failure rate of 56% (18 of 32 nests). However, only a single failure was attributable to a ghost crab predating a freshly hatched chick. This rate was comparable to abiotic or anthropogenic causes (i.e., tidal overwash, trampling, destruction by humans) and to that caused by crows, which all accounted for a single nest failure each; predation by ghost crabs was considerably less than predation by mammals (12 nest failures). Similarly, Watts & Bradshaw (1995) found a single instance of egg/nest predation by ghost crabs in 500 piping plover (*Charadrius melodus*) nests examined. Wolcott & Wolcott (1999) failed to elicit predatory behaviour of ghost crabs encountering quail eggs and chicks (used as plover surrogates), and they did not observe predation in the field. Although plover mortality could not be directly attributed to predation by crabs, ghost crabs may indirectly decrease chick survival by eliciting protective behavioural displays in parents that attract other brood predators or by displacing plover fledglings from food-rich foraging areas (Wolcott & Wolcott 1999).

In contrast to small predatory effects on some birds, in snowy plovers (*Charadrius nivosus*), breeding at Gulf Islands National Seashore (Florida), ghost crabs were the leading cause of nest failure attributable to predators, predating about 15% of nests, followed by coyotes and avian predators at less than 10% (Durkin 2012). Ghost crabs also lowered survival of early chicks in white-tailed tropicbirds (*Phaethon lepturus*) breeding on Aride Island (Seychelles), accounting for 12–24% of annual chick deaths (Ramos et al. 2005).

Consumers of ghost crabs

"When they (ghost crabs) fear some danger, they save themselves by walking sidewise into their burrows with such rapidity that this naturalist was a long time observing them before forming an idea about the species of animal which was fleeing before him; it finally took all the speed of his horse to procure several specimens of them, again after several futile attempts. One knows well that an animal which is so difficult to catch cannot serve commonly as nourishment; thus in Carolina no one makes any use of them" (Bosc 1802). Bosc's equestrian prowess was challenged by ghost crabs to the point that the dashing naturalist's deductions about their vulnerability to predators were wide of the mark: On the contrary, predation of ghost crabs—overwhelmingly by vertebrates—is widespread and common (Table 8). Most predators of ghost crabs have a predominantly terrestrial affinity, using ocean beaches and coastal dunes as foraging sites to various degrees. Thus, predation on ghost crabs may provide an important functional pathway for the spatial coupling of marine and terrestrial ecosystems where predators forage across habitat boundaries at the dune-beach interface.

Reptiles

Several species of lizards—from the families Iguanidae, Tropiduridae, and Varanidae—have been reported to consume ghost crabs. On tropical beaches in Northern Australia, coastal goannas (*Varanus panoptes*) are a major predator of ghost crabs (Blamires 2004). The lizards preferentially select beaches and dunes from amongst a broader habitat mosaic, foraging mostly near the sea. Ghost crabs make a frequent and sizeable dietary contribution (~10% of hard remains in scats) to goannas, which are prey generalists that eat anything they can subdue (Blamires 2004). Similarly, large iguanas were reported to regularly take ghost crabs on a beach in Costa Rica (Arndt 1999).

More generally, the ecological role of reptiles in coastal systems is often underappreciated (Whiting et al. 2007, Davenport 2011). Crocodiles, alligators, and snakes are known to occur and forage on beaches and dunes (Tamarack 1988, Whiting et al. 2007, Lillywhite et al. 2008) and may consume ghost crabs on ocean shores and dunes. For example, on a Georgia barrier island, American alligators (*Alligator mississippiensis*) use the ocean beach as a foraging site, where they take crabs at the surf edge (Tamarack 1988). Circumstantial evidence, such as snake tracks in or near ghost crab burrows (R. Deatsman personal communication 30 October 2010), and observations of juvenile crocodiles chasing ghost crabs on an Australian beach (S.D. Whiting personal communication 4 June 2013) also suggest that reptiles other than lizards are predators of ghost crabs.

Birds

Avian consumers of ghost crabs encompass a broad taxonomic range of species from several orders and families (Table 8). Perhaps not surprisingly, ghost crabs are prey for birds commonly associated with beaches, such as gulls, plovers, curlews, turnstones, and herons. Corvids are also known to take ghost crabs. This has possible conservation significance as there is growing concern about expanding populations of crows in beach and dune habitats, for example, in the Atlantic coast of the USA (Maslo et al. 2012). Top-level avian consumers of ghost crabs are birds of prey. There are at least three published records of kites and hawks hunting ghost crabs, but raptor predation is likely to be more widespread as this group has often been neglected in beach studies (but see Meager et al. 2012, Schlacher et al. 2013a).

Branco et al. (2010) give an interesting account of predation by owls on ghost crabs. On a Brazilian beach, crabs made up a sizeable fraction (30% by volume) of regurgitated feeding pellets of the ground-nesting owl *Speotyto cunicularia*. Because nocturnal activity of ghost crabs is frequently interpreted as an anti-predator strategy (at least in part) and because all other known avian predation events presumably occur by day, owl predation is remarkable and may possibly be more widespread.

Mammals

Mammals are the most widely recorded vertebrate consumers of ghost crabs (Table 8). Mammalian predators of *Ocypode* span a broad taxonomic range, with published records including species of the families Suidae (pigs); Canidae (dogs, wolves, foxes, jackals, coyotes); Herpestidae (mongooses); Mephitidae (skunks); Procyonidae (raccoons); Viverridae (civets and genets); Muridae (rats, mice); and Cercopithecidae (Old World monkeys) as predators.

Eleven species of canids are known to prey on ghost crabs (Table 8). Canid consumers include domestic dogs, coyotes, dingoes (Figure 8D), jackals, crab-eating foxes, and the Sechuran fox (*Lycalopex sechurae*), as well as five species of true foxes (genera *Vulpes* and *Urocyon*). Predation by coyotes (*Canis latrans*) can be intense; Rose & Polis (1998) reported that 106 burrows were excavated over a 1.5-km stretch of beach in one night by a single coyote.

Mammal predation on ghost crabs can be a pivotal mechanism in beach food webs (Prugh et al. 2009). Ghost crabs on Atlantic shores of the USA are functionally mesopredators: Raccoons control populations of the crabs, which prey on lower trophic levels encompassing invertebrates and turtle eggs. The cascading flow-on effects of lowering predation pressure on ghost crabs following raccoon removal are a widely cited example of mesopredator release. Control of raccoons (*Procyon lotor*) in Florida was intended to protect sea turtle eggs but paradoxically resulted in increased predation on the eggs because populations of ghost crabs (*Ocypode quadrata*) were released from top-down control by raccoons (Barton & Roth 2008).

Several of the canids take ghost crabs on the beach surface but will also attempt to dig them out of their burrows. Digging mammalian predators are best represented by mongooses, and there are several reports of them preying on fossorial ghost crabs. Skunks and genets may also dislodge buried crabs (Table 8).

The most spectacular example of a mammalian predator taking ghost crabs involves probable tool use: On East African beaches, yellow baboons (*Papio cynocephalus*) dig ghost crabs from their burrows either by hand or by purposefully using the bones of cuttlefish as a small shovel (Messeri 1978).

Interactions with humans

Ghost crabs have been widely used to assess the effects of human activities on beach and dune biota and ecosystems. Biological and ecological consequences of the following pressures have been measured using ghost crabs as response variables: trampling (Neves & Bemvenuti 2006, Lucrezi et al. 2009a,b); camping (Figure 9C) (Schlacher et al. 2011a); off-road vehicle traffic (Figure 9A, 9D) (Steiner & Leatherman 1981, Moss & McPhee 2006, Foster-Smith et al. 2007, Schlacher et al. 2007, Hobbs et al. 2008, Thompson & Schlacher 2008, Lucrezi & Schlacher 2010, Schlacher & Lucrezi

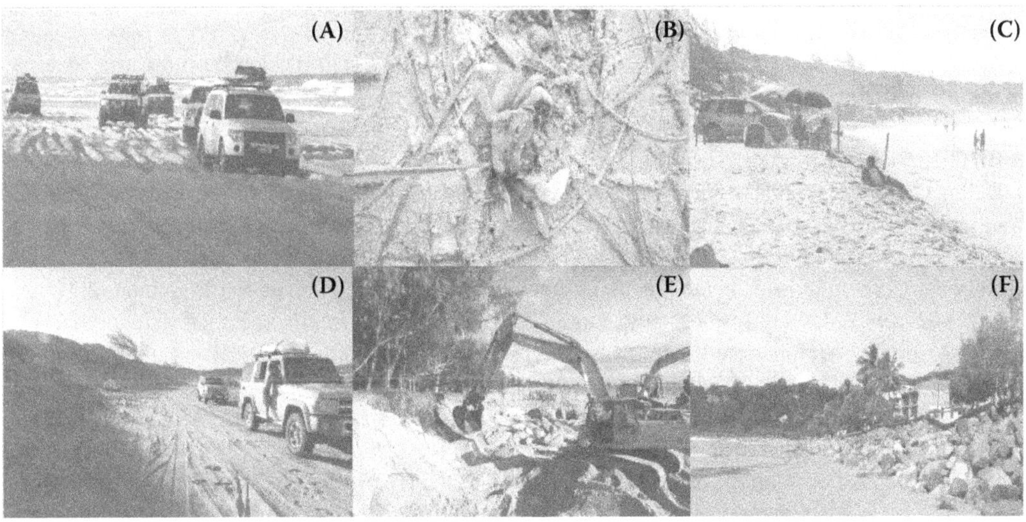

Figure 9 Typical human threats to ghost crabs. (A) Off-road vehicles driving on a beach, Queensland, Australia; (B) ghost crab crushed by off-road vehicle tyres, Queensland, Australia; (C) intense recreation on the beach, Sodwana Bay, KwaZulu-Natal (KZN), South Africa; (D) off-road vehicles departing shore, Sodwana Bay, KZN, South Africa; (E) shore armouring intervention with dozer, Mooloolaba Beach, Queensland, Australia; (F) armoured beach with slurry pump (for beach nourishment) running along rock wall, Noosa, Queensland, Australia. (Photos by S. Lucrezi and T.A. Schlacher.)

2010a,b); mechanical beach cleaning (De Araújo et al. 2008, Noriega et al. 2012); the introduction of alien vegetation (Brook et al. 2009); oil pollution (Schlacher et al. 2011b); nourishment and bulldozing (Figure 9E, 9F) (Peterson et al. 2000); armouring (Barros 2001, Lucrezi et al. 2009b, Lucrezi et al. 2010); and urbanization (Xiang & Jingming 2002, Turra et al. 2005, Souza et al. 2008, Magalhaes et al. 2009, Noriega et al. 2012).

Most impact studies have employed standardized counts of the surface openings of burrows as proxies for population densities and size distributions (e.g., Turra et al. 2005, De Araújo et al. 2008, Hobbs et al. 2008). Metrics other than burrow opening counts include species richness (Brook et al. 2009); direct measurements (as opposed to burrow counts) of densities and abundance (Xiang & Jingming 2002, Schlacher et al. 2011b); mortalities (e.g., by crushing; Schlacher et al. 2007); body condition (Berry 1976, Jackson et al. 1991, Schlacher et al. 2011a); architectural features of burrows (Lucrezi & Schlacher 2010, Schlacher & Lucrezi 2010b); and home ranges and distances travelled during surface activity (Schlacher & Lucrezi 2010b).

Typical responses of ghost crab populations to anthropogenic disturbances include significant reductions in the number and size of ghost crab burrow openings, together with substantial shifts in the distribution of ghost crab burrows across the shore (e.g., away from vehicle tracks, sea walls or infrastructure). Off-road vehicles have been shown to cause considerable mortalities of ghost crabs by directly crushing individuals (Figure 9B) both while the crabs are active on the beach surface and while they are inside their burrows (Schlacher et al. 2007). Other reported responses are reduced species richness in disturbed areas (Brook et al. 2009); external and internal body contamination from oil pollution (Berry 1976, Jackson et al. 1991); higher body condition of crabs in camping sites, presumably as a result of feeding on food scraps (Schlacher et al. 2011a); increased burrow depths and length as a result of vehicle traffic (Lucrezi & Schlacher 2010); and reduced home ranges and distances travelled by ghost crabs on beaches affected by traffic (Schlacher & Lucrezi 2010b).

In the Central Pacific, ghost crabs have cultural significance for indigenous communities of the Tokelau Islands. A traditional fable tells the story of the *kaviki* (*Ocypode ceratophthalma*), whose pursuit to find a wife seems to teach the importance of humility (Yaldwyn & Wodzicki 1979). Also in the Tokelau Islands, local fishers use the burrowing behaviour of the kaviki to make weather predictions (Elders from Atafu Atoll 2012). The fishers believe that if a kaviki piles up excavated sand quite a distance from its burrow, the weather will be fine and calm; if the kaviki piles up the sand right next to its burrow, strong winds and ocean currents are expected (Elders from Atafu Atoll 2012).

Ghost crabs are regularly harvested for food in a number of countries. On the coasts of KwaZulu-Natal and the Eastern Cape in South Africa, ghost crabs have always constituted part of the diet of indigenous communities, and collecting of ghost crabs is a common practice (Kyle et al. 1997). Despite concerns that in protected areas of coastal South Africa, such as the Maputaland Marine Reserve in KwaZulu-Natal, ghost crab numbers would be compromised by subsistence harvesting, research has demonstrated that at low levels of harvesting pressure local populations appear resilient (Kyle et al. 1997). Whilst it appears that subsidence harvesting in this locality is sustainable, this may change with increasing human population densities or changed socioeconomic circumstances in the local communities. In other areas of KwaZulu-Natal, specifically Sodwana Bay and Durban, ghost crab biomass has been shown to be reduced by up to 50% as a result of regular harvesting by local communities (Boon et al. 1999). On the Mediterranean coast of Israel, where all wildlife has legal protection and hunting is strictly regulated, ghost crabs (*Ocypode cursor*) are 'poached' by either gathering them while walking in a line along the beach or by using noose traps (Yom-Tov 2003).

In Brazil, ghost crabs are collected for medicinal purposes; the carapace and bile of *Ocypode quadrata* are used to treat flu, asthma, and fish (niquim) poisoning (Alves & Rosa 2006). Medicinal use of ghost crabs (*Ocypode ceratophthalma*) has also been reported from the Tokelau Islands (Yaldwyn & Wodzicki 1979).

Summary and conclusion

Ghost crabs are among the most conspicuous species on ocean shores and dunes: They are much larger than most other beach invertebrates, are widely distributed from tropical to temperate shores on all inhabited continents, and they construct elaborate burrows with prominent openings on the beach surface. The ecological importance of the group arises primarily from their central role in beach food webs as mesopredators and from their substantial bioturbating activities. Ghost crabs also have close connections with humans, being harvested as natural medicines and food and forming part of the cultural heritage of indigenous people.

Because ghost crabs are quintessential ecotonal species at the land-ocean transition, most biological traits of the group are primarily interpretable as adaptations to a semi-terrestrial mode of life between the sea and the land. Ghost crabs are the fastest crustaceans on land, have generally acute senses, and possess a range of anatomical, physiological, and behavioural characteristics that place them amongst the 'land crabs'. Their great trophic plasticity is remarkable: As consumers, ghost crabs feed on a wide ambit of dietary items, ranging from unicellular algae, obtained by floating feeding, to active predation on the eggs and young of turtles and shorebirds.

Despite the prominence of the group on many sandy shorelines and the long history of associated research, several aspects of their ecology have not been well addressed to date, offering opportunities for future work; some examples include (1) larval biology, dispersal, recruitment phenology, and population connectivity; (2) impacts of ghost crab predation on populations of vertebrate species with threatened conservation status (i.e., turtles, shore-nesting birds); (3) the functional role of ghost crabs as vectors of carbon transfer across ecosystem boundaries and in food webs; (4) ghost crabs as model organisms to gauge the effects of climate change on littoral species and ecosystems (e.g., changes to reproductive phenology, dispersal, survivorship, range expansions, physiological adaptations); and (5) non-lethal responses by ghost crabs to human modifications of beach and dune habitats, offering potentially novel tools to assess environmental pressures on sandy shorelines.

Acknowledgements

Thomas Wolcott shared many of his unique and wide-ranging insights on ghost crab biology with us, and we are appreciative of this. We warmly thank Peter Davie, Charles Peterson, Brooke Maslo, Dave Schoeman, Colin McLay, Mark Costello, Jenifer Dugan, and John McDermott for checking distribution maps, for providing first-hand information on ghost crab distributions and ranges or both. We also extend our gratitude to Melville Saayman and Peet van der Merwe for fully supporting the realization of this project. Sam Masters wielded her merciless editorial axe through the bramble patch of our writing and flushed out many a linguistic and grammatical ghost crab from its semantic burrow. Last, but certainly not least, we warmly thank our colleagues and students who spent many hours with us, in sometimes rather trying conditions, on beaches and dunes in pursuit of the fascinating 'ocean sprites'.

References

Aerts, K., Vanagt, T., Degraer, S., Guartatanga, S., Wittoeck, J., Fockedey, N., Cornejo-Rodriguez, M.P., Calderón, J. & Vincx, M. 2004. Macrofaunal community structure and zonation of an Ecuadorian sandy beach (Bay of Valdivia). *Belgian Journal of Zoology* **134**, 17–24.

Ahn, A., Weinstein, R.B. & Full, R.J. 1991. Increased temperature and water loss decrease locomotor performance of the ghost crab. *American Zoologist* **31**, 141A.

Alberto, R.M.F. & Fontoura, N.F. 1999. Distribução e estrutura etária de *Ocypode quadrata* (Fabricius, 1787) (Crustacea, Decapoda, Ocypodidae) em praia arenosa do litoral sul do Brasil. *Revista Brasileira de Biologia* **59**, 95–108.

Alcock, A.W. 1902. *A Naturalist in Indian Seas*. London: Murray.

Alexander, H.G.L. 1976. *An ecological study of the terrestrial decapod Crustacea of Aldabra.* PhD thesis, University of London, UK.

Alexander, H.G.L. 1979. A preliminary assessment of the role of the terrestrial decapod crustaceans in the Aldabran ecosystem. *Philosophical Transactions of the Royal Society of London B* **286**, 241–246.

Ali, A. & Ibrahim, K. 2002. Crab predation on green turtle (*Chelonia mydas*) eggs incubated on a normal beach and in turtle hatcheries. In *Proceedings of the 3rd Workshop on SEASTAR2000*, N. Arai (ed.). Kyoto, Japan: Graduate School of Informatics, Kyoto University, 95–100.

Allayie, S.A., Ravichandran, S. & Bhat, B.A. 2011. Influence of sinus and mandibular glands on the growth and behaviour of *Ocypode macrocera. Insight Endocrinology and Metabolism* **1**, 7–14.

Allen, B.L., Fleming, P.J.S., Hayward, M., Allen, L.R., Engeman, R.M., Ballard, G. & Leung, L.K.P. 2012a. Top predators as biodiversity regulators: contemporary issues affecting knowledge and management of dingoes in Australia. In *Biodiversity Enrichment in a Diverse World*, G.A. Lameed (ed.). Online. http://www.intechopen.com/books/howtoreference/biodiversity-enrichment-in-a-diverse-world/top-predators-as-biodiversity-regulators-contemporary-issues-affecting-knowledge-and-management-of-d (accessed 20 September 2012).

Allen, E.A. & Curran, H.A. 1974. Biogenic sedimentary structures produced by crabs in lagoon margin and salt marsh environments near Beaufort, North Carolina. *Journal of Sedimentary Petrology* **44**, 538–548.

Allen, L.R., Goullet, M. & Palmer, R. 2012b. The diet of the dingo (*Canis lupus dingo* and hybrids) in north-eastern Australia: a supplement to Brook and Kutt. *The Rangeland Journal* **34**, 211–217.

Allington-Jones, L., Braddy, S.J. & Trueman, C.N. 2010. Palaeoenvironmental implications of the ichnology and geochemistry of the Westbury Formation (Rhaetian), Westbury-on-Severn, South-West England. *Palaeontology* **53**, 491–506.

Al-Solamy, M.A. & Hussein, H.K. 2012. Population dynamics and reproductive aspects of the ghost crab *Ocypode saratan* at Jeddah Coast, Saudi Arabia. *Global Advanced Research Journal of Environmental Science and Toxicology* **1**, 52–59.

Alves, R.R.N. & Rosa, I.L. 2006. From cnidarians to mammals: the use of animals as remedies in fishing communities in NE Brazil. *Journal of Ethnopharmacology* **107**, 259–276.

Al-Wassia, A.H., Innes, A.J., Whiteley, N.M. & Taylor, E.W. 1989. Aerial and aquatic respiration in the ghost crab *Ocypode saratan* I. Fine structure of respiratory surfaces, their ventilation and refusion; oxygen consumption and carbon dioxide production. *Comparative Biochemistry and Physiology A: Molecular and Integrative Physiology* **94**, 755–764.

Amos, W.H. 1966. *The Life of the Seashore*. Toronto: McGraw-Hill.

Antia, E.E. 1989. Beach cusps and burrowing activity of crabs on a fine-grained sandy beach, southeastern Nigeria. *Journal of Coastal Research* **5**, 263–270.

Antunes, G.F., do Amaral, A.P.N., Ribarcki, F.P., de Fátima Willand, E., Zancan, D.M. & Vinagre, A.S. 2010. Seasonal variations in the biochemical composition and reproductive cycle of the ghost crab *Ocypode quadrata* (Fabricius 1787) in southern Brazil. *Journal of Experimental Zoology A: Ecological Genetics and Physiology* **313**, 280–291.

Apel, M. & Türkay, M. 1999. Taxonomic composition, distribution and zoogeographic relationships of the grapsid and ocypodid crab fauna of intertidal soft bottoms in the Arabian Gulf. *Estuarine, Coastal and Shelf Science* **49**, 131–142.

Armstrong, J. 1876. Notes on birds collected in the Rangoon District of the Irrawaddy Delta. In *Stray Feathers. Journal of Ornithology for India and Its Dependencies*, A.O. Hume (ed.). Calcutta: Acton, Vol. 4, 296, 303.

Arndt, R.G. 1991. Predation on hatchling diamondback terrapin, *Malaclemys terrapin* (Schoepff), by the ghost crab, *Ocypode quadrata* (Fabricius). *Florida Scientist* **54**, 215–217.

Arndt, R.G. 1994. Predation on hatchling diamondback terrapin, *Malaclemys terrapin* (Schoepff), by the ghost crab, *Ocypode quadrata* (Fabricius). II. *Florida Scientist* **57**, 1–5.

Arndt, R.G. 1999. Predation by the black iguana (*Ctenosaura similis*) on the painted ghost crab (*Ocypode gaudichaudii*) in Costa Rica. *Florida Scientist* **62**, 111–114.

Ateş, A.S., Kocataş, A., Katağan, T. & Özcan, T. 2010. An updated list of decapod crustaceans on the Turkish coast with a new record of the Mediterranean shrimp, *Processa acutirostris* Nouvel and Holthuis 1957 (Caridea, Processidae). *NorthWestern Journal of Zoology* **6**, 209–217.

Ayres, J.C. 1938. Relationship of habitat to oxygen consumption by certain estuarine crabs. *Ecology* **19**, 523–527.

Bakshi, S.K., Ray, T.K. & De, C. 1980. On the workings of some crabs on the sandy beach of Western Sundarban, Bengal Delta, India. *Journal of the Geological Society of India* **21**, 184–187.

Balss, H. 1955–1956. Decapoda. Ökologie. In *Klassen und Ordnungen des Tierriechs*, H.G. Bronn (ed.). Leipzig, Germany: Winter, Vol. 5, Part 1, 1285–1476.

Barnard, K.H. 1950. Descriptive catalogue of South African decapod Crustacea (crabs and shrimps). *Annals of the South African Museum* **38**, 1–837.

Barrass, R. 1963. The burrows of *Ocypode ceratophthalmus* (Pallas) (Crustacea, Ocypodidae) on a tidal wave beach at Inhaca Island, Moçambique. *Journal of Animal Ecology* **32**, 73–85.

Barros, F. 2001. Ghost crabs as a tool for rapid assessment of human impacts on exposed sandy beaches. *Biological Conservation* **97**, 399–404.

Barton, B.T. & Roth, J.D. 2008. Implications of intraguild predation for sea turtle nest protection. *Biological Conservation* **141**, 2139–2145.

Bent, A.C. 1942. Northern rough-winged swallow. *Smithsonian Institution United States National Museum Bulletin* **179**, 424–433.

Berry, P.P. 1976. Natal's ghost crabs. Victims of the Middle East conflict? *African Wildlife* **30**, 35–37.

Blamires, S.J. 2004. Habitat preferences of coastal goannas (*Varanus panoptes*): are they exploiters of sea turtle nests at Fog Bay, Australia? *Copeia* **2**, 370–377.

Blankensteyn, A. 2006. O uso do caranguejo maria-farinha *Ocypode quadrata* (Fabricius) (Crustacea, Ocypodidae) como indicator de impactos antropogênicos em praias arenosas da Ilha de Santa Catarina, Santa Catarina, Brasil. *Revista Brasileira de Zoologia* **23**, 870–876.

Blickhan, R. & Full, R.J. 1987. Locomotion energetics of the ghost crab II. Mechanics of the centre of mass during walking and running. *Journal of Experimental Biology* **130**, 155–174.

Blickhan, R., Full, R.J. & Ting, L. 1993. Exoskeletal strain: evidence for a trot-gallop transition in rapidly running ghost crabs. *Journal of Experimental Biology* **179**, 301–321.

Bliss, D.E. 1968. Transition from water to land in decapod crustaceans. *American Zoologist* **8**, 355–392.

Blus, L.J. & Prouty, R.M. 1979. Organochlorine pollutants and population status of least tern in South Carolina. *The Wilson Bulletin* **91**, 62–71.

Boon, R., Mann, B.Q. & Robertson, W.D. 1999. Environmental assessment: off-road vehicles at Tongaat Long Beach. Report 1, Wildlife and Environment Society of South Africa. Howick, South Africa: Wildlife and Environment Society of South Africa.

Borradaile, L.A. 1903. Land crustaceans. In *The Fauna and Geography of the Maldive and Laccadive Archipelagos*, J.S. Gardiner (ed.). Cambridge, UK: Cambridge University Press, Vol. 5, 64–100.

Bosc, L.A.G. 1802. *Historie Naturelle des Crustacés, Contenant leur Description et leurs Moeurs, avec Figures Dessinées d'aprés Nature*. Paris: Deterville, Vol. 2, plates 9–18.

Bouchard, S.S. & Bjorndal, K.J. 2000. Sea turtles as biological transporters of nutrients and energy from marine to terrestrial ecosystems. *Ecology* **81**, 2305–2313.

Braithwaite, C.J.R. & Talbot, M.R. 1972. Crustacean burrows in the Seychelles, Indian Ocean. *Palaeogeography, Palaeoclimatology, Palaeoecology* **11**, 265–285.

Branch, G.M., Griffiths, C.L., Branch, M.L. & Beckley, L.E. 1994. *Two Oceans. A Guide to the Marine Life of Southern Africa*. Cape Town: Philip.

Branco, J.O., Hillesheim, J.C., Fracasso, H.A.A., Christoffersen, M.L. & Evangelista, C.L. 2010. Bioecology of the ghost crab *Ocypode quadrata* (Fabricius, 1787) (Crustacea: Brachyura) compared with other intertidal crabs in the southwestern Atlantic. *Journal of Shellfish Research* **29**, 503–512.

Bright, D.B. & Hogue, C.L. 1972. A synopsis of the burrowing land crabs of the world and list of their arthropod symbionts and burrow associates. *Contributions in Science* **220**, 1–58.

Brook, S., Grant, A. & Bell, D. 2009. Can land crabs be used as a rapid ecosystem evaluation tool? A test using distribution and abundance of several genera from the Seychelles. *Acta Oecologica* **35**, 711–719.

Brooke, M.L. 1981. Size as a factor influencing the ownership of copulation burrows by the ghost crab (*Ocypode ceratophthalmus*). *Zeitschrift für Tierpsychologie* **55**, 63–78.

Brown, F.A. 1940. The crustacean sinus gland and chromatophore activation. *Physiological Zoology* **13**, 343–355.

Bruce-Chwatt, L.J. & Fitz-John, R.A. 1951. Mosquitoes in crab-burrows on the coast of West Africa and their control. *Journal of Tropical Medicine and Hygiene* **54**, 116–121.

Brusca, R.C. 1980. *Common Intertidal Invertebrates of the Gulf of California*. Tucson, Arizona: University of Arizona Press.

Buchholz, F. & Vetter, R.A.H. 1993. Enzyme kinetics in cold water: characteristics of N-acetyl-p-u-glucosaminidase in the Antarctic krill, *Euphausia superba,* compared with other crustacean species. *Journal of Comparative Physiology B: Biochemical, Systematic, and Environmental Physiology* **163**, 28–37.

Burger, A.E. 2005. Dispersal and germination of seeds of *Pisonia grandis*, an Indo-Pacific tropical tree associated with insular seabird colonies. *Journal of Tropical Ecology* **21**, 263–271.

Burggren, W.W., Moreira, G.S. & Santos, M.C.F. 1993. Specific dynamic action and the metabolism of brachyuran land crabs *Ocypode quadrata* (Fabricius, 1787), *Goniopsis cruentata* (Latreille, 1803) and *Cardisoma guanhumi* (Latreille, 1825). *Journal of Experimental Marine Biology and Ecology* **169**, 117–130.

Burnett, L.E. 1979. The effects of environmental oxygen levels on the respiratory function of hemocyanin in the crabs *Libinia emarginata* and *Ocypode quadrata. Journal of Experimental Zoology* **210**, 289–300.

Burrows, M. & Hoyle, G. 1973. The mechanism of rapid running in the ghost crab, *Ocypode ceratophthalma. Journal of Experimental Biology* **58**, 327–349.

Butler, J.A., Broadhurst, C., Green, M. & Mullin, Z. 2004. Nesting, nest predation and hatchling emergence of the Carolina Diamondback Terrapin, *Malaclemys terrapin centrata*, in northeastern Florida. *The American Midland Naturalist* **152**, 145–155.

Cameron, J.N. 1981. Brief introduction to the land crabs of the Palau islands: stages in the transition to air breathing. *Journal of Experimental Zoology* **218**, 1–5.

Carlton, J.T. & Hodder, J. 2003. Maritime mammals: terrestrial mammals as consumers in marine intertidal communities. *Marine Ecology Progress Series* **256**, 271–286.

Caut, S., Guirlet, E., Jouquet, P. & Girondot, M. 2006. Influence of nest location and yolkless eggs on the hatching success of leatherback turtle clutches in French Guiana. *Canadian Journal of Zoology* **84**, 908–915.

Chakrabarti, A. 1980. Burrow patterns of ghost crab *Ocypode ceratophthalma* (Pallas) as possible indicators of foreshore slopes. *AAPG Bulletin* **64**, doi: 10.1306/2F918C1B-16CE-11D7-8645000102C1865D

Chakrabarti, A. 1981. Burrow patterns of *Ocypode ceratophthalma* (Pallas) and their environmental significance. *Journal of Palaeontology* **55**, 431–441.

Chakrabarti, A. 1993. *Ocypode* burrows as predictors of ancient shoreline position: new findings from a barred tidal flat. *Indian Journal of Geology* **65**, 15–24.

Chan, B.K.K., Chan, K.K.Y. & Leung, P.C.M. 2006. Burrow architecture of the ghost crab *Ocypode ceratophthalma* on a sandy shore in Hong Kong. *Hydrobiologia* **560**, 43–49.

Chartosia, N., Kitsos, M.S., Tzomos, T.H., Mavromati, E. & Koukouras, A. 2010. Diet composition of five species of crabs (Decapoda, Brachyura) that show a gradual transition from marine to terrestrial life. *Crustaceana* **83**, 1181–1197.

Christoffers, E.W. 1986. *Ecology of the ghost crab* Ocypode quadrata *(Fabricius) on Assateague Island, Maryland and the impacts of various human uses of the beach on their distribution and abundance*. PhD thesis, Michigan State University, East Lansing, MI, USA.

Christy, J.H. 1987. Competitive mating, mate choice and mating associations of brachyuran crabs. *Bulletin of Marine Science* **41**, 177–191.

Clayton, D. 2005. Substrate (acoustic/vibrational) communication and ecology of the ghost crab *Ocypode jousseaumei* (Brachyura: Ocypodidae). *Marine and Freshwater Behaviour and Physiology* **38**, 53–70.

Clayton, D. 2008. Singing and dancing in the ghost crab *Ocypode platytarsus* (Crustacea, Decapoda, Ocypodidae). *Journal of Natural History* **42**, 141–155.

Clum, N. 2005. Ghost crab burrow construction, placement, and longevity. In *10th Symposium of the Natural History of the Bahamas, Abstracts and Program*, S.D. Buckner & T. McGrath (eds). San Salvador: Gerace Research Center, 12–13.

Coll, M., Piroddi, C., Steenbeek, J., Kaschner, K., Lasram, F.B.R., Aguzzi, J., Ballesteros, E., Bianchi, C.N., Corbera, J., Dailianis, T., Donovaro, R., Estrada, M., Froglia, C., Galil, B.S., Gasol, J.M., Gertwagen, R., Gil, J., Guilhaumon, F., Kesner-Reyes, K., Kitsos, M.S., Koukouras, A., Lampadariou, N., Laxamana, E., de la Cuadra, C.M.L.F., Lotze, H.K., Martin, D., Mouillot, D., Oro, D., Raicevich, S., Rius-Barile, J., Saiz-Salinas, J.I., Vicente, C.S., Somot, S., Templado, J., Turon, X., Vafidis, D., Villanueva, R. & Voultsiadou, E. 2013. The biodiversity of the Mediterranean Sea: estimates, patterns, and threats. *PLoS ONE* **5**, e11842. doi: 10.1371/journal.pone.0011842

Colombini, I. & Chelazzi, L. 2003. Influence of marine allochtonous input on sandy beach communities. *Oceanography and Marine Biology: Annual Review* **41**, 114–159.

Conant, T.A. 1991. *Ghost crab predation on emergent sea turtles from relocated nests on a barrier island, North Carolina*. Masters Thesis, Western Washington University, Bellingham, Washington, USA.
Cooper, R.T. 1997. Mangal-associated Brachyura (Ocypodidae, Grapsidae, Portunidae, Majidae, Xanthidae and Leucosiidae) from the north-eastern coastal islands of Abu Dhabi, United Arab Emirates. *Crustaceana* **70**, 155–179.
Cossíos, E.D. 2010. *Lycalopex sechurae* (Carnivora: Canidae). *Mammalian Species* **42**, 1–6.
Cott, H.B. 1929. Observations on the natural history of the racing crab *Ocypode ceratophthalma* from Beira. *Proceedings of the Zoological Society of London* **1929**, 755–765.
Coues, E. 1871. Notes on the natural history of Fort Macon, N.C. and vicinity. *Proceedings of the Academy of Natural Sciences of Philadelphia* **23**, 120–148.
Cowles, R.P. 1908. Habits, reactions and associations in *Ocypoda arenaria*. *Papers from the Tortugas Laboratory of the Carnegie Institution of Washington* **2**, 1–41.
Crane, J. 1939. The painted ghost crabs of Honda Bay. *Bulletin of the New York Zoological Society* **42**, 19–25.
Crane, J. 1940. Eastern Pacific expeditions of the New York Zoological Society. XXVI. On the post-embryonic development of brachyuran crabs of the genus *Ocypode*. *Zoologica* **25**, 65–82.
Crane, J. 1941. Eastern Pacific expeditions of the New York Zoological Society. XXIX. On the growth and ecology of brachyuran crabs of the genus *Ocypode*. *Zoologica* **26**, 297–310.
Dahl, E. 1953. Some aspects of the ecology and zonation of the fauna on sandy beaches. *Oikos* **4**, 1–27.
Dana, J.D. 1852. *Crustacea. United States Exploring Expedition during the years 1838, 1839, 1840, 1841, 1842 under the command of Charles Wilkes, USN*. Philadelphia: C. Sherman, Vol. 13, 58–298.
Da Rosa, L.C. & Borzone, C.A. 2008. Spatial distribution of the *Ocypode quadrata* (Crustacea: Ocypodidae) along estuarine environments in the Paranaguá Bay Complex, southern Brazil. *Revista Brasileira de Zoologia* **25**, 383–388.
Daumer, K., Jander, R. & Waterman, T.H. 1963. Orientation of the ghost-crab *Ocypode* in polarized light. *Zeitschrift für Vergleichende Physiologie* **47**, 56–76.
Davenport, J. 2011. High-trophic level consumers: trophic relationships of reptiles and amphibians of coastal and estuarine ecosystems. In *Treatise on Estuarine and Coastal Science*, E. Wolanski & D.S. McLusky (eds). Elsevier, 227–249.
Davie, P.J.F. 1994. Variations in diversity of mangrove crabs in tropical Australia. *Memoirs of the Queensland Museum* **36**, 55–58.
Davie, P.J.F. 2002. Crustacea: Malacostraca: Eucarida (Part 2). Decapoda – Anomura, Brachyura. In *Zoological Catalogue of Australia*, A. Wells & W.W.K. Houston (eds). CSIRO Publishing, vol. 19.3A, 356–358.
Day, J.H. 1969. *A Guide to Marine Life on South African Shores*. Cape Town, South Africa: A.A. Balkema.
De, C. 1998. Biological reworking of sediments by crabs: a cause for erosion of the Digha Beach, West Bengal. *Current Science* **7**, 617–620.
De, C. 2000. Neoichnological activities of endobenthic invertebrates in downdrift coastal Ganges Delta Complex, India: their significance in trace fossil interpretations and palaeoshoreline reconstructions. *Ichnos* **7**, 89–113.
De, C. 2005. Biophysical model of intertidal beach crab burrowing: application and significance. *Ichnos* **12**, 11–29.
De Araújo, C.C.V., Rosa, D.M. & Fernandes, J.M. 2008. Densidade e distribuição espacial do caranguejo *Ocypode quadrata* (Fabricius, 1787) (Crustacea, Ocypodidae) em três praias arenosas do Espírito Santo, Brasil. *Revista Biotemas* **21**, 73–80.
Demicheli, M., Martínez, A., Ortega, L., Scarabino, F., Maytía, S. & Demicheli, A. 2006. Mass stranding of *Argonauta nodosa* Lightfoot, 1786 (Cephalopoda, Argonautidae) along the Uruguayan coast (southwestern Atlantic). *Revista de Biología Marina y Oceanografía* **41**, 147–153.
Diamond, A.W. 1976. Breeding biology and conservation of Hawksbill turtles, *Eretmochelys imbricata* L., on Cousin Island, Seychelles. *Biological Conservation* **9**, 199–215.
Díaz, H. & Costlow, J.D. 1972. Larval development of *Ocypode quadrata* (Brachyura: Crustacea) under laboratory conditions. *Marine Biology* **15**, 120–131.
Díaz, H. & Rodríguez, G. 1977. The branchial chamber in terrestrial crabs: a comparative study. *The Biological Bulletin* **153**, 485–504.
Dittrich, B. 1990. Temperature dependence of the activities of trypsin-like proteases in decapod crustaceans from different habitats. *Naturwissenschaften* **77**, 491–492.

Dodd, C.K. 1988. Synopsis of the biological data on the loggerhead sea turtle *Caretta caretta* (Linnaeus 1758). U.S. Fish and Wildlife Service Biological Report **88**, 14.

Duncan, G.A. 1986. Burrows of *Ocypode quadrata* (Fabricius) as related to slopes of substrate surfaces. *Journal of Palaeontology* **60**, 384–389.

Dunham, D.W. & Gilchrist, S.L. 1988. Behavior. In *Biology of the Land Crabs*, W.W. Burggren & B.R. McMahon (eds). Cambridge: Cambridge University Press, 97–138.

Durkin, M. 2012. Distribution and dispersal of Snowy Plovers in Florida during the breeding and post-breeding period. Final report to the Edna Sussman Foundation. New York: State University of New York, College of Environmental Science and Forestry.

Dye, A.H., McLachlan, A. & Wooldridge, T. 1981. The ecology of sandy beaches in Natal. *South African Journal of Zoology* **16**, 200–209.

Eckrich, C.E. & Owens, D.W. 1995. Solitary versus arribada nesting in the olive Ridley sea turtles (*Lepidochelys olivacea*): A test of the predator-satiation hypothesis. *Herpetologica* **51**, 349–354.

Elders from Atafu Atoll. 2012. Echoes at Fishermen's Rock: Traditional Tokelau fishing. In *Knowledges of Nature 4*, A. Hooper & I. Tinielu (eds). Paris: UNESCO.

Eshky, A.A. 1985. *Aspects of the ecology, behaviour and physiology of the ghost crab* Ocypode saratan *(Forskål)*. PhD thesis, University of Glasgow, UK.

Eshky, A.A., Al-Wassia, A.H., Atkinson, R.J.A. & Taylor, A.C. 1990. Branchial ventilation in the ghost crab *Ocypode saratan* (Forskål). *Marine and Freshwater Behaviour and Physiology* **16**, 237–248.

Eshky, A.A., Atkinson, R.J.A. & Taylor, A.C. 1988. Effects of temperature on oxygen consumption and heart rate in the semi-terrestrial crab *Ocypode saratan* (Forskål). *Marine and Freshwater Behaviour and Physiology* **13**, 341–358.

Evans, S.M., Cram, A., Eaton, K., Torrance, R. & Wood, V. 1976. Foraging and agonistic behaviour in the ghost crab *Ocypode kuhlii* de Haan. *Marine and Freshwater Behaviour and Physiology* **4**, 121–135.

Ewa-Oboho, I.O. 1993. Substratum preference of the tropical estuarine crabs, *Uca tangeri* Eydoux (Ocypodidae) and *Ocypode cursor* Linne (Ocypodidae). *Hydrobiologia* **271**, 119–127.

Fales, R.R. 1976. Apparent predation on the mole crab *Emerita talpoida* (Say) by the ghost crab *Ocypode quadrata* (Fabricius). *Chesapeake Science* **17**, 65.

Farrelly, C.A. & Greenaway, P. 1994. Gas exchange through the lungs and gills in air-breathing crabs. *Journal of Experimental Biology* **187**, 113–130.

Feare, C.J., Jaquemet, S. & Le Corre, M. 2007. An inventory of Sooty Terns (*Sterna fuscata*) in the western Indian Ocean with special reference to threats and trends. *Ostrich* **78**, 423–434.

Felder, D.L. 1982. A report of the ostracode *Vargula harveyi* Kornicker & King, 1965 (Myodocopida, Cypridinidae) in the southern Bahamas and its implication in luminescence of a ghost crab, *Ocypode quadrata* (Fabricius, 1787). *Crustaceana* **42**, 222–224.

Fellows, D.P. 1966. *Zonation and burrowing behaviour of the ghost crab* Ocypode ceratophthalmus *(Pallas) and* Ocypode laevis *Dana in Hawaii*. Masters Thesis, University of Hawaii, Honolulu, Hawaii, USA.

Fellows, D.P. 1973. Behavioral ecology of the ghost crab *Ocypode ceratophthalmus* and *Ocypode cordimana* at Fanning Atoll, Line Islands. In *Fanning Island Expedition. July and August, 1972*, K.E. Chave & E.A. Kay (eds). Hawaii Institute of Geophysics, University of Hawaii, Honolulu, Hawaii, 219–241.

Fellows, D.P. 1975. On the distribution of the Hawaiian ghost crab, *Ocypode laevis* Dana. *Pacific Science* **29**, 257–258.

Fimpel, E. 1975. Phaenomene der landadaptation bei terrestrischen und semiterrestrischen Brachyura der Brasilianischen Kueste (Malacostraca, Decapoda). *Zoologische Jahrbuecher Systematik* **102**, 173–204.

Fischer, W., Bianchi, G. & Scott, W.B. (eds) 1981. *FAO species identification sheets for fishery purposes. Eastern Central Atlantic; fishing areas 34, 47 (in part)*. Canada Funds-in-Trust, Ottawa: Department of Fisheries and Oceans Canada, by arrangement with the Food and Agriculture Organization of the United Nations, vol. 6, 119–120.

Fisher, J.B. & Tevesz, M.J.S. 1979. Within-habitat spatial patterns of *Ocypode quadrata* (Fabricius) (Decapoda, Brachyura). *Crustaceana* **5S**, 31–36.

Flemister, L.J. 1958. Salt and water anatomy, constancy and regulation in related crabs from marine and terrestrial habitats. *Biological Bulletin* **115**, 180–200.

Flemister, L.J. & Flemister, S.C. 1951. Chloride ion regulation and oxygen consumption in the crab *Ocypode albicans* (Bosq). *Biological Bulletin* **101**, 259–273.

Flemister, S.C. 1959. Histophysiology of gill and kidney of crab *Ocypode albicans*. *Biological Bulletin* **116**, 37–48.

Florey, E. & Hoyle, G. 1976. The effects of temperature on a nerve-muscle system of the Hawaiian ghost crab, *Ocypode ceratophthalma* (Pallas). *Journal of Comparative Physiology A: Neuroethology, Sensory, Natural and Behavioral Physiology* **110**, 51–64.

Foster-Smith, J., Birchenough Cefas, A.C., Evans, S.M. & Prince, J. 2007. Human impacts on Cable Beach, Broome (Western Australia). *Coastal Management* **35**, 181–194.

Fowler, H.W. 1912. The Crustacea of New Jersey. Annual Report of the New Jersey State Museum. Trenton, New Jersey, 29–651.

Fowler, L.E. 1979. Hatching success and nest predation in the green sea turtle, *Chelonia mydas*, at Tortuguero, Costa Rica. *Ecology* **60**, 946–955.

Fransozo, A., Negreiros-Fransozo, M.L. & Bertini, G. 2002. Morphometric study of the ghost crab *Ocypode quadrata* (Fabricius, 1887) (Brachyura, Ocypodidae) from Ubatuba, São Paulo, Brazil. In *Modern Approaches to the Study of Crustacea*, E. Escobar-Briones & F. Alvarez (eds). Proceedings of the Crustacean Society 2000 Summer Meeting, held June 26–30, 2000, in Puerto Vallarta, Mexico. New York: Kluwer Academic/Plenum Publishers, 189–195.

Frazier, J. 1971. Observation on sea turtles at Aldabra Atoll. *Philosophical Transactions of the Royal Society of London. Series B, Biological Sciences* **260**, 373–410.

Frey, R.W. 1970. Environmental significance of recent marine Lebensspuren near Beaufort, North Carolina. *Journal of Palaeontology* **44**, 507–519.

Frey, R.W., Curran, H.A. & Pemberton, S.G. 1984. Tracemaking activities of crabs and their environmental significance: the Ichnogenus *Psilonichnus*. *Journal of Palaeontology* **58**, 333–350.

Frey, R.W. & Mayou, T.V. 1971. Decapod burrows in Holocene barrier island beaches and washover fans, Georgia. *Senckenbergiana Maritima* **3**, 53.

Frith, C.B. 1979. Feeding ecology of land birds of West Island, Aldabra Atoll, Indian Ocean: a preliminary survey. *Philosophical Transactions of the Royal Society of London. Series B, Biological Sciences* **286**, 195–210.

Full, R.J. 1987. Locomotion energetics of the ghost crab. I. Metabolic cost and endurance. *Journal of Experimental Biology* **130**, 137–153.

Full, R.J. & Tu, M.S. 1991. Mechanics of a rapid running insect: two-, four- and six-legged locomotion. *Journal of Experimental Biology* **156**, 215–231.

Full, R.J. & Weinstein, R.B. 1992. Integrating the physiology, mechanics and behavior of rapid running ghost crabs: slow and steady doesn't always win the race. *American Zoologist* **32**, 382–395.

Garassino, A., De Angeli, A., Pasini, G. & Tangocci, F. 2010. *Ocypode italica* n. sp. (Crustacea, Decapoda, Brachyura, Ocypodidae) from the Piacentian (middle Pliocene) of Tuscany (Central Italy). *Atti della Societa' Italiana di Scienze Naturali e del Museo Civico di Storia Naturale di Milano* **151**, 135–144.

Gatti, A., Bianchi, R., Rosa, C.R.X. & Mendes, S.L. 2006. Diet of two sympatric carnivores, *Cerdocyon thous* and *Procyon cancrivorus*, in a restinga area of Espirito Santo State, Brazil. *Journal of Tropical Ecology* **22**, 227–230.

Gauld, D.T. 1960. Brachyura: an annotated check-list of the Crustacea of Ghana, IV. *Journal of the West African Science Association* **6**, 68–72.

Gauld, D.T. & Buchanan, J.B. 1956. The fauna of sandy beaches in the Gold Coast. *Oikos* **7**, 93–301.

George, R.W. 1982. The distribution and evolution of the ghost crabs (*Ocypode* spp.) of Hong Kong with a description of a new species. In *Proceedings of the First International Marine Biological Workshop: The marine flora and fauna of Hong Kong and southern China, Hong Kong, 1980*, B.S. Morton & C.K. Tsang (eds). Hong Kong: Hong Kong University Press, 185–194.

George, R.W. & Knott, M.E. 1965. The *Ocypode* ghost crabs of Western Australia (Crustacea, Brachyura). *Journal of the Royal Society of Western Australia* **48**, 15–21.

Gibson-Hill, C.A. 1950. A note on the reptiles occurring on the Cocos-Keeling Islands. *Bulletin of the Raffles Museum* **22**, 206–211.

Glen, F., Broderick, A.C., Godley, B.J. & Hays, G.C. 2005. Patterns in the emergence of green (*Chelonia mydas*) and loggerhead (*Caretta caretta*) turtle hatchlings from their nests. *Marine Biology* **146**, 1039–1049.

Grant, D. 1983. Ghost crabs at Sandy Hook N.J. *Underwater Naturalist* **14**, 25.

Greenaway, P. 1988. Ion and water balance. In *Biology of the Land Crabs*, W.W. Burggren & B.R. McMahon (eds). New York: Cambridge University Press, 211–248.

Greenaway, P. 1993. Calcium and magnesium balance during molting in land crabs. *Journal of Crustacean Biology* **13**, 191–197.

Greenaway, P. & Farrelly, C.A. 1984. The venous system of the terrestrial crab *Ocypode cordimanus* (Desmarest 1825) with particular reference to the vasculature of the lungs. *Journal of Morphology* **181**, 133–142.

Griffin, D.J.G. 1968. Social and maintenance behaviour in two Australian ocypodid crabs (Crustacea: Brachyura). *Journal of Zoology* **156**, 291–305.

Gross, W.J. 1955. Aspects of osmotic regulation in crabs showing the terrestrial habit. *The American Naturalist* **847**, 205–222.

Grubb, P. 1971. Ecology of terrestrial decapod crustaceans on Aldabra. *Philosophical Transactions of the Royal Society of London* **260**, 411–416.

Guinot-Dumortier, D. & Dumortier, B. 1960. La stridulation chez les crabs. *Crustaceana* **1**, 117–155.

Haddoud, D.A. & El Gomati, H.M. 2011. A general assessment of marine turtle nesting activity on the Lybian saline coast. In *Tasks for Vegetation Science Series 46. Sabkha Ecosystems. Volume 3: Africa and Southern Europe*, M. Öztürk et al. (eds). Springer, 53–58.

Hafemann, D.R. & Hubbard, J.I. 1969. On the rapid running of ghost crabs (*Ocypode ceratophthalma*). *Journal of Experimental Zoology* **170**, 25–32.

Haig, J. 1984. Land and freshwater crabs of the Seychelles and neighbouring islands. In *Biogeography and Ecology of the Seychelles Islands*, D.R. Stoddart (ed.). The Hague: Dr W. Junk Publishers, 123–139.

Haley, S.R. 1969. Relative growth and sexual maturity of the Texas ghost crab, *Ocypode quadrata* (Fabr.) (Brachyura, Ocypodidae). *Crustaceana* **17**, 285–297.

Haley, S.R. 1972. Reproductive cycling in the ghost crab, *Ocypode quadrata* (Fabr.) (Brachyura, Ocypodidae). *Crustaceana* **23**, 1–11.

Haley, S.R. 1973. On the use of morphometric data as a guide to reproductive maturity in the ghost crab, *Ocypode ceratophthalmus* (Pallas) (Brachyura, Ocypodidae). *Pacific Science* **27**, 350–362.

Haragi, S.B., Naik, U.G. & Rathod, J.L. 2010. Brachyuran diversity in sub littoral zone of tropical estuary, Karwar, west coast of India. In Lake: *Wetlands, Biodiversity and Climate Change, 22–24 December 2010*.

Hartnoll, R.G. 1969. Mating in the Brachyura. *Crustaceana* **16**, 161–181.

Hartnoll, R.G. 1974. Variation in growth pattern between some secondary sexual characters in crabs (Decapoda Brachyura). *Crustaceana* **27**, 131–136.

Hartnoll, R.G. 1975. The Grapsidae and Ocypodidae (Decapoda: Brachyura) of Tanzania. *Journal of Zoology* **177**, 305–328.

Hartnoll, R.G. 1988a. Evolution, systematics, and geographical distribution. In *Biology of the Land Crabs*, W.W. Burggren & B.R. McMahon (eds). New York: Cambridge University Press, 6–54.

Hartnoll, R.G. 1988b. Growth and molting. In *Biology of the Land Crabs*, W.W. Burggren & B.R. McMahon (eds). New York: Cambridge University Press, 186–210.

Hayasaka, I. 1935. The burrowing activities of certain crabs and their geological significance. *American Midland Naturalist* **16**, 99–103.

Hayashi, J.H. & Cooke, I.M. 1978. Continuous recording of erythrophore membrane potential during change of stage in *Ocypode pallidula* (Crustacea). *Journal of Experimental Zoology* **204**, 163–169.

Hendrickx, M.E. 1996. Habitats and biodiversity of decapod crustaceans in the SE Gulf of California, Mexico. *Revista de Biologia Tropical* **44**, 603–617.

Henning, H.G. & Klaassen, F. 1973. Dekapode Crustaceen auf der Isla de Salamanca (Atlantik-Kueste, Kolumbien). *Mitteilungen des Instituto Colombo-Alemán Investigaciones Cientıficas* **7**, 63–84.

Henry, R.P. 1994. Morphological, behavioural, and physiological characterization of bimodal breathing crustaceans. *American Zoologist* **34**, 205–215.

Hill, G.W. 1982. Orientation of the ghost crab *Ocypode quadrata* (Fabricius) burrows as an indicator of shoreline position and wind direction. *Texas Journal of Science* **34**, 23–34.

Hill, G.W. & Hunter, R.E. 1973. Burrows of the ghost crab *Ocypode quadrata* (Fabricius) on the barrier islands, south-central Texas coast. *Journal of Sedimentary Petrology* **43**, 24–30.

Hill, R.L. & Green, D.J. 1971. Investigation of the damage by the crab *Ocypode quadrata* to the eggs of the green turtle *Chelonia mydas*. In: *Surinam Turtle Notes* – 2, R.L. Hill & D.J. Green (eds). Paramaribo: Mededeling/Stichting Natuurbehoud Suriname (STINASU), Vol. 2, 11–13.

Hines, A.H. 1986. Larval patterns in the life histories of brachyuran crabs (Crustacea, Decapoda, Brachyura). *Bulletin of Marine Science* **39**, 444–466.

Hirth, H.F. 1963. The ecology of two lizards on a tropical beach. *Ecological Monographs* **2**, 83–112.
Hitchins, P.M., Bourquin, O. & Hitchins, S. 2004. Nesting success of hawksbill turtles (*Eretmochelys imbricata*) on Cousine Island, Seychelles. *Journal of Zoology* **264**, 383–389.
Hobbs, C.H., Landry, C.B. & Perry, J.E. 2008. Assessing anthropogenic and natural impacts on ghost crabs (*Ocypode quadrata*) at Cape Hatteras National Seashore, North Carolina. *Journal of Coastal Research* **24**, 1450–1458.
Holthuis, L.B. 1961. Report on a collection of Crustacea Decapoda and Stomatopoda from Turkey and the Balkans. In *Zoologische Verhandelingen*, E.J. Brill (ed.). Leiden, vol. 47.
Horch, K. 1971. An organ for hearing and vibration sense in the ghost crab *Ocypode*. *Journal of Comparative Physiology A: Neuroethology, Sensory, Natural and Behavioral Physiology* **73**, 1–21.
Horch, K.W. 1974. Barth's myochordotonal organ as an acoustic sensor in the ghost crab, *Ocypode*. *Experientia* **30**, 630–631.
Horch, K.W. 1975. The acoustic behaviour of the ghost crab *Ocypode cordimana* Latreille, 1818 (Decapoda, Brachyura). *Crustaceana* **29**, 193–205.
Horch, K.W. & Salmon, M. 1969. Production, perception and reception of acoustic stimuli by semiterrestrial crabs (genus *Ocypode* and *Uca*, family Ocypodidae). *Forma et Functio* **1**, 1–25.
Huey, R.B. 1969. Winter diet of the Peruvian desert fox. *Ecology* **50**, 1089–1091.
Hughes, D.A. 1966. Behavioural and ecological investigations of the crab *Ocypode ceratophthalmus* (Crustacea: Ocypodidae). *Journal of Zoology* **150**, 129–143.
Hughes, D.A. 1973. On mating and the "copulation burrows" of crabs of the genus *Ocypode* (Decapoda, Brachyura). *Crustaceana* **24**, 72–79.
Imafuku, M., Habu, E. & Nakajima, H. 2001. Analysis of waving and sound-production display in the ghost crab, *Ocypode stimpsoni*. *Marine and Freshwater Behaviour and Physiology* **34**, 197–211.
Iwamoto, T. 1986. Mammals, reptiles and crabs on the Krakatau Islands: their roles in the ecosystem. *Ecological Research* **1**, 249–258.
Jackson, J.A. & Jackson, B.J.S. 1985. Status, dispersion, and population changes of the Least Tern in Coastal Mississippi. *Colonial Waterbirds* **8**, 54–62.
Jackson, L.F., Smale, M.J. & Berry, P.F. 1991. Ghost crabs of the genus *Ocypode* (Decapoda, Brachyura, Ocypodidae) of the east coast of South Africa. *Crustaceana* **61**, 280–286.
Johnson, B.A. 1987. Structure and function of the hemocyanin from a semi-terrestrial crab, *Ocypode quadrata*. *Journal of Comparative Physiology B: Biochemical, Systemic, and Environmental Physiology* **157**, 501–509.
Jones, D.A. 1972. Aspects of the ecology and behaviour of *Ocypode ceratophthalmus* (Pallas) and *O. kuhlii* de Haan (Crustacea: Ocypodidae). *Journal of Experimental Marine Biology and Ecology* **8**, 31–43.
Jones, D.S. 1988. The occurrence of *Ocypode pallidula* Jacquinot (Decapoda, Brachyura) in Australia and the Coral Sea. *Crustaceana* **54**, 33–38.
Kakati, V.S. 2005. Larval development of ghost crab *Ocypode ceratophthalma* (Pallas) under laboratory conditions. In *The Seventh Indian Fisheries Forum Proceedings*, C. Vasudevappa, Y. Basavaraju, D. Seenappa, S. Ayyappan & S.R. Reddy (eds). Mangalore, Karnataka: Asian Fisheries Society Indian Branch, 91–99.
Kazemiyan, M. 2008. Histological assessment of ghost crab's ovary (*Ocypode saratan*) in Chabahar Bay. *Pakistan Journal of Biological Sciences* **11**, 404–409.
Kazemiyan, M. & Hossein, V.G. 2008. Investigation reproduction cycle in *Ocypode saratan* (ghost crab) in sand of Park Station Beach in Chabahar Bay (Oman Sea). *Journal of Veterinary Medicine* **2**, 31–37.
Kensley, B. 1970. Some decapods crustaceans from northern south west Africa, including a new species of *Hippolyte*. *Cimbebasia A* **1**, 179–188.
Khan, S.A. & Ravichandran, S. 2007. *Brachyuran crabs*. Tamil Nadu, India: Centre of Advanced Study in Marine Biology, Annamalai University, 321–338. Online. Available http://ocw.unu.edu/international-network-on-water-environment-and-health/unu-inweh-course-1-mangroves/Brachyuran-crabs.pdf (accessed 1 February 2012).
Koepcke, H.W. & Koepcke, M. 1953. Contributión al concimiento ele la forma de vida de *Ocypode gaudichaudis* Milne Edwards et Lucas (Decapoda, Crust.). *Revista Cientifica* **55**, 157–202.
Koukouras, A., Dounas, C., Türkay, M. & Voultsiadou-Koukoura, E. 1992. Decapod crustacean fauna of the Aegean Sea: new information, check list, affinities. *Senckenbergiana Maritima* **22**, 217–244.
Kristensen, E., Penha-Lopes, G., Delefosse, M., Valdemarsen, T., Quintana, C.O. & Banta, G.T. 2012. What is bioturbation? The need for a precise definition for fauna in aquatic sciences. *Marine Ecology Progress Series* **446**, 285–302.

Kyle, R., Robertson, W.D. & Birnie, S.L. 1997. Subsistence shellfish harvesting in the Maputaland Marine Reserve in northern Kwazulu-Natal, South Africa: sandy beach organisms. *Biological Conservation* **82**, 173–182.

La Rivers, I. 1948. Some Hawaiian ecological notes. *The Wasmann Collector* **7**, 85–110.

Lawrence, J.M. 2003. Ghost crabs (*Ocypode* spp.) of Bird Island, Seychelles. *Phelsuma* **11**, 71–72.

Lazarus-Agudelo, J.F. & Cantera-Kintz, J.R. 2007. Crustáces (Crustacea: Sessilia, Stomatopoda, Isopoda, Amphipoda, Decapoda) de Bahía Málaga, Valle del Cauca (Pacífico colombiano). *Biota Colombiana* **8**, 221–239.

Leber, K.M. 1977. *Seasonal community dynamics of macrobenthos on a high energy sandy beach in North Carolina.* Masters Thesis, East Carolina University, Greenville, North Carolina, USA.

Leber, K.M. 1982. Seasonality of macroinvertebrates on a temperate, high wave energy sandy beach. *Bulletin of Marine Science* **32**, 86–98.

Lighter, F.J. 1974. A note on a behavioral spacing mechanism of the ghost crab *Ocypode ceratophthalmus* (Pallas) (Decapoda, Family Ocypodidae). *Crustaceana* **27**, 312–314.

Lighter, F.J. 1976. The social use of space in the Hawaiian ghost crab, *Ocypode ceratophthalmus. Pacific Science* **30**, 211–212.

Lighter, F.J. 1977. *The social use of space: aspects of ecology, ethology and endocrinology in the ghost crabs* Ocypode ceratophthalmus *(Pallas) and* Ocypode laevis *Dana.* PhD thesis, University of Hawaii, Hawaii, USA.

Lillywhite, H.B., Sheehy, C.M. & Zaidan, F. 2008. Pitviper scavenging at the intertidal zone: an evolutionary scenario for invasion of the sea. *BioScience* **58**, 947–955.

Lim, S.S., Yong, A.Y. & Tantichodok, P. 2011. Comparison of burrow morphology of juvenile and young adult *Ocypode ceratophthalmus* from Sai Kaew, Thailand. *Journal of Crustacean Biology* **31**, 59–65.

Limpus, C.J. 2007. A biological review of Australian Marine Turtles. 5. Flatback Turtle *Natator depressus* (Garman). Brisbane, Australia: Queensland Environmental Protection Agency.

Linsenmair, K.E. 1965. Optische signalisierung der kopulationshöhle bei der reiterkrabbe *Ocypode saratan* Forsk. (Decapoda Brachyura Ocypodidae). *Naturwissenschaften* **52**, 256–257.

Linsenmair, K.E. 1967. Konstruktion und signalfunktion der sandpyramide der reiterkrabbe, *Ocypode saratan* Forsk. (Decapoda, Brachyura, Ocypodidae). *Zeitschrift für Tierpsychologie* **24**, 403–456.

Little, G. 1967. Chromatophore responses in relation to the photoperiod and background color in the Hawaiian ghost crab, *Ocypode ceratophthalma* (Pallas). *Pacific Science* **21**, 77–84.

Loegering, J.P., Fraser, J.D. & Loegering, L.L. 1995. Ghost crab preys on a Piping Plover chick. *Wilson Bulletin* **107**, 768–769.

Lomer, C.J. 1985. Ecology of *Oryctes monoceros* in the Seychelles. *Antenna* **9**, 28–29.

Longhurst, A.R. 1958. An ecological survey of the West African marine benthos. *Colonial Office, Fishery Publications* **11**, 1–102.

Lucrezi, S. & Schlacher, T.A. 2010. Impacts of off-road vehicles (ORVs) on burrow architecture of ghost crabs (genus *Ocypode*) on sandy beaches. *Environmental Management* **45**, 1352–1362.

Lucrezi, S., Schlacher, T.A. & Robinson, W. 2009a. Human disturbance as a cause of bias in ecological indicators for sandy beaches: experimental evidence for the effects of human trampling on ghost crabs (*Ocypode* spp.). *Ecological Indicators* **9**, 913–921.

Lucrezi, S., Schlacher, T.A. & Robinson, W. 2010. Can storms and shore armouring exert additive effects on sandy beach habitats and biota? *Marine and Freshwater Research* **61**, 951–962.

Lucrezi, S., Schlacher, T.A. & Walker S.J. 2009b. Monitoring human impacts on sandy shore ecosystems: a test of ghost crabs (*Ocypode* spp.) as biological indicators on an urban beach. *Environmental Monitoring and Assessment* **152**, 413–424.

Maccarone, A.D. & Mathews, P.L. 2007. Effect of human disturbance on the abundance and spatial distribution of the Atlantic ghost crab (*Ocypode quadrata*) (Fabricius 1798) on a Texas beach. *Texas Journal of Science* **59**, 51–60.

Macnae, W. & Kalk, M. 1962. Fauna and flora of sand flats at Inhaca Island, Mocambique. *Journal of Animal Ecology* **31**, 93–128.

Magalhaes, W.F., Lima, J.B., Barros, F. & Dominguez, J.M.L. 2009. Is *Ocypode quadrata* (Fabricius, 1787) a useful tool for exposed sandy beaches management in Bahia state (northeast Brazil)? *Brazilian Journal of Oceanography* **57**, 149–152.

Manning, L.M. & Lindquist, N. 2003. Helpful habitant or pernicious passenger: interactions between an infaunal bivalve, an epifaunal hydroid and three potential predators. *Oecologia* **134**, 415–422.
Manning, R.B. & Holthuis, L.B. 1981. West African brachyuran crabs (Crustacea: Decapoda). *Smithsonian Contributions to Zoology* **306**, 1–379.
Mariappan, P., Balasundaram, C. & Schmitz, B. 2000. Decapod crustacean chelipeds: an overview. *Journal of Biosciences* **25**, 301–313.
Marks, J.S. & Hall, C.S. 1992. Tool use by Bristle-thighed Curlews feeding on albatross eggs. *The Condor* **94**, 1032–1034.
Marschhauser, S.R. 2010. *Ghost crab* (Ocypode quadrata) *abundance and depredation on Loggerhead sea turtle* (Caretta caretta) *hatchlings on Onslow Beach, North Carolina*. Masters Thesis, North Carolina State University, Raleigh, North Carolina, 1–19.
Maslo, B., Burger, J. & Handel, S.N. 2012. Modeling foraging behavior of Piping Plovers to evaluate habitat restoration success. *Journal of Wildlife Management* **76**, 181–188.
Maslo, B., Handel, S.N. & Pover, T. 2011. Restoring beaches for Atlantic Coast Piping Plovers (*Charadrius melodus*): a classification and regression tree analysis of nest-site selection. *Restoration Ecology* **19**, 194–203.
McDermott, J.J. 2009. Notes on the unusual megalopae of the ghost crab *Ocypode quadrata* and related species (Decapoda: Brachyura: Ocypodidae). *Northeastern Naturalist* **16**, 637–646.
McFadden, K.W., Sambrotto, R.N., Medellín, R.A. & Gompper, E. 2006. Feeding habits of endangered pygmy raccoons (*Procyon pygmaeus*) based on stable isotope and fecal analyses. *Journal of Mammalogy* **87**, 501–509.
McLachlan, A. 1980. Occurrence of ghost crabs *Ocypode* spp. in the Eastern Cape. *South African Journal of Zoology* **15**, 57–58.
McLachlan, A., Martin, F., Al-Habsi, H.N., Al-Shukairi, S.S. & Al-Habsi, A.M. 1998. Ecology of sandy beaches in Oman. *Journal of Coastal Conservation* **4**, 181–190.
McLachlan, A., Wooldridge, T. & Dye, A.H. 1981. The ecology of sandy beaches in southern Africa. *South African Journal of Zoology* **16**, 219–231.
McLay, C.L. 2009. New records of crabs (Decapoda: Brachyura) from the New Zealand region, including a new species of *Rochinia* A. Milne-Edwards, 1875 (Majidae), and a revision of the genus *Dromia* Weber, 1795 (Dromiidae). *Zootaxa* **2111**, 1–66.
McMahon, B.R. & Burggren, W.W. 1988. Respiration. In *Biology of the Land Crabs*, W.W. Burggren & B.R. McMahon (eds). New York: Cambridge University Press, 249–284.
Meager, J.J., Schlacher, T.A. & Nielsen, T. 2012. Humans alter habitat selection of birds on ocean-exposed sandy beaches. *Diversity and Distributions* **18**, 294–306.
Megyesi, J.L. & Griffin, C.R. 1996. Brown Noddy chick predation by Great Frigatebirds in the Northwestern Hawaiian Islands. *The Condor* **98**, 322–327.
Mendoça, V.M., Al-Kiyumi, A.A., Al-Saady, S.M., Grobler, H.J., Erzini, K., Bait Said, A.S. & Al-Rasbiy, A. 2001. Environment of the nesting and feeding grounds for endangered turtle species in Dhofar (Southern Oman). In *Proceedings of the International Conference on Fisheries, Aquaculture and Environment in the NW Indian Ocean*, M. Clareboudt, S. Goddard, H. Al-Oufi & J. McIlwain (eds). Sultan Qaboos University, Muscat, Sultanate of Oman, 151–159.
Mendoça, V.M., Al-Saady, S.M., Al-Kiyumi, A.A. & Erzini, K. 2010. Interactions between green turtles (*Chelonia mydas*) and foxes (*Vulpes vulpes arabica, V. rueppellii sabaea,* and *V. cana*) on turtle nesting grounds in the northwestern Indian Ocean: impacts of the fox community on the behavior of nesting sea turtles at the Ras Al Hadd Turtle Reserve, Oman. *Zoological Studies* **49**, 437–452.
Menezes, C., Paise, G., Levy, G. & Oliveira, L. 2007. Distribuição especial e profundidae de tocal de maria-farinha *Ocypode quadrata* (Crustacea: Decapoda) na praia de Tucuruçá, Cananéia. Projeto Orientado 03.04. In *Livro do curso campo "Ecologia da Mata Atlântica" 2007*, G. Machado, P.I.K.L. Prado & A.A. Oliveira (eds). São Paulo, Brazil: Universidade de São Paulo, 1–3.
Mense, D.J., Posey, M.H., West, T. & Kincheloe, K. 1995. Settlement of brachyuran postlarvae along the North Carolina coast. *Bulletin of Marine Science* **57**, 793–806.
Messeri, P. 1978. Some observations on a littoral troop of yellow baboons. *Monitore Zoologico Italiano Supplement* **12**, 69.
Milne, L.J. & Milne, M.J. 1946. Notes on the behaviour of the ghost crab. *The American Naturalist* **80**, 362–380.

Monod, T. 1956. Hippidea et Brachyura ouest-africains. *Memoirs de L'Institut Francois d'Afrique Noire* **45**, 1–674.

Montgomery, S.K. 1931. Report on the Crustacea Brachyura of the Percy Sladen Trust expedition to the Abrolhos Islands under the Leadership of Professor W.J. Dakin, D.Sc., F.L.S., in 1913; along with other crabs from Western Australia. *Linnean Society Journal— Zoology* **3**, 405–466.

Morris, S. & Bridges, C.R. 1985. An investigation of hemocyanin oxygen affinity in the semi-terrestrial crab *Ocypode saratan* Forsk. *Journal of Experimental Biology* **117**, 119–132.

Morris, S. & Bridges, C.R. 1994. Properties of respiratory pigments in bimodal breathing animals: air and water breathing by fish and crustaceans. *American Zoologist* **34**, 216–228.

Moss, D. & McPhee, D.P. 2006. The impacts of recreational four-wheel driving on the abundance of the ghost crab (*Ocypode cordimanus*) on subtropical sandy beaches in SE Queensland. *Coastal Management* **34**, 133–140.

Murphy, R.C. 1924. The Whitney South Sea Expedition—a sketch of the bird life of Polynesia. *Natural History* **24**, 539–553.

Myint, M. 2001. *Psilonichnus quietis* isp. nov. from the Eocene Iwaki Formation, Shiramizu Group, Joban Coal Field, Japan. *Ichnos* **8**, 1–14.

Nara, M. 1998. The fossil *Psilonichnus* in the Holocene deposit of the Choshi Peninsula, Japan, and its formative environment. In *The 105th Annual Meeting of the Geological Society of Japan*, Vol. 151, Abstracts.

Negreiros-Fransozo, M.L., Fransozo, A. & Bertini, G. 2002. Reproductive cycle and recruitment period of *Ocypode quadrata* (Decapoda, Ocypodidae) at a sandy beach in southeastern Brazil. *Journal of Crustacean Biology* **22**, 157–161.

Nesbitt, E.A. & Campbell, K.A. 2006. The paleoenvironmental significance of *Psilonichnus*. *Palaios* **21**, 187–196.

Neto de Carvalho, C. & Baucon, A. 2010. Jurassic sex in the beach? *Macanopsis* from the Kimmeridgian of Praia do Salgado (Portugal). In *Workshop on Crustacean Bioturbation— Fossil and Recent*, J. de Gibert, F. Muñiz, E. Mayoral & Z. Belaústegui (eds). Lepe, Spain, May 31-June 4, 71–75.

Neves, F.M. & Bemvenuti, C.E. 2006. The ghost crab *Ocypode quadrata* (Fabricius, 1787) as a potential indicator of anthropic impact along the Rio Grande do Sul coast, Brazil. *Biological Conservation* **133**, 431–435.

Norcross-Nu'u, Z. & Brown, D. 2006. Marine monitoring for small-scale beach replenishment at Maalaea, Maui, Hawaii. *Shore and Beach* **74**, 32–41.

Noriega, R., Schlacher, T.A. & Smeuninx, B. 2012. Reductions in ghost crab populations reflect urbanization of beaches and dunes. *Journal of Coastal Research* **28**, 123–131.

Olive, P.J.W. 1992. The adaptive significance of seasonal reproduction in marine invertebrates: the importance of distinguishing between models. *Invertebrate Reproduction and Development* **22**, 165–174.

Olive, P.J.W. 1995. Annual breeding cycles in marine invertebrates and environmental temperature: probing the proximate and ultimate causes of reproductive synchrony. *Journal of Thermal Biology* **20**, 79–90.

Palmer, J.D. 1971. Comparative studies of circadian locomotory rhythms in four species of terrestrial crabs. *The American Midland Naturalist* **85**, 97–107.

Pearse, A.S. 1929. The ecology of certain estuarine crabs at Beaufort, NC. *Journal of the Elisha Mitchell Scientific Society* **44**, 230–237.

Pearse, A.S., Humm, H.J. & Wharton, G.W. 1942. Ecology of sand beaches at Beaufort, NC. *Ecological Monographs* **12**, 135–190.

Pearse, A.S. & Wharton, G.W. 1938. The oyster "Leech," *Stylochus inimicus* Palombi, associated with oysters on the coasts of Florida. *Ecological Monographs* **4**, 605–656.

Perry, M.J., Tait, J., Hu, J., White, S.C. & Medler, S. 2009. Skeletal muscle fiber types in the ghost crab, *Ocypode quadrata*: implications for running performance. *Journal of Experimental Biology* **212**, 673–683.

Peters, H.M. 1955. Die Winkebärde von *Uca* and *Minuca* (Brachyura) in vergleichend-ethologischer, -ökologischer und -morphologisch-anatomischer betrachtung. *Zeitschrift für Morphologie und Ökologie der Tiere* **43**, 425–500.

Peterson, C.H., Bishop, M.J., Johnson, G.A., D'Anna, L.M. & Manning, L.M. 2006. Exploiting beach filling as an unaffordable experiment: benthic intertidal impacts propagating upwards to shorebirds. *Journal of Experimental Marine Biology and Ecology* **338**, 205–221.

Peterson, C.H., Fegley, S.R., Voss, C.M., Marschhauser, S.R. & Van Dusen, B.M. 2013. Conservation implications of density-dependent predation by ghost crabs on hatchling sea turtles running the gauntlet to the sea. *Marine Biology* **160**, 629–640.

Peterson, C.H., Hickerson, D.H.M. & Johnson, G.G. 2000. Short-term consequences of nourishment and bulldozing on the dominant large invertebrates of a sandy beach. *Journal of Coastal Research* **16**, 368–378.

Peterson, R.T. & Fisher, J. 1956. *Wild America*. London: Collins Press.

Phillips, A.M. 1940. The ghost crab—Adventures investigating the life of a curious and interesting creature that lives on our doorstep, the only large crustacean of our North Atlantic coast that passes a good part of his life on land. *Natural History* **46**, 36–41.

Phillips, N.J. 1987. The breeding biology of White-tailed Tropicbirds *Phaethon lepturus* at Cousin Island, Seychelles. *Ibis* **129**, 10–24.

Pilcher, N. 1999. The Hawksbill turtle, *Eretmochelys imbricata*, in the Arabian Gulf. *Chelonian Conservation and Biology* **3**, 312–317.

Polis, G.A. & Hurd, S.D. 1996. Linking marine and terrestrial food web: allochtonous input from the ocean supports high secondary productivity on small islands and coastal land communities. *The American Naturalist* **147**, 396–423.

Popper, A.N., Salmon, M. & Horch, K.W. 2001. Acoustic detection and communication by decapod crustaceans. *Journal of Comparative Physiology A: Neuroethology, Sensory, Natural and Behavioral Physiology* **187**, 83–89.

Portell, R.W. 2002. Occurrence and significance of the ghost crab *Ocypode quadrata* (Fabricius, 1787) from the Upper Pleistocene to Holocene Anastasia Formation of Florida. Paper no. 39 in *North-Central Section (36th) and Southeastern Section (51st), Joint Annual Meeting, April 3–5, 2002*, The Geological Society of America.

Portell, R.W., Turner, R.L. & Beerensson, J.L. 2003. Occurrence of the Atlantic ghost crab *Ocypode quadrata* from the Upper Pleistocene to Holocene Anastasia Formation of Florida. *Journal of Crustacean Biology* **23**, 712–722.

Poupin, J. 2010. Biodiversité de l'Indo-Pacifique tropical français: 2514 espèces de crustacés décapodes et stomatopodes. Rapport scientifique de l'Institut de Recherche de l'Ecole Navale, Octobre 2010.

Poupin, J., Zubia, M., Gravier-Bonnet, N., Chabanet, P. & Malay, M. 2012. *New observations on the crustacean fauna of Europa Island, Mozambique Channel (Crustacea, Decapoda)*. Publication Project for Western Indian Ocean Journal of Marine Science. Online. Available http://decapoda.ecole-navale.fr/pdf/poupin-etal-decapoda-europa-island.pdf (accessed 28 May 2012).

Powers, L.W. 1977. A catalogue and bibliography to the crabs (Brachyura) of the Gulf of Mexico. *Contributions in Marine Science* **20**, 1–190.

Powers, L.W. & Bliss, D.E. 1983. Terrestrial adaptations. In *The Biology of Crustacea*, F.J. Vernberg & W.B. Vernberg (eds). New York: Academic Press, Vol. 8, 271–333.

Pretzmann, G. 1975. Verhaltensstudien an Stranddekapoden bei Bandarabass (Str. v. Hormo). *Anzeiger der Akademie der Wissenshaften in Wien* **2**, 14–18.

Pritchard, P.C.H. & Marquez, R.E.N.E. 1973. Kemp's Ridley Turtle or Atlantic Ridley. IUCN Monograph 2. Morges, Switzerland: International Union for Conservation of Nature and Natural Resources.

Prugh, L.R., Stoner, C.J., Clinton, W.E., Bean, W.T., Ripple, W.J., Laliberte, A.S. & Brashares, J.S. 2009. The rise of the mesopredator. *BioScience* **59**, 779–791.

Radwanski, A. 1977. Burrows attributable to the ghost crab *Ocypode* from the Korytnica basin (Middle Miocene; Holy Cross Mountains, Poland). *Acta Geologica Polonica* **27**, 217–225.

Rafinesque, C.S. 1815. *Analyse de la nature, ou tableau de l'univers et des corps organisés*. Palermo: L'Imprimerie de Jean Barravecchia.

Raja Bai Naidu, K.G. 1951. Some aspects in the development and bionomics of *Ocypoda platytarsis*. *Proceedings of the Indian Academy of Sciences, Section B* **33**, 32–40.

Raja Bai Naidu, K.G. 1954. The post-larval development of the shore crabs *Ocypoda platytarsis* Milne Edwards and *Ocypoda cordimana* Desmarest. *Proceedings of the Indian Academy of Sciences, Section B* **40**, 89–101.

Ramirez-Llodra, E. 2002. Fecundity and life-history strategies in marine invertebrates. *Advances in Marine Biology* **43**, 87–170.

Ramos, J.A., Bowler, J., Betts, M., Pacheco, C., Agombar, J., Bullock, I. & Monticelli, D. 2005. Productivity of White-tailed Tropicbird on Aride Island, Seychelles. *Waterbirds* **28**, 405–410.

Rao, K.R. 1966. Studies on the influence of environmental factors on growth in the crab, *Ocypode macrocera* H. Milne Edwards. *Crustaceana* **11**, 257–276.

Rao, K.R. 1968. The pericardial sacs of *Ocypode* in relation to the conservation of water, molting, and behavior. *American Zoologist* **8**, 561–567.

Rao, K.R. 1969. The effects of temperature on the responses of chromatophores in the crab, *Ocypode macrocera. Journal of Comparative Physiology* **65**, 233–241.

Rathbun, M.J. 1899. Notes on the Crustacea of the Tres Marias Islands. *North American Fauna* **14**, 73–75.

Rathbun, M.J. 1900. Synopses of North-American invertebrates. XI. The Catometopous or grapsoid crabs of North America. *The American Naturalist* **34**, 583–592.

Rathbun, M.J. 1921. The brachyuran crabs collected by the American Museum Congo Expedition, 1909–1915. (Ecological and other notes by Herbert Lang). *Bulletin of the American Museum of Natural History* **43**, 379–468.

Ratnaswamy, M.J. 1995. *Raccoon depredation of sea turtle nests at Canaveral National Seashore, Florida: implications for species management and conservation.* PhD thesis, University of Georgia, Athens, Georgia, USA.

Rebelo, R., Barbosa, C., Granadeiro, J.P., Indjai, B., Novais, B., Rosa, G.M. & Catry, P. 2012. Can leftovers from predators be reliably used to monitor marine turtle hatchling sex-ratios? The implications of prey selection by ghost crabs. *Marine Biology* **159**, 613–620.

Rittschof, D. 1992. Chemosensation in the daily life of crabs. *American Zoologist* **32**, 363–369.

Robertson, J.R. & Pfeiffer, W.J. 1982. Deposit-feeding by the ghost crab *Ocypode quadrata* (Fabricius). *Journal of Experimental Marine Biology and Ecology* **56**, 165–177.

Rodriguez, G. 1963. The intertidal estuarine communities of Lake Maracaibo, Venezuela. *Bulletin of Marine Science* **13**, 197–218.

Rose, A.B. 2001. Predators of nesting little terns *Sterna albifrons* at Foster, New South Wales. *Australian Bird Watcher* **19**, 79–87.

Rose, M.D. & Polis, G.A. 1998. The distribution and abundance of coyotes: the effects of allochthonous food subsidies from the sea. *Ecology* **79**, 998–1007.

Rosenberg, J. & Langer, H. 2001. Ultrastructural changes of rhabdoms of the eyes of *Ocypode* species in relation to different regimes of light and dark adaptation. *Journal of Crustacean Biology* **21**, 345–353.

Sabine, J.B., Meyers, J.M. & Schweitzer, S.H. 2005. A simple, inexpensive video camera setup for the study of avian nest activity. *Journal of Field Ornithology* **76**, 293–297.

Sabine, J.B., Schweitzer, S.H. & Meyers, J.M. 2006. Nest fate and productivity of American Oystercatchers, Cumberland Island National Seashore, Georgia. *Waterbirds* **29**, 308–314.

Saigusa, M., Okochi, T. & Ikei, S. 2003. Nocturnal occurrence, and synchrony with tidal and lunar cycles, in the invertebrate assemblage of a subtropical estuary. *Acta Oecologica* **24**, S191–S204.

Sakai, K. & Türkay, M. 2013. Revision of the genus *Ocypode* with the description of a new genus, *Hoplocypode* (Crustacea: Decapoda: Brachyura). *Memoirs of the Queensland Museum–Nature* **56**, 665–793.

Salmon, M. 1983. Acoustic 'calling' by fiddler and ghost crabs. *Australian Museum Memoir* **18**, 63–76.

Salmon, M., Horch, K.W. & Hyatt, G.W. 1977. Barth's myochordotonal organ as a receptor for auditory and vibrational stimuli in fiddler crabs (*Uca pugilator* and *U. minax*). *Marine and Freshwater Behaviour and Physiology* **4**, 187–194.

Sandon, H. 1937. Differential growth in the crab *Ocypoda. Proceedings of the Zoological Society of London* **107**, 397–414.

Santos, M.C.F. & Moreira, G.S. 1999. Time course of osmoionic compensations to acute salinity exposure in the ghost crab *Ocypode quadrata* (Fabricius, 1787). *Journal of Experimental Marine Biology and Ecology* **235**, 91–104.

Santos, M.C.F., Moreira, G.S. & Brotto, R.O. 1989. Osmo-ionic regulation in the Brazilian ghost crab *Ocypode quadrata* (Fabricius, 1787) (Crustacea, Brachyura). In *Topics in Marine Biology*, J.D. Ros (ed.). *Scientia Marina* **53**, 691–694.

Savazzi, E. 1985. Functional morphology of the cuticular terraces in burrowing terrestrial brachyuran decapods. *Lethaia* **18**, 147–154.

Schenk, S.C. & Wainwright, P.C. 2001. Dimorphism and the functional basis of claw strength in six brachyuran crabs. *Journal of Zoology* **255**, 105–119.

Schlacher, T.A., de Jager, R. & Nielsen, T. 2011a. Vegetation and ghost crabs in coastal dunes as indicators of putative stressors from tourism. *Ecological Indicators* **11**, 284–294.
Schlacher, T.A., Holzheimer, A., Stevens, T. & Rissik, D. 2011b. Impacts of the 'Pacific Adventurer' oil spill on the macrobenthos of subtropical sandy beaches. *Estuaries and Coasts* **34**, 937–949.
Schlacher, T.A. & Lucrezi, S. 2010a. Experimental evidence that vehicle traffic changes burrow architecture and reduces population density of ghost crabs on sandy beaches. *Vie et Milieu—Life and Environment* **60**, 313–320.
Schlacher, T.A. & Lucrezi, S. 2010b. Compression of home ranges in ghost crabs on sandy beaches impacted by vehicle traffic. *Marine Biology* **157**, 2467–2474.
Schlacher, T.A., Strydom, S. & Connolly, R.M. 2013a. Multiple scavengers respond rapidly to pulsed carrion resources at the land-ocean interface. *Acta Oecologica* **48**, 7–12.
Schlacher, T.A., Strydom, S., Connolly, R.M. & Schoeman, D. 2013b. Donor-control of scavenging food webs at the land-ocean interface. *PLoS ONE* **8**, e68221. doi: 10.1371/journal.pone.0068221
Schlacher, T.A., Thompson, L.M.C. & Price, S. 2007. Vehicles versus conservation of invertebrates on sandy beaches: mortalities inflicted by off-road vehicles on ghost crabs. *Marine Ecology-Evolutionary Perspective* **28**, 354–367.
Schmidt, T.S. & Burney, C.M. 1998. Effects of ghost crab (*Ocypode quadrata*) invasion on loggerhead sea turtle (*Caretta caretta*) nests at Hillsboro Beach, Florida. In *Proceedings of the 17th Annual Sea Turtle Symposium, 4–8 March 1997*, S.P. Epperly & J. Braun (eds). Orlando, Florida, 279–280.
Schober, U.M. & Christy, J.H. 1993. Sand disposal of the painted ghost crab *Ocypode gaudichaudii* (Decapoda: Ocypodidae): a possible role in courtship. *Marine Biology* **116**, 53–60.
Schöne, H. 1968. Agonistic and sexual display in aquatic and semi-terrestrial brachyuran crabs. *American Zoologist* **8**, 641–654.
Schöne, H. & Schöne, H. 1961. Eyestalk movements induced by polarized light in the ghost crab, *Ocypode quadrata. Science* **134**, 675–676.
Schulte, S.A. 2012. *Ecology and population dynamics of American Oystercatchers* (Haematopus palliatus). PhD thesis, North Carolina State University, Raleigh, North Carolina, USA.
Seike, K. & Nara, M. 2007. Occurrence of bioglyphs on *Ocypode* crab burrows in a modern sandy beach and its palaeoenvironmental implications. *Palaeogeography, Palaeoclimatology, Palaeoecology* **252**, 458–463.
Seike, K. & Nara, M. 2008. Burrow morphologies of the ghost crabs *Ocypode ceratophthalma* and *O. sinensis* in foreshore, backshore, and dune subenvironments of a sandy beach in Japan. *Journal of the Geological Society of Japan* **11**, 591–596.
Sherfy, J.A. 1986. *Aspects of the foraging behavior of the painted ghost crab* (Ocypode gaudichaudii)*: the influence of density-dependent agonistic interactions, predation and reproductive biology*. PhD thesis, University of California, Berkeley, CA, USA.
Shuchman, E. & Warburg, M.R. 1978. Dispersal, population structure and burrow shape of *Ocypode cursor. Marine Biology* **49**, 255–263.
Sikes, D.S. & Slowik, J. 2010. Terrestrial arthropods of pre- and post-eruption Kasatochi Island, Alaska, 2008–2009: a shift from plant-based to a necromass-based food web. *Arctic, Antarctic, and Alpine Research* **42**, 297–305.
Simms, C.N., Clarke, M. & Campbell, A.C. 2002. Ghost crab predation of loggerhead turtle hatchlings in the Sinai Region of Egypt. In *Proceedings of the 20th Annual Symposium of Sea Turtle Biology and Conservation*, A. Mosier, A. Foley & A.B. Brost (eds). NOAA Technical Memorandum NMFS-SEFC-477, Silver Spring, MD, 202–203.
Simões, N., Apel, M. & Jones, D.A. 2001. Intertidal habitats and decapod faunal assemblages (Crustacea: Decapoda) of Socotra Island, Republic of Yemen. *Hydrobiologia* **449**, 81–97.
Smith, D.A.S. 1975. Polymorphism and selective predation in *Donax faba* Gmelin (Bivalvia: Tellinacea). *Journal of Experimental Marine Biology and Ecology* **17**, 205–219.
Smith, J.I.D. 1992. *The feeding ecology of brahminy kites,* Haliastur indus *from South-east Queensland.* BSc (Hons) thesis. University of Queensland, Brisbane, Australia.
Smith, R.C., Godley, B.J. & Broderick, A.C. 1996. The effect of predation by the ghost crab *Ocypode cursor* on eggs and hatchlings of marine turtles in N. Cyprus. In *Proceedings of the 16th Annual Symposium on Sea Turtle Biology and Conservation*, R. Byles & Y. Fernandez (eds). NOAA Technical Memorandum NMFS-SEFC-412, Silver Spring, MD, 126–127.
Smith, S.I. 1873. The megalops stage of *Ocypoda. American Journal of Science* **6**, 67–68.

Smith, S.I. 1880. Occasional occurrence of tropical and subtropical species of decapod Crustacea on the coast of New England. *Transactions of the Connecticut Academy of Arts and Sciences* **4**, 254–267.

Souza, J.R.B., Lavoie, N., Bonifàcio, P.H. & da Rocha, C.M.C. 2008. Distribution of *Ocypode quadrata* (Fabricius, 1787) on sandy beaches of northeastern Brazil. *Atlântica, Rio Grande* **30**, 139–145.

Spagna, J.C., Goldman, D.I., Lin, P.-C., Koditschek, D.E. & Full, R.J. 2007. Distributed mechanical feedback in arthropods and robots simplifies control of rapid running on challenging terrain. *Bioinspiration and Biomimetics* **2**, 9–18.

Specht, J. 1985. Crabs as disturbance factors in tropical archaeological sites. *Australian Archaeology* **21**, 11–18.

Sprunt, A., Jr. 1948. The tern colonies of the Dry Tortugas Keys. *Auk* **65**, 1–19.

Stancyk, S.E. 1982. Non human predators of sea turtles and their control. In *Biology and Conservation of Sea Turtles*, K.A. Bjorndal (ed.). Washington: Smithsonian Institution Press, 139–152.

Steiner, A.J. & Leatherman, S.P. 1981. Recreational impacts on the distribution of ghost crab *Ocypode quadrata* Fab. *Biological Conservation* **20**, 111–122.

Stephenson, D.G. 1965. Fossil burrows on the coast of Kenya. *Nature* **207**, 850–851.

Stevens, M., Rong, C.P. & Todd, P.A. 2013. Colour change and camouflage in the horned ghost crab *Ocypode ceratophthalmus*. *Biological Journal of the Linnean Society* **109**, 257–270.

Storch, V. & Welsch, U. 1975. Über bau und function der keimen und lungen von *Ocypode ceratophthalma* (Decapoda: Crustacea). *Marine Biology* **29**, 363–371.

Strachan, P.H., Smith, R.C., Hamilton, D.A.B., Taylor, A.C. & Atkinson, R.J.A. 1999. Studies on the ecology and behaviour of the ghost crab, *Ocypode cursor* (L.) in northern Cyprus. *Scientia Marina* **63**, 51–60.

Suvatti, C. 1950. *Fauna of Thailand*. Bangkok, Thailand: Department of Fisheries, 2: 1–1100.

Takahashi, S. 1932. On the burrows of *Ocypoda ceralophthalma* Fabricius. *Kwagaku* **2**, 329–335.

Takeda, S., Matsumasa, M., Kikuchi, S., Poovachiranon, S. & Murai, M. 1996. Variation in the branchial formula of semiterrestrial crabs (Decapoda: Brachyura: Grapsidae and Ocypodidae) in relation to physiological adaptations to the environment. *Journal of Crustacean Biology* **16**, 472–486.

Tamarack, J.L. 1988. Georgia's coastal island alligators, variation in habitat and prey availability. In *Proceedings of the 8th Working Meeting of the Crocodile Specialist Group*. Gland, Switzerland: IUCN Publications, 105–118.

Thompson, L.M.C. & Schlacher, T.A. 2008. Physical damage to coastal dunes and ecological impacts caused by vehicle tracks associated with beach camping on sandy shores: a case study from Fraser Island, Australia. *Journal of Coastal Conservation* **12**, 67–82.

Titgen, R.H. 1982. *The systematics and ecology of the decapods of Dubai, and their zoogeographic relationships to the Arabian Gulf and the Western Indian Ocean*. PhD thesis, Texas A & M University, Texas, USA.

Tomillo, P.S., Paladino, F.V., Suss, J.S. & Spotila, J.R. 2010. Predation of leatherback turtle hatchlings during the crawl to the water. *Chelonian Conservation and Biology* **9**, 18–25.

Trevallion, A., Ansell, A.D., Sivadas, P. & Narayanan, B. 1970. A preliminary account of two sandy beaches in South-West India. *Marine Biology* **6**, 268–279.

Trivedi, J.N., Gadhavi, M.K. & Vachhrajani, K.D. 2012. Diversity and habitat preference of brachyuran crabs in the Gulf of Kutch, Gujarat, India. *Arthropods* **1**, 13–23.

Trott, T.J. 1987a. Chemoreception in the painted ghost crab *Ocypode gaudichaudii* H. Milne Edwards and Lucas (Brachyura: Ocypodidae): Implications for foraging. *Zoologischer Anzeiger* **218**, 295–303.

Trott, T.J. 1987b. The prevalence of left-handedness in the painted ghost crab *Ocypode gaudichaudii* H. Milne Edwards & Lucas (Decapoda Brachyura, Ocypodidae). *Crustaceana* **52**, 213–215.

Trott, T.J. 1988. Note on the foraging activities of the painted ghost crab *Ocypode gaudichaudii* H. Milne Edwards & Lucas in Costa Rica (Decapoda, Brachyura). *Crustaceana* **55**, 217–219.

Trott, T.J. 1998. On the sex ratio of the painted ghost crab *Ocypode gaudichaudii* H. Milne Edwards & Lucas, 1843 (Brachyura, Ocypodidae). *Crustaceana* **71**, 47–56.

Trott, T.J. 1999. Gustatory responses of ghost crab *Ocypode quadrata* to seawater extracts and chemical fractions of natural stimuli. *Journal of Chemical Ecology* **25**, 375–388.

Trott, T.J. & Robertson, J.R. 1982. Chemoreception and feeding stimulants of the ghost crab *Ocypode quadrata* (Fabricius). *American Zoologist* **22**, 853.

Trott, T.J. & Robertson, J.R. 1984. Chemical stimulants of cheliped flexion behaviour by the Western Atlantic ghost crab *Ocypode quadrata* (Fabricius). *Journal of Experimental Marine Biology and Ecology* **78**, 237–252.

Tsai, J.R. & Lin, H.C. 2012. A shift to an ion regulatory role by gills of a semi-terrestrial crab, *Ocypode stimpsoni*. *Zoological Studies* **51**, 606–618.

Türeli, C., Duysak, Ö., Akamca, E. & Kiyagi, V. 2009. Spatial distribution and activity pattern of the ghost crab, *Ocypode cursor* (L., 1758) in Yumurtalik Bay, North-Eastern Mediterranean-Turkey. *Journal of Animal and Veterinary Advances* **8**, 165–171.

Türkay, M., Sakai, K. & Apel, M. 1996. The *Ocypode* ghost crabs (Crustacea: Decapoda: Brachyura) of the Arabian Peninsula and adjacent regions. *Fauna of Saudi Arabia* **15**, 99–117.

Turra, A., Gonçalves, M.A.O. & Denadai, M.R. 2005. Spatial distribution of the ghost crab *Ocypode quadrata* in low-energy tide-dominated sandy beaches. *Journal of Natural History* **39**, 2163–2177.

Tweedie, M.W.F. 1950. Notes on grapsoid crabs from the Raffles Museum. *Bulletin of the Raffles Museum* **23**, 310–324.

Utashiro, M. & Horii, Y. 1965. Some knowledges on the ecology of *Ocypode stimpsoni* Ortmann and on its burrows—Biological study of "Lebensspuren" Part 6. *Memoirs, Takada Branch, Faculty of Education, Niigata University* **9**, 121–141.

Vader, W. 1982. Pied Wagtails catching young ghost crabs. *Ostrich* **53**, 205.

Valero-Pacheco, E., Alvarez, F., Abarca-Arenas, L.G. & Escobar, M. 2007. Population density and activity pattern of the ghost crab, *Ocypode quadrata*, in Veracruz, Mexico. *Crustaceana* **80**, 313–325.

Vannini, M. 1975. Researches on the coast of Somalia. The shore and the dune of Sar Uanle. 5. Description and rhythmicity of digging behaviour in *Caenobita rugosus*. *Monitore Zoologico Italiano Supplement* **6**, 233–242.

Vannini, M. 1976. Researches on the coast of Somalia. The shore and the dune of Sar Uanle. 10. Sandy beach decapods. *Monitore Zoologico Italiano Supplement* **8**, 255–286.

Vannini, M. 1980a. Notes on the behaviour of *Ocypode ryderi* Kingsley (Crustacea, Brachyura). *Marine and Freshwater Behaviour and Physiology* **7**, 171–183.

Vannini, M. 1980b. Researches on the coast of Somalia. The shore and the dune of Sar Uanle. 27. Burrows and digging behaviour in *Ocypode* and other crabs (Crustacea Brachyura). *Monitore Zoologico Italiano Supplement* **13**, 12–14.

Vannini, M. & Valmori, P. 1981. Researches on the coast of Somalia. The shore and the dune of Sar Uanle. 31. Ocypodidae and Gecarcinidae (Decapoda, Brachyura). *Monitore Zoologico Italiano* **14**, 199–226.

Veloso, V.G., Cardoso, R.S. & Fonseca, D.B. 1997. Adaptations and biology of exposed sandy beaches macrofauna with emphasis in the intertidal species of Fluminense littoral. In *Oecologia Brasiliensis, vol. 3: Ecologia de praias arenosas do litoral Brasileiro*, R.S. Absalão & A.M. Esteves (eds). Rio de Janeiro: Programa de Pós-Graduação em Ecologia, Instituto de Biologia, UFRJ, 135–154.

Vernberg, F.J. 1956. Study of the oxygen consumption of excised tissues of certain marine decapod Crustacea in relation to habitat. *Physiological Zoology* **29**, 227–234.

Vernberg, W.B. & Vernberg, F.J. 1968. Physiological diversity in metabolism in marine and terrestrial Crustacea. *American Zoologist* **8**, 449–458.

Vinagre, A.S., do Amaral, A.P.N., Ribarcki, F.P., da Silveira, E.F. & Perico, E. 2007. Seasonal variation of energy metabolism in ghost crab *Ocypode quadrata* at Siriu Beach (Brazil). *Comparative Biochemistry and Physiology A: Molecular and Integrative Physiology* **146**, 514–519.

Vogt, G. 2013. Abbreviation of larval development and extension of brood care as key features of the evolution of freshwater Decapoda. *Biological Reviews* **88**, 81–116.

Von Prahl, H., Ramos, G.E. & Rios, R. 1990. The crabs (Crustacea: Decapoda: Brachyura) of the Pacific coast of Colombia. *Revista de Ciencias* **2**, 23–35.

Warrick, S.F. & Wilcox, E.D. (eds.) 1981. Big river: the natural history of an endangered northern California estuary. Environmental Field Program, University of California, Santa Cruz.

Watts, B.D. & Bradshaw, D.S. 1995. Ghost crab preys on Piping Plover eggs. *Wilson Bulletin* **107**, 767–768.

Weihrauch, D., Morris, S. & Towle, D.W. 2004. Ammonia excretion in aquatic and terrestrial crabs. *Journal of Experimental Biology* **207**, 4491–4504.

Weinstein, R.B. 1995. Locomotor behaviour of nocturnal ghost crabs on the beach: focal animal sampling and instantaneous velocity from three-dimentional motion analysis. *Journal of Experimental Biology* **198**, 989–999.

Weinstein, R.B. 1998. Performance limits of low-temperature, continuous locomotion are exceeded when locomotion is intermittent in the ghost crab. *Physiological Zoology* **71**, 274–284.

Weinstein, R.B. 2001. Terrestrial intermittent exercise: common issues for human athletics and comparative animal locomotion. *American Zoologist* **41**, 219–228.

Weinstein, R.B. & Full, R.J. 1992. Foraging ghost crabs move intermittently. *American Zoologist* **32**, 45A.

Weinstein, R.B. & Full, R.J. 1994. Thermal dependence of locomotor energetics and endurance capacity in the ghost crab, *Ocypode quadrata. Physiological Zoology* **67**, 855–872.

Weinstein, R.B., Full, R.J. & Ahn, A.N. 1994. Moderate dehydration decreases locomotor performance of the ghost crab, *Ocypode quadrata. Physiological Zoology* **67**, 873–891.

Wellins, C.A., Rittschof, D. & Wachowiak, M. 1989. Location of volatile odor sources by ghost crab *Ocypode quadrata* (Fabricius). *Journal of Chemical Ecology* **15**, 1161–1169.

Whiteley, N.M., Innes, A.J., Al-Wassia, A.H. & Taylor, E.W. 1990. Aerial and aquatic respiration in the ghost crab *Ocypode saratan* II. Respiratory gas exchange and transport in the haemolymph. *Marine Behaviour and Physiology* **16**, 261–273.

Whiting, S.D. Long, J.L., Hadden, K.M. Lauder, A.D.K. & Koch, A.U. 2007. Insights into size, seasonality and biology of a nesting population of the Olive Ridley turtle in northern Australia. *Wildlife Research* **34**, 200–210.

Williams, A.B. 1965. Marine decapod crustaceans of the Carolinas. *U.S. Fish and Wildlife Service Fishery Bulletin* **65**, 225–227.

Williams, J.A. 1983. Environmental regulation of burrow depth distribution of the sand-beach amphipod *Talitrus saltator. Estuarine, Coastal and Shelf Science* **16**, 291–298.

Wingate, D.B. 1982. Successful reintroduction of the Yellow-crowned Night-Heron as a nesting resident on Bermuda. *Colonial Waterbirds* **5**, 104–115.

Witherington, B., Hirama, S. & Mosier, A. 2011. Sea turtle responses to barriers on their nesting beach. *Journal of Experimental Marine Biology and Ecology* **401**, 1–6.

Wolcott, D.L. & Wolcott, T.G. 1999. High mortality of Piping Plovers on beaches with abundant ghost crabs: correlation, not causation. *Wilson Bulletin* **111**, 321–329.

Wolcott, T.G. 1976. Uptake of soil capillary water by ghost crabs. *Nature* **264**, 756–757.

Wolcott, T.G. 1978. Ecological role of ghost crabs, *Ocypode quadrata* (Fab.) on an ocean beach: scavengers or predators? *Journal of Experimental Marine Biology and Ecology* **31**, 67–82.

Wolcott, T.G. 1984. Uptake of soil interstitial water: mechanisms and ecological significance in the ghost crab *Ocypode quadrata* and two gecarcinid land crabs. *Physiological Zoology* **57**, 161–184.

Wolcott, T.G. 1988. Ecology. In *Biology of the Land Crabs*, W.W. Burggren & B.R. McMahon (eds). Cambridge: Cambridge University Press, 55–96.

Wolcott, T.G. 1992. Water and solute balance in the transition to land. *American Zoologist* **32**, 428–437.

Wolcott, T.G. & Wolcott, D.L. 1985. Extrarenal modification of urine for ion conservation in ghost crabs, *Ocypode quadrata* (Fabricius). *Journal of Experimental Marine Biology and Ecology* **91**, 93–107.

Wolcott, T.G. & Wolcott, D.L. 2001. Role of behavior in meeting osmotic challenges. *American Zoologist* **41**, 795–805.

Wooldridge, T., Dye, A.H. & McLachlan, A. 1981. The ecology of sandy beaches in Transkei. *South African Journal of Zoology* **16**, 210–218.

Wright, H.O. 1968. Visual displays in brachyuran crabs: field and laboratory studies. *American Zoologist* **8**, 655–665.

Xiang, G. & Jingming, X. 2002. Impact of Rizhao coastal area exploitation on inter-tidal habitats and zoobenthic communities. *Studia Marina Sinica* doi: CNKI:SUN:HKJK.0.2002-00-007

Yaldwyn, J.C. & Wodzicki, K. 1979. Systematics and ecology of the land crabs (Decapoda: Coenobitidae, Grapsidae and Gecarcinidae) of the Tokelau Islands, Central Pacific. *Atoll Research Bulletin* **235**, 1–53.

Yasué, M. & Dearden, P. 2006. The potential impact of tourism development on habitat availability and productivity of Malaysian plovers *Charadrius peronii. Journal of Applied Ecology* **43**, 978–989.

Yom-Tov, Y. 2003. Poaching of Israeli wildlife by guest workers. *Biological Conservation* **110**, 11–20.

Yong, A.Y.P., Lim, S.S.L., Kaenphet, A. & Tantichodok, O. 2011. Evidence of precision engineering in the excavation of *Ocypode ceratophthalmus* burrows on the west and east coasts of Thailand. *Crustaceana* **84**, 749–761.

Ziese, M. 1985. Further records of the crab species *Ocypode cursor* (Linnaeus 1758) for the eastern part of the Mediterranean. *Senckenbergiana Biologica, Frankfurt* **66**, 123–125.

Oceanography and Marine Biology: An Annual Review, 2014, **52**, 257-314
© Roger N. Hughes, David J. Hughes, and I. Philip Smith, Editors
Taylor & Francis

CITIZEN SCIENTISTS AND MARINE RESEARCH: VOLUNTEER PARTICIPANTS, THEIR CONTRIBUTIONS, AND PROJECTION FOR THE FUTURE

MARTIN THIEL[1,2*], MIGUEL ANGEL PENNA-DÍAZ[1], GUILLERMO LUNA-JORQUERA[1,2], SONIA SALAS[3,2], JAVIER SELLANES[1,2] & WOLFGANG STOTZ[1,2]

[1]*Facultad de Ciencias del Mar, Universidad Católica del Norte, Larrondo 1281, Coquimbo, Chile*
E-mail: thiel@ucn.cl
[2]*Centro de Estudios Avanzados en Zonas Áridas (CEAZA), Coquimbo, Chile*
[3]*Department of Psychology, University of La Serena, La Serena, Chile*

The field of citizen science is flourishing, and although terrestrial projects are more visible, in recent years thousands of volunteers have actively participated in marine research activities. These volunteers (also termed 'citizen scientists') may have experience in the research in which they are participating, but they have no formal degree in marine science or related topics. The participation of large numbers of volunteers with variable educational or professional backgrounds poses particular challenges for the professional scientists coordinating such research. Knowledge about the structure of these projects, the research activities conducted by citizen scientists, and quality control of data collected by volunteers is essential to identify their contribution to marine science. We examined 227 published studies in which professional scientists collaborated with volunteers in a wide range of marine investigations. Most studies focused on a diverse assemblage of animals, followed by flora and other topics (e.g., contamination or beach dynamics). Seabirds, marine mammals, turtles, and fishes were the most commonly studied animals, but several studies also dealt with marine invertebrates. Many of the studied taxa were commercially important, emblematic, or endangered species. Surveys of invasive species took advantage of the extensive spatial scale that can be covered by large numbers of volunteers. As would be expected, the research activities of citizen scientists were concentrated in easily accessible coastal habitats, including sandy beaches, estuaries, coral reefs, and seagrass beds. Hot spots of marine citizen science projects (CSPs) were found not only in North America and Europe, but also in the Indo-West Pacific region. Contributions made by citizen scientists were equally based on incidental observations as on standardized surveys. Some of the research projects had been active for more than a decade, but most were midterm programmes, lasting a few years or less. Volunteer participants came from a wide range of demographic backgrounds. Usually, the participants were adults of both sexes, but a few studies considered either only men or only women (mainly in small fishing communities). Whereas several studies were based on schoolchildren as volunteers, no study worked specifically with senior citizens. The educational level of participants, often not explicitly mentioned in the publications, was also diverse. Some projects selected participants based on their experiences, skills, or profession, but in the majority of the studies, there was either no selection or no information was provided, suggesting that any interested citizen could participate. The preparation of participants ranged from brief written or oral instructions to extensive (weeks) training sessions with professional experts. In general, training effort increased with the complexity of the tasks conducted by volunteers, a crucial element being the adjustment of simple methodologies to the capabilities of participants. Studies for which volunteers needed to identify many different species and estimate their abundances were considered the most complex tasks, and subsequent analysis of such studies by professional scientists must consider inherent bias or shortcomings. About half of the examined studies included some type of quality control to ensure that the data collected by citizen scientists met the standards of rigorous scientific studies. Several authors emphasized that data quality increased with the duration of project participation. Efforts therefore should be made to retain experienced volunteers over time, which is facilitated when volunteers perceive that their efforts lead to something of practical use, such as publications, conservation initiatives, management decisions, or policy actions. Participants seemed to value personal satisfaction and public recognition, but learning about the ocean was also important. The coordinators of marine CSPs often collaborated with organizations such as conservation groups, birdwatchers, dive associations, or fishermen's cooperatives to recruit volunteers, but media campaigns, personal communication, social media, and functional websites were also important. Some studies were based on small numbers of participants (e.g., artisanal fishermen); others involved thousands of volunteers (e.g., coral or litter surveys). Volunteer-generated data contributed information about population dynamics, health, or distribution of marine organisms and supported long-term monitoring programmes of marine protected areas, harmful algal blooms, or marine litter, among others. In general, the contribution of citizen scientists greatly enhances research capacity, providing an increased workforce over extensive spatial and intensive temporal scales at comparatively moderate costs. Citizen science is able to make significant contributions to marine science, where professional scientific activities are limited by the available human resources. Considering the vastness of the oceans and the diversity of habitats, communities, and species, proper understanding of this realm requires intensive research activities over time and space. This recognition should lead to increased consideration of citizen science as a powerful tool for the generation and spread of scientific knowledge. Furthermore, sharing knowledge between volunteer participants and professional scientists improves communication, trust, and capacity building, facilitating efficient collaboration in much-needed conservation initiatives.

Introduction

> Upon the morning of arrival, the expected naturalist finds upon his table a collection of preserved material ... or a dish of fresh seawater with some brilliantly coloured squirming inhabitant of the subtropical Gulf of Naples, brought that morning by a barefooted fisherman. ... Every morning at 10 o'clock the ten or more fishermen of the station, with buckets or baskets full of glass jars poised gracefully on their heads, march into the court and receiving room of the station with their prizes from the sea—scarlet starfishes, orange feather stars, red and black sea-cucumbers, sea-urchins, squirming bits of red coral or a wriggling, creeping octopus.
>
> **—Charles A. Kofoid (1910)**

This colourful description of the daily functioning of the Naples Marine Station by the eminent zoologist Charles Kofoid shows how citizens (here fishermen) are supporting marine research. These fishermen are not mere sampling assistants. Based on their experiences, they recognize which organisms could be of interest to the marine biologists at the station (e.g., because they have never been seen before or because they appear in atypical locations or at unusual times). From their conversations with the marine biologists, they are aware of the specific research interests, and they catch and bring the organisms the scientists are working on. Consequently, these fishermen, who closely collaborate with the professional scientists, play an active and autonomous role in marine science.

Marine scientists often depend on the help of other people who lack scientific training. Ship crews are essential for successful research in the oceans. Fishermen are keen observers of marine life and consult scientists with questions, report unusual events, or (as in the example) provide material for study by professional scientists. Also, laypeople have 'fed' marine scientists with observations or samples in the past and continue to do so more than ever before: Thousands of volunteers roam the coasts of the world's oceans during weekends or vacations, providing a wealth of data that can be of scientific value. Much of this information is scattered over personal web pages, blogs, or home computers, but scientists are increasingly recognizing its scientific value and encourage enthusiastic volunteers to report their findings on central data platforms or to participate in planned surveys. This review is a first attempt to explore the extensive literature on volunteer participation in marine research.

Historical context

Observation of nature was crucial for the first humans, enabling them to learn about phenomena relevant for survival. The acquired knowledge improved foraging, hunting, or fishing and aided protection from the rigors of nature. This knowledge led to the development of agriculture and improved fishing technology, promoting the acquisition of food, and later through industrialization and accumulation of capital, made society increasingly insulated against the direct effects and variability of nature (Cohen 1977, Tudge 1998, Zohany & Hopf 2000, Bellwood 2005). For much of human society, this meant that knowledge about nature was no longer important for individual survival, Plato's republic, among others, becoming based on the division of labour (Silvermintz 2010). Nature for many people became something distant, to which they did not feel connected, and something to enjoy simply for recreation. The study of nature remained increasingly in the hands of specialists, highly trained scientists. Nevertheless, professions such as agriculture, fisheries, and aquaculture, as well as some human cultures, still rely directly on nature, maintaining a need for people to observe and learn directly about properties of nature relevant to its management (Berkes et al. 2000). Currently, this knowledge is termed 'traditional ecological knowledge' (TEK) or local ecological knowledge (LEK). TEK has been gained through lifetime observations and experience, often related to some professional activity (e.g., fishermen, skippers, divers, etc.). It is typically produced by direct observation and verbally transmitted, being distinguished from 'scientific knowledge', which is generated by a formal methodology for gathering and analysing data and transmitted by writing (Rutherford & Ahlgren 1991).

As negative effects of humankind on nature increase, conservation and management become progressively more urgent. This requires acquisition of knowledge on a scale that can only be met by bringing people closer to nature and involving them directly in decision or conservation/management processes. In this context, TEK has become valuable once again but is now acquired by specialists and integrated with scientific knowledge. Non-specialists, especially those who observe nature through curiosity, become increasingly involved by recording data and by making results accessible to laypeople, thereby enhancing public awareness of nature, often for conservation purposes. This new involvement of non-professional scientists is termed 'citizen science'. In this context, citizen science is understood as any involvement of volunteer participants in the generation of scientific knowledge, that is, observation, sampling, collection of data, and analysis of data, which enter the normal scientific production cycle, ideally leading to reports supporting management decisions and scientific publications (Bonney et al. 2009).

The field of citizen science is flourishing, and there are many different programmes in which citizens participate in the collection of scientific data (Silvertown 2009, Dickinson et al. 2010, Shirk et al. 2012). Here, we make no attempt to identify the first scientific publication that depended on observations by untrained citizens, but we are aware of the fact that there were many instances during the past centuries in which progress in marine science depended on the attentive eye of interested citizens. In this review, we evaluate studies published over the past 30 years. During this

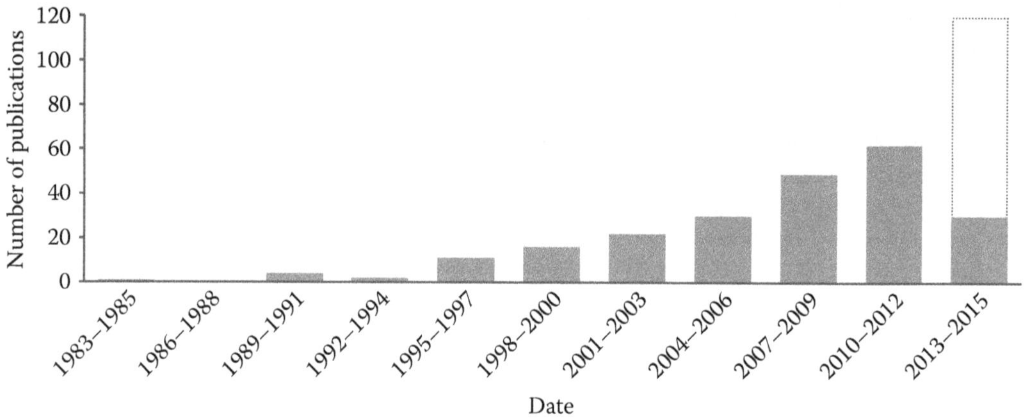

Figure 1 Number of scientific publications on marine topics that are based entirely or partly on information contributed by citizen scientists per 3-year period. The dotted line for the period 2013–2015 is the projection for that 3-year period based on 30 publications until 30 September 2013.

period, the number of new studies that incorporated volunteer-collected data increased continuously and is a trend likely to continue well into the future (Figure 1). It appears that marine citizen science is a healthy field and that many people are keen to actively support scientific research in the oceans.

Marine citizen scientists: a diverse group of many trades

Citizens without marine scientific background support research in various forms. They transport scientists to specific locations (e.g., the staff of research vessels); they launch scientific equipment (e.g., mariners releasing drifter buoys or weather balloons); they enable scientists to collect data of marine resources (e.g., fishermen allowing scientists to take measurements of their catch); they carry scientific equipment on their vessels (e.g., sensors or samplers such as the continuous plankton recorder); they take samples and pass them on to scientists (e.g., fishermen taking water samples or preserving fish or parts thereof); they actively record observations (e.g., measuring water temperature or clarity); or they engage in particular programmes in which they take standardized data that are then evaluated and interpreted by professional scientists (Figure 2). In the most advanced cases, citizens can participate in data evaluation, interpretation, and publication. As will become evident, there are many examples of volunteer participation in marine science (Figure 2). Reflecting the diversity of research topics in which they participate, marine citizen scientists range from inexperienced beach strollers reporting incidental observations (e.g., Casale et al. 2011, Hong et al. 2013) to advanced citizen scientists with special training who aid professional scientists in data collection (e.g., Uychiaoco et al. 2005, Edgar et al. 2009). In certain cases, through activities that in general start as a 'weekend hobby' (e.g., birdwatching, shell collecting), citizens can become highly specialized in their fields and gain a solid reputation as experts, even among the professionally trained scientific community.

Here, we consider those cases for which volunteers (citizen scientists) collect data and record these in a rigorous form. The volunteers may have received specific training for the research in which they are participating, but they had no formal degree in marine science or related topics (but possibly in other academic fields). The data contributed by citizens were not necessarily collected for scientific purposes, but rigorous data recording is required for a study to be considered in this review. For example, catch data collected by recreational fishermen within the context

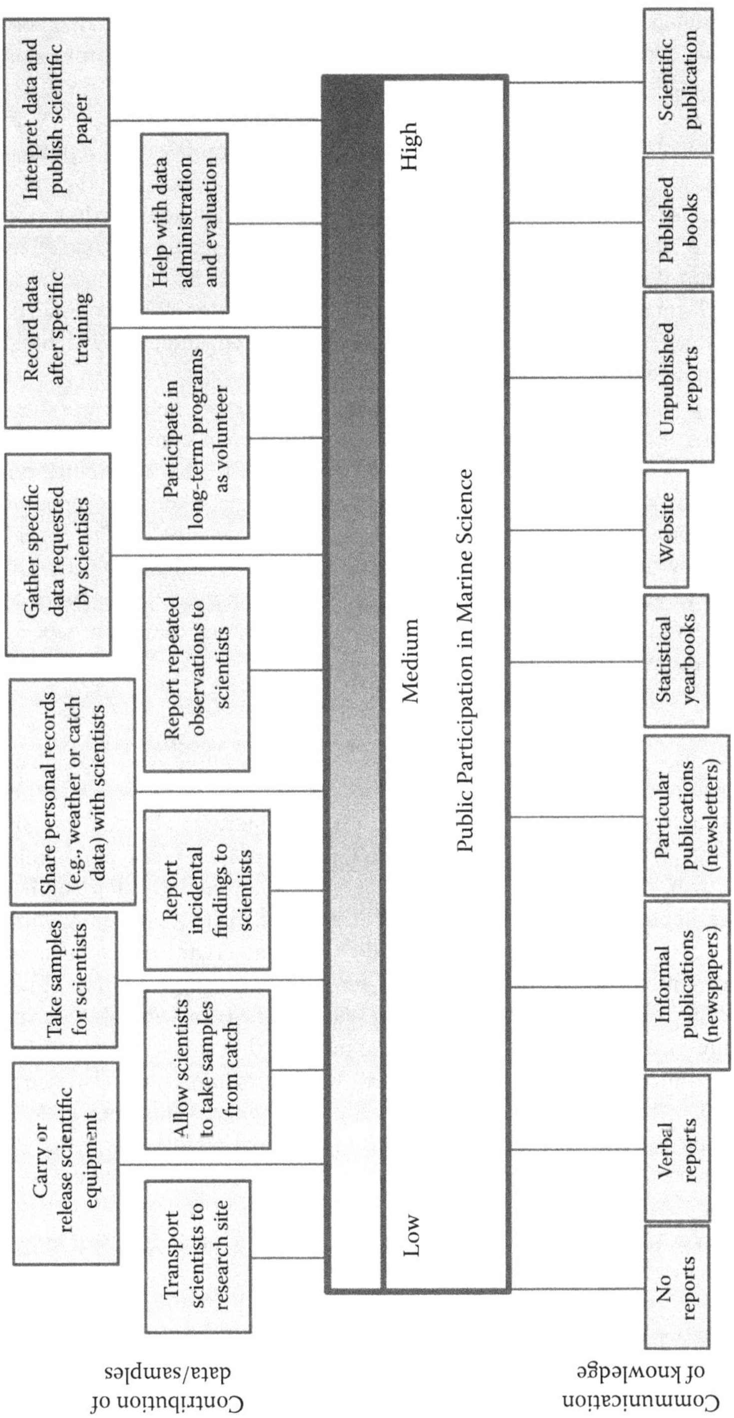

Figure 2 Different levels of volunteer participation in marine sciences.

of a competition would be carefully recorded by citizens without specific scientific training—but when such data are made available to scientists, they can provide relevant information (e.g., Godoy et al. 2010).

A number of data are generated within the fisheries context for which the line between professional data recording and citizen science is especially vague. For example, professional scientists collate a large number of catch data provided by fishermen to determine the status of exploited natural resources. However, we do not consider such studies in our review because they are conducted within the framework of resource monitoring, often published in professional reports or databases, and therefore do not strictly represent citizen science. On the other hand, studies in which scientists mark fish or birds to determine their movements and fishermen voluntarily report findings of marked fish (Moser & Shepherd 2009, Björnsson et al. 2011) or ringed birds drowned in fishing nets (e.g., Lyngs & Kampp 1996, Österblom et al. 2002) are considered citizen science because the fishermen contribute information that helps to answer a scientific question. We recognize that the dividing line is diffuse, but we decided to limit our review to published studies in which data collection by volunteers was motivated by a specific scientific question.

In general, any private person who voluntarily contributes data, samples, or other information and who has no formal education in the field of research is herein considered a citizen scientist. Fishermen (whether recreational or professional) who voluntarily share their logbooks with scientists to respond to a scientific question act as citizen scientists. Even fishermen who receive a financial reward for providing knowledge, samples, work, or ship time can be considered citizen scientists because they are free to refuse participation. However, fishermen who are obliged by administrative rules to provide catch data or samples of the catch are not considered citizen scientists because the motivation may not be scientific curiosity, but rather the risk of receiving a fine by not delivering the required information.

The role of citizen scientists in marine research

Volunteer participation in marine research, as in other research fields (Silvertown 2009, Dickinson et al. 2010), is on the rise (see also Figure 1). The research questions citizen scientists are helping to answer are diverse. Some popular, well-known programmes focus on shorebirds, marine mammals, turtles, fish, corals, and seagrasses (Figure 3; Hodgson 1999, Pattengill-Semmens & Semmens 2003, Bart et al. 2007, Moore et al. 2009, Finn et al. 2010, Cornwell & Campbell 2012). However, citizens not only are interested in the study of marine organisms but also are concerned about marine conservation—this is impressively underlined by the large number of people participating in beach cleanups and volunteers helping to investigate the types, abundances, and sources of marine litter (Figure 3). Many other programmes are motivated by conservation, often focusing on marine vertebrates. If there is a bias in taxa studied by volunteers (targeting mostly large, visible, or charismatic species), there also might be a bias in the habitats where volunteers work. Accessibility can be a challenge for marine research, and it could thus be expected that projects with volunteer participants will be concentrated in particular habitats, most likely beaches and nearshore, shallow-water environments.

One of the main advantages of collecting scientific data with large teams of volunteers is that observers can spread out over extensive areas, sometimes exceeding national borders. However, this also poses challenges (Bonney et al. 2009, Dickinson et al. 2010, Finn et al. 2010). Can anyone interested participate, or should there be a selection process? Who are the people who voluntarily offer their time and efforts for marine research programmes? What motivates them to participate in these studies? Clearly, the answers to these questions will help better focus research activities involving citizen scientists.

Scientific studies also require careful planning and logistic support. Volunteers need to be recruited, and ideally they should receive some basic training. Collection of scientific data also

Figure 3 Citizen scientists working in the marine environment. (Upper left courtesy of Martin Stock; upper right courtesy of Ivan Hinojosa.)

requires some tools and equipment, ranging from simple recording sheets to expensive laboratory equipment. Data need to be centralized and managed in easily accessible databases. And, at the end of each study there should be a thorough analysis of the collected data; this will be the task of professional scientists. Which scientists collaborate with volunteers? Are they drawn from specific disciplines of marine science? Do they work for universities or for government institutions?

Scientific research also usually depends on limited funding. Although it is often suggested that collaborating with citizen scientists might reduce costs, some basic financial supported is needed. For better planning of citizen science, it is important to identify the sources of current financial support and to consider future potential sources.

To be admitted to the holy grail of scientific knowledge, data need to meet the high standards of scientific rigor. Quality control of data collected by volunteers with no or little scientific training is thus of utmost importance (e.g., Foster-Smith & Evans 2003). Is validation of volunteer-contributed data a central part of marine citizen science research, and if so, what type of validation? Knowing the type and quality of data collected by citizen scientists will help to better judge the contribution that volunteer participation can make to marine science. Our review collates a wide range of marine studies to answer these questions.

The baseline: studies considered in this review

Communicating knowledge is an inherent part of the scientific process. Information contributed by citizens only becomes part of the scientific knowledge if it is published. Therefore, we only consider those studies published in the primary scientific literature (edited books or journals). Such publications must have at least one clear reference (in the main text or in the acknowledgements) to citizens contributing part or all of the data reported. We also surveyed whether initiatives had a public website, but projects for which results were only reported through a website without any scientific publication are not considered here. The literature was scanned independently by all coauthors, and we examined the degree of saturation (i.e., the point at which, after independent searches by all coauthors, the cumulative number of publications in the database tended towards an asymptote). This exhaustive literature review produced a total of 227 publications (Table 1). Although we recognize that there are additional publications on marine citizen science projects (CSPs), we are confident that the database we used is sufficiently representative to provide a first overview of the contribution of volunteer participants to marine science.

Table 1 Synthesis of the database of marine citizen science publications (n = 227)

Study subject	Country	Political extension	Habitat	Objective	Reference
Fauna	Argentina	Regional	Subtidal	Conservation	Negri et al. 2012
Fauna	Australia	Regional	Intertidal	Conservation	Finn et al. 2007
Fauna	Australia	National	Coral reef	Conservation	Edgar et al. 2009
Fauna	Australia	Regional	Subtidal	Conservation	Huveneers et al. 2009
Fauna	Australia	Regional	Shoreline	Population	Holmberg et al. 2009
Fauna	Australia	Regional	Urban area	Social	Koss & Kingsley 2010
Fauna	Australia	Regional	Coral reef	Social	Whatmough et al. 2011
Fauna	Australia	Regional	Subtidal	Social	Hammerton et al. 2012
Fauna	Australia	Regional	Subtidal	Conservation	Jaine et al. 2012
Fauna	Australia	Regional	Subtidal	Conservation	Hassell et al. 2013
Fauna	Bahamas	Regional	Subtidal	Invasive species	Green et al. 2012
Fauna	Brazil	Regional	Beach	Conservation	Parente et al. 2004
Fauna	Brazil	Local	Coral reef	Conservation	Rosa et al. 2006
Fauna	Brazil	Regional	Beach	Conservation	Meirelles et al. 2009
Fauna	Brazil	Regional	Pelagic	Conservation	Marigo & Giffoni 2010
Fauna	Canada	Regional	Beach	Conservation	Stacey et al. 1989
Fauna	Canada	Regional	Beach	Conservation	Baird & Guenther 1995
Fauna	Canada	Regional	Coastal waters	Biology	Ford et al. 1998
Fauna	Canada	Regional	Coastal waters	Conservation	Willis & Baird 1998
Fauna	Canada	Regional	Intertidal & subtidal	Conservation	Baird et al. 2002
Fauna	Canada	Regional	Pelagic	Conservation	Wiber et al. 2004

Table 1 (Continued) Synthesis of the database of marine citizen science publications ($n = 227$)

Study subject	Country	Political extension	Habitat	Objective	Reference
Fauna	Canada	Regional	Subtidal	Conservation	Martin & James 2005
Fauna	Canada	Regional	Continental shelf	Conservation	James et al. 2006
Fauna	Canada	Regional	Intertidal & subtidal	Climate change	Berkes et al. 2007
Fauna	Canada	Regional	Supratidal	Modelling	Norris et al. 2007
Fauna	Canada	Local	Salt marshes	Social	Peloquin & Berkes 2009
Fauna	Canada	Regional	Subtidal	Management	MacKenzie & Cox 2013
Fauna	Canada	Regional	Continental shelf	Management	Cadigan & Brattey 2006
Fauna	Caribbean	Regional	Intertidal	Population	Aiken et al. 2001
Fauna	Cayman Islands	Regional	Subtidal	Biology	Whaylen et al. 2004
Fauna	Cayman Islands	Regional	Subtidal	Biology	Semmens et al. 2006
Fauna	Cayman Islands	Regional	Beach	Population	Bell et al. 2009
Fauna	Chile	Regional	Rocky reef	Conservation	Godoy et al. 2010
Fauna	Chile	Regional	Intertidal & supralittoral	Social	Roura 2012
Fauna	Chile	Local	Fjords	Management	Suazo et al. 2013
Fauna	Colombia	National	Intertidal	Population	Johnston-González et al. 2008
Fauna	Costa Rica	Regional	Intertidal	Social	Campbell & Smith 2006
Fauna	Ecuador	Regional	Subtidal	Population	Acuña-Marrero et al. 2013
Fauna	Fiji	Local	Coral reef	Methodology	Léopold et al. 2009
Fauna	France	Regional	Beach	Distribution	Jung et al. 2009
Fauna	France	Regional	Intertidal	Modelling	Peltier et al. 2012
Fauna	Germany	Regional	Rocky intertidal	Conservation	Landschoff et al. 2013
Fauna	Germany & Netherlands	International	Intertidal	Contamination	Camphuysen et al. 1999
Fauna	Greece	Local	Intertidal	Conservation	Margaritoulis 1983
Fauna	Greece	Local	Subtidal	Population	Arvanitidis et al. 2011
Fauna	Greece	National	Intertidal to subtidal	Invasive species	Zenetos et al. 2013
Fauna	Grenada	Local	Subtidal	Social	Grant & Berkes 2007
Fauna	Guinea	Regional	Subtidal	Conservation	Le Fur et al. 2011
Fauna	Honduras	Local	Coral reef	Conservation	Biggs & Olden 2011
Fauna	Iceland	Regional	Continental shelf	Management	Björnsson et al. 2011

Continued

Table 1 (Continued) Synthesis of the database of marine citizen science publications (n = 227)

Study subject	Country	Political extension	Habitat	Objective	Reference
Fauna	Indonesia	Regional	Coral reef	Biology	Bell 2007
Fauna	Indonesia & Jamaica	International	Subtidal	Population	Crabbe et al. 2004
Fauna	International	International	Coastal waters	Population	Lyngs & Kampp 1996
Fauna	International	International	Subtidal	Conservation	Österblom et al. 2002
Fauna	International	International	Coral reefs	Conservation	Pattengill-Semmens & Semmens 2003
Fauna	International	International	Intertidal & supratidal	Population	Milton 2003
Fauna	International	International	Coral reef	Management	Bruno & Selig 2007
Fauna	International	International	Subtidal	Population	Moreno et al. 2007
Fauna	International	International	Coral reef	Modelling	Stallings 2009
Fauna	International	International	Coral reef	Biology	Holt et al. 2010
Fauna	International	Global	Coral reef	Conservation	Selig & Bruno 2010
Fauna	International	Global	Intertidal & subtidal	Distribution	Condon et al. 2012
Fauna	International	Global	Coral reef	Conservation	Marshall et al. 2012
Fauna	International	Global	Subtidal	Modelling	Stuart-Smith et al. 2013
Fauna	International	International	Subtidal	Conservation	Ward-Paige et al. 2010
Fauna	International	International	Intertidal	Modelling	Peltier et al. 2013
Fauna	Italy	National	Subtidal	Population	Goffredo et al. 2004
Fauna	Italy	Regional	Subtidal	Distribution	Boero et al. 2009
Fauna	Italy	National	Subtidal	Population	Bramanti et al. 2011
Fauna	Italy	National	Coastal waters	Biology	Boero et al. 2013
Fauna	Italy	National	Intertidal	Population	Casale et al. 2011
Fauna	Italy	Regional	Subtidal	Conservation	Azzurro et al. 2013a
Fauna	Jamaica	National	Subtidal	Conservation	Crabbe 2012
Fauna	Kenya	Regional	Subtidal	Social	Obura 2001
Fauna	Kenya & Tanzania	International	Subtidal	Social	Obura et al. 2002
Fauna	Korea	Local	Wetland	Conservation	Jang et al. 2013
Fauna	Madagascar	Regional	Subtidal	Conservation	Brenier et al. 2011
Fauna	Madagascar	Regional	Intertidal & subtidal	Conservation	Humber et al. 2011
Fauna	Maldives	National	Subtidal	Modelling	Davies et al. 2012
Fauna	Mexico	Regional	Subtidal	Conservation	Sáenz-Arroyo et al. 2005
Fauna	Mexico	Regional	Intertidal & subtidal	Conservation	Pabody et al. 2009

Table 1 (Continued) Synthesis of the database of marine citizen science publications ($n = 227$)

Study subject	Country	Political extension	Habitat	Objective	Reference
Fauna	Mexico	National	Coastal waters	Modelling	Fujitani et al. 2012
Fauna	Mexico	Local	Subtidal	Conservation	Cudney-Bueno & Basurto 2009
Fauna	Mexico	Regional	Subtidal	Management	Moreno-Baez et al. 2010
Fauna	Netherlands Antilles	Regional	Coral reef	Population	Auster et al. 2005
Fauna	New Zealand	Regional	Pelagic	Management	Starr & Vignaux 1997
Fauna	New Zealand	National	Pelagic	Management	Starr 2010
Fauna	Norway	Regional	Subtidal	Management	Kleiven et al. 2012
Fauna	Papua New Guinea & Australia	International	Coral reef	Conservation	Almany et al. 2010
Fauna	Peru	Regional	Intertidal	Population	Alfaro-Shigueto et al. 2011
Fauna	Philippines	Local	Coral reef	Conservation	Uychiaoco et al. 2005
Fauna	Philippines	Local	Coral reef	Conservation	Yasué et al. 2010
Fauna	Philippines	Local	Coastal waters	Management	O'Donnell et al. 2012
Fauna	Philippines	Regional	Subtidal	Conservation	Yasué et al. 2012
Fauna	Puerto Rico	Regional	Beach	Conservation	Mignucci-Giannoni et al. 1999
Fauna	Puerto Rico, British & US Virgin Islands	International	Beach	Conservation	Mignucci-Giannoni et al. 2000
Fauna	Scotland	Regional	Intertidal & supratidal	Ecology	Foster-Smith & Evans 2003
Fauna	Scotland	Regional	Subtidal	Conservation	Moore 2003
Fauna	Scotland	Regional	Intertidal	Climate change	MacLeod et al. 2005
Fauna	Scotland	National	Coastal waters	Population	Cheney et al. 2013
Fauna	Solomon Islands	Local	Mangroves	Conservation	Aswani & Weiant 2004
Fauna	Solomon Islands	Local	Subtidal	Conservation	Aswani & Hamilton 2004
Fauna	Solomon Islands	Local	Coral reef	Conservation	Hamilton et al. 2012
Fauna	South Africa	Regional	Intertidal & subtidal	Conservation	Beckley et al. 1997
Fauna	South Africa	Regional	Rocky reef	Management	Wheeler et al. 2008
Fauna	South Africa	National	Subtidal	Modelling	Bernard et al. 2013
Fauna	Spain	Regional	Supratidal cliff	Modelling	Lopez-Darias et al. 2011
Fauna	Spain	National	Subtidal	Conservation	Azzurro et al. 2013b
Fauna	Sudan	Local	Rocky reef	Population	Hussey et al. 2013

Continued

Table 1 (Continued) Synthesis of the database of marine citizen science publications ($n = 227$)

Study subject	Country	Political extension	Habitat	Objective	Reference
Fauna	Tanzania	Regional	Coral reef	Population	Darwall & Dulvy 1996
Fauna	Thailand	Regional	Coastal waters	Population	Theberge & Dearden 2006
Fauna	Thailand	Local	Coral reef	Conservation	Worachananant et al. 2007
Fauna	Thailand	Local	Intertidal	Biology	Wells et al. 2008
Fauna	Thailand	National	Subtidal	Population	Ward-Paige & Lotze 2011
Fauna	United Kingdom	Regional	Estuaries	Distribution	Gill et al. 2001
Fauna	United Kingdom	Regional	Shoreline	Population	Bristow et al. 2001
Fauna	United Kingdom	Regional	Pelagic	Methodology	Marrs et al. 2002
Fauna	United Kingdom	Local	Coastal waters	Conservation	Pierpoint et al. 2009
Fauna	United Kingdom	Local	Seagrass	Conservation	Garrick-Maidment et al. 2010
Fauna	United Kingdom	National	Pelagic	Distribution	Witt et al. 2012
Fauna	United Kingdom & France	International	Pelagic	Management	Rochet et al. 2008
Fauna	United Kingdom, France, Germany, & Norway	International	Mud flats	Contamination	Evans et al. 2000
Fauna	USA	Regional	Intertidal & supratidal	Conservation	Casey & Kohler 1992
Fauna	USA	Local	Intertidal & supratidal	Conservation	Colwell & Cooper 1993
Fauna	USA	Regional	Coral reef	Conservation	Schmitt & Sullivan 1996
Fauna	USA	Regional	Subtidal	Conservation	Pattengill-Semmens & Semmens 1998
Fauna	USA	Local	Urban area	Social	Pflugh et al. 1999
Fauna	USA	Regional	Subtidal	Management	Lucy & Davy 2000
Fauna	USA	National	Subtidal	Conservation	Loftus et al. 2000
Fauna	USA	Regional	Intertidal	Conservation	Ellis & Cowan 2001
Fauna	USA	Regional	Beach	Biology	Smith et al. 2002
Fauna	USA	Regional	Subtidal	Conservation	Semmens et al. 2004
Fauna	USA	Regional	Rocky intertidal	Climate change	Osborn et al. 2005
Fauna	USA	Regional	Pelagic	Modelling	Walsh et al. 2005
Fauna	USA	Regional	Continental shelf	Management	Shepherd et al. 2006
Fauna	USA	Regional	Beach	Conservation	Harris et al. 2006
Fauna	USA	Regional	Intertidal	Conservation	Delaney et al. 2008
Fauna	USA	Regional	Intertidal	Oceanography	Parrish et al. 2007
Fauna	USA	Regional	Intertidal	Population	Avens & Goshe 2007

Table 1 (Continued) Synthesis of the database of marine citizen science publications ($n = 227$)

Study subject	Country	Political extension	Habitat	Objective	Reference
Fauna	USA	Regional	Beach	Conservation	Moore et al. 2009
Fauna	USA	Regional	Continental shelf	Management	Moser & Shepherd 2009
Fauna	USA	Local	Intertidal	Biology	Vargo et al. 2009
Fauna	USA	Regional	Shoreline & wetlands	Population	Bower 2009
Fauna	USA	Local	Subtidal	Social	Baker & Oeschger 2009
Fauna	USA	Regional	Subtidal	Management	Schroeter et al. 2009
Fauna	USA	Local	Beach	Conservation	Smith 2010
Fauna	USA	Regional	Subtidal	Conservation	Culver et al. 2010
Fauna	USA	Regional	Subtidal	Management	Johnson 2010
Fauna	USA	Regional	Pelagic	Management	Sampson 2011
Fauna	USA	Regional	Rocky reef & kelp forest	Conservation	Gillett et al. 2012
Fauna	USA	Regional	Intertidal	Conservation	Cornwell & Campbell 2012
Fauna	USA	Regional	Subtidal	Biology	Whitney et al. 2012
Fauna	USA	Regional	Subtidal	Conservation	Ruttenberg et al. 2012
Fauna	USA	Regional	Supratidal	Population	Wilson et al. 2013
Fauna	USA	Regional	Intertidal	Biology	Gannon et al. 1997
Fauna	USA	Regional	Intertidal	Conservation	McFee et al. 2006
Fauna	USA	Regional	Subtidal	Conservation	Wolfe & Pattengill-Semmens 2013a
Fauna	USA	Regional	Subtidal	Modelling	Wolfe & Pattengill-Semmens 2013b
Fauna	USA	Local	Subtidal	Conservation	Rogers-Bennett et al. 2013
Fauna	USA	National	Urban area	Biology	Anderson 2013
Fauna	USA & Australia	International	Coral reef	Management	Beeden et al. 2012
Fauna	USA & Canada	International	Beach	Population	Bart et al. 2007
Fauna	USA & Canada	International	Beach	Modelling	Hamel et al. 2009
Fauna	USA & Canada	Regional	Beach	Population	Howe et al. 1989
Fauna	USA, Jamaica, & Caribbean	International	Pelagic	Conservation	Ward-Paige et al. 2011
Fauna	Uruguay	National	Intertidal	Conservation	Vélez-Rubio et al. 2013
Contamination	Canada	Local	Beach	Contamination	Ross et al. 1991
Contamination	Chile	National	Intertidal	Contamination	Bravo et al. 2009
Contamination	Chile	National	Intertidal	Social	Eastman et al. 2013
Contamination	Chile	National	Intertidal	Contamination	Hidalgo-Ruz & Thiel 2013

Continued

Table 1 (Continued) Synthesis of the database of marine citizen science publications ($n = 227$)

Study subject	Country	Political extension	Habitat	Objective	Reference
Contamination	Greece	Regional	Beach	Contamination	Kordella et al. 2010
Contamination	Greece	National	Intertidal	Contamination	Kordella et al. 2013
Contamination	Japan & Russia	International	Beach	Contamination	Kusui & Noda 2003
Contamination	Korea	National	Intertidal	Contamination	Hong et al. 2013
Contamination	Netherlands	Regional	Beach	Contamination	Camphuysen 1998
Contamination	New Zealand	Regional	Beach	Contamination	Gregory 1991
Contamination	Poland	Regional	Beach	Conservation	Jóźwiak 2005
Contamination	Scotland	Regional	Beach	Conservation	Storrier et al. 2007
Contamination	United Kingdom	Regional	Beach	Contamination	Rees & Pond 1995
Contamination	United Kingdom	Local	Beach	Contamination	Williams et al. 1999
Contamination	USA	Local	Estuary	Contamination	Jackson et al. 1997
Contamination	USA	Local	Beach	Contamination	Ribic 1998
Contamination	USA	Regional	Beach	Contamination	Moore et al. 2001
Contamination	USA	Regional	Beach	Conservation	Roletto et al. 2003
Contamination	USA	Regional	Intertidal	Contamination	Ribic et al. 2010
Contamination	USA	National	Intertidal	Contamination	Ribic et al. 2012
Contamination	USA	Regional	Intertidal	Contamination	Rosevelt et al. 2013
Contamination	USA	Regional	Coastal dunes to intertidal	Contamination	Martin 2013
Contamination	USA	Regional	Subtidal	Contamination	Carson et al. 2013
Contamination	USA & Puerto Rico	International	Intertidal	Contamination	Ribic et al. 2011
Contamination	USA & Canada	International	Urban area	Water quality	Kimball et al. 2009
Fauna & flora	Australia	Regional	Subtidal	Conservation	Mckenzie et al. 2000
Fauna & flora	Australia	Regional	Coral reef	Conservation	Monk et al. 2008
Fauna & flora	Australia	National	Subtidal	Conservation	Edgar & Stuart-Smith 2009
Fauna & flora	Australia	Regional	Rocky reef & rocky intertidal	Conservation	Koss et al. 2009
Fauna & flora	Belize	Local	Coral reef & seagrass	Conservation	Mumby et al. 1995
Fauna & flora	Belize	Local	Coral reef	Validation	Mumby et al. 1996
Fauna & flora	Chile	Local	Subtidal	Conservation	Rojas-Nazar et al. 2012
Fauna & flora	Fiji	Regional	Subtidal	Conservation	Harding et al. 2006
Fauna & flora	International	Global	Coral reef	Conservation	Hodgson 1999
Fauna & flora	International	Global	Subtidal	Methodology	Campbell et al. 2007
Fauna & flora	International	Global	Coral reef	Ecology	Bruno et al. 2009

Table 1 (Continued) Synthesis of the database of marine citizen science publications ($n = 227$)

Study subject	Country	Political extension	Habitat	Objective	Reference
Fauna & flora	Italy	National	Subtidal	Conservation	Goffredo et al. 2010
Fauna & flora	Solomon Islands	Local	Subtidal	Conservation	Aswani & Lauer 2006a
Fauna & flora	Solomon Islands	Local	Subtidal	Conservation	Aswani & Lauer 2006b
Fauna & flora	Tanzania	National	Coral reef & mangrove	Conservation	Wagner 2005
Fauna & flora	United Kingdom	Regional	Subtidal	Conservation	Holt et al. 2013
Fauna & flora	United Kingdom	International	Subtidal	Climate change	Bull et al. 2013
Fauna & flora	USA	Regional	Intertidal	Social	Pearse et al. 2003
Fauna & flora	USA	Regional	Intertidal & subtidal	Conservation	Cohen et al. 2011
Fauna & flora	USA	Regional	Intertidal	Population	Cox et al. 2012
Fauna & contamination	Britain, France, Germany, & Norway	International	Intertidal	Contamination	van Franeker et al. 2011
Fauna & contamination	International	International	Intertidal	Contamination	Camphuysen & Heubeck 2001
Fauna & contamination	USA	Local	Supratidal	Contamination	Lindborg et al. 2012
Fauna & contamination	USA & Canada	International	Intertidal	Contamination	Avery-Gomm et al. 2012
Flora	Australia	Local	Seagrass	Conservation	Mellors et al. 2008
Flora	Australia	Regional	Soft sediment	Management	Finn et al. 2010
Flora	International	Global	Subtidal	Oceanography	Bhat & Matondkar 2004
Flora	Portugal	Regional	Coastal waters	Biology	Assis et al. 2009
Flora	USA	Regional	Coastal waters	Contamination	Lewitus & Holland 2003
Flora	USA	Regional	Coastal waters	Contamination	Heil & Steidinger 2009
Fauna, flora, & contamination	USA	Local	Seagrass	Conservation	Lewis III et al. 1998
Fauna, flora, & contamination	USA	Regional	Subtidal	Contamination	Whereat et al. 2004
Fauna, flora, & contamination	USA	Regional	Shoreline	Contamination	Kirkpatrick et al. 2008
Oceanography	International	Global	Oceanic	Climate change	Kennedy et al. 2011
Oceanography	USA	Regional	Intertidal & supratidal	Oceanography	Hill et al. 2002
Geology	United Kingdom	Local	Coastal dunes	Conservation	Evans et al. 2008

Note: Studies are listed in decreasing order of subject frequency.

Marine research topics supported by volunteers

Main topics and taxa

Of the 227 publications analysed, 84.6% were dedicated to, or at least included, the study of fauna. Other common topics, but in lower proportion, were flora (12.8%) and contamination (14.1%). Examples of topics seldom (<2%) chosen were beach morphology (Hill et al., 2002) and social studies evaluating the perception, behaviour, or knowledge of volunteers about certain subjects but still providing biological or ecological information (e.g., Pflugh et al. 1999, Koss & Kingsley 2010, Moreno-Baez et al. 2010, Roura 2012).

Table 2 synthesizes main topics and taxa selected in those projects dealing with fauna and flora. Taxa of scientific interest were investigated in 195 studies, ranging from micro- (e.g., Bhat & Matondkar 2004, Heil & Steidinger 2009) or macroalgae (Assis et al. 2009) to seagrasses (McKenzie et al. 2000, Mellors et al. 2008, Finn et al. 2010) and most major taxa of animals. Bell (2007) studied morphological variations in sponges with the aid of volunteer divers within the framework of a conservation programme. Growth and resilience of coral reefs were studied in Indonesia and Jamaica with the help of volunteer undergraduate students (Crabbe et al. 2004). Another example of this type is CoralWatch (Marshall et al. 2012), a project with a worldwide coverage and monitoring mainly coral colour through volunteer observations. Study of arthropods has focused mainly on commercial species, such as lobsters *Jasus edwardsii* in New Zealand (Starr & Vignaux 1997, Starr 2010) and *Nephrops norvegicus* in the Clyde Sea (Marrs et al. 2002) but has also included invasive species such as *Carcinus maenas* and *Hemigrapsus* spp. (Delaney et al. 2008, Landschoff et al. 2013). One particular study addressed over 2 years the spatial and temporal extent of the spawning of the horseshoe crab *Limulus polyphemus* in Delaware Bay; the species

Table 2 Principal taxa studied in marine citizen science programmes

Taxa		Number of studies
Chordata	Actinopterygii	54
	Aves	28
	Mammalia	23
	Chondrichthyes	14
	Testudines	11
	Tunicata	2
Echinodermata	Echinoidea	1
Arthropoda	Crustacea	10
	Chelicerata	1
Mollusca	General	2
	Bivalvia	1
	Gastropoda	2
Cnidaria	Anthozoa	9
	Scyphozoa	2
Porifera	General	1
Plantae	Angiosperma	2
Heterokontophyta	General	1
Dinoflagellata	General	5
Diverse taxa	General	28
	Total	**195**

Note: Some studies focus on several taxa simultaneously.

has been monitored by a network of volunteers and motivated mainly by the population decrease of this species due to habitat destruction by anthropogenic activities (Smith et al. 2002). Among vertebrates, seabirds are one of the preferred subjects of marine citizen scientists (Bower 2009). For example, Camphuysen (1998) reported on stranded birds that have been surveyed since the 1960s by volunteers in the Netherlands to determine mortality caused by oil spills. A similar study was performed by Roletto et al. (2003) in central California, which also allowed assessment of the ecological effects of the strong 1997/1998 El Niño event. Mortality of birds because of by-catch in fishing gear was monitored by beach visitors in North America (Hamel et al. 2009) and with support from fishermen in Europe (Lyngs & Kampp 1996, Österblom et al. 2002). Beached bird surveys performed with the aid of volunteers were also used to monitor the presence of brevetoxin, the toxin associated with red tides caused by the toxic dinoflagellate *Karenia brevis* in Florida (Vargo et al. 2009). Fishes are another popular group, and among the most emblematic examples is the REEF (Reef Environmental Education Foundation) Fish Survey Project, a monitoring programme performed with the aid of volunteer divers and snorkelers, which started locally in Florida in 1993 and then was extended worldwide, including by 2001 over 40,000 surveys conducted by REEF members (Pattengill-Semmens & Semmens 2003).

For cetaceans, a good example is the multipurpose Stranded Whale and Dolphin Program of British Columbia, Canada, which was started in 1987 (Stacey et al. 1989). This programme not only considers the range and relative abundance information provided by strandings, but also seeks to maximize the use of carcasses for scientific and educational purposes. For the same area (British Columbia), incidental mortality of the grey whale *Eschrichtius robustus* associated with commercial fishing operations was assessed through information provided by fishermen and stranding reports from volunteers (Baird et al. 2002).

Projects addressing taxa of commercial importance were considered in 56 publications in our database, including mainly bony fishes, but also other taxa (e.g., algae, arthropods, molluscs, echinoderms, birds). Fish tags recovered by fishermen enabled monitoring of the distribution and movement of the black sea bass *Centropristis striata* in the north-western Atlantic (Moser & Shepherd 2009), and observations of volunteer divers at Cayman Islands were used to detect spawning aggregations of the grouper *Epinephelus striatus* (Whaylen et al. 2004).

A comparatively large proportion of initiatives dealt with emblematic species (52 studies). An example of an emblematic species that has been targeted by citizen scientists is the whale shark *Rhincodon typus*. Beckley et al. (1997) used information by professional divers to create a database of sightings and strandings of this species in South Africa. Sightings of sharks by ecotourism operators in the Andaman Sea were used for the same purpose (Theberge & Dearden 2006). A major collaborative study included tagging of sharks by scientists and subsequent observations of tagged animals by both tourists and tour operators at the Ningaloo Marine Park, Australia (Holmberg et al. 2009). This study found a decline in reported shark sizes over the past decade, but a coincidental slight increase in resightings of individual sharks suggested an increased residency rate. A similar approach allowed accurate photo identification and monitoring of sharks using pictures taken by tourists at the Maldives Islands (Davies et al. 2012). This project confirmed a high residency rate of sharks, at least at the study site, and highlighted the usefulness of this approach in places where data collected by tourists are available online and when research funding is limited.

Marine invasive species have been studied in relatively few (12) CSPs (for microalgae, see Whereat et al. 2004). The presence of colonial tunicates *Didemnum vexillum* and related species was confirmed with the aid of volunteers in Sitka, Alaska (Cohen et al. 2011). A major initiative involving more than 1000 volunteers, including high school students, helped to determine the distribution of invasive crabs, *Carcinus maenas* and *Hemigrapsus sanguineus*, in the intertidal zone of seven states along the East Coast of the United States (Delaney et al. 2008). The presence of invasive lionfish *Pterois* spp. in Caribbean coral reef habitats has been assessed using sightings by divers and snorkelers (Biggs & Olden 2011, Green et al. 2012).

Conservation status of focal organisms ranged from 'least concern' or 'near threatened' (33 studies) to 'vulnerable' and 'critically endangered' (43 studies). Studies of a critically endangered emblematic taxon are exemplified by work on leatherback turtles *Dermochelys coriacea* (Martin & James 2005, James et al. 2006) in Nova Scotia and the Canadian Atlantic, where a network of volunteer fishermen and tour boat operators helped to gather more than 800 geo-referenced records of this species.

Habitats and depth ranges

Probably because of easier accessibility, the most frequently studied habitats by marine citizen scientists were beaches (10.7%), the intertidal zone (22.6%), or some combinations of these shallow-water habitats, including particular habitats like estuaries and wetlands (6.3%), totalling 100 references for these coastal habitats (Figure 4). Coral reefs were also a popular subject for volunteers (10.7%), as well as other subtidal areas (28.6%) that can be directly reached using relatively simple methodologies such as scuba diving. Some emblematic coral reef studies provide excellent examples of worldwide initiatives supported by citizen scientists, for example, the global assessments of human effects on corals (Hodgson 1999) and the analysis of the effectiveness of marine protected areas (MPAs) in the conservation of coral habitats (Selig & Bruno 2010).

Other marine habitats targeted by voluntary studies include mudflats (e.g., Evans et al. 2000), mangroves (e.g., Aswani & Weiant 2004), salt marshes (e.g., Peloquin & Berkes 2009), seagrass beds (e.g., Lewis et al. 1998, McKenzie et al. 2000, Mellors et al. 2008, Garrick-Maidment et al. 2010), estuaries and fjord regions (e.g., Suazo et al. 2013). A few studies considered marine-related issues within urban areas, for example, public perception of risk from contaminated crabs and fish (Pflugh et al. 1999). As expected, projects on habitats beyond the reach of scuba divers, such as those involving the continental shelf, slope, or open ocean, accounted for only 6.4% of all studies. Such studies were generally based on logbooks and observations of amateur and professional fishermen.

In terms of depth range, about 57.7% of studies were performed on the seashore or in areas shallower than 10 m; 28.4% corresponded to depths accessible by regular scuba diving (i.e., <40 m). The remaining (13.9%) studies concerned deeper environments (>40 m), aiming mainly at deep-water fisheries, such as for rock lobsters *Jasus edwardsii* inhabiting depths down to 200 m around

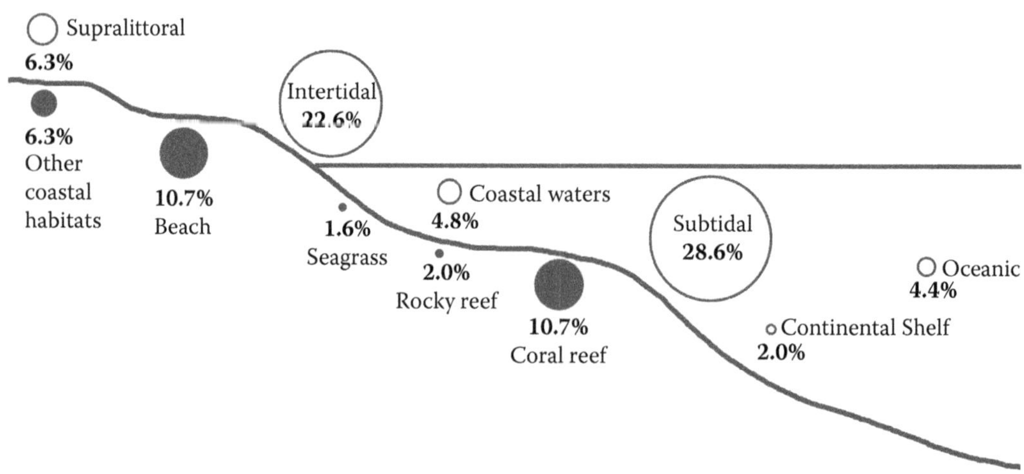

Figure 4 Frequency of marine studies with volunteer participation in particular habitats and depth ranges.

New Zealand (Starr & Vignaux 1997) and the black sea bass in the north-western Atlantic (Moser & Shepherd 2009).

Spatial coverage of studies

The most prolific country in terms of marine citizen science is the United States, followed by the United Kingdom, Canada, and Australia. The remaining studies are distributed over 55 different countries, but few of these have more than one study (Figure 5). Here, it is also interesting to note that citizen science in the marine environment is still an infrequently used approach in developing countries. There are generally few examples from South America, Africa, and Asia. From South America, most studies are from Chile and Brazil; for Africa, examples are from Kenya, Tanzania, Madagascar, and South Africa, whereas studies from Asian countries in our database are from Thailand and Indonesia.

In terms of the geographic extent of each initiative, most (122) were regional studies, and 40 were local studies (Table 3). The remaining 65 projects had wider geographical coverage, with 28 of national extent, 28 international, and 9 global. In general, international and global citizen science ventures focused on coral reef monitoring, fisheries, seabirds, and contamination. An example of a worldwide project studying human impacts on corals is CoralWatch (Marshall et al. 2012). For shorebirds, a

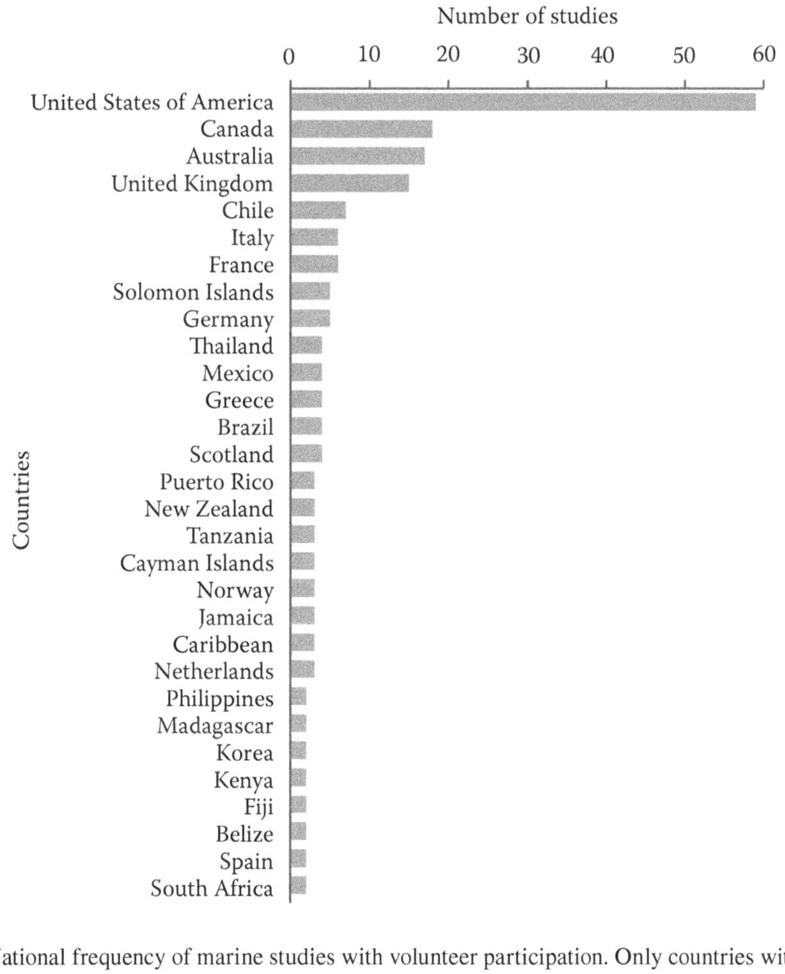

Figure 5 National frequency of marine studies with volunteer participation. Only countries with more than one study are shown.

Table 3 Number of studies with different spatial and temporal coverage

	Number of studies
Spatial coverage	
Local	40
Regional	122
National	28
International	28
Global	9
Total	**227**
Temporal coverage (days)	
1–10	10
11–100	26
101–365	35
366–3650	74
>3650	58
No data	24
Total	**227**

good example of an international initiative is the East Asian-Australasian Flyway long-term monitoring, involving more than 50 threatened bird species (Milton 2003). A massive international initiative involving more than 3000 volunteers in Mediterranean countries is the recreational monitoring of marine biodiversity within the programme Divers for the Environment: Mediterranean Underwater Biodiversity Project (Goffredo et al. 2010), targeting more than 60 taxa from algae to animals.

Figure 6 summarizes the distribution of marine citizen science efforts around the globe using the biogeographic classification for the world's coastal and shelf areas proposed by Spalding et al.

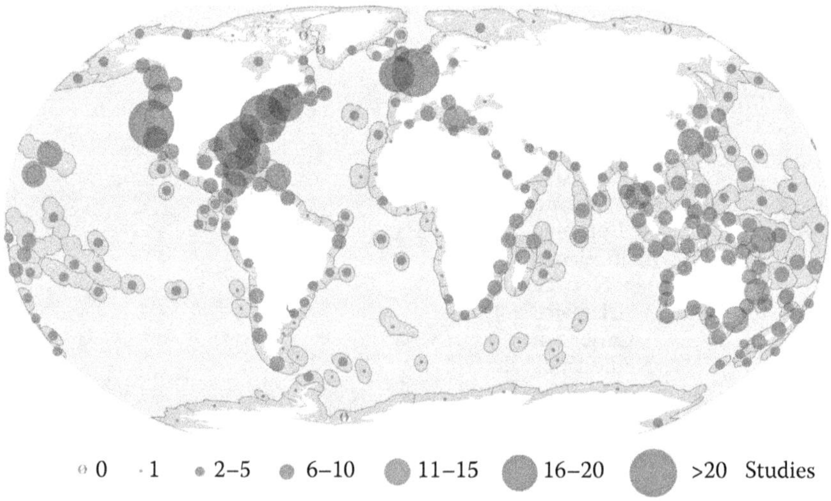

Figure 6 Geographical distribution of marine studies with volunteer participation in marine coastal ecoregions. Each ecoregion is characterized by one dot that is placed approximately in the center of the respective ecoregion. (Classification of ecoregions based on Spalding, M.D., Fox, H.E., Allen, G.R., Davidson, N., Ferdaña, Z.A., Finlayson, M., Halpern, B.S., Jorge, M.A., Lombana, A., Lourie, S.A., Martin, K.D., McManus, E., Molnar, J., Recchia, C.A. & Robertson, J. *BioScience* **57**, 573–583, 2007.)

(2007). There are hot spots of marine CSPs not only in North America and Europe but also in the Indo-West Pacific.

Methodology and duration of research programmes

Methodologies employed by marine citizen scientists are rather heterogeneous, but surveys, transects, or opportunistic sightings (incidental observations) are the most commonly reported approaches. Surveys generally comprise sightings of live (e.g., macroinvertebrates, reef fishes) or dead organisms (e.g., stranded seabirds, turtles, cetaceans) in planned areas or transects. These standardized surveys differ fundamentally from chance encounters (incidental observations), which in general are fortuitous and therefore unplanned (e.g., recovery of tagged fishes by fishermen, whale shark sightings by tour operators, finding of ringed birds by beach visitors). Additional methods include, for example, bibliographic surveys, as well as database and logbook analyses. Interviews and questionnaires are frequently used in studies of information provided by fishermen or tour operators. One particular approach recorded the behaviour of more than 50 Antarctic tourists and recovered biological data from information and pictures published in the blogs of these tourists (Roura 2012). In other studies, volunteers adopted a more active role in terms of sampling, for example, volunteers collecting water samples in the case of the long-term monitoring of red tides in South Carolina and in the Gulf of Mexico (Lewitus & Holland 2003 and Heil & Steidinger 2009, respectively) or recreational anglers tagging and releasing fishes in the United States (Loftus et al. 2000).

Sampling frequencies and duration of programmes vary widely (Table 3). At least 58 of the studies in our database covered a time span of more than one decade, the longest being more than 150 years of meteorological records (Kennedy et al. 2011). This is followed by a study with 55 years of cetacean stranding data (MacLeod et al. 2005) and the 53-year record of sharks and rays provided by divers to the magazine *Sport Diving* (Whatmough et al. 2011). Other examples include the 40-year study of seabirds in the North Sea affected by polyisobutylene and plastic contamination (Camphuysen et al. 1999, van Franeker et al. 2011); the 42-year record of stranded cetaceans on the French Atlantic coasts (Peltier et al. 2012); the 41 years of sightings of bottlenose dolphins along the Scottish coast (Cheney et al. 2013); and the 28 years of observations initiated by undergraduate students and subsequently including high school students, who monitored several sites on the rocky intertidal shorelines of the Monterey Bay National Marine Sanctuary, California (Osborn et al. 2005).

The majority of studies (102) were midterm monitoring programmes generally lasting a few years, and 37 studies were either projects run for a short time (weeks to months) or single opportunities (Table 3). An example of a citizen science study that was performed over only a couple of days is the volunteer survey of marine environments near Sitka, Alaska, for several invasive invertebrates (Cohen et al. 2011). The idea of this initiative was mainly to create awareness among the local population about these species. Experienced monitors helped in the training of volunteers in invertebrate identification. Laminated field handouts were distributed, and 10 sites were surveyed by groups of 2 to 10 volunteers during the two survey days. The main result was the confirmed occurrence of the colonial tunicate *Didemnum vexillum* in Alaska, representing about a 1000-km northward extension of its previously known range. The authors emphasized the power of a citizen science network that was large for detection of these invasive species (Cohen et al. 2011), a benefit also seen in other studies (e.g., Crall et al. 2010, Landschoff et al. 2013).

Characterization of volunteers participating in research

Volunteer participants included a diverse range of people differing in sex, age, educational level, scientific background, selection process, type of involvement (volunteer/hired), skills, experience, and incentive. Most studies were open to participants of all ages and educational levels, but usually there was a predominance of adult participants who had already passed or were in the final stage of their school education.

Demographic characteristics of volunteer participants

Of the 227 studies we reviewed, 206 made no reference to the gender of participants and most likely did not select participants according to gender (Table 4); 14 considered only men as volunteers, 1 only women, and 6 included mixed groups. For example, Alfaro-Shigueto et al. (2011) personally recruited and interviewed a group composed of six (male) fishermen to validate the information on a new record of the marine otter *Lontra felina* north of its current distribution in Chimbote, Peru. Another case is provided by Peloquin & Berkes (2009), who examined how (male) Cree indigenous people deal with uncertainties in the process of hunting geese in the Canadian Eastern Subarctic. Through sharing knowledge and scientific information hunters build up a collective understanding that leads to sustainable and adaptive strategies for hunting; for instance, space and time rotation for hunting geese is one of the noticeable practices used by this community. Aswani & Weiant (2004) developed and evaluated an interesting programme in the Solomon Islands, where shellfish collection is done by women. Using a participatory approach, the researchers and the women from one of the islands established an MPA, improving shellfish biomass by providing spatial and temporal refuges for marine invertebrates. This initiative also enhanced local environmental awareness,

Table 4 Demographic characteristics of volunteer participants in marine science studies

Sex	Number of studies	Percentage
Male	14	6.2
Female	1	0.4
Mixed	6	2.6
Not defined	206	90.7
Total	**227**	
Age	Frequency	Percentage
<18	25	10.5
≥18	149	62.3
Not defined	65	27.2
Total	**239**	
Educational level	Frequency	Percentage
Primary-secondary	17	7.5
High school	9	4.0
Superior	9	4.0
Post-graduate	1	0.4
Illiterate	2	0.9
Heterogeneous	12	5.3
Not defined	191	84.1
Total	**241**	

Note: Because of overlap in age and educational level in some categories, the total number of studies is larger than the number of studies in the database (Table 1). Percentages are based on all 227 studies reviewed.

maintaining a long-term commitment and community participation, by offering women a potential source for business.

Studies involved participants of all ages (Table 4), ranging from schoolchildren to older adults (10 to 80 years old), but most worked with adult volunteers above 18 years of age, whereas others involved the 'school-age category' (10 to 18 years old). No study specifically worked with senior citizens, even though they represent an important fraction of the volunteers collaborating in conservation programmes (O'Brien et al. 2010). Sixty-five studies gave no information about the age of volunteers.

Educational background of volunteer participants

Only 36 studies (22.0%) specified the educational level of participants, but as among these studies there were several with participants of variable educational background, the sum of all categories adds up to 50; 191 studies did not identify the level of education of the volunteer participants (Table 4). It is interesting to note that three studies exclusively employed volunteers with a higher educational level (Kirkpatrick et al. 2008, Beeden et al. 2012, Azzurro et al. 2013a); at the other extreme, there was one case of a study supported by mostly illiterate fishermen (Le Fur et al. 2011). Azzurro et al. (2013a) selected 30 highly trained biology graduate students for a study of diel rhythms in rocky-reef fishes. On the other hand, Grant & Berkes (2007) used a qualitative approach to conclude that fishermen (some of them without formal education) were able to structure a decision-making process (using applied and scientific knowledge) for finding and catching fishes.

In CSPs with elementary and high school students, as well as with citizens in general, surveys of litter or of seagrass were used for teaching purposes, providing training opportunities to address environmental problems or sustain biological monitoring initiatives (Jackson et al. 1997, Mellors et al. 2008, Bravo et al. 2009). Some studies also were based on partnerships with organizations that promote involvement of volunteer school groups (Kimball et al. 2009).

It appears that the educational level of participants was not mentioned in a large number of studies mainly because information was submitted by hundreds of diverse volunteers from a wide range of educational levels (Holt et al. 2010). However, because data quality might be influenced by the educational level of participants, information on educational background should be included in future CSPs to rigorously evaluate this potential relationship.

Selection of volunteer participants

About a third of reviewed studies did not mention whether volunteers were recruited through a selection process, about a quarter had no selection process, and the rest used a wide range of procedures to select participants (Table 5). Whereas some studies were conducted following a rigorous selective training programme (20 weeks for beginners), requiring marks of 95% or above on tests measuring identification skills (Bower 2009), others (e.g., Ellis & Cowan 2001) considered 'hands-on training in the field with a handbook' to be sufficient. In some cases, selection was done through a process of media announcements, assigning groups into teams based on their own interests and geographical location (Hill et al. 2002). Taking into account the diversity of different studies, it can be inferred that the need for participant selection depends on the type and context of the respective projects and the questions and tasks involved.

Most studies (97.8%) worked with unpaid volunteers, and only in a few (2.2%) did the citizen scientists receive some payment, mainly because work took a considerable amount of time and because participants belonged to an economically vulnerable population (Aswani & Weiant 2004). In some studies (3.5%), the participants were mixed groups of volunteers and paid citizen scientists.

Table 5 Criteria for selecting participants in marine citizen science programmes (n = 93 studies)

Kind of selection	Frequency	Percentage
Profession	46	20.3
Experience	20	8.8
Local/native	10	4.4
Education level	9	4.0
Special license	5	2.2
Personal motivation	4	1.8
Skills	2	0.9
Gender	1	0.4
Not defined	1	0.4
Total	**98**	

Note: Percentages are based on all 227 studies reviewed; some studies used several criteria to select participants.

According to Tulloch & Szabo (2012), unpaid volunteers are more responsive to intrinsic motivation (working for its own sake), whereas paid participants are driven by extrinsic forces (perceived personal benefits). In this sense, it would be important to know the motivation of participants prior to selection, thereby optimizing returns in terms of work accuracy and efficiency (Campbell & Smith 2006).

Skill and experience of volunteer participants

In the majority (58%) of reviewed studies, volunteer participants had no profession related to the marine environment, but in 42%, some volunteers were professionals with marine-related jobs (e.g., fishermen, divers, tour operators, and others) (Figure 7). Unsurprisingly, there seems to be a close relationship between the questions or tasks of respective studies and the professional occupation of volunteers (e.g., fishermen usually helped answer questions about the biology of fisheries resources). On the other hand, in studies for which volunteers had no professional relationship with the oceans, the questions and tasks could be highly diverse. For example, scientific volunteers from different backgrounds provided visual and physical evidence of predation of a group of killer

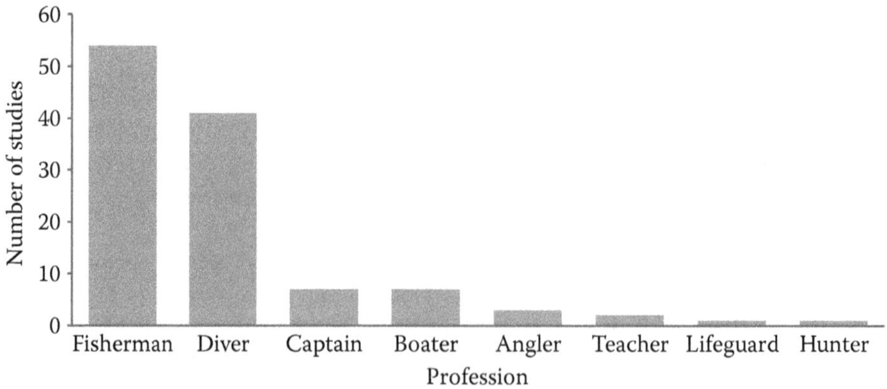

Figure 7 Frequency of studies undertaken by citizen scientists according to professional background.

Table 6 Experience levels of citizen scientists participating in marine studies

Experience	Number of studies	Percentage
Beginner	109	48.0
Basic experience	97	42.7
Advanced experience	139	61.2
Expert	159	70.0
Master	25	11.0
Not determined	16	7.0
Total	**545**	

Note: Beginner is a first-time participant; participants with basic experience have participated in one or two surveys and have some minimal experience; participants with advanced experience have participated in repeated surveys and gained more experience; experts have extensive experience that may be work related (e.g., fishermen, boat skippers, or dive instructors); masters have the maximum possible level of experience and often have received special training for the citizen science research they are conducting. Percentages are based on all 227 studies reviewed.

whales in one study (Ford et al. 1998), and citizen scientists of another study identified microalgae and estimated their abundances (Whereat et al. 2004). According to Greenwood (2007), it is possible that amateurs could be less experienced than professionals, but because amateurs are mostly local participants, it is assumed that they are more familiar with the study area, thereby compensating for differences in experience.

Many (55.9%) of the reviewed studies used volunteer groups that included participants with various levels of relevant experience, ranging from beginners to professionals. Most of these studies used volunteers with advanced or expert experience, whereas only a few worked exclusively with beginners, who represent a particular challenge for training (Table 6).

In a study of the home range and site fidelity of volunteers, Tulloch & Szabo (2012) found that participants with 'high site fidelity' (generally locals) achieved high species detection rates, beneficial for local long-term monitoring programmes. On the other hand, so-called roaming volunteers, generally tourists far from home, were more suitable for surveying threatened species.

Motivation of volunteer participants

The motives of volunteers participating in marine science studies are as diverse as their demographic and educational backgrounds. In most studies (84.6%), citizen scientists considered personal satisfaction and public recognition as most rewarding (Figure 8). Knowledge gained during participation (52.0%) also appeared to be a frequent incentive. Financial support, mixed purposes, and education were emphasized as incentives in a few studies (Figure 8).

Volunteer participation and contact with the marine environment seem to produce cognitive, emotional, and social benefits. Participants declared that working as a volunteer produces active learning and in turn activates memory processes (Koss & Kingsley 2010). Also, feelings of emotional satisfaction and sense of achievement are mentioned. Participation in CSPs also generated a positive sense of contributing to environmental protection, increasing self- and social awareness (Campbell & Smith 2006). Psychosocial characteristics of volunteers should be taken into account when planning future CSPs to achieve more reliable information that matches with realistic project goals.

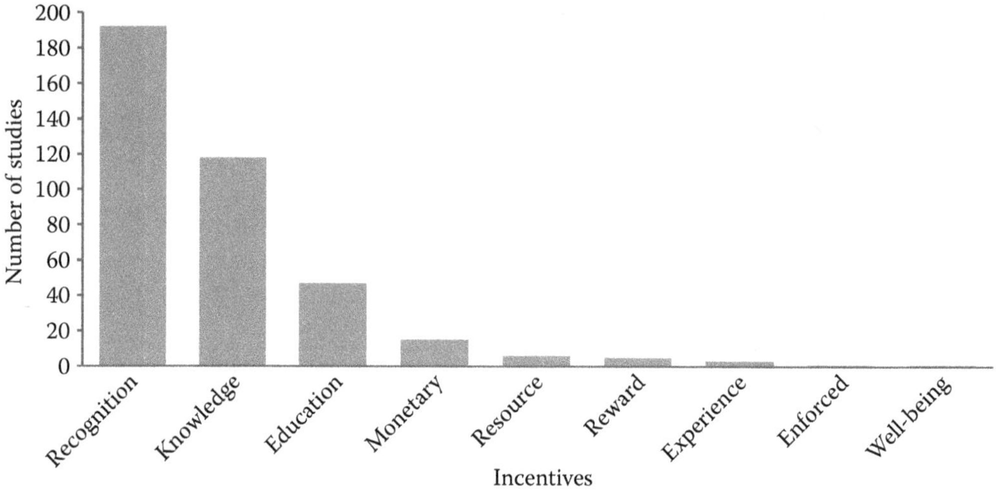

Figure 8 Motivation of volunteers participating in marine science studies.

Organization and logistics of volunteer groups

Invitation and communication

Gathering a large number of volunteers to participate in a scientific research programme is a challenge in itself, and diverse approaches are used to invite volunteers, including posters, newspaper articles, websites, personal conversations, and others (Table 7). Most commonly, researchers collaborated with organized volunteer groups such as conservation groups, diver associations, or birdwatchers. A typical example is the study by Johnston-González et al. (2008), who collaborated with the Colombian National Network of Birdwatchers to conduct surveys of shorebirds in Colombia. This network is part of a wider effort involving birdwatchers from many other countries in the neotropical region. Word of mouth (personal conversations) is another approach for recruiting

Table 7 Communication channels used to recruit citizen scientists

Calls for participation	Number of studies	Percentage
Organized volunteer groups	76	33.5
Personal conversations	34	15.0
Internet and websites	13	5.7
School program	7	3.1
Newspapers and magazines	5	2.2
Dive schools	4	1.8
Local communities	3	1.3
Posters, flyers, and newsletters	3	1.3
TV and radio	1	0.4
Others	2	0.9
Not mentioned	86	37.9
Total	**234**	

Note: Several studies use various communication channels. Percentages are based on all 227 studies reviewed.

volunteers (Table 7). For example, Almany et al. (2010) engaged in intensive conversations with the members of a fishing community and thereby obtained their agreement to participate in research on the connectivity of fish populations between coral reefs. In general, personal conversations are an important way of approaching fishermen (e.g., Hamilton et al. 2012, Suazo et al. 2013). When working with native people, personal communication also plays an important role, often involving multiple meetings, and even sharing the routine life of people (Le Fur et al. 2011). In urbanized settings or large-scale studies, e-mail and phone contacts can also be used (Mellors et al. 2008, Bravo et al. 2009).

Interestingly, electronic media are not often mentioned as tools for inviting participation, but given the rapidly increasing social relevance of these media, their use in CSPs is most likely underestimated in our database. This inference is supported by the relatively large number of organizations collaborating with citizen scientists that maintain a functional website (Table 8). Surprisingly, a large proportion of the examined studies did not report how volunteers were invited (Table 7), but because many of these were of local and regional extent, it is likely that local media (newspapers and radio programmes) and personal conversations played an important role in recruitment. Many of these last programmes also have been active for many years, suggesting that the local community is well aware of their existence.

Table 8 Organizations collaborating with citizen scientists that maintain an active website (as of 15 June 2014)

Organization/programme/network/survey	Website
Aquatic Mammal Rescue Program	http://www4.icmbio.gov.br/cma/index.php
Basking Shark Watch (BSW)	http://www.mcsuk.org/sightings/baskingshark.php
Beachwatch	http://farallones.org/
Caribbean Stranding Network	http://manatipr.org/sobre-nosotros/rcv/
Cetacean stranding network, United Kingdom, France, Denmark, Germany, Belgium	http://www.walvisstrandingen.nl
Christmas Bird Count	http://birds.audubon.org/christmas-bird-count
Científicos de la Basura	http://www.cientificosdelabasura.cl
Coastal Observation and Seabird Survey Team (COASST)	http://depts.washington.edu/coasst/
Coastwatch	http://www.coastwatch.org/Coastwatch.org/Home.html
COMBER	http://www.comber.hcmr.gr
Coral Cay Conservation	http://www.coralcay.org/
Coralwatch	http://www.coralwatch.org
Cousteau Society	http://www.cousteau.org
Earthwatch	http://www.earthwatch.org
ECOCEAN	http://www.whaleshark.org
Ellenic Network on Aquatic Invasive Species (ELNAIS)	https://services.ath.hcmr.gr/
Fish and Wildlife Research Institute (FWRI)	http://isurus.mote.org/niehsredtidestudy/INDEX1.htm
Independent	http://www.roatanmarinepark.com/lionfish/
International Shorebird Survey	http://ebird.org/content/iss
Jellyfish Working Group	http://www.jellywatch.org/blooms
LEMM	http://www.mammiferes-marins-bretagne.fr/php/observation/index.htm
Merri Marine Sanctuary Program	http://www.peopleandparks.org
Met Office	http://www.metoffice.gov.uk/hadobs/
Meteo Meduse reporting network	http://meteomeduse.focus.it/
Mobile Manatee Sighting Network (MMSN)	http://manatee.disl.org/

Continued

Table 8 (Continued) Organizations collaborating with citizen scientists that maintain an active website (as of 15 June 2014)

Organization/programme/network/survey	Website
National Audubon Society Christmas Bird Count (CBC)	http://www.audubon.org
National Sea Turtle Strandings and Salvage Network (STSSN)	http://www.sefsc.noaa.gov/species/turtles/strandings.htm
Nova Scotia Leatherback Turtle Working Group (NSLTWG)	http://www.freetheleatherback.com/
Opihi Program	http://www.hawaii.edu/gk-12/opihi/classroom_home.shtml
Our Sea of East Asia Network (OSEAN)	http://www.osean.net
Partnership for Interdisciplinary Studies of Coastal Oceans (PISCO)	http://www.piscoweb.org/
Port Townsend Marine Science Center (PTMSC)	http://www.ptmsc.org
Pro Delphinus	http://www.prodelphinus.org
Project Seahorse	http://seahorse.fisheries.ubc.ca/
Red de rescate y varamiento de tortugas marinas	http://www.karumbe.org/
Reef Environmental Education Foundation (REEF)	http://www.reef.org
Reef Life Survey	http://reeflifesurvey.com/
Reefcheck	http://reefcheck.org/
Sea Watch Foundation	http://www.seawatchfoundation.org.uk/
Seagrass Watch (SGW)	http://seagrasswatch.org/
SEANET	http://www.tufts.edu/vet/seanet/
Seasearch Project	http://www.seasearch.co.uk
Seawatchers	http://www.observadoresdelmar.es
SO COOL	http://coolgate.mote.org/beachconditions/
Sound Citizen	http://soundcitizen.org
South Carolina Harmful Algal Bloom Program	http://www.chbr.noaa.gov/pmn/
The Marine Science Group	http://www.marinesciencegroup.org
The Sea Horse Trust	http://www.theseahorsetrust.org/
University of Delaware Citizen Monitoring Program	http://citizen-monitoring.udel.edu/links.shtml
University of Maine	http://www.geology.um.maine.edu/beach

In some cases, citizen scientists were initially unaware that their data were used for scientific studies (e.g., recreational divers who volunteered catch data many years after recording them) (Godoy et al. 2010). A special case in this category is the study of Roura (2012), who reviewed the blogs of tourists who had visited Antarctica, collecting from them data on human impacts. Similarly, in the Gulf of California, where boaters report their destinations to a volunteer radio rescue service that tracks each boat until it safely returns to harbour, destination records were evaluated to determine the area use of recreational anglers before and after the establishment of an MPA (Fujitani et al. 2012). Those citizens were unaware that their records had contributed to science. This example also highlights that incorporation of volunteer data raises ethical questions (see also Riesch & Potter 2014): What information can be used, and do citizens have to be explicitly informed about the potential use of their data in scientific publications?

Communication with volunteers is not only crucial at the beginning of a study but also is important during the sampling period and should extend until at least the publication of results. Diverse media are utilized to maintain contact with citizen scientists throughout project development, although rarely mentioned. Newsletters can serve this purpose, and although printed newsletters were important in the past, currently they seem to be produced exclusively in electronic format, enabling their distribution to large numbers of participants via e-mail. Some studies also have an interactive website that allows citizen scientists to do basic analyses and generate simple graphics, which are of great value for keeping participants informed about the results of their efforts. Efficient communication with volunteer participants is especially important for long-term projects in which citizen scientists participate repeatedly in surveys. Camphuysen & Heubeck (2001, p. 449) nicely underscore the significance of frequent communication with citizen scientists: "The longest running surveys have tended to be those where a degree of contact is maintained between organizers and participants, either personally or through the circulation of reports, analyses and publications so that individual volunteers see that their relatively small contribution helps build a larger, meaningful picture."

Number of participants

The numbers of volunteers participating in marine science studies range from a handful of people to hundreds of participants (Figure 9). A few studies are based on small numbers of volunteers, who were often selected according to their background (e.g., small communities of fishing people) or training experience. For example, the study by Hamilton et al. (2012) was based on four participants from a small fishing community, all of whom were experienced divers and fishermen. On a small island (0.5-km diameter) in the Great Barrier Reef, a small group of volunteers from the local dive centre (dive instructors and boat skippers) helped to collect sighting data and behavioural observations on manta rays (Jaine et al. 2012). On the other hand, studies on beach litter often involved hundreds and in several cases even thousands of participants. For example, Kordella et al. (2013) engaged more than 15,000 volunteers in a litter survey of 80 beaches from Greece that was part of a wider initiative called Clean Up the Med. Long-term monitoring programmes of marine litter on beaches along the US coasts were also supported by hundreds of volunteers (e.g., Ribic et al. 2010, 2011, 2012).

Training of participants

Gathering scientific data with volunteers often requires some basic training, which many (41.4%) of the reviewed studies provided. Training methods were highly variable, ranging from brief instructions at the beginning of the surveys to long-term training, in some cases taking several days or weeks of intense training in the field (see also previous discussion). Some studies (26.9%) explicitly mentioned that no special training was provided, and the rest (31.7%) provided no information. However, in most of the latter cases it can be inferred that volunteers received written instructions, either on paper or via a website from which data sheets could be downloaded. For example, participants of litter surveys generally received simple data sheets with some basic written and oral instructions (Jóźwiak 2005, Bravo et al. 2009, Ribic et al. 2011). Also, particularly in studies

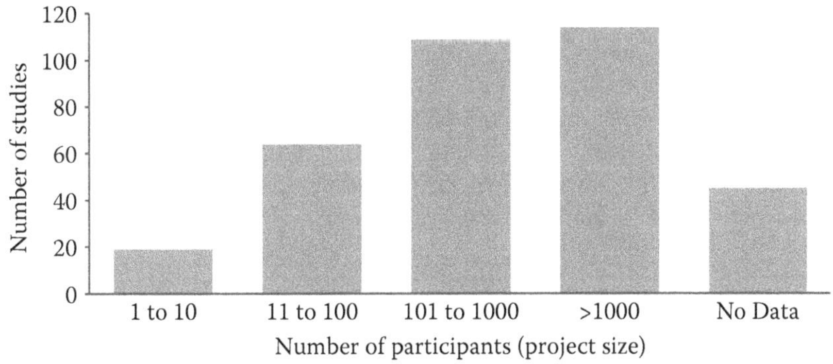

Figure 9 Frequency of marine science studies according to number of voluntary participants (citizen scientists).

of single, easily recognizable species, volunteers often began surveys on their own with only brief written instructions and record cards (Bramanti et al. 2011). Extensive presurvey workshops and presentations were often employed in studies where volunteers needed to distinguish different species and count or estimate their abundance (Whereat et al. 2004, Osborn et al. 2005, Mellors et al. 2008). In one study on the diet of gulls, volunteers were intensively trained in bolus dissection and identification of the food remains (Lindborg et al. 2012).

In addition to instructions or training, in many studies (28.2%) professional scientists accompanied volunteers during the research activities. Surprisingly, in many studies (48.5%) volunteers conducted the research without professional assistance or supervision. This could be because collection of data or samples was relatively simple, as is the case, for example, with surveys of stranded animals for which volunteers report findings to professionals, who on arrival identify the species and take samples (e.g., of tissues or bones), or when beach visitors report stranded turtles, and scientists then recover and process the carcasses for further analyses (Avens & Goshe 2007, Avens et al. 2009). Such processing of samples by professional scientists is a common feature of studies of stranded animals (e.g., Camphuysen 1998, van Franeker et al. 2011).

In general, training effort increases with the complexity of the tasks conducted by volunteer participants. People who make incidental observations usually receive no training or instructions (e.g., a beach visitor who finds a ringed bird and depends exclusively on the information provided on the ring). An interesting strategy was chosen by a bird-ringing association in the United States; in this study, a full postal address and a toll-free phone number were stamped on larger rings, substantially increasing reports from the general public (Tautin et al. 1999). In other projects, volunteers completed more complex tasks, in some cases resembling the work of professional scientists. Here, citizen scientists were often selected based on their previous experiences and skills, and they received substantial training before collecting data in a planned and standardized sampling design (e.g., Edgar & Stuart-Smith 2009).

Tools and materials provided to volunteers

Even the most basic scientific research requires some minimal materials, such as data sheets, sampling devices, or other aids. In many studies, the citizen scientists received data sheets and other written support materials from the organizers (Table 9). Educational materials (e.g., booklets and identification charts) were also commonly provided. In some studies, the organizers provided more expensive equipment (GPS, cameras, binoculars, microscopes, etc.) to the volunteers (Table 9). For example, in a study monitoring red tide microalgae, the trained participants received a microscope to identify and estimate the density of algal cells in water samples that they took at preassigned sampling locations close to their homes (Whereat et al. 2004). In another study of red tide algae, lifeguards received a personal digital assistant that allowed them to submit observations to a web-based platform in real time (Kirkpatrick et al. 2008). Baker & Oeschger (2009) took an interesting approach to obtain new information for fisheries management: They provided prepaid mobile phones to recreational anglers, who recorded important variables, including fishing effort and species captured. Provision of expensive tools, however, is uncommon in CSPs. In most studies, only basic sampling equipment (quadrats, vials, data sheets) are provided to the volunteer participants. Often, the volunteers use their own equipment; for example, divers used their own dive gear or birdwatchers use personal binoculars and spotting scopes.

Institutional and legal framework

A wide range of different institutions coordinates projects that engage citizens in scientific activity in the marine realm (Table 10). The majority of studies are supported by independent groups, and citizen science organizations (e.g., REEF, Coastwatch, Seagrass Watch) coordinate one-quarter of

Table 9 Materials and tools provided to participants in marine citizen science studies

Materials and tools	Number of studies	Percentage
Data sheets (diaries, paper, record books)	49	21.6
Didactic materials (comics, posters, pictures)	17	7.5
Species identification guides/charts	12	5.3
Questionnaire	12	5.3
Global positioning system device	10	4.4
Camera	8	3.5
Quadrats	8	3.5
Diving equipment	8	3.5
Sampling guides/instructions	7	3.1
Tags	7	3.1
Software/online platform	6	2.6
Binoculars	4	1.8
Personal computer	3	1.3
Cell phone	2	0.9
Smartphone application	2	0.9
Sampling kit	2	0.9
Microscopes	2	0.9
Fishing nets	1	0.4
Other measuring devices (rulers, thermometers, weights, sieves)	13	5.7
Other materials (buckets, magnifying glasses, ropes)	6	2.6
No data	116	51.1
Total	**295**	

Note: Percentages are based on all 227 studies reviewed.

Table 10 Institutions organizing and supporting marine citizen science studies

Institutions	Number of studies	Percentage
Non-governmental organizations	44	19.4
Governmental organizations	13	5.7
Long-term research programmes	34	15.0
Research groups	11	4.8
Reporting networks	15	6.6
Specific research projects	13	5.7
Surveys	10	4.4
Independent studies	96	42.3
Total	**236**	

Note: Percentages are based on all 227 studies reviewed.

the CSPs. Forty-nine reviewed studies were conducted within the framework of larger research programmes and networks (Table 10). Some of these networks span many different countries across the entire globe (Hodgson 1999, Bruno et al. 2009, Marshall et al. 2012).

The comparatively large number of studies that deal with endangered or emblematic species (see various previous sections) or are being conducted in MPAs suggests that the presence of certified professionals is needed or that permits are required for the research activities. A large number of studies are conducted under the auspices of government organizations (Table 11), which likely

Table 11 Administrative support and legal framework of marine citizen science studies

Legislative support	Number of studies	Percentage
Government	88 (3)	38.8
Non-governmental organization	72 (1)	31.7
No special permission needed	64 (37)	28.2
Special permission obtained	2	0.9
No data	17	7.5
Total	**243**	

Note: Percentages are based on all 227 studies reviewed. Numbers in parentheses represent those studies for which the category was inferred based on the study description.

facilitates the issuing of permits where needed. For many studies, no special permits are needed, for example, those on marine litter on public beaches. A large proportion of studies did not explicitly mention whether permits were needed or obtained for the research with citizen scientists (Table 11).

Coordination and funding of research activities

The coordinators (principal investigators) of marine CSPs have diverse professional backgrounds but usually include biologists (Table 12). In all reviewed studies, at least some of the organizers had passed the higher educational system. Overall, little information could be obtained about the organizers or coordinators of marine citizen science initiatives (for coordinators of terrestrial CSPs, see Riesch & Potter 2014). Given that the coordinators are the driving force behind CSPs, it appears a timely task to examine their experiences and motivations. Furthermore, because the coordinators are the people who know their projects best, and who can identify the things that work and

Table 12 Professions of coordinators or organizers of marine citizen science studies

Profession	Number of studies	Percentage
Biologist	223	98.2
Conservationist	130	57.3
Manager	37	16.3
Anthropologist	18	7.9
Ornithologist	16	7.0
Environmentalist	3	1.3
Ecologist	3	1.3
Oceanographer	3	1.3
Mathematician	2	0.9
Geologist	1	0.4
Data analyst	1	0.4
Meteorologist	1	0.4
Psychologist	1	0.4
Sociologist	1	0.4
Taxonomist	1	0.4
Teacher	1	0.4
Total	**442**	

Note: Percentages are based on all 227 studies reviewed.

Table 13 Financial support of marine citizen science studies

Funding sources	Number of studies	Percentage
Government	117	51.5
Non-governmental organization	127	55.9
Private	150	66.1
No data	14	6.2
Total	**408**	

Note: Percentages are based on all 227 studies reviewed. Some studies have multiple funding sources.

those that do not, utilizing such first-hand knowledge would also be highly instructive for future citizen science initiatives.

Financial support for CSPs typically comes from governmental institutions but also to a large extent from non-governmental organizations (Table 13). Private funding support is also common and usually refers to citizen scientists using their own equipment or paying for paper, gasoline, and the like out of their pockets.

Evaluation of volunteer-contributed data

Data management and processing

Data obtained by volunteers are submitted in various forms, mainly as hard copy to local coordinators or as electronic data to publicly accessible websites (see, e.g., Table 8). Usually, data are checked before inclusion in larger databases. To maximize efficiency of storage, retrieval, and analysis of data, it is important to standardize the way in which data will be registered (Crall et al. 2010, Kelling 2012, Pulsifer et al. 2012); this refers to variables such as sampling date, geographical coordinates, and species names or any other categorical variable.

In most studies (52.9%), it was not explained how data were processed after field surveys, but from the publications we reviewed, it can be inferred that data were available as hard copy or in digital format. In the majority of studies, initial data recording was made on hard copies, but in many cases data were subsequently transferred to digital files (Table 14). In some cases, visual information (videos, photographs) was also stored.

An increasingly common way of communicating in CSPs is through the use of smartphones and their applications. The use of these devices improves the quality of the survey process (Crall et al.

Table 14 Storage media for the data generated in marine citizen science studies

Storage medium	Number of studies	Percentage
Hard copies	190 (123)	83.7
Digitalized files	152 (121)	67.0
Pictures	22	9.7
Video	6	2.6
Audio	2	0.9
Total	**372**	

Note: Numbers of studies represent the total of studies in each category, and numbers in parentheses represent those studies for which the category was inferred based on the study description. Percentages are based on all 227 studies reviewed.

2010), taking advantage of mobile Internet connectivity, the possibility of geotagging and adding comments to photos or video, plus the immediate integration of data into a mapping program or web page (e.g., Boero et al. 2013, Martin 2013). The smartphone is a promising tool for the next generation of CSPs, but efficient peer review or expert validation after initial uploading of the records appears particularly important to guarantee data quality.

Quality control of marine citizen science data

In their highly valuable 'Guide to Citizen Science', Tweddle et al. (2012) suggest two complementary processes for assessing data collected by volunteers: validation and verification. Validation is defined as the process of checking if specific information satisfies a defined criterion and can therefore be interpreted correctly so that its subsequent use will be free of misinterpretation (e.g., Fowler et al. 2013). Verification is defined as the process designed for checking data accuracy by means of statistical methods. In most studies, however, no distinction is made between the two processes (validation and verification), and the procedure for checking the quality and reliability of the data at any stage of the study is interchangeably called quality control, validation, or verification. Hitherto, we primarily examine the methods, rather than the concepts, used to measure the validity and accuracy of data collected by volunteers.

Evaluation of data usually is conducted by professional scientists (information taken from the affiliation of authors of the corresponding publications). Wherever possible, statistical methods should be used to measure accuracy of the data collected by volunteers using a set, or even a subset, of data generated independently by a scientist or trainer with certified skills. Because no filter or other quality measure can account for all potential error and bias, advanced statistical tools can also be used when evaluating the data collected by volunteers (Bird et al. 2013).

Of the 227 studies we examined in this review, 55.1% included some type of verification mechanism to ensure the quality of information. For example, Starr & Vignaux (1997) validated the reliability of logbook data on catches of rock lobsters *Jasus edwardsii* through comparison with a catch-sampling programme (for further examples in the fisheries context, see, e.g., Marrs et al. 2002, Starr 2010). In the majority of cases, the volunteers were trained to apply sampling methods and to recognize species. Delaney et al. (2008) pointed out that education of participants is an important component of data quality. In their study, education was a significant predictor of the ability of participants to correctly identify crab species and sex: Although young students (grade 3) were able to differentiate between species of crab with 80% accuracy, seventh-grade students achieved greater than 95% accuracy. In addition, Jiguet (2009) reported that data quality increased with experience of the volunteers; also, availability and familiarity with protocols seemed to improve identification of collected material. In some cases, volunteers received intensive training and were accepted for participation only after passing a final evaluation (e.g., Bower 2009). In contrast, data provided by birdwatchers from ornithological organizations (e.g., Christmas Bird Counts, USA) were considered sufficiently reliable without any validation step. If questionnaires and interviews are used to obtain important information from citizens (e.g., fishermen), it has to be assumed that answers are given in good faith (Baird et al. 2002).

There seems to be an implicit consensus among scientists that training significantly increases the quality of the data stemming from volunteer surveys. An example is given by a study in which community variables from reefs in MPAs and adjacent fished zones were compared using a similarity index (Edgar & Stuart-Smith 2009). The ratio of similarity indices taken by volunteers and trainers was regressed against the number of training dives to assess whether performance of volunteers improved with training (Figure 10). Data produced by volunteers that fell within 10% of the mean similarity of data produced by the trainers were considered to be of a standard adequate for use in the study. Also, a minimum threshold of data quality collected by volunteers may be used to ensure acceptable accuracy. For example, in a study on abundance and diversity of crabs, counts made by

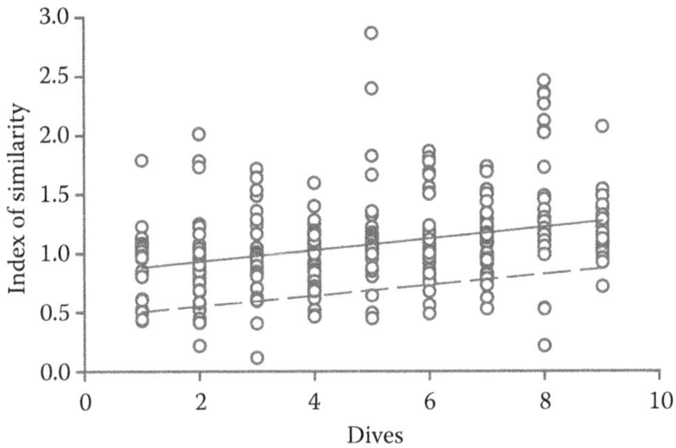

Figure 10 An example of validation process in a study supported by citizen scientists. The performance of the volunteers in estimating macroinvertebrate abundance increased with the number of training dives. Only data within 10% (i.e., Index of Similarity > 0.9) of the mean similarity produced by trainers (dashed line) were considered for subsequent analyses. (From Edgar, G.J. & Stuart-Smith, R.D. *Marine Ecology Progress Series* **388**, 51–62, 2009.)

volunteers were excluded from analysis if they differed by more than 3.5% from counts made by scientists (Culver et al. 2010).

Of the 125 studies with validation, 51 used a statistical test to verify the accuracy of data taken by volunteers. In certain cases, Student's *t* test (for paired or independent samples depending on the grade of independence of the replicates) is useful to test for significant differences between data from volunteers and from a control (Hidalgo-Ruz & Thiel 2013). The most common tests used were the correlation coefficients and the Kappa coefficient. For example, in the study by Evans et al. (2000), volunteers were trained for 2 days before sampling occurred. A randomly selected subsample was independently analysed by a scientific expert, and correspondence between the two sources of data (experts and volunteers) was evaluated by correlation analysis (see Figure 11A and 11B). Goffredo et al. (2010) used a similar validation process for a study of habitat variables, species richness, and abundance. Volunteer data, expressed as the Marine Biodiversity Index, were compared with those collected by a control diver (Figure 11D).

Different statistical methods can also be applied at several stages of the verification process, as was done by Goffredo et al. (2010), who used a correlation test (Spearman rank and Cronbach's alpha) and Czekanowski proportional similarity index. The Czekanowski index is an objective and simple estimate of the area of intersection between two frequency distributions, in which the value of the index ranges from 1 (highest similarity) to 0 (no similarity). The levels of accuracy are typically reported in terms of percentage, ranging between 75% and 95%. The use of the confidence interval around the mean or around the line of the best-fit equation (Figures 11C and 11D) enables the detection of outliers and their exclusion from analysis if they are identified as mistakes (Hidalgo-Ruz & Thiel 2013). The least accuracy is typically observed in studies that require the taxonomic identification of species (e.g., Bell 2007). Estimation of abundance is less prone to errors by volunteers, and accuracy commonly reaches high values of agreement between volunteers and professional scientists (95%; Delaney et al. 2008). In a study on reef fishes, experts consistently recorded higher densities than volunteer divers, but overall effects (higher fish densities in a no-take marine reserve) observed by both groups were similar (Hassell et al. 2013).

In the case of categorical variables (i.e., sex, presence or absence of species), logistic regression can be used to obtain a measure of accuracy by comparing data collected by volunteers and those validated by scientists or trainers, as in work by Delaney et al. (2008), who collaborated with

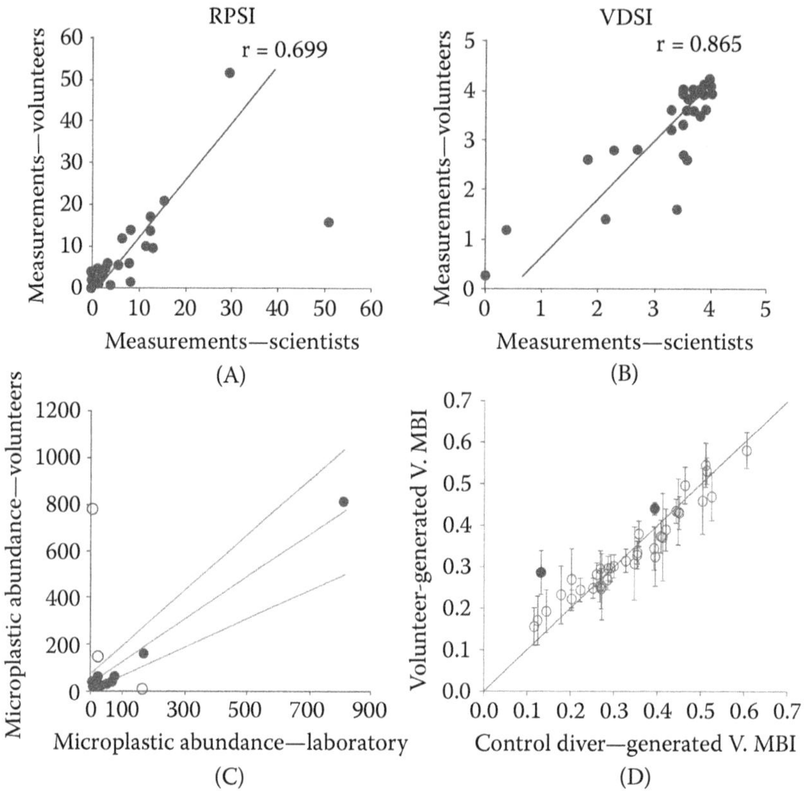

Figure 11 Correlations between data by volunteer participants and by professional scientists. (A) and (B) Data from a study of imposex in *Nucella lapillus*. (Modified after Evans, S.M., Birchenough, A.C. & Fletcher, H. *Marine Pollution Bulletin* **40**, 220–225, 2000.) (A) RPSI (Relative Penis Size Index) is the length of the penis in both males and those females possessing one. (B) VDSI (Vas Deferens Sequence Index) is a six-point score of imposex estimates based on the development of the vas deferens (a score > 5 corresponds to a sterile animal). (C) In a study of microplastic abundance, volunteer data were contrasted with recounts made in the laboratory by a scientist. The regression analysis revealed three data points that fell outside the 95% confidence interval; after removing these outliers (open dots), the correlation coefficient increased from 0.681 to 0.984. (Modified after Hidalgo-Ruz, V. & Thiel, M. *Marine Environmental Research* **459**, 121–134, 2013.) (D) Comparison of marine biodiversity index (V.MBI) values calculated from volunteer data with those from control divers (original values were transformed to positive numbers). In 36 of 38 trials (94.7%), the values generated by volunteers were not significantly different from those of control divers. Black dots indicate volunteer-generated values that are significantly different from values obtained by control divers. (Modified after Goffredo, S., Pensa, F., Neri, P., Orlandi, A., Gagliardi, M.S., Velardi, A., Piccinetti, C. & Zaccanti, F. *Ecological Applications* **20**, 2170–2187, 2010.)

volunteers to assess the presence/absence and sex of an invasive and a native species of crab. Also, error matrices are used in the validation of studies on the conservation status of widely distributed organisms (e.g., kelps; Monk et al. 2008, Assis et al. 2009) and on the diversity of taxa for which identification requires specialized knowledge (e.g., sponges; Bell 2007). In such cases, validation is made by randomly selecting a number of sampling sites or quadrats surveyed by scientists. Error matrices for data collected by volunteers and scientists are used to compute the Kappa coefficient, measuring proportional agreement between the two sets of data (e.g., Assis et al. 2009). For example, Monk et al. (2008) used the Kappa coefficient in a study in which data from a community-based monitoring programme were integrated with a geographical information system to create maps of the marine habitat. In general, verification through error matrices is recommended when data include different types of factors, such as abiotic variables and species identity.

Recommendations for future studies with volunteer participation

It seems that in most citizen science studies of the abundance of marine organisms, quality lies within the range of variation for data collected by professional scientists (Foster-Smith & Evans 2003, Cox et al. 2012). For counts of one or more species or of specific objects (e.g., microplastics or litter), a variety of statistical methods is available to test for the accuracy. However, the strongest concern about data quality is principally related to the ability of volunteers to identify species correctly, especially in studies involving diversity or species richness at several community levels. Thus, it is necessary to devote great effort to training volunteers in taxonomic identification to validate data and promote confidence, either for purely basic or applied purposes.

In estimating the abundance of fishes, invertebrates, and sessile species, a variety of methods has been used, including direct counts and comparatively subjective methods (e.g., coverage or relative abundance), with the latter subjected to the influence of the educational level, experience, and specific skills of the volunteers. Also, it seems that there is a risk of sampling bias resulting from sampling efforts that are unevenly distributed in space and time. The use of standardized sampling protocols, such as transect or quadrat methods, may reduce bias and increase accuracy in data collected by volunteers.

In measuring accuracy, it is possible that data are not always independent, and some type of pseudoreplication is present. Another common problem for statistical analysis is disparate sample size among groups. Randomization routines and generation of null models would help to overcome such difficulties, and consultation of a statistician is recommendable. For those cases in which the assumptions of normality and homoscedasticity of variance are not fulfilled, even after transformation, it is possible to apply a permutation test that is generally more powerful than a traditional non-parametric test. Scientists can also employ advanced statistical methods for evaluating data from CSPs (Bird et al. 2013).

To improve the quality of data collected by volunteers, a minimal number of steps for reducing sources of inaccuracy, error, and variation can be implemented during the design phase of the study. We recommend considering the following aspects:

1. *Simplicity of the task*: Several studies emphasized that the research questions investigated by citizen scientists should be straightforward and adjusted to the skills and capabilities of the participating volunteers. This aspect was nicely synthesized by Riesch & Potter (2014, p. 112), who emphasized that CSP coordinators should "make sure the questions are simple and easy enough that there is little that can go wrong". For example, species to be surveyed should be selected according to the capabilities of the participating volunteers to guarantee correct identification (Goffredo et al. 2010).
2. *Training and support materials*: Correct identification of species, items, or categories (e.g., type of plastic, sex, adult or juvenile) can be improved by training, which should be adequately implemented before volunteers collect data. In multispecies studies, it will be of great help to register all species or items present that are recorded by citizen scientists after repeated surveys. Although it is possible that an exhaustive list of species is unavailable, a reference list (as complete as possible) might be compiled. Based on such a reference list, it is possible to rapidly detect species or items (or categories) recorded by volunteers that are highly unusual and hence should be rechecked by professionals or be categorized as erroneous records.

3. *Consistency*: If data are collected by different groups of citizen scientists at the same place (quadrats or transects), it is desirable that the separate sets of data have high similarity. Here, it is assumed that sources of error are the same for all the cases. Consistency can be measured using correlation methods or the Czekanowski index for data expressed as frequency distributions.
4. *Representativeness*: Studies relying on incidental reports from volunteers may not be representative for the entire range or temporal presence of a species. These shortcomings can be overcome by complementing data sets generated by volunteers with the coordinator's own systematic observations or measurements, other sources or pieces of information, or improving the probability of encounter. For example, Björnsson et al. (2011) showed that mark-recapture studies of cod *Gadus morhua* around Iceland could be improved by double tagging each individual, which increased the reporting rate by the commercial fleet.
5. *Abundance estimation*: Abundance assessments based on direct counts are less prone to produce errors than quadrat estimates of per cent coverage. Correct estimations are influenced by the ability in estimating coverage, but also depend on the correct identification of species, and again, training can enhance accuracy. In other cases, abundance is quantified using an ordinal scale (e.g., high, medium, low), which needs prior clear definition of the criteria for each category. In such cases, unusual values should be scrutinized for accuracy.
6. *Accuracy estimation*: The statistical test needed for estimating accuracy depends on the type of data collected, and although the majority of tests used are relatively straightforward (e.g., correlation methods and error matrices), it is wise to seek expert help in the selection of an appropriate method. Although a high level of accuracy is desirable, complete similarity between data generated by volunteers and the control (i.e., scientists, trainers, or experts) can be difficult to achieve under certain circumstances, depending on the skills of participants, and conditions in which data are collected (e.g., diving depth, weather, equipment availability).

If verification is made using photographs, authors usually do not include any statistical evaluation to obtain a quantitative assessment of correct identification, but efforts could be made to obtain some assessment of accuracy. For example, a chi-square procedure may be applied to test for significant deviation from a hypothesized expected proportion (e.g., 95% correct vs. 5% incorrect) of a number of photographs verified by an expert.

The slogan 'science by the people' promoted by Silvertown (2009) sounds not only distinctive but also necessary in the present and future scenarios of dramatic changes caused by humans in the marine environment. Citizens are increasingly interested in participating in the process of testing sound hypotheses related to the conservation of marine resources and biodiversity. However, as also emphasized by Silvertown, 'data collected by the public must be validated in some way'. Clearly, from our review, validation has not yet become a standard procedure in marine CSPs, a situation that perhaps can be improved with the aid of journal editors during the peer review process. On the other hand, the high quality of information in those studies for which the accuracy of data collected by volunteers has been properly measured is encouraging. This is especially so because certain studies included a number of community variables, abundance estimates, and a large number of species, some of which are difficult to identify even for experts. Thus, in planning volunteer-supported studies, we strongly recommend implementation of the suggestions given and proper consideration of the basic principles of experimental design.

Principal contributions to scientific knowledge

Roughly two-thirds of the reviewed studies addressed a scientific question; the rest dealt with general aspects of citizen science, analysing its potential contribution to professional science

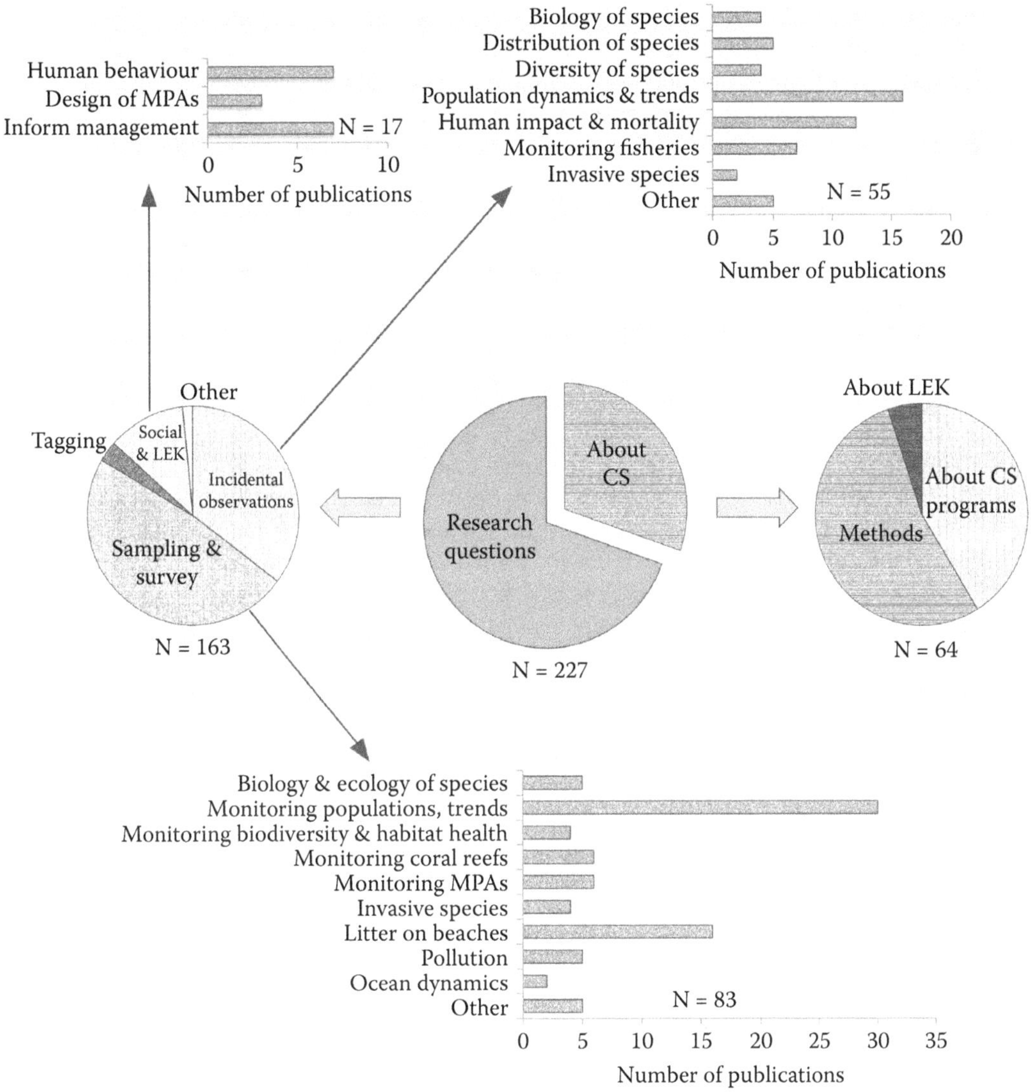

Figure 12 Main areas in which citizen science contributes to improved knowledge in marine sciences.

(Figure 12, pie chart at the centre of the figure). In the first group, we considered all those publications that reported and analysed new or novel information; these were dominated by two main types of contribution (Figure 12, pie chart at the left of the figure): (1) incidental observations, in which volunteers recorded and reported random observations or sightings made during work or recreation, and (2) surveys or samplings, in which volunteers joined an organized workforce to make observations or gather data about organisms or phenomena covering a specific area or time frame, often employing standardized methodologies normally used in scientific studies (transects, quadrats, etc.). In addition to those two categories, we distinguished a small fraction of studies in which volunteers reported tagged organisms and also those in which citizens contributed to science by sharing with the researcher (e.g., through interview or questionnaire) their own TEK (Figure 12). In addition, we included a few studies of human behaviour, which in various forms were based on volunteer participation.

Incidental observation by citizens can be considered analogous to remote sensing, extensively used in oceanography or meteorology. In this context, citizens act as 'probes' to record data to be collated and analysed by researchers. Participation of large numbers of volunteers thus enhances research capacity by contributing information about species or events over long temporal or extensive geographic scales (Theberge & Dearden 2006). This is of particular importance for studies of rare and endangered species (as, for example, manatees; Pabody et al. 2009) and for collecting considerable amounts of data at low cost (Goffredo et al. 2010). Such information, when assembled in long-term databases, has provided information mainly on the population dynamics, trends, or health of diverse taxa (Figure 12) (e.g., Norris et al. 2007, Bramanti et al. 2011, Casale et al. 2011, Jaine et al. 2012, Peltier et al. 2012, Whitney et al. 2012).

Incidental observations of stranded organisms reported by citizens contributed importantly to the study of mortality, especially anthropogenic causes (Figure 12), of diverse taxa (Camphuysen 1998, Meirelles et al. 2009, Parente et al. 2004, Whitney et al. 2012). Also, monitoring of fisheries for management purposes benefitted from the recording and reporting of catches by fishermen (e.g., Obura 2001, Obura et al. 2002, Sáenz-Arroyo et al. 2005, Godoy et al. 2010, Brenier et al. 2011, Humber et al. 2011). This type of incidental contribution is probably strongly underrepresented in our database, as it has only relatively recently become common practice: Almost all fisheries assessments are currently based on catch data delivered by those involved in the activity, with the contributor no longer specifically acknowledged. Fishermen may contribute information on a voluntary basis, but in many cases, they are also required by law to provide or allow independent observers to quantify catch data (MacKenzie & Cox 2013).

Contribution to basic biological and ecological information was less frequent through incidental than planned observation (Figure 12). This was because of various shortcomings of opportunistic samplings that reduced the representativeness of data. For example, incomplete geographic coverage and effort, bias introduced by the difficulty of detecting less-conspicuous species, failure to record all occasions or sites in which the target species was not observed, among other reasons, led to problems in interpreting the information (Mignucci-Giannoni et al. 1999, James et al. 2006, Jung et al. 2009, Goffredo et al. 2010, Bird et al. 2013). Even the reporting behaviour of citizens may change over time, as revealed for ring recovery rates for ducks and water birds (Robinson et al. 2009, Guillemain et al. 2011) (Figure 13A and 13B). There may also be regional differences in reporting behaviour, such as was found for fishermen reporting rings of seabirds drowned in fishing gear (Lyngs & Kampp 1996).

Complementing incidental observations of volunteers with other sources of information, or own records, samplings or sample analysis, overcame some of these problems and enabled useful knowledge to be gained about the biology (Ford et al. 1998, Avens & Goshe 2007, Casale et al. 2011), distribution (Semmens et al. 2004, Boero et al. 2009, Alfaro-Shigueto et al. 2011, Biggs & Olden 2011), and diversity of species (Stacey et al. 1989, Willis & Baird 1998, Marigo & Giffoni 2010) (Figure 12). Notable in the context of complementing diverse sources of information are the publications of Condon et al. (2012), questioning the current paradigm regarding the rise of gelatinous zooplankton in the world oceans, and Norris et al. (2007) concerning the diet and population dynamics of a threatened bird species. In studies of biological processes such as reproductive behaviour or growth of turtles, volunteers can make helpful contributions by collecting and delivering skeletal parts to the researcher (Avens & Goshe 2007, Casale et al. 2011), or in the case of the diet of a whale, alerting and guiding the researcher to feeding events (Ford et al. 1998). Even harmful algal blooms (HABs) are monitored with the aid of incidental observations, by laboratory analysis of stranded birds found by volunteers (Vargo et al. 2009).

The standardized samplings and surveys through which volunteers contribute to marine research enable long-term monitoring programmes to be maintained for diverse purposes (Figure 12), such as monitoring the population health, size, and trends in distribution over time (Howe et al. 1989, Milton 2003, Bruno & Selig 2007, Bower 2009, Holmberg et al. 2009, Pierpoint et al. 2009, Stallings

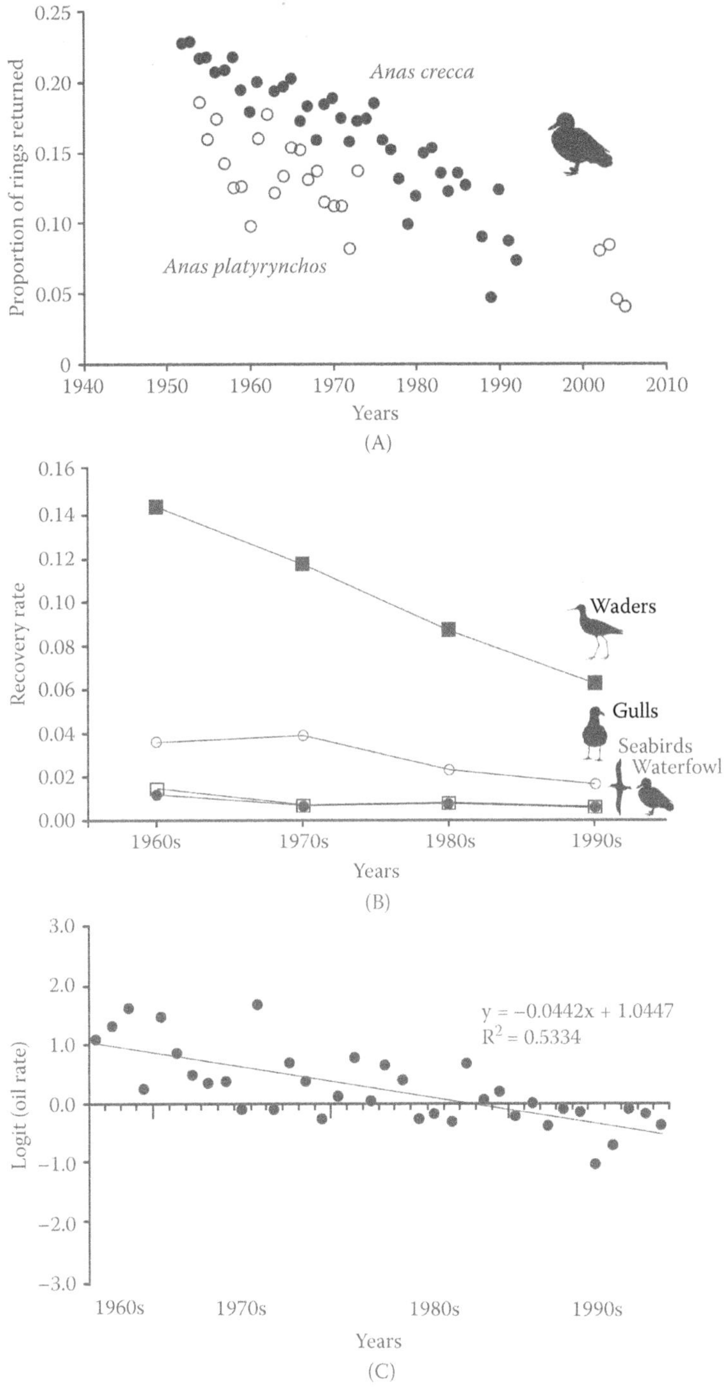

Figure 13 Temporal trends of data collected by citizen scientists. Ring recovery rates of diverse species of (A) ducks and (B) aquatic birds. (Modified from Guillemain, M., Devineau, O., Gauthier-Clerc, M., Hearn, R., King, R., Simon, G., & Grantham, M. *Journal of Ornithology* **152**, 55–61, 2011, and Robinson, R.A., Grantham, M.J., & Clark, J.A. *Ringing & Migration* **24**, 266–272, 2009.) (C) Changes in oiling rates of stranded birds in the Netherlands, showing decline of oil pollution over a period of more than 30 years. (Modified from Camphuysen, C.J., & Heubeck, M. *Environmental Pollution* **112**, 443–461, 2001.)

2009, Ward-Paige et al. 2011, Witt et al. 2012, Hussey et al. 2013) for conservation and management. Volunteers also make important contributions to the monitoring of MPAs (Worachananant et al. 2007, Monk et al. 2008, Edgar et al. 2009, Edgar & Stuart-Smith 2009, Yasué et al. 2010), coral reefs, and biodiversity in general (Hodgson 1999, Crabbe et al. 2004, Bruno et al. 2009, Goffredo et al. 2010, Selig & Bruno 2010, Jang et al. 2013), delivering basic information necessary for the management of these areas or habitats. Many initiatives enlist and enable divers and other marine enthusiasts to become active ocean stewards and citizen scientists, teaching how to collect meaningful marine life data that will contribute to the preservation of ocean environments.

Volunteer participation enables sampling of large areas and coverage of a wide range of habitats, depths, and times of the year (Rees & Pond 1995, Ellis & Cowan 2001, Almany et al. 2010, Stuart-Smith et al. 2013), producing highly valuable information about species or habitat health that can be used for management decisions and conservation planning (Ward-Paige & Lotze 2011). Also notable in this context is Whereat et al.'s (2004) monitoring programme for HABs, in which the participation of volunteers living along the coast of Delaware in the United States enabled high sampling frequency to be maintained and the inclusion of sites not sampled by the state. Such a network of volunteer observers guarantees extensive geographic and temporal coverage that facilitates detection of the normally patchy and ephemeral occurrence of algal blooms, contributing to an effective early warning system for HABs in the inland bays of Delaware.

CSPs contribute significantly to the study of litter on beaches through sampling and surveys. Volunteer participation enables litter surveys to be conducted simultaneously over extensive geographic scales, covering the coasts of entire countries (Bravo et al. 2009, Ribic et al. 2010, 2012, Kordella et al. 2013). Collection of data at many different sites has shown that in general plastic is the predominant litter (Ribic 1998, Moore et al. 2001, Kusui & Noda 2003, Bravo et al. 2009, Martin 2013), that its main sources are terrestrial (Ross et al. 1991), and by temporal comparison that it is related to national development (Jóźwiak 2005). Volunteers also contribute to the study of the biological effects of litter, including entanglement and ingestion by organisms (for birds, see, e.g., van Franeker et al. 2011, Lindborg et al. 2012). A major contribution of CSPs is the creation of public awareness of the problem of litter, largely through the involvement of thousands (10,000 to 15,000) of volunteers participating in nationwide beach-cleaning campaigns (Ribic et al. 2010, 2012, Kordella et al. 2013, Martin 2013). The results of these studies have even generated changes in local policies. For example, the finding that styrofoam was the main litter around Monterey Bay (USA) has led several municipalities around the bay to ban styrofoam takeout containers (Rosevelt et al. 2013).

Extensive spatial and temporal coverage is also attained through volunteers in the monitoring of pollution, mainly by the analysis of stranded birds (Camphuysen et al. 1999, Roletto et al. 2003, van Franeker et al. 2011). This approach has revealed, for example, that oil pollution in the North Sea has diminished over a period of more than 30 years (Figure 13C) (Camphuysen & Heubeck 2001). Remarkable in this context because of its geographic extent, covering several countries around the North Sea and English Channel, is the evaluation of pollution by tributyltin (TBT) by the determination of the occurrence of imposex in the dogwhelk *Nucella lapillus* (Figure 14) (Evans et al. 2000). Citizen scientists can aid in monitoring remote areas and so complement scientific surveys, something of increasing importance in the face of climate change (Berkes et al. 2007). Intense temporal sampling contributes to detection of phenomena that may be missed by intermittent surveys that normal research practices are constrained to maintain (Bristow et al. 2001). Since the recovery of communities within MPAs may take decades, long-term monitoring programmes, which can be supported by volunteer participation, are essential (Edgar et al. 2009).

As many volunteer programmes contribute to management and conservation initiatives, the involvement of volunteers also has positive impacts on the environmental awareness of participants (Trumbull et al. 2000, Foster-Smith & Evans 2003, Wagner 2005, Evans et al. 2008, Shirk

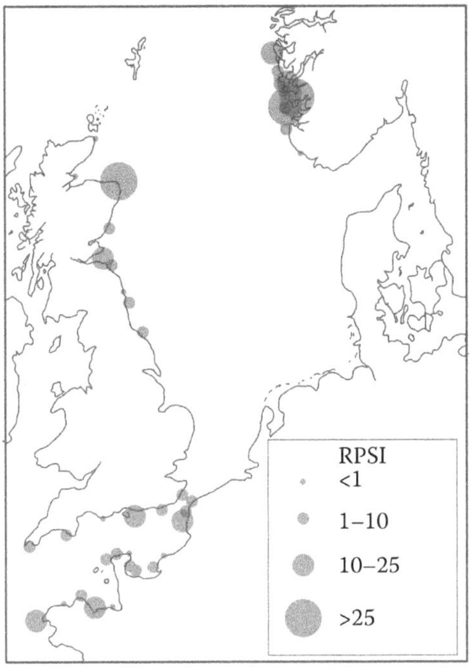

Figure 14 Proportion of individuals with imposex in the dogwhelk *Nucella lapillus* around the North Sea and English Channel. (Modified from Evans, S.M., Birchenough, A.C. & Fletcher, H. *Marine Pollution Bulletin* **40**, 220–225, 2000.). Reproduction is impaired at values of RPSI > 25.

et al. 2012). Such awareness contributes to long-term community support, reinforcing cooperation for conservation, and thereby also improves effectiveness of management (Lucy & Davy 2000, Cudney-Bueno & Basurto 2009, Almany et al. 2010) and fosters coproduction of conservation practices (Cornwell & Campbell 2012). The two-way flow of knowledge, inherent to these programmes, improves communication, the building of trust and work capacity among the general public, scientists, and administrators, promoting the mutual understanding necessary for collective action (Johnson 2010). Consideration of LEK or TEK is also something that greatly contributes to the last (Rochet et al. 2008).

LEK and TEK are probably much underrepresented in our database as they represent a different type of contribution by citizen scientists. Mostly, LEK and TEK are used to complement scientific information in the context of conservation, for example, in the design of MPAs (Aswani & Lauer 2006a,b, Rojas-Nazar et al. 2012) and informing management (Aswani & Hamilton 2004, Moreno et al. 2007, Rochet et al. 2008, Moreno-Baez et al. 2010, Le Fur et al. 2011, Cornwell & Campbell 2012, Suazo et al. 2013) (Figure 12). Rarely does TEK or LEK feed directly into scientific knowledge, but as shown by Rochet et al. (2008) for bottom communities of the English Channel, each can be used to verify scientific knowledge (e.g., of ecosystem changes). As pointed out by Le Fur et al. (2011), LEK is used as a supplementary source of knowledge or as a basis to guide new scientific research, and in the context of management or conservation, it may often be the only knowledge available. Depending on the subject, TEK/LEK may be equally valid as (e.g., on fish diet) or be a satisfactory proxy for (e.g., regarding trophic networks) information produced by scientific professionals. TEK/LEK can thus serve to obtain wide-ranging knowledge in a context of limited financial resources for scientific research by professionals (Le Fur et al. 2011). It must be emphasized, though, that this only applies to special cases, and that citizen science cannot replace professional scientific research.

Tagging or mark-and-recapture studies are common in fisheries science and ornithological research and generally depend on the help of volunteers for the recovery of the tag or marked animals. Nevertheless, this is often not specifically acknowledged in publications, and it is likely that this type of contribution by citizen scientists is grossly underrepresented in our database. Nevertheless, four works illustrate the crucial contribution of volunteers in studying the movements of sharks (Casey & Kohler 1992) and fishes (Shepherd et al. 2006, Moser & Shepherd 2009), as well as reproductive behaviour of turtles (Margaritoulis 1983). The general public has always been instrumental in the recovery of bird rings (Robinson et al. 2009, Guillemain et al. 2011); interestingly, these authors noted a decrease in recovery rates over recent decades, which they attributed to changing reporting behaviours (Figure 13).

Regarding the one-third of publications in our database that analysed the potential of citizen science, half described diverse CSPs; the other half analysed methodological aspects and the quality and usefulness of the information produced, and a small fraction examined the use of LEK (Figure 12, pie chart at the right of the figure). The information gained through these programmes is mostly considered equivalent to that produced by scientific researchers (Williams et al. 1999, Foster-Smith & Evans 2003, Pearse et al. 2003, Koss et al. 2009, Finn et al. 2010), but also less precision is often recognized (Mumby et al. 1995, Pattengill-Semmens & Semmens 1998, Uychiaoco et al. 2005, Delaney et al. 2008, Wells et al. 2008, Gillett et al. 2012).

In general, CSPs can become powerful tools when appropriately guided by experienced scientists (Evans et al. 2000), and to maintain their persistence over time, proper feedback is necessary so that volunteers feel and realize that their contributions lead to something scientifically significant (Camphuysen & Heubeck 2001). A report, or better still a publication, should be a minimum requirement, but an important aim should be to bring knowledge generated by CSPs into policy, as occurred for litter around Monterey Bay (Rosevelt et al. 2013), or into palpable management and conservation practices (Shirk et al. 2012).

Outreach activities

CSPs have received extensive media coverage during recent years. Furthermore, the organizations that work with volunteers often use public platforms to invite participation in their projects. It thus could be expected that many of the studies resulting from these projects engaged in outreach activities or closely collaborated with the public media. Because this review is based on scientific publications, however, the probability of outreach and media coverage being reported is low, and indeed few studies mentioned outreach activities.

Important channels for informing the general public about the activities and results of CSPs are the press and electronic media. Nine of the examined studies mentioned publications in newspapers, and 11 studies reported that their results were published in newsletters or magazines, but it is likely that the true number is substantially higher. A large number of newspapers and weekly magazines have published articles about CSPs during recent years, reflecting the great interest of the general public in being part of science (Figure 15).

Ninety-four studies used websites, which were either presented directly or were easily accessed via the Internet. We tracked all available websites, and the majority of these were functional (for a selection of specific websites see Table 8), providing an important tool for the general public and for volunteer participants. Social media (blogs, Facebook, Twitter) are frequently used, and many entries about CSPs can be found on the Internet. However, such entries were rarely mentioned in the examined publications. Most international programmes (e.g., CoralWatch, REEF, Seagrass Watch; see Table 8) not only have complete websites with abundant information for volunteer participants but also have direct links to their Facebook and Twitter accounts. Blogs maintained by individuals or by organizations collaborating with citizen scientists often receive many entries from visitors; in

BBC News Sport Weather Capital Culture Autos

NEWS EDUCATION & FAMILY

Home UK Africa Asia Europe Latin America Mid-East US & Canada Business Health Sci/Environ

England Northern Ireland Scotland Wales UK Politics Education

23 February 2013 Last updated at 08:26 GMT

Seafarer science: Sailors asked to help measure plankton

By Judith Burns
BBC News education reporter

Scientists are enlisting sailors and fishermen to help with what they hope will be the world's biggest study of plankton in the oceans.

Plankton are microscopic organisms, key to the marine food chain, but research suggests they are in decline.

Researchers at Plymouth University's Marine Institute want seafarers to help measure the quantity of plankton.

The results will "help us understand how life in the oceans is changing", said project leader, Dr Richard Kirby.

Scientists fear that the population of phytoplankton in the world's oceans is being affected by rising sea temperatures, with a paper published in the journal Nature in 2010 controversially suggesting that it had declined by as much as 40% since the 1950s due to climate change.

PLYMOUTH UNIVERSITY

Scientists want seafarers to help collect data on

LEARNING FROM THE BEST: Sunshine Coast Grammar School students get a few tips from University of Queensland and CoralWatch scientist Dave Logan.

Students get a learned hand

IN DEEP
Ian McKinnon

THE diving industry breathed a sigh of relief following a very busy Easter weekend. It is hoped the spell of good weather will continue.
SURFACE CONDITIONS: Light winds with a slight chop
UNDERWATER VISIBILITY: 10-12 Metres
WATER TEMPERATURE: 24 Degrees

survey results suggest that our coral is in good health.
Having returned from a recent trip to the Galapagos Islands and Tahiti, Marine Studies teacher and Scuba

T20 // medio ambiente

TENDENCIAS | LATERCERA | Sábado 6 de abril de 2013

Detectives científicos rastrean la basura en las playas de Chile

Más de 1.000 alumnos de una treintena de escuelas llevan años visitando playas a lo largo del país para estudiar la contaminación. Su trabajo ha dado origen a varios estudios que muestran el impacto de la basura en el ecosistema marino y, ahora, comenzarán a evaluar el efecto de la basura que llega desde los ríos hasta la zona costera.

TEXTO Ricardo Acevedo

Playas contaminadas

Más de 40 playas a lo largo de Chile han sido analizadas en estudios realizados entre 2007 y 2012 con la participación de más de mil estudiantes de 40 colegios del país. Los principales contaminantes son microplásticos, colillas de cigarro y pellets.

Figure 15 Newspaper articles of citizen scientists contributing to marine research.

particular, those entries on emblematic species (e.g., sea turtles, whales, or manatees) may be used in conservation programmes (Mignucci-Giannoni et al. 2000, Pabody et al. 2009).

Outreach activities have multiple effects. They are assumed to contribute to environmental awareness among the general public, even though this variable is rarely measured (Finn et al. 2010). In the context of CSPs, 'getting out the word' is of crucial importance for recruiting volunteers who contribute incidental observations or participate in planned surveys. The number of sightings reported to the Manatee Monitoring Network significantly increased during months when media coverage in local newspapers or television programmes was high (Pabody et al. 2009). Also, the success of long-term projects, which are well known among specific groups (e.g., recreational divers, local communities, birdwatchers), demonstrates that efficient outreach facilitates volunteer recruitment. Whereas new or short-term projects mostly depend on beginners, long-term studies increasingly rely on experienced volunteers, who have been shown to generate better and more reliable data than beginners (Jiguet 2009, Schmeller et al. 2009). Retention of these experienced participants is highly desirable and depends to a large extent on the feedback they receive from project coordinators (Camphuysen & Heubeck 2001, Finn et al. 2010).

Volunteer-based research as an opportunity for marine science

Overall, the studies examined herein are positive about the participation of citizen scientists. In many works, the authors declared this approach as a win-win situation: (1) scientists gain, for example, by enhancing the spatiotemporal coverage of their studies, and (2) the general public and stakeholders (e.g., tour operators, fishermen) become familiar with the research results and with the professional investigators (Goffredo et al. 2010), leading to better acceptance of the scientific information to which they themselves have contributed (Starr 2010).

In the marine realm, many habitats offer limited accessibility to both professional scientists and the general public. Logistic impediments limit the number of people and the times they can visit the ocean. Improving the observational and reporting capacities of those who regularly move in the marine environment thus seems to be an important advantage of CSPs. In this context, two aspects were repeatedly mentioned in the studies examined: (1) Increasing the numbers of citizen scientists enhances the spatial and temporal coverage of studies; and (2) training and accumulated experience of citizen scientists enhance the quality of the data they are collecting. Both these aspects represent important challenges. Coordinating large numbers of citizen scientists requires efficient communication via the traditional media (newspaper, radio, television), via the electronic media (e-mail, websites, smartphones, etc.), and via personal communication (direct, phone, letter). Training of citizen scientists requires manpower (professional scientists), the necessary infrastructure (offices, training labs, cars, and boats), and some basic tools (e.g., dive equipment, cameras, etc.). Also, data management and preparation for analysis require professionals with basic scientific education.

Although all this costs money, the benefits can be enormous. Large numbers of citizen scientists can generate data that allow describing the large-scale distribution and temporal dynamics of endangered species, fisheries resources, harmful species, litter, and pollution. This information can be used for much more reliable management of marine organisms and habitats, something that is of utmost importance with the ever-increasing use of the marine environment for shipping, aquaculture, generation of energy, construction, and recreation.

Professionals required for the coordination of functional CSPs do not need to be highly qualified scientists with years of post-graduate studies. Professionals with a basic degree in marine sciences can become efficient coordinators of CSPs. We suggest that the cost-benefit analysis of such a marine citizen science programme would be positive, as has already been demonstrated for numerous terrestrial programmes (Dickinson and Bonney 2012).

Marine CSPs collaborate with stakeholders (e.g., fishermen, tour operators) and with leisure seekers (e.g., divers, birdwatchers, beachcombers). The benefits that these two types of participants obtain from collaborating with professional scientists may differ: Stakeholders will be interested in economic benefits (higher or at least a stable income), and leisure seekers who participate in scientific projects are looking for recognition and education. Whether CSPs can satisfy the expectations of stakeholders is not always predictable, and it is less controllable by the coordinators. However, the expectations of leisure seekers can be relatively easily fulfilled when certain steps (e.g., acknowledge their contributions and provide regular feedback) are established and maintained over time.

In synthesis, citizen science is able to make significant contributions to marine sciences where professional scientific activities are limited in their potential. Considering the vastness of the oceans, the extensiveness of the coastlines, and the diversity of habitats, communities, and species, a proper understanding of this realm requires intensive research activities over time and space. This recognition should lead to an increased consideration of citizen science as a powerful tool for the generation of scientific knowledge. But, beyond its contribution to science, citizen science initiatives should also be promoted because of their benefits in creating awareness of the challenges facing the world's oceans.

Acknowledgements

We are especially grateful to Enrique Martinez and Carlos Gaymer, who share our enthusiasm and provided initial input to this review; because of other time commitments, they could not be part of this time-consuming task, but we are looking forward to future endeavours with them. Every effort was made to generate a representative set of published studies, and we apologize to those authors whose publications we unintentionally overlooked: We salute all colleagues who engage in marine citizen science, whether their results have been published or not—there is still so much to learn about the oceans, and all help is needed, including that of interested citizens who aid in data collection and foremost in spreading the word. Our activities with volunteer participants in marine research have been supported over the years by the Chilean Science Foundation through the program EXPLORA-CONICYT. Last but not least, we are extremely grateful to all volunteers for their participation and inspiration. This review is dedicated to them.

References

Acuña-Marrero, D., Zimmerhackel, J.S., Mayorga J. & Hearn, A. 2013. First record of three shark species, *Odontaspis ferox*, *Mustelus albipinnis* and *Centrophorus squamosus*, from the Galápagos Islands. *Marine Biodiversity Records* **6**, e87.

Aiken, J.J., Godley, B.J., Broderick, A.C., Austin, T., Ebanks-Petrie, G. & Hays, G.C. 2001. Two hundred years after a commercial marine turtle fishery: the current status of marine turtles nesting in the Cayman Islands. *Oryx* **35**, 145–151.

Alfaro-Shigueto, J., Valqui, J. & Mangel, J.C. 2011. New record of the marine otter *Lontra felina* (Molina, 1782) north to its current distribution. *Ecología Aplicada* **10**, 87–91.

Almany, G.R., Hamilton, R.J., Williamson, D.H., Evans, R.D., Jones, G.P., Matawai, M., Potuku, T., Rhodes, K.L., Russ, G.R. & Sawynok, B. 2010. Research partnerships with local communities: two case studies from Papua New Guinea and Australia. *Coral Reefs* **29**, 567–576.

Anderson, P.A. 2013. Acoustic characterization of seahorse tank environments in public aquaria: a citizen science project. *Aquacultural Engineering* **54**, 72–77.

Arvanitidis, C., Faulwetter, S., Chatzigeorgiou, G., Penev, L., Banki, O., Dailianis, T., Pafilis, E., Kouratoras, M., Chatzinikolaou, E., Fanini, L., Vasileiadou, A., Pavloudi, C., Vavilis, P., Koulouri, P. & Dounas, C. 2011. Engaging the broader community in biodiversity research: the concept of the COMBER pilot project for divers in ViBRANT. *ZooKeys* **150**, 211–229.

Assis, J., Tavares, D., Tavares, J., Cunha, A., Alberto, F. & Serrao, E.A. 2009. Findkelp, a GIS-based community participation project to assess Portuguese kelp conservation status. *Journal of Coastal Research* **56**, 1469–1473.

Aswani, S. & Hamilton, R.J. 2004. Integrating indigenous ecological knowledge and customary sea tenure with marine and social science for conservation of bumphead parrotfish (*Bolbometopon muricatum*) in the Roviana Lagoon, Solomon Islands. *Environmental Conservation* **31**, 69–83.

Aswani, S. & Lauer, M. 2006a. Benthic mapping using local aerial photo interpretation and resident taxa inventories for designing marine protected areas. *Environmental Conservation* **33**, 263–273.

Aswani, S. & Lauer, M. 2006b. Incorporating fishermen's local knowledge and behavior into geographical information systems (GIS) for designing marine protected areas in Oceania. *Environmental Conservation* **65**, 81–102.

Aswani, S. & Weiant, P. 2004. Scientific evaluation in women's participatory management: monitoring marine invertebrate refugia in the Solomon Islands. *Human Organization* **63**, 301–319.

Auster, P.J., Semmens, B.X. & Barber, K. 2005. Pattern in the co-occurrence of fishes inhabiting the coral reefs of Bonaire, Netherlands Antilles. *Environmental Biology of Fishes* **74**, 187–194.

Avens, L. & Goshe L.R. 2007. Comparative skeletochronological analysis of Kemp's ridley (*Lepidochelys kempii*) and loggerhead (*Caretta caretta*) humeri and scleral ossicles. *Marine Biology* **152**, 1309–1317.

Avens, L., Taylor, J.C., Goshe, L.R., Jones, T.T. & Hastings, M. 2009. Use of skeletochronological analysis to estimate the age of leatherback sea turtles *Dermochelys coriacea* in the western North Atlantic. *Endangered Species Research* **8**, 165–177.

Avery-Gomm, S., O'Hara, P.D., Kleine, L., Bowes, V., Wilson, L.K. & Barry, K.L. 2012. Northern fulmars as biological monitors of trends of plastic pollution in the eastern North Pacific. *Marine Pollution Bulletin* **64**, 1776–1781.

Azzurro, E., Aguzzi, J., Maynou, F., Chiesa, J.J. & Savini D. 2013a. Diel rhythms in shallow Mediterranean rocky-reef fishes: a chronobiological approach with the help of trained volunteers. *Journal of the Marine Biological Association of the United Kingdom* **93**, 461–470.

Azzurro, E., Broglio, E., Maynou, F. & Bariche, M. 2013b. Citizen science detects the undetected: the case of *Abudefduf saxatilis* from the Mediterranean Sea. *Management of Biological Invasions* **4**, 167–170.

Baird, R.W. & Guenther, T.J. 1995. Account of harbour porpoise (*Phocoena phocoena*) strandings and bycatches along the coast of British Columbia. *International Whaling Commission Special Issues* **16**, 159–168.

Baird, R.W., Stacey, P.J., Duffus, D.A. & Langelier, K.M. 2002. An evaluation of gray whale (*Eschrichtius robustus*) mortality incidental to fishing operations in British Columbia, Canada. *Journal of Cetacean Research and Management* **4**, 289–296.

Baker, M.S., Jr. & Oeschger, I. 2009. Description and initial evaluation of a text message based reporting method for marine recreational anglers. *Marine and Coastal Fisheries: Dynamics, Management, and Ecosystem Science* **1**, 461–470.

Bart, J., Brown, S., Harrington, B. & Morrison, R.I.G. 2007. Survey trends of North American shorebirds: population declines of shifting distributions? *Journal of Avian Biology* **38**, 73–82.

Beckley, L.E., Cliff, G., Smole, M.J. & Compagno, L.J. 1997. Recent strandings and sightings of whale sharks in South Africa. *Environmental Biology of Fishes* **50**, 343–348.

Beeden, R., Maynard, J.A., Marshall, P.A., Heron, S.F. & Willis, B.L. 2012. A framework for responding to coral disease outbreaks that facilitates adaptive management. *Environmental Management* **49**, 1–13.

Bell, C.D., Blumenthal, J.M., Austin, T.J., Ebanks-Petrie, G., Broderick, A.C. & Godley, B.J. 2009. Harnessing recreational divers for the collection of sea turtle data around the Cayman Islands. *Tourism in Marine Environments* **5**, 245–257.

Bell, J.J. 2007. The use of volunteers for conducting sponge biodiversity assessments and monitoring using a morphological approach on Indo-Pacific coral reefs. *Aquatic Conservation: Marine and Freshwater Ecosystems* **17**, 133–145.

Bellwood, P. 2005. *First Farmers: The Origins of Agricultural Societies*. New York: Blackwell.

Berkes, F., Berkes, M.K. & Fast, H. 2007. Collaborative integrated management in Canada's North: the role of local and traditional knowledge and community-based monitoring. *Coastal Management* **35**, 143–162.

Berkes, F., Colding, J. & Folke, C. 2000. Rediscovery of traditional ecological knowledge as adaptive management. *Ecological Applications* **10**, 1251–1262.

Bernard, A.T.F., Götz, A., Kerwath, S.E & Wilke, C.G. 2013. Observer bias and detection probability in underwater visual census of fish assemblages measured with independent double-observers. *Journal of Experimental Marine Biology and Ecology* **443**, 75–84.

Bhat, S.R. & Matondkar, S.G.P. 2004. Algal blooms in the seas around India—networking for research and outreach. *Current Science* **87**, 1079–1083.

Biggs, C.R. & Olden, J.D. 2011. Multi-scale habitat occupancy of invasive lionfish (*Pterois volitans*) in coral reef environments of Roatan, Honduras. *Aquatic Invasions* **6**, 347–353.

Bird, T.J., Bates, A.E., Lefcheck, J.S., Hill, N.A., Thomson, R.J., Edgar, G.J., Stuart-Smith, R.D., Wotherspoon, S., Krkosek, M., Stuart-Smith, J.F., Pecl, G.T., Barrett, N. & Frusher, S. 2013. Statistical solutions for error and bias in global citizen science datasets. *Biological Conservation* **173**, 144–154.

Björnsson, B., Karlsson, H., Thorskinsson, V. & Solmundsson, J. 2011. Should all fish in mark-recapture experiments be double-tagged? Lessons learned from tagging coastal cod (*Gadus morhua*). *Journal of Marine Science* **68**, 603–610.

Boero, F., Belmonte, G., Bracale, R., Fraschetti, S., Piraino, S. & Zampardi, S. 2013. A salp bloom (Tunicata, Thaliacea) along the Apulian coast and in the Otranto Channel between March–May 2013. *F1000Research* **2**, 1–6.

Boero, F., Putti, M., Trainito, E., Prontera, E., Piraino, S. & Shiganova, T.A. 2009. First records of *Mnemiopsis leidyi* (Ctenophora) from the Ligurian, Thyrrhenian and Ionian Seas (Western Mediterranean) and first record of *Phyllorhiza punctata* (Cnidaria) from the western Mediterranean. *Aquatic Invasions* **4**, 675–680.

Bonney, R., Cooper, C.B., Dickinson, J., Kelling, S., Phillips, T., Rosenberg, K.V. & Shirk, J. 2009. Citizen science: a developing tool for expanding science knowledge and scientific literacy. *BioScience* **59**, 977–984.

Bower, J.L. 2009. Changes in marine bird abundance in the Salish Sea: 1975 to 2007. *Marine Ornithology* **37**, 9–17.

Bramanti, L., Vielmini, I., Rossi, S., Stolfa, S. & Santangelo, G. 2011. Involvement of recreational scuba divers in emblematic species monitoring: the case of Mediterranean red coral (*Corallium rubrum*). *Journal for Nature Conservation* **19**, 312–318.

Bravo, M., Gallardo, M., Luna-Jorquera, G., Núñez, P., Vásquez, N. & Thiel, M. 2009. Anthropogenic debris on beaches in the SE Pacific (Chile): results from a national survey supported by volunteers. *Marine Pollution Bulletin* **58**, 1718–1726.

Brenier, A., Ferraris, J. & Mahafina, J. 2011. Participatory assessment of the Toliara Bay reef fishery, southwest Madagascar. *Madagascar Conservation & Development* **6**, 60–67.

Bristow, T., Glanville, N. & Hopkins, J. 2001. Shore-based monitoring of bottlenose dolphins (*Tursiops truncatus*) by trained volunteers in Cardigan Bay, Wales. *Aquatic Mammals* **27**, 115–120.

Bruno, J.F. & Selig, E.R. 2007. Regional decline of coral cover in the Indo-Pacific: timing, extent and subregional comparisons. *PloS ONE* **2**, e711.

Bruno, J.F., Sweatman, H., Precht, W.F., Selig, E.R. & Schutte, V.G.W. 2009. Assessing evidence of phase shifts from coral to macroalgal dominance on coral reefs. *Ecology* **90**, 1478–1484.

Bull, J.C., Mason, S., Wood, C. & Price, A.R.G. 2013. Benthic marine biodiversity patterns across the United Kingdom and Ireland determined from recreational diver observations: a baseline for possible species range shifts induced by climate change. *Aquatic Ecosystem Health & Management* **16**, 20–30.

Cadigan, N.G. & Brattey, J. 2006. Reporting and shedding rate estimates from tag recovery experiments on Atlantic cod (*Gadus morhua*) in coastal Newfoundland. *Canadian Journal of Fisheries and Aquatic Sciences* **63**, 1944–1958.

Campbell, L.M. & Smith, C. 2006. What makes them pay? Values of volunteer tourists working for sea turtle conservation. *Environmental Management* **38**, 84–98.

Campbell, M.L., Gould, B. & Hewitt, C.L. 2007. Survey evaluations to assess marine bioinvasions. *Marine Pollution Bulletin* **55**, 360–378.

Camphuysen, K.C.J. 1998. Beached bird surveys indicate decline in chronic oil pollution in the North Sea. *Marine Pollution Bulletin* **36**, 519–526.

Camphuysen, K.C.J., Barreveld, H., Dahlmann, G. & van Franeker, J.A. 1999. Seabirds in the North Sea demobilized and killed by polyisobutylene (C4H8)n(PIB). *Marine Pollution Bulletin* **38**, 1171–1176.

Camphuysen, C.J. & Heubeck, M. 2001. Marine oil pollution and beached bird surveys: the development of a sensitive monitoring instrument. *Environmental Pollution* **112**, 443–461.

Carson, H.S., Lamson, M.R., Nakashima, D., Toloumu, D., Hafner, J., Maximenko, N. & McDermid, K.J. 2013. Tracking the sources and sinks of local marine debris in Hawai'i. *Marine Environmental Research* **84**, 76–83.

Casale, P., Mazaris, A.D. & Freggi, D. 2011. Estimation of age at maturity of loggerhead sea turtles *Caretta caretta* in the Mediterranean using length-frequency data. *Endangered Species Research* **13**, 123–129.

Casey, J.G. & Kohler, N.E. 1992. Tagging studies on the short fin mako shark (*Isurus oxyrinchus*) in the Western North Atlantic. *Marine and Freshwater Research* **43**, 45–60.

Cheney, B., Thompson, P.M., Ingram, S.N., Hammond, P.S., Stevick, P.T., Durban, J.W., Culloch, R.M., Elwen, S.H., Mandleberg, L., Janik, V.M., Quick, N.J., Islas-Villanueva, V., Robinson, K.P., Costa, M., Eisfeld, S.M., Walters, A., Phillips, C., Weir, C.R., Evans, P.G.H., Anderwald, P., Reid, R.J., Reid, J. & Wilson, B. 2013. Integrating multiple data sources to assess the distribution and abundance of bottlenose dolphins *Tursiops truncatus* in Scottish waters. *Mammal Review* **43**, 71–88.

Cohen, C.S., McCann, L., Davis, T., Shaw, L. & Ruiz, G. 2011. Discovery and significance of the colonial tunicate *Didemnum vexillum* in Alaska. *Aquatic Invasions* **6**, 263–271.

Cohen, M.N. 1977. *The Food Crisis in Prehistory: Overpopulation and the Origins of Agriculture*. New Haven, Connecticut: Yale University Press.

Colwell, M.A. & Cooper, R.J. 1993. Estimates of coastal shorebird abundance: the importance of multiple counts. *Journal of Field Ornithology* **64**, 293–301.

Condon, R., Graham, W., Duarte, C., Pitt, K., Lucas, C., Haddock, S., Sutherland, K., Robinson, K., Dawson, M., Decker, M., Mills, C., Purcell, J., Malej, A., Mianzan, H., Uye, S., Gelcich, S. & Madin, L. 2012. Questioning the rise of gelatinous zooplankton in the world's oceans. *Bioscience* **62**, 160–169.

Cornwell, M.L. & Campbell, L.M. 2012. Co-producing conservation and knowledge: citizen-based sea turtle monitoring in North Carolina, USA. *Social Studies of Science* **42**, 101–120.

Cox, T.E., Philippoff, J., Baumgartner, E. & Smith, C.M. 2012. Expert variability provides perspective on the strengths and weaknesses of citizen-driven intertidal monitoring program. *Ecological Applications* **22**, 1201–1212.

Crabbe, M.J., Karaviotis, S. & Smith, D.J. 2004. Monitoring growth of hard corals as performance indicators for coral reefs. *Journal of Biological Education* **38**, 113–117.

Crabbe, M.J.C. 2012. From citizen science to policy development on the coral reefs of Jamaica. *International Journal of Zoology* **2012**, 1–6.

Crall, A.W., Newman, G.J., Jarnevich, C.S., Stohlgren, T.J., Waller, D.M. & Graham, J. 2010. Improving and integrating data on invasive species collected by citizen scientists. *Biological Invasions* **12**, 3419–3428.

Cudney-Bueno, R. & Basurto, X. 2009. Lack of cross-scale linkages reduces robustness of community-based fisheries management. *PloS ONE* **4**, e6253

Culver, C.S., Schroeter, S.C., Page, H.M. & Dugan, J.E. 2010. Essential fishery information for trap-based fisheries: development of a framework for collaborative data collection. *Marine and Coastal Fisheries* **2**, 98–114.

Darwall, W.R.T. & Dulvy, N.K. 1996. An evaluation of the suitability of non-specialist volunteer researchers for coral reef fish surveys. Mafia Island, Tanzania—a case study. *Biological Conservation* **78**, 223–231.

Davies, T.K., Stevens, G., Meekan, M.G., Struve, J. & Rowcliffe, J.M. 2012. Can citizen science monitor whale-shark aggregations? Investigating bias in mark-recapture modeling using identification photographs sourced from the public. *Wildlife Research* **39**, 696–704.

Delaney, D.G., Sperling, C.D., Adams, C.S. & Leung, B. 2008. Marine invasive species: validation of citizen science and implications for national monitoring networks. *Biological Invasions* **10**, 117–128.

Dickinson, J.L. & Bonney, R. (eds) 2012. *Citizen Science: Public Participation in Environmental Research*. Ithaca, New York: Cornell University Press.

Dickinson, J.L., Zuckerberg, B. & Bonter, D.N. 2010. Citizen science as an ecological research tool: challenges and benefits. *Annual Review of Ecology, Evolution, and Systematics* **41**, 149–172.

Eastman, L.B., Núñez, P., Crettier, B. & Thiel, M. 2013. Identification of self-reported user behavior, education level, and preferences to reduce littering on beaches—a survey from the SE Pacific. *Ocean & Coastal Management* **78**, 18–24.

Edgar, G.J., Barrett, N.S. & Stuart-Smith, R.D. 2009. Exploited reefs protected from fishing transform over decades into conservation features otherwise absent from seascapes. *Ecological Applications* **19**, 1967–1974.

Edgar, G.J. & Stuart-Smith, R.D. 2009. Ecological effects of marine protected areas on rocky reef communities—a continental-scale analysis. *Marine Ecology Progress Series* **388**, 51–62.

Ellis, S.L. & Cowan, D.F. 2001. Volunteer-based monitoring of juvenile American lobster, *Homarus americanus*. *Marine Freshwater Research* **52**, 1103–1112.

Evans, S.M., Birchenough, A.C. & Fletcher, H. 2000. The value and validity of community based research: TBT contamination of the North Sea. *Marine Pollution Bulletin* **40**, 220–225.

Evans, S.M., Gebbels, S. & Stockill, J. M. 2008. 'Our shared responsibility': participation in ecological projects as a means of empowering communities to contribute to coastal management processes. *Marine Pollution Bulletin* **57**, 3–7.

Finn, P.G., Catterall, C.P. & Driscoll, P.V. 2007. Determinants of preferred intertidal feeding habitat for eastern curlew: a study at two spatial scales. *Austral Ecology* **32**, 131–144.

Finn, P.G., Udy, N.S., Baltais, S.J., Price, K. & Coles, L. 2010. Assessing the quality of seagrass data collected by community volunteers in Moreton Bay Marine Park, Australia. *Environmental Conservation* **37**, 83–89.

Ford, J.K.B., Ellis, G.M., Barrett-Lennard, L.G., Morton, A.B., Palm, R.S. & Balcomb, K.C., III. 1998. Dietary specialization in two sympatric populations of killer whales (*Orcinus orca*) in coastal British Columbia and adjacent waters. *Canadian Journal of Zoology* **76**, 1456–1471.

Foster-Smith, J. & Evans, S.M. 2003. The value of marine ecological data collected by volunteers. *Biological Conservation* **113**, 199–213.

Fowler, A., Whyatt, J.D., Davies, G. & Ellis, R. 2013. How reliable are citizen-derived scientific data? Assessing the quality of contrail observations made by the general public. *Transactions in GIS* **17**, 488–506.

Fujitani, M.L., Fenichel, E.P., Torre, J. & Gerber, L.R. 2012. Implementation of a marine reserve has a rapid but short-lived effect on recreational angler use. *Ecological Applications* **22**, 597–605.

Gannon, D.P., Read, J.A., Craddock, J.E. & Mead J.G. 1997. Stomach contents of long-finned pilot whales (*Globicephala melas*) stranded on the U.S. Mid-Atlantic coast. *Marine Mammal Science* **13**, 405–418.

Garrick-Maidment, N., Strewhella, S., Hatcher, J., Collins, K.J. & Mallinson, J.J. 2010. Seahorse tagging project, Studland Bay, Dorset U. K. *Marine Biodiversity Records* **3**, 1–4.

Gill, J.A., Sutherland, W.J. & Norris, K. 2001. Depletion models can predict shorebird distribution at different spatial scales. *Proceedings of the Royal Society of London. Series B: Biological Sciences* **268**, 369–376.

Gillett, D.J., Pondella, D.J., Freiwald, J., Schiff, K.C., Caselle, J.E., Shuman, C. & Weisberg, S.B. 2012. Comparing volunteer and professionally collected monitoring data from the rocky subtidal reefs of Southern California, USA. *Environmental Monitoring and Assessment* **184**, 3239–3257.

Godoy, N., Gelcich, S., Vásquez, J.A. & Castilla, J.C. 2010. Spearfishing to depletion: evidence from temperate reef fishes in Chile. *Ecological Applications* **20**, 1504–1511.

Goffredo, S., Pensa, F., Neri, P., Orlandi, A., Gagliardi, M.S., Velardi, A., Piccinetti, C. & Zaccanti, F. 2010. Unite research with what citizens do for fun: 'recreational monitoring' of marine biodiversity. *Ecological Applications* **20**, 2170–2187.

Goffredo, S., Piccinetti, C. & Zaccanti, F. 2004. Volunteers in marine conservation monitoring: a study of the distribution of seahorses carried out in collaboration with recreational scuba divers. *Conservation Biology* **18**, 1492–1503.

Grant, S. & Berkes, F. 2007. Fisher knowledge as expert system: a case from the longline fishery of Grenada, the Eastern Caribbean. *Fisheries Research* **84**, 162–170.

Green, S.J., Akins, J.L., Maljkovic, A. & Côté, I.M. 2012. Invasive lionfish drive Atlantic coral reef fish declines. *PloS ONE* **7**, e32596.

Greenwood, J.J. 2007. Citizens, science and bird conservation. *Journal of Ornithology* **148**, 77–124.

Gregory, M.R. 1991. The hazards of persistent marine pollution drift plastics and conservation islands. *Journal of the Royal Society of New Zealand* **21**, 83–100.

Guillemain, M., Devineau, O., Gauthier-Clerc, M., Hearn, R., King, R., Simon, G. & Grantham, M. 2011. Changes in ring recovery rates over the last 50 years: shall we continue to ring ducks? *Journal of Ornithology* **152**, 55–61.

Hamel, N.J., Burger, A.E., Charleton, K., Davidson, P., Lee, S., Bertram, D.F. & Parrish, J.K. 2009. Bycatch and beached birds: assessing mortality impacts in coastal net fisheries using marine bird strandings. *Marine Ornithology* **37**, 41–60.

Hamilton, R.J., Giningele, M., Aswani, S. & Ecochand, J.L. 2012. Fishing in the dark—local knowledge, night spearfishing and spawning aggregations in the western Solomon Islands. *Biological Conservation* **145**, 246–257.

Hammerton, Z., Dimmock, K., Hahn, C., Dalton, S.J. & Smith, S.D. 2012. Scuba diving and marine conservation: collaboration at two Australian subtropical destinations. *Tourism in Marine Environments* **8**, 1–2.

Harding, S., Comley, J. & Raines, P. 2006. Baseline data analysis as a tool for predicting the conservation value of tropical coastal habitats in the Indo-Pacific. *Ocean & Coastal Management* **49**, 696–705.

Harris, R.J., Tseng, F.S., Pokras, M.A., Suedmeyer, B.A., Bogart, J.S.H., Prescott, R.L. & Newman, S.H. 2006. Beached bird surveys in Massachusetts: the Seabird Ecological Assessment Network (SEANET). *Marine Ornithology* **34**, 115–122.

Hassell, N.S., Williamson, D.H., Evans, R.D. & Russ, G.R. 2013. Reliability of non-expert observer estimates of the magnitude of marine reserve effects. *Coastal Management* **41**, 361–380.

Heil, C.A. & Steidinger, K.A. 2009. Monitoring, management and mitigation of *Karenia* blooms in the eastern Gulf of Mexico. *Harmful Algae* **8**, 611–617.

Hidalgo-Ruz, V. & Thiel, M. 2013. Distribution and abundance of small plastic debris on beaches in the SE Pacific (Chile): a study supported by a citizen science project. *Marine Environmental Research* **459**, 121–134.

Hill, H.H., Kelley, J.T., Belknap, D.F. & Dickson, S.M. 2002. Co-measurement of beaches in Maine, USA: volunteer profiling of beaches and annual meetings. *Journal of Coastal Research* **SI 36**, 374–380.
Hodgson, G. 1999. A global assessment of human effects on coral reefs. *Marine Pollution Bulletin* **38**, 345–355.
Holmberg, J., Norman, B. & Arzoumanian, Z. 2009. Estimating population size, structure, and residency time for whale sharks *Rhincodon typus* through collaborative photo-identification. *Endangered Species Research* **7**, 39–53.
Holt, B.G., Côté, I.M. & Emerson, B.C. 2010. Signatures of speciation? Distribution and diversity of *Hypoplectrus* (Teleostei: Serrasidae) color morphotypes. *Global Ecology and Biogeography* **19**, 432–441.
Holt, B.G., Rioja-Nieto, R., MacNeil, M.A., Lupton, J. & Rahbek, C. 2013. Comparing diversity data collected using a protocol designed for volunteers with results from a professional alternative. *Methods in Ecology and Evolution* **4**, 383–392.
Hong, S., Lee, J., Jang, Y.C., Kim, Y.J., Kim, H.J., Han, D., Hong, S.H., Kang, D. & Shim, W.J. 2013. Impacts of marine debris on wild animals in the coastal area of Korea. *Marine Pollution Bulletin* **66**, 117–124.
Howe, M.A., Geissler, P.H. & Harrington, B.A. 1989. Population of North American shorebirds based on the international shorebird survey. *Biological Conservation* **49**, 185–199.
Humber, F., Godley, B.J., Ramahery, V. & Broderick, A.C. 2011. Using community members to assess artisanal fisheries: the marine turtle fishery in Madagascar. *Animal Conservation* **14**, 175–185.
Hussey, N., Stroh, N., Klaus, R., Chekchak, T. & Kessel, S. 2013. Scuba diver observations and placard tags to monitor grey reef sharks, *Carcharhinus amblyrhynchos*, at Sha'ab Rumi, the Sudan: assessment and future directions. *Journal of the Marine Biological Association of the United Kingdom* **93**, 299–308.
Huveneers, C., Luo, K., Otway, N.M. & Harcourt, R. G. 2009. Assessing the distribution and relative abundance of wobbegong sharks (Orectolobidae) in New South Wales, Australia, using recreational scuba-divers. *Aquatic Living Resources* **22**, 255–264.
Jackson, N.L., Cerrato, M.L. & Elliot, N. 1997. Geography and fieldwork at the secondary school level: an investigation of anthropogenic litter on an estuarine shoreline. *Journal of Geography* **96**, 301–306.
Jaine, F.R.A., Couturier, L.I.E., Weeks, S.J., Townsend, K.A., Bennett, M.B., Fiora, K. & Richardson, A.J. 2012. When giants turn up: sighting trends, environmental influences and habitat use of the manta ray *Manta alfredi* at a coral reef. *PloS ONE* **7**, e46170.
James, M.C., Sherrill-Mix, S.A., Martin, K. & Myers, R.A. 2006. Canadian waters provide critical foraging habitat for leatherback sea turtles. *Biological Conservation* **133**, 347–357.
Jang, Y.C., Hong, S.Y., Kim, Y.H. & Lee, B.K. 2013. The birds of Changpo Bay, Gyeongnam Province, Korea. *Korean Journal of Ornithology* **20**, 1–12.
Jiguet, F. 2009. Method-learning caused first-time observer effect in a newly-started breeding bird survey. *Bird Study* **56**, 253–258.
Johnson, T.R. 2010. Cooperative research and knowledge flow in the marine commons: lessons from the north-east United States. *International Journal of the Commons* **4**, 251–272.
Johnston-González, R., Peña-Herrera, V. & Castillo, L.F. 2008. El censo neotropical de aves acuáticas y el conocimiento de aves playeras neotropicales en Colombia. *Ornitologia Neotropical* **19**, 1–8.
Jóźwiak, T. 2005. Tendencies in the numbers of beverage containers on the Polish coast in the decade from 1992 to 2001. *Marine Pollution Bulletin* **50**, 87–90.
Jung, J.L., Stephan, E., Louis, M., Alfonsi, E., Liret, C., Carpentier, F.G. & Hassani, S. 2009. Harbour porpoises (*Phocoena phocoena*) in north-western France: aerial survey, opportunistic sightings and strandings monitoring. *Journal of the Marine Biological Association of the United Kingdom* **89**, 1045–1050.
Kelling, S. 2012. Using bioinformatics in citizen science. In *Citizen Science: Public Participation in Environmental Research*, J.L. Dickinson et al. (eds). Ithaca, New York: Cornell University Press, 58–68.
Kennedy, J.J., Rayner, N.A., Smith, R.O., Parker, D.E. & Saunby, M. 2011. Reassessing biases and other uncertainties in sea surface temperature observations measured *in situ* since 1950, part 2: biases and homogenization. *Journal of Geophysical Research: Atmospheres (1984–2012)* **116**, 1–16.
Kimball, B., Myers-Pigg, A., Clay, T., Neibauer, J. & Keil, R. 2009. Sound Citizen: students and citizens conducting environmental science together. In *Proceedings from the NAAEE Research Symposium, Portland, Oregon*, USA, October 2009, 1–4.
Kirkpatrick, B., Currier, R., Nierenberg, K., Reich, A., Backer, L.C., Stumpf, R., Fleming, L. & Kirkpatrick, G. 2008. Florida red tide and human health: a pilot beach conditions reporting system to minimize human exposure. *Science of the Total Environment* **402**, 1–8.

Kleiven, A.R., Olsen, E.M. & Vølstad, J.H. 2012. Total catch of a red-listed marine species is an order of magnitude higher than official data. *PloS ONE* **7**, e31216.
Kordella, S., Geraga, M., Papatheodorou, G., Fakiris, E. & Mitropoulou, I.M. 2013. Litter composition and source contribution for 80 beaches in Greece, Eastern Mediterranean: a nationwide voluntary clean-up campaign. *Aquatic Ecosystem Health & Management* **16**, 111–118.
Kordella, S., Papatheodorou, G., Geraga, M., Fakiris, E. & Mitropoulou, A. 2010. Litter composition and source identification for 80 beaches in Greece. *Rapport Commission International Mer Méditerranée* **39**, 764 only.
Koss, R.S. & Kingsley, J.Y. 2010. Volunteer health and emotional wellbeing in marine protected areas. *Ocean & Coastal Management* **53**, 447–453.
Koss, R.S, Miller, K., Wescott, G., Bellgrove, A., Boxshall, A., McBurnie, J., Bunce, A., Gilmour, P. & Lerodiaconou, D. 2009. An evaluation of sea search as a citizen science programme in marine protected areas. *Pacific Conservation Biology* **15**, 116–127.
Kofoid C.A. 1910. *The Biological Stations of Europe*. US Bureau of Education Bulletin 4. Washington, District of Columbia: US Government Printing Office.
Kusui, T. & Noda, M. 2003. International survey on the distribution of stranded and buried litter on beaches along the Sea of Japan. *Marine Pollution Bulletin* **47**, 175–179.
Landschoff, J., Lackschewitz, D., Kesy, K. & Reise, K. 2013. Globalization pressure and habitat change: Pacific rocky shore crabs invade armored shorelines in the Atlantic Wadden Sea. *Aquatic Invasions* **8**, 77–87.
Le Fur, J., Guilavogui, A. & Teitelbaum, A. 2011. Contribution of local fishermen to improving knowledge of the marine ecosystem and resources in the Republic of Guinea, West Africa. *Canadian Journal of Fisheries and Aquatic Sciences* **68**, 1454–1469.
Léopold, M., Cakacaka, A., Meo, S., Sikolia, J. & Lecchini, D. 2009. Evaluation of the effectiveness of three underwater reef fish monitoring methods in Fiji. *Biodiversity & Conservation* **18**, 3367–3382.
Lewis, R.R., III, Clark, P.A., Fehring, W.K., Greening, H.S., Johansson, R.O. & Paul, R.T. 1998. The rehabilitation of the Tampa Bay estuary, Florida, USA, as an example of successful integrated coastal management. *Marine Pollution Bulletin* **37**, 468–473.
Lewitus, A.J. & Holland, F.A. 2003. Initial results from a multi-institutional collaboration to monitor harmful algal blooms in South Carolina. *Environmental Monitoring and Assessment* **81**, 361–371.
Lindborg, V.A., Ledbetter, J.F., Walat, J.M. & Moffett, C. 2012. Plastic consumption and diet of glaucous-winged gulls (*Larus glaucescens*). *Marine Pollution Bulletin* **64**, 2351–2356.
Loftus, A.J., Waldon, J., Fay, V., Davy, K. & Lucy, J. 2000. Overview of angler-based tagging programs and management issues. *Fisheries Management* **25**, 8–13.
Lopez-Darias, M., Luzardo, J., Martínez, R., González, D., García, E.A. & Cabrera, J. 2011. Poaching vs. patrolling: effects on conservation of Cory's Shearwater *Calonectris diomedea borealis* colonies. *Bird Conservation International* **21**, 342–352.
Lucy, J. & Davy, K. 2000. Benefits of angler-assisted tag and release programs. *Fisheries* **25**, 18–23.
Lyngs, P. & Kampp, K. 1996. Ringing recoveries of razorbills *Alca torda* and guillemots *Uria aalge* in Danish waters. *Dansk Ornitologisk Forenings Tidsskrift* **90**, 119–132.
MacKenzie, C.J. & Cox, S.P. 2013. Building legitimacy of the recreational fishing sector in mixed commercial-recreational fisheries. *Ocean & Coastal Management* **75**, 11–19.
MacLeod, C.D., Bannon, S.M., Pierce, G.J., Schweder, C., Learmonth, J.A., Herman, J.S. & Reid, R.J. 2005. Climate change and the cetacean community of North-West Scotland. *Biological Conservation* **124**, 477–483.
Margaritoulis, D. 1983. There is an inter-nesting interval of turtles in Zakynthos loggerheads. In *Adaptions to Terrestrial Environments*, N.S. Margaris et al. (eds). New York: Plenum Press, 135–144.
Marigo, J. & Giffoni, B.B. 2010. Sightings and bycatch of small pelagic cetaceans, new information registered by volunteer fishermen of São Paulo, Brazil. *Brazilian Journal of Oceanography* **58**, 71–75.
Marrs, S.J., Tuck, I.D., Atkinson, R.J.A., Stevenson, T.D.I. & Hall, C. 2002. Position data loggers and logbooks as tools in fisheries research: results of a pilot study and some recommendations. *Fisheries Research* **58**, 109–117.
Marshall, N.J., Kleine, D.A. & Dean, A.J. 2012. CoralWatch: education, monitoring, and sustainability through citizen science. *Frontiers in Ecology and the Environment* **10**, 332–334.
Martin, J.M. 2013. Marine debris removal: one year of effort by the Georgia Sea Turtle-Center-Marine Debris Initiative. *Marine Pollution Bulletin* **74**, 165–169.

Martin, K. & James, M.C. 2005. Conserving sea turtles in Canada: successful community-based collaboration between fishers and scientists. *Chelonian Conservation and Biology* **4**, 899–907.

McFee, W.E., Hopkins-Murphy, S.R. & Schwacke, L.H. 2006. Trends in bottlenose dolphin (*Tursiops truncatus*) strandings in South Carolina, USA, 1997–2003: implications for the southern North Carolina and South Carolina management units. *Journal of Cetacean Research and Management* **8**, 195–201.

McKenzie, L.J., Lee Long, W.J., Coles, R.G. & Roder, C.A. 2000. Seagrass-watch: community based monitoring of seagrass resources. *Biología Marina Mediterranea* **7**, 393–396.

Meirelles, A.C.O., Monteiro-Neto, C., Martins, A.M.A., Costa, A.F., Barros, H.M.D.R. & Alves, M.D.O. 2009. Cetacean strandings on the cost of Ceara, north-eastern Brazil. (1992–2005). *Journal of the Marine Biological Association of the United Kingdom* **89**, 1083–1090.

Mellors, J.E., Mackenzie, L.J. & Coles, R.G. 2008. Seagrass-watch engaging Torres Strait Islanders in marine habitat monitoring. *Continental Shelf Research* **28**, 2339–2349.

Mignucci-Giannoni, A.A., Montoya-Ospina, R.A., Jiménez-Marrero, N.M., Rodríguez-López, M.A., Williams, E.H., Jr. & Bonde, R.K. 2000. Manatee mortality in Puerto Rico. *Environmental Management* **25**, 189–198.

Mignucci-Giannoni, A.A., Pinto-Rodriguez, B., Velasco-Escudero, M., Montoya-Ospina, R.A., Jimenez-Marrero, N.M., Rodriguez-López, M.A., Williams, E.H., Jr. & Odell, D.K. 1999. Cetacean strandings in Puerto Rico and the Virgins Islands. *Journal of Cetacean Research and Management* **1**, 191–198.

Milton, D. 2003. Threatened shorebird species of the East Asian-Australasian Flyway: significance for Australian wader study groups. *Wader Study Group Bulletin* **100**, 105–110.

Monk, J., Jerodiaconou, D., Bellgrove, A. & Laureson, L. 2008. Using community-based monitoring with GIS to create habitat maps for a marine protected area in Australia. *Journal of the Marine Biological Association of the United Kingdom* **88**, 865–871.

Moore, E., Lyday S., Roletto, J., Litle, K., Parrish, J.K., Nevins, H., Harvey, J., Mortenson, J., Greig, D., Piazza, M., Hermance, A., Lee, D., Adams, D., Allen, S. & Kell, S. 2009. Entanglements of marine mammals and seabirds in central California and the north-west coast of the United States 2001–2005. *Marine Pollution Bulletin* **58**, 1045–1051.

Moore, P.G. 2003. Seals and fisheries in the Clyde Sea area (Scotland): traditional knowledge informs science. *Fisheries Research* **61**, 51–61.

Moore, S.L., Gregorio, D., Carreon, M., Weisberg, S.B. & Leecaster, M.K. 2001. Composition and distribution of beach debris in Orange County, California. *Marine Pollution Bulletin* **42**, 241–245.

Moreno, G., Dagorn, L., Sancho, G. & Itano, D. 2007. Fish behavior from fisher's knowledge: the case study of tropical tuna around drifting fish aggregating devices (DFADs). *Canadian Journal of Fisheries and Aquatic Sciences* **64**, 1517–1528.

Moreno-Baez, M., Orr, B.J., Cudney-Bueno, R. & Shaw, W.W. 2010. Using fishers local knowledge to aid management at regional scales: spatial distribution of small-scale fisheries in the Northern Gulf of California, Mexico. *Bulletin of Marine Science* **86**, 339–353.

Moser, J. & Shepherd, G.R. 2009. Seasonal distribution and movement of black sea bass (*Centropristis striata*) in the North West Atlantic as determined from a mark-recapture experiment. *Journal of Northwest Atlantic Fishery Science* **40**, 17–28.

Mumby, P.Y., Clarke, K.R. & Harborne, A.R. 1996. Weighting species abundance estimates for marine resource assessment. *Aquatic Conservation: Marine and Freshwaters Ecosystems* **6**, 115–120.

Mumby, P.Y., Harborne, A.R., Raines, P.S. & Ridley, J.M. 1995. A critical assessment of data derived from coral cay conservation volunteers. *Bulletin of Marine Science* **56**, 737–751.

Negri, M.F., Denuncio, P., Panebianco, M.V. & Cappozzo, H.L. 2012. Bycatch of Franciscana dolphins *Pontoporia blainvillei* and the dynamic of artisanal fisheries in the species' southernmost area of distribution. *Brazilian Journal of Oceanography* **60**, 149–158.

Norris, D.R., Arcese, P., Preikshot, D., Bertram, D.F. & Kyser, T.K. 2007. Diet reconstruction and historic population dynamics in a threatened seabird. *Journal of Applied Ecology* **44**, 875–884.

O'Brien, L., Townsend, M. & Ebden, M. 2010. 'Doing something positive': volunteers' experiences of the well-being benefits derived from practical conservation activities in nature. *Voluntas* **21**, 525–545.

Obura, D.O. 2001. Participatory monitoring of shallow tropical marine fisheries by artisanal fishers in Diani, Kenya. *Bulletin of Marine Science* **69**, 777–791.

Obura, D.O., Wells, S., Church, J. & Horrill, C. 2002. Monitoring of fish and fish catches by local fishermen in Kenya and Tanzania. *Marine Freshwater Research* **53**, 215–222.

O'Donnell, K.P., Molloy, P.P. & Vincent, A.C.J. 2012. Comparing fisher interviews, logbooks, and catch landings estimates of extraction rates in a small-scale fishery. *Coastal Management* **40**, 594–611.

Osborn, D.A., Pearse, J.S. & Roe, C.A. 2005. Monitoring rocky intertidal shorelines: a role for the public in resource management. In *California and the World Ocean '02, Conference Proceedings American Society of Civil Engineers, Reston, VA*, 624–636.

Österblom, H., Fransson, T. & Olsson, O. 2002. Bycatches of common guillemot (*Uria aalge*) in the Baltic Sea gillnet fishery. *Biological Conservation* **105**, 309–319.

Pabody, C.M., Carmichael, R.H., Rice, L. & Ross, M. 2009. A new sighting network adds to 20 years historical data on fringe West Indian Manatee (*Trichechus manatus*) population in Alabama Waters. *Gulf of Mexico Science* **27**, 52–61.

Parente, C.L., Vergara-Parente, J.E. & Pinto de Lima, R. 2004. Strandings of Antillean manatees, *Trichechus manatus manatus*, in northeastern Brazil. *Latin American Journal of Aquatic Mammals* **3**, 69–75.

Parrish, J.K., Bond, N., Nevins, H., Mantua, N., Loeffel, R., Peterson, W.T. & Harvey, J.T. 2007. Beached birds and physical forcing in the California current system. *Marine Ecology Progress Series* **352**, 275–288.

Pattengill-Semmens, C.V. & Semmens, B.X. 1998. Fish census data generated by non-experts in the Flower Garden Banks National Marine Sanctuary. *Journal of Gulf Mexico Science* **2**, 196–207.

Pattengill-Semmens, C.V. & Semmens, B.X. 2003. Conservation and management applications of the REEF volunteer fish monitoring program. *Environmental Monitoring and Assessment* **81**, 43–50.

Pearse, J.S., Osborn, D.A. & Roe, C.A. 2003. Assessing sanctuary shorelines: a role for volunteers, particularly high school students, in resource management. California Sea Grant College Program. San Diego, California: University of California, San Diego.

Peloquin, C. & Berkes, F. 2009. Local knowledge, subsistence harvests, and social-ecological complexity in James Bay. *Human Ecology* **37**, 533–545.

Peltier, H., Baggøe, H.J., Camphuysen, K.C.J., Czeck, R., Dabin, W., Daniel, P., Deaville, R., Haelters, J., Jauniaux, T., Jensen, L.F., Jepson, P.D., Keijil, G.O., Siebert, U., van Canneyt, O. & Ridoux, V. 2013. The stranding anomaly as population indicator: the case of harbour porpoise *Phocoena phocoena* in North-Western Europe. *PloS ONE* **8**, e62180.

Peltier, H., Dabin, W., Daniel, P., Van Canneyt, O., Dorémus, G., Huon, M. & Ridoux, V. 2012. The significance of stranding data as indicators of cetacean populations at sea: modelling the drift of cetacean carcasses. *Ecological Indicators* **18**, 278–290.

Pflugh, K.K., Lurig, L., Von Hagen, L.A., Von Hagen, S. & Burger, J. 1999. Urban angler's perception of risk from contaminated fish. *The Science of the Total Environment* **228**, 203–218.

Pierpoint, C., Allan, L., Arnold, H., Evans, P., Perry, S., Wilberforce, L. & Baxter, J. 2009. Monitoring important coastal sites for bottlenose dolphin in Cardigan Bay, UK. *Journal of the Marine Biological Association of the United Kingdom* **89**, 1033–1043.

Pulsifer, P., Gearheard, S., Huntington, H.P., Parsons, M.A., McNeave, C. & McCann, H.S. 2012. The role of data management in engaging communities in Arctic research: overview of the Exchange for Local Observations and Knowledge of the Arctic (ELOKA). *Polar Geography* **35**, 271–290.

Rees, G. & Pond, K. 1995. Marine litter monitoring programs—a review of methods with special reference to national surveys. *Marine Pollution Bulletin* **30**, 103–108.

Ribic, C.A. 1998. Use of indicator items to monitor marine debris on a New Jersey beach from 1991 to 1996. *Marine Pollution Bulletin* **36**, 887–891.

Ribic, C.A., Sheavly, S.B. & Rugg, D.J. 2011. Trends in marine debris in the US Caribbean and the Gulf of Mexico 1996–2003. *Journal of Integrated Coastal Zone Management* **11**, 7–19.

Ribic, C.A., Sheavly, S.B., Rugg, D.J. & Erdmann, E.S. 2010. Trends and drivers of marine debris on the Atlantic coast of the United States 1997–2007. *Marine Pollution Bulletin* **60**, 1231–1242.

Ribic, C.A., Sheavly, S.B., Rugg, D.J. & Erdmann, E.S. 2012. Trends in marine debris along the US Pacific Coast and Hawai'i 1998–2007. *Marine Pollution Bulletin* **64**, 994–1004.

Riesch, H. & Potter, C. 2014. Citizen science as seen by scientists: methodological, epistemological and ethical dimensions. *Public Understanding of Science*, **23**, 107–120.

Robinson, R.A., Grantham, M.J. & Clark, J.A. 2009. Declining rates of ring recovery in British birds. *Ringing & Migration* **24**, 266–272.

Rochet, M.J., Prigent, M., Bertrand, J.A., Carpentier, A., Coppin, F., Delpech, J.P., Fontenelle, G., Foucher, E., Mahé, K., Rostiaux, E. & Trenkel, V.M. 2008. Ecosystem trends: evidence for agreement between fisher's perceptions and scientific information. *ICES Journal of Marine Science: Journal du Conseil* **6**, 1057–1068.

Rogers-Bennett, L., Hubbard, K.E. & Juhasz, C.I. 2013. Dramatic declines in red abalone populations after opening a '*de facto*' marine reserve to fishing: testing temporal reserves. *Biological Conservation* **157**, 423–431.

Rojas-Nazar, U., Gaymer, C.F., Squeo, F.A., Garay-Flühmann, R. & López, D. 2012. Combining information from benthic community analysis and social studies to establish no-take zones within a multiple uses marine protected area. *Aquatic Conservation: Marine and Freshwater Ecosystems* **22**, 74–86.

Roletto, J., Mortenson, J., Harrold, J., Hall, J. & Grella, L. 2003. Beached bird surveys and chronic oil pollution in central California. *Marine Ornithology* **31**, 21–28.

Rosa, J.L., Sampaio, C.L.S. & Barros, A.T. 2006. Collaborative monitoring of the ornamental trade of seahorses and pipefishes (Teleostei: Syngnathidae) in Brazil: Bahia State as a case study. *Neotropical Ichthyology* **4**, 247–252.

Rosevelt, C., Los Huertos, M., Garza, C. & Nevins, H.M. 2013. Marine debris in central California: quantifying type and abundance of beach litter in Monterey Bay, CA. *Marine Pollution Bulletin* **71**, 299–306.

Ross, J.B., Parker, R. & Strickland, M. 1991. A survey of shoreline litter in Halifax Harbour 1989. *Marine Pollution Bulletin* **22**, 245–248.

Roura, M.R. 2012. Being there: examining the behavior of Antarctic tourists through their blogs. *Polar Research* **31**, 1–23.

Rutherford, F.J. & Ahlgren, A. 1991. *Science for ALL Americans*. New York: Oxford University Press.

Ruttenberg, B.I., Schofield, P.J., Akins, J.L., Acosta, A., Feeley, M. W., Blondeau, J., Smith, S.G. & Ault, J.S. 2012. Rapid invasion of Indo-Pacific lionfishes (*Pterois volitans* and *Pterois miles*) in the Florida Keys, USA: evidence from multiple pre- and post-invasion datasets. *Bulletin of Marine Science* **88**, 1051–1059.

Sáenz-Arroyo, A., Roberts, C.M., Torre, J., Cariño-Olvera, M. & Enríquez-Andrade, R.R. 2005. Rapidly shifting environmental baselines among fishers of the Gulf of California. *Proceedings of the Royal Society B: Biological Sciences* **272**, 1957–1962.

Sampson, D.B. 2011. The accuracy of self-reported fisheries data: Oregon trawl logbook fishing locations and retained catches. *Fisheries Research* **112**, 59–76.

Schmeller, D.S., Henry, P.Y., Julliard, R., Gruber, B., Clobert, J., Dziock, F., Lengyel, S., Nowicki, P., Déri, E., Budrys, E., Kull, T., Tali, K., Bauch, B., Settele, J., van Swaay, C., Kobler, A., Babij, V., Papastergiadou, E. & Henle, K. 2009. Advantages of volunteer-based biodiversity monitoring in Europe. *Conservation Biology* **23**, 307–316.

Schmitt, E.F. & Sullivan, K.M. 1996. Analysis of a volunteer method for collecting fish presence and abundance data in the Florida Keys. *Bulletin of Marine Science* **59**, 404–416.

Schroeter, S.C., Gutiérrez, N.L., Robinson, M., Hilborn, R. & Halmay, P. 2009. Moving from data poor to data rich: a case study of community-based data collection for the San Diego red sea urchin fishery. *Marine and Coastal Fisheries* **1**, 230–243.

Selig, E.R. & Bruno, J.F. 2010. A global analysis of the effectiveness of marine protected areas in preventing coral loss. *PloS ONE* **5**, e9278.

Semmens, B.X., Buhle, E.R., Salomon, A.K. & Pattengill-Semmens, C.B. 2004. A hotspot of non-native marine fishes: evidence for the aquarium trade as an invasion pathway. *Marine Ecology Progress Series* **266**, 239–244.

Semmens, B.X., Luke, K.E., Bush, P.G., McCoy, C.M.R. & Johnson, B.C. 2006. Isopod infestation of post spawning Nassau grouper around Little Cayman Island. *Journal of Fish Biology* **69**, 933–937.

Shepherd, G.R., Moser, J., Deuel, D. & Carlsen, P. 2006. The migration patterns of bluefish (*Pomatomus saltatrix*) along the Atlantic coast determined from tag recoveries. *Fishery Bulletin* **104**, 559–570.

Shirk, J.L., Ballard, H.L., Wilderman, C.C., Phillips, T., Wiggins, A., Jordan, R., McCallie, E., Minarchek, M., Lewenstein, B.V., Krasny, M.E. & Bonney, R. 2012. Public participation in scientific research: a framework for deliberate design. *Ecology and Society* **17**, 29.

Silvermintz, D. 2010. Plato's supposed defense of the division of labor: a reexamination of the role of job specialization in the republic. *History of Political Economy* **42**, 747–772.

Silvertown, J. 2009. A new dawn for citizen science. *Trends in Ecology & Evolution* **24**, 467–471.

Smith, B.S. 2010. Patterns of nonbreeding snowy plover (*Charadrius alexandrinus*), piping plover (*C. melodus*), and red knot (*Calidris canutus*) distribution in northwest Florida. *Florida Field Naturalist* **38**, 43–54.

Smith, D.R., Pooler, P.S., Swan, B.L., Michels, S.F., Hall, W.R., Himchak, P.J. & Millard, M.J. 2002. Spatial and temporal distribution of horseshoe crab (*Limulus polyphemus*) spawning in Delaware Bay: implications for monitoring. *Estuaries* **25**, 115–125.

Spalding, M.D., Fox, H.E., Allen, G.R., Davidson, N., Ferdaña, Z.A., Finlayson, M., Halpern, B.S., Jorge, M.A., Lombana, A., Lourie, S.A., Martin, K.D., McManus, E., Molnar, J., Recchia, C.A. & Robertson, J. 2007. Marine ecoregions of the world: a bioregionalization of coastal and shelf areas. *BioScience* **57**, 573–583.

Stacey, P.J., Baird, R.W. & Langelier, K.M. 1989. Stranded whale and dolphin program 1988 report. *Wildlife Veterinary Report* **2**, 10–11.

Stallings, C.D. 2009. Fishery-independent data reveal negative effect of human population density on Caribbean predatory fish communities. *PloS ONE* **4**, e5333.

Starr, P. 2010. Fisher-collected sampling data: lessons from the New Zealand experience. *Marine and Coastal Research* **2**, 47–59.

Starr, P.J. & Vignaux, M. 1997. Comparison of data from voluntary logbook and research catch-sampling programmes in the New Zealand lobster fishery. *Marine Freshwater Research* **48**, 1075–1080.

Storrier, K.L., McGlashan, D.J., Bonellie, S. & Velander, K. 2007. Beach litter deposition at a selection of beaches in the Firth of Forth, Scotland. *Journal Coastal Research* **23**, 813–822.

Stuart-Smith, R.D., Bates, A.E., Lefcheck, J.S., Duffy, J.E., Baker, S.C., Thomson, R.J., Stuart-Smith, J.F., Hill, N.A., Kininmonth, S.J., Airoldi, L., Becerro, M.A., Campbell, S.J., Dawson, T.P., Navarrete, S.A., Soler, G.A., Strain, E.M.A, Willis, T.J. & Edgar G.J. 2013. Integrating abundance and functional traits reveals new global hotspots of fish diversity. *Nature* **501**, 539–542.

Suazo, C.G., Schlatter, R.P., Arriagada, A. M., Cabezas, L. & Ojeda, J. 2013. Fishermen's perceptions of interactions between seabirds and artisanal fisheries in the Chonos archipelago, Chilean Patagonia. *Oryx* **47**, 184–189.

Tautin, J., Metras, L. & Smith, G. 1999. Large-scale studies of marked birds in North America. *Bird Study* **46**, S271–S278.

Theberge, M.M. & Dearden, P. 2006. Detecting a decline in whale shark *Rhincodon typus* sightings in the Andaman Sea, Thailand, using ecotourist operator-collected data. *Oryx* **40**, 337–342.

Trumbull, D.J., Bonney, R., Bascom, D. & Cabral, A. 2000. Thinking scientifically during participation in a citizen-science project. *Science Education* **84**, 265–275.

Tudge, C. 1998. *Neanderthals, Bandits and Farmers: How Agriculture Really Began*. London: Weidenfeld & Nicolson.

Tulloch, A.I.T. & Szabo, J.K. 2012. A behavioural ecology approach to understand volunteer surveying for citizen science datasets. *EMU* **112**, 313–325.

Tweddle, J.C., Robinson, L.D., Pocock, M.J.O. & Roy, H.E. 2012. Guide to citizen science: developing, implementing and evaluating citizen science to study biodiversity and the environment in the UK. Natural History Museum and NERC Centre for Ecology & Hydrology for UK-EOF, London, United Kingdom. 36 pp.

Uychiaoco, A.J., Arceo, H.O., Green, S.J., de la Cruz, M.J., Gaite, P.A. & Aliño, P.M. 2005. Monitoring and evaluation of reef protected areas by local fishers in the Philippines: tightening the adaptive management cycle. *Biodiversity and Conservation* **14**, 2775–2794.

van Franeker, J.A., Blaize, C., Danielsen, J., Fairclough, K., Gollan, J., Guse, N., Hansen, P.L., Heubeck, M., Jensen, J.K., le Guillou, G., Olsen, B., Olsen, K.O., Pedersen, J., Stienen, E.W.M. & Turne, D.M. 2011. Monitoring plastic ingestion by the northern fulmar *Fulmarus glacialis* in the North Sea. *Environmental Pollution* **159**, 2609–2615.

Vargo, G.A., Atwood, K., van Deventer, M. & Harris, R. 2009. Beached birds surveys on shell key, Pinellas County, Florida. *Florida Field Naturalist* **34**, 21–27.

Vélez-Rubio, G., Estrades, A., Fallabrino, A. & Tomás, J. 2013. Marine turtle threats in Uruguayan waters: insights from 12 years of stranding data. *Marine Biology* **160**, 2797–2811.

Wagner, G.M. 2005. Participatory monitoring of changes in coastal and marine biodiversity. *Indian Journal of Marine Sciences* **34**, 136–146.

Walsh, W.A., Ito, R.Y., Kawamoto, K.E. & McCracken, M. 2005. Analysis of logbook accuracy for blue marlin (*Makaira nigricans*) in the Hawaii-based longline fishery with a generalized additive model and commercial sales data. *Fisheries Research* **75**, 175–192.

Ward-Paige, C.A. & Lotze, H.K. 2011. Assessing the value of recreational divers for censusing elasmobranchs. *PloS ONE* **6**, e25609.

Ward-Paige, C.A., Mora, C., Lotze, H.K., Pattengill-Semmens, C., McClenachan, L., Arias-Castro, E. & Myers, R.A. 2010. Large-scale absence of sharks on reefs in the greater-Caribbean: a footprint of human pressures. *PloS ONE* **5**, e11968.

Ward-Paige, C.A., Pattengill-Semmens, C., Myers, R.A. & Lotze, H.K. 2011. Spatial and temporal trends in yellow stingray abundance: evidence from diver surveys. *Environmental Biology of Fishes* **90**, 263–276.

Wells, F.E., Chalermwat, K., Chitramvong, Y., Kakhai, N., Putchakarn, S. & Sanpanich, K. 2008. Assessment of three techniques for measuring the biodiversity of mollusks on rocky intertidal shorelines in Eastern Thailand. *The Raffles Bulletin of Zoology, Supplement* **18**, 259–264.

Whatmough, S., van Putten, I. & Chin, A. 2011. From hunters to nature observers: a record of 53 years of diver attitudes towards sharks and rays and marine protected areas. *Marine and Freshwater Research* **62**, 755–763.

Whaylen, L., Pattengill-Semmens, C.V., Semmens, B.X., Bush, P.G. & Boardman, M.R. 2004. Observations of a Nassau grouper, *Epinephelus striatus*, spawning aggregation site in Little Cayman, Cayman Islands, including multi-species spawning information. *Environmental Biology of Fishes* **70**, 305–313.

Wheeler, M., Chambers, F.M.J., Sims-Castley, R., Cowling, R.M. & Schoeman, D.S. 2008. From beans to breams: how participatory workshops can contribute to marine conservation planning. *African Journal of Marine Science* **30**, 475–487.

Whereat, E.B., Farrell, J.G. & Humphries, E.M. 2004. Volunteer phytoplankton monitoring in the inland bays of Delaware, USA. In: Steidinger,K.A., Landsberg, J.H., Tomas, C.R. & Vargo, G.E. (Eds.). 2004. *Harmful Algae* 2002. Florida Fish and Wildlife Conservation Commission, Florida Institute of Oceanography, and Intergovernmental Oceanographic Commission of UNESCO, St. Petersburg, Florida, USA. 367–368.

Whitney, N.M., Pyle, R.L., Holland, K.N. & Barcz, J.T. 2012. Movements, reproductive seasonality, and fisheries interactions in the whitetip reef shark (*Triaenodon obesus*) from community-contributed photographs. *Environmental Biology of Fishes* **93**, 121–136.

Wiber, M., Berkes, F., Charles, A. & Kearney, J. 2004. Participatory research supporting community-based fishery management. *Marine Policy* **28**, 459–468.

Williams, A.T., Pond, K., Tudor, D.T., Jansen, H. & Liu, H.B. 1999. The robustness of litter transect data collection by different survey groups. In *Proceedings of the MEDCOAST* **99**, 715–725.

Willis, P.M. & Baird, R.W. 1998. Sightings and stranding of beaked whales on the west coast of Canada. *Aquatic Mammals* **24**, 21–25.

Wilson, S., Anderson, E.M., Wilson, A.S.G., Bertram, D.F. & Arcese, P. 2013. Citizen science reveals an extensive shift in the winter distribution of migratory western grebes. *PloS ONE* **8**, e65408.

Witt, M.J., Hardy, T., Johnson, L., McClellan, C.M., Pikesley, S.K., Ranger, S., Richardson, P.B., Solandt, J.-L., Speedie, C., Williams, R. & Godley, B.J. 2012. Basking sharks in the northeast Atlantic: spatio-temporal trends from sightings in UK waters. *Marine Ecology Progress Series* **459**, 121–134.

Wolfe, J.R. & Pattengill-Semmens, C.V. 2013a. Fish population fluctuation estimates based on fifteen years of reef volunteer diver data for the Monterey peninsula, California. *California Cooperative Oceanic Fisheries Investigations Reports* **54**, 141–154.

Wolfe, J.R. & Pattengill-Semmens, C.V. 2013b. Estimating fish populations from reef citizen science volunteer diver order-of-magnitude surveys. *California Cooperative Oceanic Fisheries Investigations Reports* **54**, 127–140.

Worachananant, S., Carter, R.W. & Hockings, M. 2007. Impacts of the 2004 tsunami on Surin Marine National Park, Thailand. *Coastal Management* **35**, 399–412.

Yasué, M., Kaufman, L. & Vincent, A.C.J. 2010. Assessing ecological changes in and around marine reserves using community perceptions and biological surveys. *Aquatic Conservation: Marine and Freshwater Ecosystems* **20**, 407–418.

Yasué, M., Nellas, A. & Vincent, A.C.J. 2012. Seahorses helped drive creation of marine protected areas, so what did these protected areas do for the seahorses? *Environmental Conservation* **39**, 183–193.

Zenetos, A., Koutsogiannopoulos, D., Ovalis, P. & Poursanidis, D. 2013. The role played by citizen scientists in monitoring marine alien species in Greece. *Cahiers de Biologie Marine* **54**, 419–426.

Zohany, D. & Hopf, M. 2000. *Domestication of Plants in the Old World: The Origin and Spread of Cultivated Plants in West Asia, Europe, and the Nile Valley*. New York: Oxford University Press.

Oceanography and Marine Biology: An Annual Review, 2014, **52**, 315-336

Taylor & Francis

THE CLAW HYPOTHESIS: A NEW PERSPECTIVE ON THE ROLE OF BIOGENIC SULPHUR IN THE REGULATION OF GLOBAL CLIMATE

TAMARA K. GREEN & ANGELA D. HATTON

Department of Biogeochemistry & Earth Science, Scottish Association for Marine Science, Oban, Argyll PA37 1QA, UK
E-mail: 09002154@uhi.ac.uk; angela.hatton@sams.ac.uk

In 1987, R.J. Charlson et al. (*Nature* **326**, 655–661) proposed that biogenic sulphur, primarily dimethyl sulphide (DMS), a by-product of dimethyl sulphoniopropionate (DMSP) produced by marine organisms, could act as cloud condensation nuclei (CCN) in the atmosphere. Higher levels of CCN would lead to increased cloud cover, with consequent changes to the radiation budget on Earth, thereby establishing a climatic feedback loop reliant on marine organisms. This CLAW hypothesis inspired a large body of research intended to test key aspects of the proposed feedback mechanism. We consider here the key components of the hypothesis as outlined by the original authors: light, temperature, salinity, DMS-CCN correlation, and the palaeo record. Literature reviewed here indicates that it is time to modify the CLAW hypothesis in light of recent advances in knowledge, particularly the emerging importance of microbial activity in the control of water column DMS concentrations and increased awareness of the importance of dimethylsulphoxide (DMSO) as a source and sink of DMS.

Introduction: The CLAW hypothesis and its importance

In 1987, Charlson, Lovelock, Andreae, and Warren published a paper proposing a feedback system whereby dimethyl sulphide (DMS) produced by marine phytoplankton is oxidised in the atmosphere to form sulphate aerosols, thus acting as a precursor to cloud condensation nuclei (CCN) (Charlson et al. 1987). Because cloud albedo, and therefore Earth's radiation budget, is sensitive to CCN density, it was suggested that production of DMS could have a role in climate regulation. This proposed mechanism has become known as the CLAW hypothesis, an acronym for the surnames of the four authors of the paper.

Dimethyl sulphide is not produced directly by marine algae; its precursor is dimethyl sulphoniopropionate (DMSP), a compatible tertiary sulphonium solute produced by several classes of marine algae and some classes of higher plants (Reed 1983, Van Diggelen et al. 1986, Keller et al. 1989, Pakulski & Kiene 1992, Hanson et al. 1994, Mulholland & Otte 2000, Stefels 2000). Globally, the main producers of DMSP (Table 1) are the classes Dinophyceae (dinoflagellates); Prymnesiophyceae (prymnesiophytes, including coccolithophores); Chrysophyceae; and Bacillariophyceae (diatoms) (Keller et al. 1989). Typical DMSP concentration values are 50–400 mM, and this can account for 50–100% of the total cellular organic sulphur (Matrai & Keller 1994, Keller et al. 1999a,b, Stefels 2000).

Although it is still unclear why organisms produce DMSP, there is evidence for several possible roles. Some studies indicate that DMSP may have an antioxidant role, helping organisms cope with high ultraviolet (UV) radiation by scavenging the harmful hydroxyl radical (Sunda et al. 2002,

Table 1 Intracellular concentration of DMSP in marine phytoplankton

Organism	Cytosolic DMSP (mM)	Reference
Hymenomonas carterae	120	Vairavamurthy et al. 1985
Gymnodinium nelsoni	280	Dacey & Wakeham 1986
Platymonas subcordiformis	170	Dickson & Kirst 1986
Phaeocystis sp.	71–169	Stefels & van Boekel 1993
Melosira numuloides	264	Keller et al. 1989
Chrysameoba sp.	596	Yoch 2002
Ochromonas sp.	529	Yoch 2002
Prorocentrum sp. IIB_2b_1	1082[b]	Yoch 2002
Emiliana huxleyi BT6	166	Yoch 2002
Diatoms (from seawater)	38 ± 18	Turner et al. 1998
>10-µm flagellates (from seawater)	125	Belviso et al. 1990
	650 ± 429	Turner et al. 1998
Dinoflagellates (from seawater)	640	Belviso et al. 1990
	70 ± 30	Turner et al. 1998

Source: Adapted from D.C. Yoch. *Applied and Environmental Microbiology* **68**, 5804–5815, 2002. A reviewer of this article highlighted that "seawater osmolarity is about 0.7 M, and DMSP is never the only osmolyte, so its intracellular concentrations can't be >1M."

Vallina & Simó 2007). It also may be produced as a cryoprotectant (Welsh 2000, Lee et al. 2001), and there is limited evidence that DMSP may act as either a methyl donor in metabolic reactions (Ishida 1968), as a precursor of the phospholipid phosphatidylsulphocholine (Kates & Volcani 1996), as an overflow mechanism for excess reducing power under conditions of unbalanced growth (Stefels 2000), and as an antigrazing agent (Wolfe et al. 1997). The most likely role for DMSP, however, is that it is produced as an osmoregulant, enabling algae to regulate their osmotic potential in response to changing salinities (Malin & Kirst 1997, Stefels 2000).

How DMSP is ultimately transformed to DMS is discussed further in this review. Suffice to say it is now known that higher concentrations of DMS in seawater drive a net flux of DMS to the atmosphere, where it is oxidised by the hydroxyl radical to form sulphur dioxide (SO_2) and methanesulphonate (MSA) (Hatton 2002, Quinn & Bates 2011). Sulphur dioxide is usually oxidised to sulphate, so the end result of DMS oxidation yields two major compounds: MSA and non-sea-salt sulphate (nss-SO_4^{2-}). These sulphur aerosols, once in the atmosphere, act as CCN. The CLAW hypothesis proposed that DMS-derived sulphate makes up the majority of these CCN in the form of MSA or nss-SO_4^{2-}. The hypothesis further proposed that increased DMS emissions would result in an increase in CCN, cloud albedo, and cloud droplet concentrations, with a subsequent decrease in the amount of solar radiation reaching Earth's surface (Charlson et al. 1987, Quinn & Bates 2011). A reduction in solar radiation reaching Earth would alter the populations of DMSP-producing phytoplankton, potentially setting up a climate feedback loop between cloud albedo and surface ocean DMS concentrations (Charlson et al. 1987). The original authors further posited that this feedback loop could be exploited to counteract the effects of increasing atmospheric CO_2, thus alleviating the effects of global warming.

Charlson et al. (1987) highlighted that further research would be required to test the hypothesis. Since that time, hundreds of papers addressing aspects of the proposed mechanism have been published, but there remains no consensus regarding its validity, and no review paper has yet addressed the CLAW hypothesis in its entirety. Most important is the need to understand the climatic factors affecting DMS emission, which must include controls on those phytoplankton species capable of producing DMSP (Charlson et al. 1987). The relationship between DMS concentrations and CCN is also of paramount importance, as is the influence of CCN on cloud cover. The original authors suggested the possibility that Earth's climate has been regulated in the past by this feedback loop, which would make consideration of palaeo ice and sedimentary records of considerable interest. Given the potentially significant impact of this trace biogas, the production and cycling of DMS is important especially in light of the continuing issue of global climate change. The aim of this review is to address the key points highlighted in the CLAW hypothesis (light, temperature, salinity, DMS-CCN relationship, and the palaeo record), reviewing key papers in each area and assessing the validity of the hypothesis in the light of current evidence. We also aim to assess whether the production of DMS is likely to ameliorate global warming as suggested by the original authors.

Testing the predictions of the CLAW hypothesis

The CLAW hypothesis (Charlson et al. 1987) includes the following key arguments:

- DMS is a breakdown product of DMSP, a compatible solute produced by marine algae in response to osmoregulatory demands. Production of DMSP is controlled primarily by light, temperature, and salinity; the "warmest, most saline and most intensely illuminated" regions of the oceans exhibit the highest DMS fluxes.
- The primary sources of marine boundary layer CCN are DMS oxidation products.
- If DMS exerts a biogenic feedback control on cloud cover (through its role as a source of CCN), the palaeo record should show evidence of a correlation between past atmospheric DMS concentrations and global climate.

The following sections address each of these points in turn and examine whether these predictions are supported by the results of research carried out since the hypothesis was first proposed.

Factors governing DMSP production: light

Central to CLAW is the suggestion that fluctuating light levels alter the species composition and abundance of certain DMSP-producing organisms, setting up a climate feedback loop between DMS production and cloud albedo (Charlson et al. 1987, Quinn & Bates 2011). As noted previously, the CLAW hypothesis predicted that the "warmest, most saline and most intensely illuminated" regions of the oceans will show the highest rate of DMS emission to the atmosphere. This argument was supported by empirical evidence from Andreae (1986) that the largest flux of DMS comes from the tropical and equatorial oceans. CLAW proposes that the effect of DMS in such regions is to contribute to increased cloud albedo, so reducing the input of heat into the low-latitude oceans (Charlson et al. 1987). The original authors considered the main effect of light on DMS production to be exerted through changes in the abundance and distribution of photosynthetic phytoplankton. There are, however, other complicating factors at play that affect the rate of DMSP and DMS production and subsequent cycling.

Light/dark experiments and their effect on DMSP production

It has been observed that the ratios of DMSP and DMS to chlorophyll-a are highest near the surface, decreasing with depth (Belviso et al. 1993, Simó et al. 1995, Dacey et al. 1998), leading to the suggestion that DMSP production could be related to high irradiation. Experiments on the effects of light on DMSP production by macroalgae showed that although DMSP production was stimulated after several weeks of incubation, short-day incubations of 6 h light and 18 h dark did not result in any change in algal cell DMSP content (Karsten et al. 1990, 1991, 1992). Long-day incubations of 18 h light and 6 h dark, however, resulted in doubling of DMSP production. This suggests that sudden changes in light are not directly associated with DMSP production, but it is the long-term exposure to different light levels that is of more importance. Moreover, work by Sunda et al. (2002) showed that, under lab conditions at least, many phytoplankton species produce more DMS and DMSP in response to elevated oxidant levels caused by UV light exposure.

Solar irradiance and DMS concentrations

Positive correlations have been reported between seasonal variations in DMS concentration and local solar irradiance at the sea surface (Toole & Siegel 2004) as well as between DMS concentration and the solar radiation dose (SRD) (the average radiation in the surface mixed layer) (Vallina & Simó 2007). Perhaps the greatest support for a correlation between DMSP/DMS concentrations and solar irradiance comes from Vallina & Simó (2007), who not only found a significant positive correlation between surface DMS concentrations and the SRD, but also found that the SRD accounted for 81–94% of the variance in the monthly surface DMS concentrations in the northwestern Mediterranean Sea. While this makes a compelling case for the importance of light in DMS cycling, it should be noted that although about 26,400 data points represent a great deal of information, they were gathered over a 30-year period and may not provide an accurate representation of global-scale relationships at a single time point. This highlights a potential area for future research efforts. Other authors have provided similar support; Miles et al. (2009) reported positive correlations between DMS concentration and local solar irradiance at the sea surface.

It is not only effects on phytoplankton species composition that makes irradiance levels of such importance in global DMS cycling. Solar radiation also exerts significant control over two major DMS loss pathways, and the loss of DMS is a factor of equal importance as production in controlling surface water concentrations.

Photochemical removal of DMS Once in the water column, any DMS not ventilated to the atmosphere can either be consumed by the biota (Kiene & Bates 1990, Kiene 1992, Wolfe & Kiene 1993, Ledyard & Dacey 1994, Kiene et al. 2000, Miles et al., 2012) or photo-oxidised to DMSO (dimethylsulphoxide; Brimblecombe & Shooter 1986, Kieber et al. 1996, Brugger et al. 1998, Hatton 2002). Here, we consider the effect of light on the photo-oxidation of DMS to DMSO, which represents a significant sink for DMS.

The photo-oxidation of DMS to DMSO is well established; Brimblecombe & Shooter (1986) showed that DMS not only could be rapidly destroyed by intense UV radiation, but it also could be photolysed by visible light in the presence of photosensitisers such as humic acid. Although Brimblecombe & Shooter (1986) did not measure DMSO directly, they did report that DMSO and singlet oxygen would be formed, with the photosensitiser absorbing light and reacting with dissolved molecular oxygen. Kieber et al. (1996) found that, on calm days, DMS photo-oxidation was of similar importance to microbial degradation, although their results showed that a smaller fraction of DMS (14%) underwent this pathway. Hatton (2002) found that in the northern North Sea DMS photolysis is a significant removal pathway when compared with atmospheric ventilation and bacterial consumption. Furthermore, Kieber et al. (1996) showed that photochemical removal of DMS occurs in the UVB range; Hatton (2002) found that UVA/visible light can also photochemically remove DMS.

It has also been proposed that solar irradiance plays an important role in mediating ambient seawater DMS concentrations beyond the effects of irradiance on primary production and photochemical removal (Miles et al. 2012), with a positive correlation between DMS and underwater irradiance, suggestive of repression of the bacterial metabolism of DMS (Kieber et al. 1996, Simó & Pedrós-Alió 1999, Miles et al. 2012). Although Miles et al. (2012) concluded that photo-destruction of DMS was a minor process, they did find that release of DMS under stress was the dominant light-related process.

It can be reasonably assumed that interglacial conditions are cloudier than glacial conditions; less water is locked up in the cryosphere. This has implications for CLAW; reduced light conditions should favour production of less DMSP as an antioxidant, with increased light conditions favouring the production and release of more DMSP. However, the evidence presented in the literature is far from conclusive, and although it is clear that light does have some effect on DMSP/DMS production and cycling, the CLAW hypothesis does not adequately explain why the "most intensely illuminated" regions of the world's oceans should exhibit the highest DMS emissions.

Factors governing DMSP production: temperature

A central tenet of the CLAW hypothesis is that the "warmest, most saline and most intensely illuminated" regions of the oceans have the greatest rate of DMS emission to the atmosphere. Because DMSP, the DMS precursor, is produced by phytoplankton, this seems counterintuitive; it is known that these regions are also the most oligotrophic of the world's oceans. Indeed, the original authors of the theory stressed that DMS emission is only weakly correlated with chlorophyll-a and ^{14}C uptake—the standard measures of phytoplankton productivity. Supporting evidence was provided by Andreae (1986), who concluded that the calculated flux of DMS from tropical oceans is approximately the same (2.2 mmol m^{-2} yr^{-1}) as from temperate oceans (2.4 mmol m^{-2} yr^{-1}). Charlson et al. (1987) further suggested that this emission rate is largely dependent on DMS removal as much as production, with microbial and photochemical processes responsible for removal, whereas production is affected more by phytoplankton species composition than biomass. Charlson et al. (1987, p. 656) suggested that "some algal groups, such as the coccolithophorids, which are most abundant in tropical, oligotrophic waters, have the highest rate of DMS excretion per unit biomass". This could potentially explain why these areas emit more DMS than temperate zones where diatoms are usually dominant. It is not clear from the formulation of the CLAW hypothesis, however, what specific effect temperature has on DMS production other than indirectly through its influence on phytoplankton community structure. Since publication of the original paper in 1987, researchers have examined how temperature controls DMS production, both by considering DMSP production at the cellular level and by effects on the distribution of DMSP-producing organisms. There has been much speculation regarding the possible roles of DMSP, which include (among others) the idea that it could act as a cryoprotectant (Kirst et al. 1991, Lee & De Mora 1999) or as an antioxidant under times of temperature-induced oxidative stress (Sunda et al. 2002, Yost et al. 2010).

DMSP as a cryoprotectant

If DMSO or DMSP were to act as a cryoprotectant, it has been calculated that 2 mol l^{-1} DMSO would be needed to depress the freezing point of pure water by 2°C, as is the case with salt in 35-psu seawater (Crown-Zellerbach 1966, Lee & De Mora 1999). However, Lee et al. (2001) showed that ice diatoms had less intracellular DMSO than their pelagic counterparts. This would seem to suggest that DMSO is not produced as a cryoprotectant, although the authors did highlight that DMSO is a neutral osmolyte, similar to carbohydrate osmolytes such as sugars and polyols, which also depress freezing points (Yancey 1994). Similarly, when DMSP is described as having cryoprotective properties, it is as a compatible solute that offers protection under times of stress induced by salinity/temperature changes or desiccation (Lee et al. 2001). It is possible, therefore, that DMSO offers protection against extreme temperatures in a similar manner, without forcing organisms to produce large and energetically expensive quantities of it during times of stress. Indeed, if DMSP and DMSO are produced by marine algae as a cryoprotectant, a negative correlation would be expected between temperature and intracellular concentration. Van Rijssel & Gieskes (2002) observed increased DMSP accumulation in *Emiliana huxleyi* cells at low temperatures, which they observed to be inconsistent with the feedback suggested by the CLAW hypothesis. It should be borne in mind, however, that although a feedback pathway is implied by Charlson et al. (1987), nothing is stated by the authors regarding its sign (positive or negative). It is reasonable to assume that the lower temperatures characteristic of glacial periods would be accompanied by less cloud cover, with more water locked up as ice and snow. This brings into question the conclusions drawn by Van Rijssel & Gieskes (2002); lower temperatures would result in increased exposure to light, therefore requiring organisms to produce more DMSP as an antioxidant defence, meaning that DMSP production is altered in response to light and temperature. It is possible, then, based on this evidence, that lower temperatures and increased levels of DMSP actually support the predictions of the CLAW hypothesis.

DMSP as a temperature-induced antioxidant

McLenon & DiTullio (2012) analysed the dinoflagellate coral symbiont *Symbiodinium microadriaticum* for changes in DMSP production under different temperature regimes. Increased temperature and light can cause oxidative stress by overwhelming the cell's antioxidant systems with added reactive oxygen species (ROS) produced during stress (McLenon & DiTullio 2012). Superoxide dismutases, catalase, and ascorbate peroxidase are all used to break down ROS (Dykens & Shick 1982, Lesser et al. 1990, Fridovich 1995), but Sunda et al. (2002) have suggested that the production and turnover of DMSP could result in four different antioxidant compounds that operate as a "highly effective antioxidant system", also known as the 'antioxidant cascade'. McLenon & DiTullio (2012) found that the oxidative stress imparted by elevated temperatures led to a 49% decrease in cellular DMSP in *Symbiodinium microadriaticum* as well as an 8-fold increase in ROS. If DMSP were produced to act as an antioxidant in response to extreme temperature changes, an increase in cellular DMSP accompanied by a decrease in cellular ROS should have been observed. This is in direct contrast to the findings of Van Rijssel & Gieskes (2002) and would seem to refute the prediction of the CLAW hypothesis. It is noteworthy, however, that organisms living on coral reefs and those living in polar seas deal with very different stresses. Production of DMSP will undoubtedly reflect this, and it may be unrealistic to expect a consistent pattern in species from such widely contrasting environments.

DMSP-producing organisms in polar waters and sea ice

Irrespective of why DMSP is produced, let us reconsider the regions of the oceans Charlson et al. (1987) predicted would show the greatest rate of DMS emission to the atmosphere (the "warmest, most saline and most intensely illuminated regions"). Their conclusion was based on data reported by Andreae (1986), but since then the presence of DMSP has been confirmed in both the Arctic (Levasseur et al. 1994) and Antarctic sea ice (Kirst et al. 1991) with the detection of abundant DMSP-producing algae (Trevena et al. 2003, Gambaro et al. 2004). Measuring DMS in sea ice had previously proved difficult because thawing of the ice and subsequent osmotic shock resulted in cell lysis and the artificial conversion of DMSP into DMS (Stefels et al. 2012). These difficulties have been overcome by development of new techniques, such as acidification of samples (Trevena & Jones 2006), dry crushing, and addition of stable isotopes (Tison et al. 2010, Stefels et al. 2012). It is now known that sea ice holds extremely high concentrations of DMSP, making the polar regions important sources of DMS to the atmosphere, either through sea-ice melting and associated phytoplankton blooms (Levasseur et al. 1994, Curran & Jones 2000, Trevena & Jones 2006, 2012, Tison et al. 2010) or through direct ventilation of DMS to the atmosphere (Zemmelink et al. 2008, Nomura et al. 2012). Although there are relatively few measurements of direct ventilation, fluxes of 0.1–5.3 μmol DMS m^{-2} d^{-1} have been recorded by Nomura et al. (2012) from landfast multiyear ice off Dronning Maud Land in the Southern Ocean, and Zemmelink et al. (2008) reported fluxes up to 11 μmol DMS m^{-2} d^{-1} from multiyear ice in the Weddell Sea. Whether from direct venting or from sea-ice melting, DMS fluxes to the polar atmosphere not only are more significant than the original CLAW authors suggested, but also exhibit extreme variability both regionally and seasonally (Turner et al. 1995, Curran & Jones 2000).

Another important source of DMS comes from pelagic organisms inhabiting polar waters; Davison et al. (1996) reported atmospheric concentrations of DMS from polar regions and areas south of the Falkland Islands of 3–714 ng S m^{-3}; equatorial regions saw atmospheric concentrations of 3–46 ng S m^{-3}. Galí & Simó (2010) reported fluxes of 0.5–22.5 μmol DMS m^{-2} d^{-1} in the Greenland Sea and Arctic Ocean during July 2007. Global DMS climatology compiled by Lana et al. (2011), building on previous work by Kettle et al. (1999) and Kettle & Andreae (2000), revealed that higher latitudes saw higher DMS concentrations than equatorial zones. This calls into question the validity of the conclusions drawn by Charlson et al. (1987), specifically the lack of a consistent relationship between temperature and DMSP or DMS production. It is worth bearing in mind, however, that although this climatology (Lana et al. 2011) used 47,313 DMS seawater concentration measurements from both hemispheres over 30 years, these represent only 1577 data points per year for the global ocean, suggesting more research is required to provide a comprehensive global synthesis.

The monthly climatology for DMS fluxes compiled by Lana et al. (2011) shows elevated DMS fluxes in regions other than those Charlson et al. (1987) suggested would exhibit the highest DMS fluxes. This certainly brings up questions about the original statement made by the authors of the CLAW hypothesis. In light of knowledge gained from research conducted since 1987, it appears that the CLAW hypothesis needs revision with respect to the role temperature plays in the global biogeochemical cycle of DMS.

Factors governing DMSP production: salinity

It has been known for several decades that DMSP is produced as an osmoregulant by marine algae (Vairavamurthy et al. 1985), and this role is a key element of the CLAW hypothesis. Because the principal precursor to DMS is DMSP, the effect of salinity changes on DMSP production is considered here.

Because many unicellular algae lack cell walls or have a cell wall with low elasticity, they are not able to develop high-turgor pressures in the cell (Stefels 2000). Changes to external salinity and water potential will change the osmotic potential of the cell, meaning cells must try to recover their original volume (Stefels 2000). This is achieved by adjusting the osmotic potential of the cell by active transport of ions and by the accumulation/degradation of low molecular weight organic solutes such as DMSP. Both processes are metabolically regulated, but degradation proceeds at a slower rate (Kirst 1989) and has higher energy costs (Stefels 2000).

Pelagic marine organisms live in environments of low water potential, with salinities varying between 34 and 37 psu. The challenge for marine organisms in such environments is to produce organic solutes at concentrations that are high enough to be osmotically active and metabolically compatible; DMSP, as a tertiary sulphonium compound, is one such compatible solute, although Kirst (1996) concluded that DMSP may be less effective in this role than betaine, proline, or glycerol. It has also been observed that changes in water potential do not result in rapid synthesis/degradation of DMSP (Stefels 2000). Studies looking at DMSP production following hyperosmotic shock indicate that generally under such conditions, DMSP is produced slowly, if at all (Reed 1983, Dickson & Kirst 1986, Edwards et al. 1987, Stefels et al. 1996). A few studies (Dickson et al. 1982, Vairavamurthy et al. 1985, Stefels 2000), however, have demonstrated rapid adaptation of intracellular DMSP content. During hypo-osmotic shock, DMSP may be rapidly released from some cells (Dickson & Kirst 1986). Rapid release of intracellular DMSP has been reported for a benthic diatom (Van Bergeijk et al. 2002, 2003) and for the dinophyte *Heterocapsa triquetra* (Niki et al. 2007). It is interesting to note that in the laboratory study by Niki et al. (2007), the activity of DMSP lyase, the enzyme by which DMSP is broken down into DMS and acrylate, was not altered, suggesting that the production of DMS under such conditions is not enzymatic but rather is a result of DMSP release directly from the cell. Niki et al. (2007) concluded that low-salinity shock increases DMS production by stimulating DMSP production and release and decreases bacterial DMSP consumption. These results suggest that low-salinity conditions increase the algal contribution to total DMS production. Similar results were obtained from *Hymenomonas carterae*, in which DMS output increased with salinity up to 52 psu (Vairavamurthy et al. 1985), and from *Ulva lactuca* (Dickson et al. 1980), *Scrippsiella trochoidea*, and *Prorocentrum minimum* (Zhuang et al. 2011). Zhuang et al. (2011) suggested that increased production of DMS in line with increasing salinity might be representative of upregulation of DMSP in the cell because of osmotic stress. It is worth mentioning here that algal cells can have more than one compatible solute, so, for instance, diatoms can switch between glycine betaine and DMSP, depending on nitrogen availability (Keller et al. 1999a,b).

Some studies reported no change in intracellular DMSP levels caused by salinity changes (Van Diggelen et al. 1986, Edwards et al. 1987, Colmer et al. 1996), so it would seem that the use of DMSP as an osmoregulant is not universal. It is more likely that its role is taxon specific, acting as an osmoregulant in certain species and as an antioxidant in others (or playing another role entirely). Considering implications for the CLAW hypothesis, the original authors only provided one possible explanation for DMSP production (osmoregulatory). They further suggested that reduced ocean area and altered wind regimes during glacial periods could be responsible for a decrease in the flux of DMS to the atmosphere, and that ecological changes discussed in the CLAW hypothesis could favour algae with higher DMS output rates (Charlson et al. 1987). Hendy et al. (2002) reported that sea-surface salinities were higher during glacial than during interglacial periods, which could provide a plausible mechanism for this DMS/climate feedback loop; higher salinities during glacial periods stimulate an increase in the production of DMSP and DMS that compensates for the decrease in ocean area, leading to an increase in CCN and altered cloud albedo. This still relies, however, on the existence of a positive correlation between DMS and CCN population.

Atmospheric DMS and CCN population

Possibly the most crucial aspect of the CLAW hypothesis is the proposed relationship between DMS and CCN production. The original authors suggested that because the albedo of clouds (and therefore Earth's radiation budget) is sensitive to CCN density, higher DMS emissions will increase the number of CCN, thereby providing a plausible mechanism for the biological regulation of Earth's climate.

Identification of sources that contribute to marine CCN requires analysis of particles with diameters less than 300 nm (Quinn & Bates 2011). Clouds of liquid water droplets only form in the presence of CCN; Pruppacher & Klett (1978) found that the concentration of potential CCN particles in the unpolluted marine atmosphere varied from 30 to 200 cm^{-3} depending on aerosol content, meteorological conditions, and supersaturation. Koehler (1936) argued that most CCN are composed of water-soluble materials, making nss-SO_4^{2-} a viable source of CCN. It was further argued by Charlson et al. (1987) that sea-salt particle concentrations at cloud height are not usually more than 1 cm^{-3}; consequently, they cannot be the main CCN; this was supported by evidence from Pruppacher & Klett (1978) and Hobbs (1971). It is this evidence that forms the central argument of CLAW and is possibly the strongest argument for a DMS/climate feedback loop provided by the original authors.

Correlations between DMS concentrations and marine CCN

Since the CLAW hypothesis was proposed in 1987, its strongest supporting evidence has come from Cape Grim, Tasmania, where measurements commenced in 1976. At this remote sampling site in the Southern Ocean, concentrations of DMS, nss-SO_4^{2-}, MSA, and CCN all peak in the summer and are at a minimum in the winter (Ayers & Gras 1991, Ayers et al. 1997). Similar measurements have been made over the Tropical South Atlantic (Andreae et al. 1995) and north-eastern Pacific (Hegg et al. 1991), where 40–50% of CCN variance could be explained by DMS (Quinn & Bates 2011). Several other studies have recorded positive correlations between DMS, its atmospheric oxidation products, sulphates, and CCN (Prospero et al. 1991, Andreae et al. 1995, Clarke et al. 1998,

Andreae et al. 1999, Ayers & Gillett 2000). However, these studies focused mainly on the Southern Hemisphere or the Tropical South Atlantic and lacked the temporal and spatial coverage necessary to confirm DMS-CCN coupling as a globally relevant climate process (Vallina & Simó 2008).

Evidence for alternative sources of marine boundary layer CCN

Although data from the South Atlantic provided evidence for the atmospheric chemistry component of CLAW, namely, the conversion of sulphate and MSA into CCN, they do not indicate a sensitivity of CCN to changes in DMS emissions. In other words, it is likely that there are other sources of CCN in the remote marine boundary layer.

With respect to alternative potential CCN sources, there are a small number of measurements that provide evidence of a role for sea salt; Campuzano-Jost et al. (2003) recorded the presence of sodium in particles with diameters larger than 500 nm off the coast of Oahu, Hawaii. Furthermore, they found that sulphate was the dominant constituent at smaller diameters. More compelling evidence for sea salt as CCN came from a study by Twohy & Anderson (2008); residual particles (i.e., CCN) of evaporated cloud droplets were analysed, and 60% of them were found to be sea salt. There is now growing evidence that sea salt can be a significant source of CCN in the marine boundary layer (O'Dowd & Smith 1993, Murphy et al. 1998, Yoon & Brimblecombe 2002, Clarke et al. 2006, Smith 2007, Peter et al. 2008), which is inconsistent with the claims made in the CLAW hypothesis that DMS fluxes lead to an increase in CCN. Yoon & Brimblecombe (2002) proposed that the main control on marine boundary layer CCN was in fact wind speed (and not temperature, as suggested under the CLAW hypothesis), which exerts a strong influence on sea-salt particle production, as well as on the DMS flux.

Organic sources for marine boundary layer CCN

Sea salt is not the sole contributor to marine boundary layer CCN; organic matter is also capable of acting as an aerosol after injection into the atmosphere through bubble bursting (Leck & Bigg 2005, 2008, Bigg 2007, Facchini et al. 2008, Hawkins & Russell 2010, Russell et al. 2010). Novakov & Penner (1993) showed at a marine site that sulphates were about 37% of CCN and organics were 63%; most CCN recorded over the north-eastern Atlantic of the United States were not sulphates (Saxena et al. 1995). Organic matter in marine air has been shown to be similar to organic matter in ocean surface waters and includes particulates such as phytoplankton, bacteria, viruses, DNA/RNA, organic detritus, and fragments of larger organisms (Leck & Bigg 2005, Bigg 2007, Facchini et al. 2008, Quinn & Bates 2011). Organic acids have also been measured in marine air (Yu 2000), with formic and acetic acids constituting the most abundant acids in the troposphere (Keene et al. 1995). Yu (2000) reported that organic acids are likely to make a significant contribution to CCN because of biomass burning, making them potentially of more importance over or near continental land masses. Exopolymer gels in particular have attracted significant interest (Decho 1990, Bigg 2007, Leck & Bigg 2008) as they are insoluble, thermally stable, highly surface active, and hydrated. These properties allow them to readily sequester dissolved organic matter and turn particulate matter into aggregates, making them well suited to act as CCN. Evidence for exopolymer aggregates as CCN has been provided by O'Dowd et al. (2004), who observed that during summer phytoplankton blooms in the north-eastern Atlantic organic matter contributed 63% to the submicrometre aerosol mass, decreasing to 15% in winter, the majority of which was insoluble. A further study in the northern Atlantic and Arctic Oceans (Russell et al. 2010) found that organic matter accounted for 15–47% of the submicrometre particle mass, suggesting that organics play a significant role in the formation of marine CCN in these regions. Indeed, several studies have found that in the Northern Hemisphere the majority of atmospheric aerosols and CCN do not come from the oxidation of DMS, but from continental sources (Charlson et al. 1992, Husar et al.1997, Andreae et al. 2003). So, it would appear that DMS oxidation can be a major source of CCN in unpolluted marine air far from continental aerosol sources (as in the Southern Ocean), but closer to land the dominant contribution to CCN is organic matter (including acids from biomass burning and vegetation) delivered into the atmosphere following bubble bursting.

A recent study by Woodhouse et al. (2010) found a low sensitivity of CCN density to changes in the sea-air flux of DMS. They found that in a globally warmed scenario in which DMS flux increases by about 1% relative to present day, CCN abundance would increase by about a mere 0.1%. Woodhouse et al. (2010) further suggested that because future seawater DMS concentrations are not expected to change significantly, DMS would be expected to play only a weak role in climate regulation. This has important implications for the CLAW hypothesis, which claims an approximate doubling of CCN abundance would be needed to counteract the warming arising from a doubling of atmospheric CO_2 (Charlson et al. 1987).

DMSP/DMS cycling, the role of microbes, and the DMS 'summer paradox'

It has been noted previously in this review that DMSP is produced for a wide range of reasons, but we have not yet considered how DMSP is transformed into DMS. Because the bulk of DMS emissions comes from DMSP in phytoplankton cells, the processes that release them are critical for understanding global DMS production (Yoch 2002). Although recent studies show that DMS can also be produced via reduction of DMSO (Spiese et al. 2009), ultimately the source for any dimethylated sulphur compound in the marine environment is DMSP. It is therefore essential to understand how dimethylated sulphur compounds are cycled and transformed if we are to gain a more accurate understanding of global DMS production.

Production of DMS via different pathways

DMS can be produced via three different pathways: through direct release of DMSP from the algal cell, by the reduction of DMSO, or by the enzymatic cleavage of dissolved DMSP (Stefels et al. 2007). The dominant fate for about 75% of dissolved DMSP is through assimilation by bacteria (Kiene et al. 2000); the remaining fraction can be cleaved by DMSP lyase-like enzymes (Todd et al. 2007, 2009, 2011, Curson et al. 2008). Bacteria are crucial not only in the enzymatic cleavage of DMSP to DMS (Belviso et al. 1990) but also in the reduction of DMSO to DMS, and they can utilise DMS and DMSP as a carbon or energy source. Figure 1 shows the pathways involved in the biogeochemical cycling of dimethylated sulphur compounds, demonstrating the complexity of the system and the role played by microbes. Biotic control of DMS production and cycling is, therefore, heavily influenced by bacteria. Recent studies have shown that bacterial metabolism of DMS can proceed via two pathways: assimilation or oxidation to DMSO. The former represents only about 2% of the DMS consumed by natural bacterial populations in the Sargasso Sea (del Valle et al. 2007), North Sea (Zubkov et al. 2002), and Gulf of Mexico (Vila-Costa et al. 2006). DMS oxidation to DMSO might be the major pathway by which DMS is removed from surface waters (del Valle et al. 2007, 2009). Whereas this does not conflict with the claims made in the CLAW hypothesis, the microbial role in global DMS production, removal, and cycling is probably much greater than envisaged by Charlson et al. in 1987. A key facet of the CLAW hypothesis is that marine phytoplankton produce DMSP as an osmoregulant, which is then converted to DMS via biotic and abiotic processes. Had the role of microbes in DMS production and removal been more fully appreciated in 1987, it is likely the conclusions reached would have been very different.

DMS and the summer paradox

To further explain the role of bacteria in DMS cycling, we need to understand the temporal decoupling between maximum concentrations of chlorophyll, DMS, and DMSP in subtropical and temperate

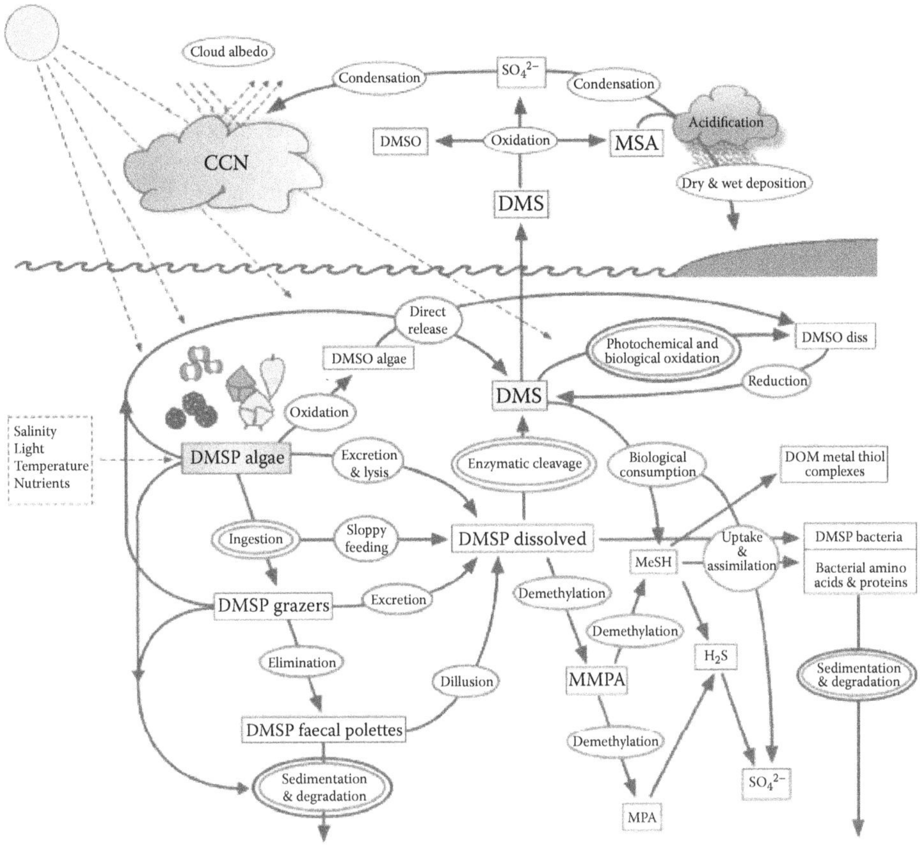

Figure 1 Schematic representation of the processes and pools involved in the marine biogeochemical cycling of DMSP and DMS. CCN, cloud condensation nuclei; DOM, dissolved organic material; DMSO, dimethylsulphoxide; MeSH, methanethiol; MPA, mercaptopropionate; MMPA, methylmercaptopropionate; MSA, methanesulphonic acid. (Figure and descriptive text from J. Stefels, M. Steinke, S. Turner, G. Malin, & S. Belviso. *Biogeochemistry* **83**, 245–275, 2007.)

latitudes (Dacey et al. 1998, Vila-Costa et al. 2008, Archer et al. 2009), a process known as the 'summer DMS paradox' (Simó & Pedrós-Alió 1999). Attempts to model this phenomenon (Toole et al. 2008, Vallina et al. 2008, Le Clainche et al. 2010) suggested that elevated UV radiation leads to a direct exudation of DMS by phytoplankton during the summer months. Although this explanation fits the role of DMSP as an antioxidant, it fails to account for the 2-month lag between peak levels of DMSP and DMS. More recently, Polimene et al. (2012) suggested that phytoplanktonic exudation of DMS is the main source of DMS for most of the year, and bacterially mediated DMS production is the pivotal process in triggering the temporal decoupling between DMSP and DMS. It is possible that early findings (Andreae 1986) showing DMS emissions to be highest in the "warmest, most saline and most intensely illuminated" waters were, in fact, an early observation of the summer paradox. As yet, however, modelling studies of the summer paradox are the main source of information on this phenomenon, with little in the way of bacterial assemblage and DMS consumption/production data available. Additional experimental work is therefore vital to gain a more complete understanding.

The palaeoclimatic record: evidence of a past role for DMS in climate regulation?

In their 1987 paper, the authors of the CLAW hypothesis made reference to glacial periods and the establishment of the proposed climatic feedback loop. In their concluding remarks, they suggested that the proposed feedback might have regulated Earth's climate in the past, either by counteracting the increasing luminosity of the sun or by counteracting the influence of increased greenhouse gas concentrations (Charlson et al. 1987). Palaeoclimatic data from ice cores in Greenland and Antarctica have increased enormously since the publication of CLAW in 1987; here, we examine whether the palaeo record provides evidence that the processes envisaged in CLAW have been important influences on climate prior to anthropogenic interference in atmospheric chemistry.

Methanesulphonate deposition as a proxy for past levels of DMS production

One of the key atmospheric oxidation products of DMS is MSA; MSA and the ratio of MSA/nss-SO_4^{2-} can be useful tracers for biogenic sulphur (Saltzman et al. 1983). The Greenland Ice Sheet Project 2 (GISP2) ice core indicated that over the last 110 Kyr the glacial atmosphere over Greenland had reduced concentrations of biogenic sulphur (DMS) compared with the present, suggesting that biogenic sulphur emissions must have been lower during glacial periods (Saltzman et al. 1997). The GISP2 ice core shows distinct climatic variability in MSA flux, with major increases during the warm periods around 75, 40–55, and 0–10 kyr before present (Saltzman et al. 1997). Furthermore, the largest change in MSA accumulation occurred between the last glacial maximum and the Holocene, for which a notable increase in ^{18}O was also observed. There was, however, a slow increase in MSA at the beginning of the Holocene, suggesting that the MSA record reflects responses to changes in climate and atmospheric/ocean circulation (Saltzman et al. 1997). In fact, MSA continued to increase for several thousand years after the end of the glacial/interglacial transition when the isotope record clearly shows that the climate had stabilised rapidly (Saltzman et al. 1997), suggesting there may have been changes in ocean biota occurring over longer timescales, producing a lag in DMS levels. Another possible explanation, however, is that changes in atmospheric transport processes affected the depositional regime. In a further study on the Siple Dome ice core in West Antarctica, Saltzman et al. (2006) found that not only were MSA levels low during the last glacial maximum, as they were in GISP2, but also there was a several-thousand-year lag relative to deglacial warming, as observed in GISP2. A similar picture has been found for the Renland ice core (Hansson & Saltzman 1993) and the GRIP ice core in Greenland (Legrand et al. 1997), for which MSA levels were lower during glacial than in interglacial periods.

Vostok ice core data and MSA deposition

The data from Greenland and West Antarctica are in contrast to results obtained from the Vostok ice core in eastern Antarctica, for which levels of MSA and MSA/nss-SO_4^{2} were higher during glacials and lower during interglacials (Saigne & Legrand 1987, Legrand et al. 1991). Given that isotopic temperature profiles for Greenland and Antarctica show roughly parallel climatic development (Johnsen & Dansgaard 1992), it appears that the role of marine biogenic sulphur in climate forcing is different from that originally proposed.

The contrast in findings from the ice core records in Northern and Southern Hemispheres raises important questions for the CLAW hypothesis. Whereas the Greenland cores, particularly GISP2, show an increase in biogenic sulphur emissions during interglacials, implying a negative feedback of the kind proposed in CLAW, this is in contrast to the Antarctic cores, for which colder conditions saw higher levels of biogenic sulphur. A possible explanation for the Vostok results is that during glacial periods greater areas of continental shelf were exposed and additional dust loading led to increased input of iron to the Southern Ocean (Martin & Fitzwater 1988). The importance of iron in the Southern Ocean has been demonstrated with *in situ* fertilisation experiments (Martin et al.

1994, Boyd et al. 2000), which suggest that the additional input would have stimulated phytoplankton growth during colder times, thereby increasing the flux of DMS and marine sulphate aerosol to the atmosphere, which in turn would lead to increased cloudiness. It has been suggested that in this scenario the climatic forcing would be in the direction of cooling, with the sulphur/dust system exerting positive feedback and stabilising the glacial climate (Legrand et al. 1988, Watson & Liss 1998). However, dust loading in ice cores has been reported for the Northern Hemisphere high latitudes (Yang et al. 1997), for which concomitant increased DMS emissions were not reported. Another possible explanation for elevated MSA levels in the Vostok ice core during glacial periods might lie in the effect of glacial conditions on global ocean salinity. Hendy et al. (2002) reported that ocean salinities were higher during glacial periods, which seems fairly intuitive with large volumes of water locked up in the cryosphere. Given the role of DMSP as an osmoregulant, it is possible that increased salinity during glacial periods stimulates the production and release of DMSP that eventually ends up as MSA deposited on Antarctic ice sheets.

Although cogent and valid explanations have been proposed to account for the asymmetry between Greenland and Antarctic ice core MSA deposits, the CLAW hypothesis in its current form suggests that it is predominantly temperature that governs DMS production. For this reason, palaeodata do not fully support the CLAW hypothesis.

Conclusions

On the basis of the evidence gathered since 1987 and summarised here, it is clear that the CLAW hypothesis is an oversimplification of a complicated ocean–atmosphere system. That is not to say that it does not hold true in certain respects; it is clear from the Cape Grim data that increased DMS emissions during summer do in fact lead to an increased CCN population, which regulates the local climate on seasonal timescales. But, CLAW fails to acknowledge other possible sources of CCN; the implication by Charlson et al. (1987) is that DMS forms the majority of CCN, whereas organic matter, sea salt, and nss-SO_4^{2-} have all been subsequently shown to contribute to the CCN population to varying degrees, sometimes significantly.

Further problems arise when considering the reasons put forward for the production of DMSP (the precursor to DMS); it was suggested by the original authors that DMSP is produced as an osmoregulant, and although this is true for certain species, it oversimplifies the role of this compound, which has since been shown to have roles as varied as an antioxidant, a cryoprotectant, and acting as an overflow mechanism for excess sulphur. Furthermore, DMSP synthesis in all organisms is not necessarily impacted by changes to salinity or light. Some species do show an increase in DMSP production under elevated light/salinity conditions, whereas others either show no change at all or even show a decrease in production.

These results indicate that CLAW was too simplistic in its assumptions about the functional role of DMSP and therefore the pattern of response of DMSP production to changing environmental conditions. The climate feedback loop itself hinges on the premise that DMS directly contributes to, and correlates with, CCN population, thereby helping regulate Earth's climate. This also has been shown not to be the case, with model studies suggesting that a 1% increase in DMS would only lead to a 0.1% increase in CCN. This is in stark contrast to the numbers offered by Charlson et al. (1987), who claimed a 50% increase in DMS emissions would be enough to offset the recent doubling of carbon dioxide as a result of anthropogenic emissions.

Clearly, the CLAW hypothesis cannot now be accepted as a complete explanatory model of global climate regulation. The ice core record provides clear evidence of this; whereas the Northern Hemisphere sites provide support for the CLAW hypothesis, the eastern Antarctic ice cores (Vostok) do not. If DMS emissions were driven by light and temperature in the way predicted by CLAW, we would expect all ice cores to exhibit the same correlation between DMS and climate (as indicated by the oxygen isotope record). The absence of this consistent pattern implies that the global sulphur cycle is affected by factors not accounted for by CLAW or that there is hemispheric asymmetry in the sulphur cycle.

When proposed, the CLAW hypothesis was genuinely visionary, combining atmospheric chemistry and physics, biochemistry, climate science, and marine biology into one plausible feedback loop. More than two decades later, it is becoming increasingly clear that CLAW has not stood the test of time and, in a sense, has masterminded its own downfall by inspiring a generation of research that has allowed us to more fully appreciate the complexity and interrelationships of marine, atmospheric, and terrestrial biogeochemistry. So, although we cannot rule out a marine biogenic influence on Earth's climate, the CLAW hypothesis, in its original form, must be rejected as a comprehensive explanatory model.

References

Andreae, M.O. 1986. The ocean as a source of atmospheric sulfur compounds. In *The Role of Air-Sea Exchange in Geochemical Cycling*, P. Buat-Menard (ed.). Dordrecht, the Netherlands: Reidel, 331–362.

Andreae, M.O., Andreae, T.W., Meyerdierks, D. & Thiel, C. 2003. Marine sulfur cycling and the atmospheric aerosol over the springtime North Atlantic. *Chemosphere* **52**, 1321–1343.

Andreae, M.O., Elbert, W., Cai, Y., Andreae, T.W. & Gras, J. 1999. Non-sea-salt sulfate, methanesulfonate, and nitrate aerosol concentrations and size distributions at Cape Grim, Tasmania. *Journal of Geophysical Research: Atmospheres (1984–2012)* **104**, 21695–21706.

Andreae, M.O., Elbert, W. & deMora, S.J. 1995. Biogenic sulfur emissions and aerosols over the tropical South Atlantic 3, Atmospheric dimethylsulfide, aerosols and cloud condensation nuclei. *Journal of Geophysical Research* **100**, 11335–11356.

Archer, S.D., Cummings, D.G., Llewellyn, C.A. & Fishwick, J.R. 2009. Phytoplankton taxa, irradiance and nutrient availability determine the seasonal cycle of DMSP in temperate shelf seas. *Marine Ecology Progress Series* **394**, 111–124.

Ayers, G.P., Cainey, J.M., Gillett, R.W. & Ivey, J.P. 1997. Atmospheric sulphur and cloud condensation nuclei in marine air in the Southern Hemisphere. *Philosophical Transactions of the Royal Society of London. Series B: Biological Sciences* **352**, 203–211.

Ayers, G.P. & Gillett, R.W. 2000. DMS and its oxidation products in the remote marine atmosphere: implications for climate and atmospheric chemistry. *Journal of Sea Research* **43**, 275–286.

Ayers, G.P. & Gras, J.L. 1991. Seasonal relationship between cloud condensation nuclei and aerosol methanesulphonate in marine air. *Nature* **353**, 834–835.

Belviso, S., Buat-Ménard, P., Putaud, J.P., Nguyen, B.C., Claustre, H. & Neveux, J. 1993. Size distribution of dimethylsulfoniopropionate (DMSP) in areas of the tropical northeastern Atlantic Ocean and the Mediterranean Sea. *Marine Chemistry* **44**, 55–71.

Belviso, S., Kim, S.K., Rassoulzadegan, F., Krajka, B., Nguyen, B.C., Mihalopoulos, N. & Buat-Menard, P. 1990. Production of dimethylsulfonium propionate (DMSP) and dimethylsulfide (DMS) by a microbial food web. *Limnology and Oceanography* **35**, 1810–1821.

Bigg, E.K. 2007. Sources, nature and influence on climate of marine airborne particles. *Environmental Chemistry* **4**, 155–161.

Boyd, P.W., Watson, A.J., Law, C.S., Abraham, E.R., Trull, T., Murdoch, R., Bakker, D.C.E., Bowie, A.R., Buesseler, K.O., Chang, H., Charette, M., Croot, P., Downing, K., Frew, R., Gall, M., Hadfield, M., Hall, J., Harvey, M., Jameson, G., LaRoche, J., Liddicoat, M., Ling, R., Maldonada, M.T., McKay, R.M., Nodder, S., Pickmere, S., Pridmore, R., Rintoul, S.R., Safi, K., Sutton, P., Strzepek, R., Tanneberger, K., Turner, S., Waite, A. & Zeldis, J. 2000. A mesoscale phytoplankton bloom in the polar Southern Ocean stimulated by iron fertilization. *Nature* **407**, 695–702.

Brimblecombe, P. & Shooter, D. 1986. Photo-oxidation of dimethylsulphide in aqueous solution. *Marine Chemistry* **19**, 343–353.

Brugger, A., Slezak, D., Obernosterer, I. & Herndl, G.J. 1998. Photolysis of dimethylsulfide in the northern Adriatic Sea: dependence on substrate concentration, irradiance and DOC concentration. *Marine Chemistry* **59**, 321–331.

Campuzano-Jost, P., Clark, C.D., Maring, H., Covert, D.S., Howell, S., Kapustin, V., Clarke, K.A., Saltzman, E.S. & Hynes, A.J. 2003. Near-real-time measurement of sea-salt aerosol during the SEAS campaign: comparison of emission-based sodium detection with an aerosol volatility technique. *Journal of Atmospheric and Oceanic Technology* **20**, 1421–1430.

Charlson, R.J., Lovelock, J.E., Andreae, M.O. & Warren, S.G. 1987. Oceanic phytoplankton, atmospheric sulphur, cloud albedo and climate. *Nature* **326**, 655–661.

Charlson, R.J., Schwartz, S.E., Hales, J.M., Cess, R.D., Coakley, J.A., Jr., Hansen, J.E. & Hofmann, D.J. 1992. Climate forcing by anthropogenic aerosols. *Science* **255**, 423–430.

Clarke, A.D., Davis, D., Kapustin, V.N., Eisele, F., Chen, G., Paluch, I., Lenschhow, D., Bandy, A.R., Thornton, D., Moore, K., Mauldin, L., Tanner, D., Litchy, M., Carroll, M.A., Collins, J. & Albercook, G. 1998. Particle nucleation in the tropical boundary layer and its coupling to marine sulfur sources. *Science* **282**, 89–92.

Clarke, A.D., Owens, S.R. & Zhou, J. 2006. An ultrafine sea-salt flux from breaking waves: implications for cloud condensation nuclei in the remote marine atmosphere. *Journal of Geophysical Research: Atmospheres (1984–2012)* **111**, D06202.

Colmer, T.D., W-M, F.T., Läuchli, A. & Higashi, R.M. 1996. Interactive effects of salinity, nitrogen and sulphur on the organic solutes in *Spartina alterniflora* leaf blades. *Journal of Experimental Botany* **47**, 369–375.

Crown-Zellerbach. 1966. Dimethyl sulfoxide (DMSO) technical bulletin. Camas, Washington: Crown-Zellerbach.

Curran, M.A. & Jones, G.B. 2000. Dimethyl sulfide in the Southern Ocean: seasonality and flux. *Journal of Geophysical Research: Atmospheres (1984–2012)* **105**, 20451–20459.

Curson, A.R.J., Rogers, R., Todd, J.D., Brearley, C.A. & Johnston, A.W.B. 2008. Molecular genetic analysis of a dimethylsulfoniopropionate lyase that liberates the climate-changing gas dimethylsulfide in several marine α-proteobacteria and *Rhodobactersphaeroides*. *Environmental Microbiology* **10**, 757–767.

Dacey, J.W., Howse, F.A., Michaels, A.F. & Wakeham, S.G. 1998. Temporal variability of dimethylsulfide and dimethylsulfoniopropionate in the Sargasso Sea. *Deep-Sea Research I* **45**, 2085–2104.

Dacey, J.W. & Wakeham, S.G. 1986. Oceanic dimethylsulfide production during zooplankton grazing on phytoplankton. *Science* **233**, 1314–1316.

Davison, B., O'Dowd, C., Hewitt, C.N., Smith, M.H., Harrison, R.M., Peel, D.A., Wolf, E., Mulvaney, R., Schwikowski, M. & Baltenspergert, U. 1996. Dimethyl sulfide and its oxidation products in the atmosphere of the Atlantic and Southern Oceans. *Atmospheric Environment* **30**, 1895–1906.

Decho, A.W. 1990. Microbial exopolymer secretions in ocean environments: their role(s) in food webs and marine processes. *Oceanography and Marine Biology: an Annual Review* **28**, 73–153.

del Valle, D.A., Kieber, D.J. & Kiene, R.P. 2007. Depth-dependent fate of biologically-consumed dimethylsulfide in the Sargasso Sea. *Marine Chemistry* **103**, 197–208.

del Valle, D.A., Kieber, D.J., Toole, D.A., Brinkley, J. & Kiene, R.P. 2009. Biological consumption of dimethylsulfide (DMS) and its importance in DMS dynamics in the Ross Sea, Antarctica. *Limnology and Oceanography* **54**, 785–798.

Dickson, D.M.J., Jones, R.W. & Davenport, J. 1982. Osmotic adaptation in *Ulva lactuca* under fluctuating salinity regimes. *Planta* **155**, 409–415.

Dickson, D.M. J. & Kirst, G.O. 1986. The role of β-dimethylsulphoniopropionate, glycine betaine and homarine in the osmoacclimation of *Platymonas subcordiformis*. *Planta* **167**, 536–543.

Dickson, D.M., Wyn Jones, R.G. & Davenport, J. 1980. Steady state osmotic adaptation in *Ulva lactuca*. *Planta* **150**, 158–165.

Dykens, J.A. & Shick, J.M. 1982. Oxygen production by endosymbiotic algae controls superoxide dismutase activity in their animal host. *Nature* **297**, 579–580.

Edwards, D.M., Reed, R.H., Chudek, J.A., Foster, R. & Stewart, W.D.P. 1987. Organic solute accumulation in osmotically-stressed *Enteromorpha intestinalis*. *Marine Biology* **95**, 583–592.

Facchini, M.C., Rinaldi, M., Decesari, S., Carbone, C., Finessi, E., Mircea, M., Fuzzi, S., Ceburnis, D., Flanagan, R., Nilsson, E.D., de Leeuw, G., Martino, M., Woeltjen, J. & O'Dowd, C.D. 2008. Primary submicron marine aerosol dominated by insoluble organic colloids and aggregates.*Geophysical Research Letters* **35**, L17814.

Fridovich, I. 1995. Superoxide radical and superoxide dismutases. *Annual Review of Biochemistry* **64**, 97–112.

Galí, M. & Simó, R. 2010. Occurrence and cycling of dimethylated sulfur compounds in the Arctic during summer receding of the ice edge. *Marine Chemistry* **122**, 105–117.

Gambaro, A., Moret, I., Piazza, R., Andreoli, C., Da Rin, E., Capodaglio, G., Barbante, C. & Cescon, P. 2004. Temporal evolution of DMS and DMSP in Antarctic coastal seawater. *International Journal of Environmental Analytical Chemistry* **84**, 401–412.

Hanson, A.D., Rivoal, J., Paquet, L. & Gage, D.A. 1994. Biosynthesis of 3-dimethylsulfoniopropionate in *Wollastonia biflora* (L.) DC. *Plant Physiology* **105**, 103–110.

Hansson, M.E. & Saltzman, E.S. 1993. The first Greenland ice core record of methanesulfonate and sulfate over a full glacial cycle. *Geophysical Research Letters* **20**, 1163–1166.

Hatton, A.D. 2002. Influence of photochemistry on the marine biogeochemical cycle of dimethylsulphide in the northern North Sea. *Deep Sea Research II* **49**, 3039–3052.

Hawkins, L.N. & Russell, L.M. 2010. Polysaccharides, proteins, and phytoplankton fragments: four chemically distinct types of marine primary organic aerosol classified by single particle spectromicroscopy. *Advances in Meteorology* **2010**, article ID. 612132. doi: 101029/91JD01309

Hegg, D.A., Ferek, R.J., Hobbs, P.V. & Radke, L.F. 1991. Dimethyl sulfide and cloud condensation nucleus correlations in the northeast Pacific Ocean. *Journal of Geophysical Research: Atmospheres (1984–2012)* **96**, 13189–13191.

Hendy, E.J., Gagan, M.K., Alibert, C.A., McCulloch, M.T., Lough, J.M. & Isdale, P.J. 2002. Abrupt decrease in tropical Pacific sea surface salinity at end of Little Ice Age. *Science* **295**, 1511–1514.

Hobbs, P.V. 1971. Simultaneous airborne measurements of cloud condensation nuclei and sodium-containing particles over the ocean. *Quarterly Journal of the Royal Meteorological Society* **97**, 263–271.

Husar, R.B., Prospero, J.M. & Stowe, L.L. 1997. Characterization of tropospheric aerosols over the oceans with the NOAA advanced very high resolution radiometer optical thickness operational product. *Journal of Geophysical Research* **102**, 16889–16916.

Ishida, Y. 1968. Physiological studies on the evolution of dimethyl-sulfide. *Memoirs of the College of Agriculture Kyoto University* **94**, 47–82.

Johnsen, S.J. & Dansgaard, W. 1992. On flow model dating of stable isotope records from Greenland ice cores. The last deglaciation: absolute and radiocarbon chronologies. *NATO ASI Series* **Ser. I**, 13–24.

Karsten, U., Kirst, G.O. & Wiencke, C. 1992. Dimethylsulphoniopropionate (DMSP) accumulation in green macroalgae from polar to temperate regions: interactive effects of light versus salinity and light versus temperature. *Polar Biology* **12**, 603–607.

Karsten, U., Wiencke, C. & Kirst, G.O. 1990. The effect of light intensity and day length on the β-dimethylsulphoniopropionate (DMSP) content of marine green macroalgae from Antarctica. *Plant, Cell & Environment* **13**, 989–993.

Karsten, U., Wiencke, C. & Kirst, G.O. 1991. Growth pattern and β-dimethylsulphoniopropionate (DMSP) content of green macroalgae at different irradiances. *Marine Biology*, **108**, 151–155.

Kates, M. & Volcani, B.E. 1996. Biosynthetic pathways for phosphatidylsulfocholine, the sulfonium analoque of phosphatidylcholine, in diatoms. In *Biological and Environmental Chemistry of DMSP and Related Sulfonium Compounds,* R.P. Kiene et al. (eds). New York: Plenum Press, 109–119.

Keene, W.C., Mosher, B.W., Jacob, D.J., Munger, J.W., Talbot, R.W., Artz, R.S., Maben, J.R., Daube, B.C. & Galloway, J.N. 1995. Carboxylic acids in clouds at a high-elevation forested site in central Virginia. *Journal of Geophysical Research* **100**, 9321–9334.

Keller, M.D., Bellows, W.K. & Guillard, R.R.L. 1989. Dimethyl sulfide production in marine phytoplankton. In *Biogenic Sulfur in the Environment,* E.S. Saltzman & W.J. Cooper (eds). Washington, District of Columbia: American Chemical Society, 167–182.

Keller, M.D., Kiene, R.P., Matrai, P.A. & Bellows, W.K. 1999a. Production of glycine betaine and dimethylsulphoniopropionate in marine phytoplankton I. Batch cultures. *Marine Biology* **135**, 237–248.

Keller, M.D., Kiene, R.P., Matrai, P.A. & Bellows, W.K. 1999b. Production of glycine betaine and dimethylsulphoniopropionate in marine phytoplankton II. N-Limited chemostat cultures. *Marine Biology* **135**, 249–257.

Kettle, A.J. & Andreae, M.O. 2000. Flux of dimethylsulfide from the oceans: a comparison of updated data seas and flux models. *Journal of Geophysical Research* **105**, 793–808.

Kettle, A.J., Andreae, M.O., Amouroux, D., Andreae, T.W., Bates,T.S., Berreshiem, H., Bingemer, H., Boniforti, R., Curran, M.A.J., DiTullio, G.R., Helas, G., Jones, G.B., Keller, M.D., Kiene, R.P., Leck, C., Levasseur, M., Malin, G., Maspero, M., Matria, P., McTaggart, A.R., Mihalopoulos, N., Nguyen, B., Novo, A., Putuad, J.P., Rapsomanikis, S., Roberts, G., Schebeske, G., Sharma, S., Simó, R., Staubes, R., Turner, S. & Uher, G. 1999. A global database of sea surface dimethylsulfide (DMS) measurements and a procedure to predict sea surface DMS as a function of latitude, longitude, and month. *Global Biogeochemical Cycles* **13**, 399–444.

Kieber, D.J., Jiao, J., Kiene, R.P. & Bates, T.S. 1996. Impact of dimethylsulfide photochemistry on methyl sulfur cycling in the equatorial Pacific Ocean. *Journal of Geophysical Research* **101**, 3715–3722.

Kiene, R.P. 1992. Dynamics of DMS and DMSP in oceanic water samples. *Marine Chemistry* **37**, 29–52.

Kiene, R.P. & Bates, T.S. 1990. Biological removal of dimethylsulphide from seawater. *Nature* **345**, 702–705.

Kiene, R.P., Linn, L.J. & Bruton, J.A. 2000. New and important roles for DMSP in marine microbial communities. *Journal of Sea Research* **43**, 209–224.

Kirst, G.O. 1989. Salinity tolerance of eukaryotic marine algae. *Annual Review of Plant Biology* **41**, 21–53.

Kirst, G.O. 1996. Osmotic adjustment in phytoplankton and macroalgae. In *Biological and Environmental Chemistry of DMSP and Related Sulfonium Compounds,* R.P. Kiene et al. (eds). New York: Plenum Press, 121–129.

Kirst, G.O., Thiel, C., Wolff, H., Nothnagel, J., Wanzek, M. & Ulmke, R. 1991. Dimethylsulfoniopropionate (DMSP) in ice algae and its possible biological role. *Marine Chemistry* **35**, 381–388.

Koehler, H. 1936. The nucleus in and the growth of hygroscopic droplets. *Transactions of the Faraday Society* **32**, 1152–1161.

Lana, A., Bell, T.G., Simó, R., Vallina, S.M., Ballabrera-Poy, J., Kettle, A.J., Dachs, J., Bopp, L., Saltzman, E.S., Stefels, J., Johnson, J.E. & Liss, P.S. 2011. An updated climatology of surface dimethylsulfide concentrations and emission fluxes in the global ocean. *Global Biogeochemical Cycles* **25**, GB1004.

Le Clainche, Y., Vézina, A., Levasseur, M., Cropp, R.A., Gunson, J.R., Vallina, S., Vogt, M., Lancelot, C., Allen, J.I., Archer, S.D., Bopp, L., Deal, C., Elliott, S., Jin, M., Malin, G., Schoeman, V., Simó, R., Six, K.D. & Stefels, J. 2010. A first appraisal of prognostic ocean DMS models and prospects for their use in climate models. *Global Biogeochemical Cycles* **24**, GB3021.

Leck, C. & Bigg, E.K. 2005. Source and evolution of the marine aerosol—a new perspective. *Geophysical Research Letters* **32**, L19803.

Leck, C. & Bigg, E.K. 2008. Comparison of sources and nature of the tropical aerosol with the summer high Arctic aerosol. *Tellus B* **60**, 118–126.

Ledyard, K.M. & Dacey, J.W.H. 1994. Dimethylsulfide production from dimethyl-sulfoniopropionate by a marine bacterium. *Marine Ecology Progress Series* **110**, 95–103.

Lee, P.A. & De Mora, S.J. 1999. Intracellular dimethylsulfoxide (DMSO) in unicellular marine algae: speculations on its origin and possible biological role. *Journal of Phycology* **35**, 8–18.

Lee, P.A., De Mora, S.J., Gosselin, M., Levasseur, M., Bouillon, R.C., Nozais, C. & Michel, C. 2001. Particulate dimethylsulfoxide in Arctic sea-ice algal communities: the cryoprotectant hypothesis revisited. *Journal of Phycology* *37*, 488–499.

Legrand, M.R., Delmas, R.J. & Charlson, R.J. 1988. Climate forcing implications from Vostok ice-core sulphate data. *Nature* **334**, 418–420.

Legrand, M., Feniet-Saigne, C., Saltzman, E.S., Germain, C., Barkov, N.I. & Petrov, V.N. 1991. Ice-core record of oceanic emissions of dimethyl sulphide during the last climate cycle. *Nature* **350**, 144–146.

Legrand, M., Hammer, C., de Angelis, M., Savarino, J., Delmas, R., Clausen, H. & Johnsen, S.J. 1997. Sulfur-containing species (methanesulfonate and SO_4) over the last climatic cycle in the Greenland Ice Core Project (central Greenland) ice core. *Journal of Geophysical Research: Oceans* **102**, 26663–26679.

Lesser, M.P., Stochaj, W.R., Tapley, D.W. & Shick, J.M. 1990. Bleaching in coral reef anthozoans: effects of irradiance, ultraviolet radiation, and temperature on the activities of protective enzymes against active oxygen. *Coral Reefs* **8**, 225–232.

Levasseur, M., Gosselin, M. & Michaud, S. 1994. A new source of dimethylsulfide (DMS) for the Arctic atmosphere: ice diatoms. *Marine Biology* **121**, 381–387.

Malin, G. & Kirst, G.O. 1997 Algal production of dimethyl sulphide and its atmospheric role. *Journal of Phycology* **33**, 889–896.

Martin, J.H., Coale, K.H., Johnson, K.S., Fitzwater,S.E., Gordon, R.M., Tanner, S. J, Hunter, C.N., Elrod, V.A., Nowicki, L., Coley, T.L., Barber, R.T., Lindley, S., Watson, A.J., Van Scoy, K., Law, C.S., Liddicoat, M.I., Ling, R., Stanton, T., Stockel, J., Collins, C., Anderson, A., Bidigare, R., Ondrusek, M., Latasa, M., Millero, F.J., Lee, K., Yao, W., Zhang, J.Z., Friederich, G., Sakamoto, C., Chavez, F., Buck, K., Kolber, Z., Greene, R., Falkowski, P., Chisholm, S.W., Hoge, F., Swift, R., Yungel, J., Turner, S., Nightingale, P., Hatton, A., Liss, P. & Tindale, N.W. 1994. Testing the iron hypothesis in ecosystems of the equatorial Pacific Ocean. *Nature* **371**, 123–129.

Martin, J.H. & Fitzwater, S.E. 1988. Iron deficiency limits phytoplankton growth in the north-east Pacific sub-arctic. *Nature* **331**, 341–343.

Matrai, P.A. & Keller, M.D. 1994. Total organic sulfur and dimethylsulfoniopropionate in marine phytoplankton: intracellular variation. *Marine Biology* **119**, 61–68.

McLenon, A.L. & DiTullio, G.R. 2012. Effects of increased temperature on dimethylsulfoniopropionate (DMSP) concentration and methionine synthase activity in *Symbiodinium microadriaticum*. *Biogeochemistry* **110**, 17–29.

Miles, C.J., Bell, T.G. & Lenton, T.M. 2009. Testing the relationship between the solar radiation dose and surface DMS concentrations using *in situ* data. *Biogeosciences* **6**, 1927–1934.

Miles, C.J., Bell, T.G. & Suntharalingam, P. 2012. Investigating the inter-relationships between water attenuated irradiance, primary production and DMS(P). *Biogeochemistry* **110**, 201–213.

Mulholland, M.M. & Otte, M.L. 2000. Effects of varying sulphate and nitrogen supply on DMSP and glycine betaine levels in *Spartina anglica*. *Journal of Sea Research* **43**, 199–207.

Murphy, D.M., Anderson, J.R., Quinn, P.K., Mclnnes, L.M., Brechtel F.J., Kreidenweis, S.M., Middlebrook, A.M., Posfai, M., Thomson, D.S. & Buseck, P.R. 1998. Influence of sea-salt on aerosol radiative properties in the Southern Ocean marine boundary layer. *Nature* **392**, 62–65.

Niki, T., Shimizu, M., Fujishiro, A. & Kinoshita, J. 2007. Effects of salinity downshock on dimethylsulfide production. *Journal of Oceanography* **63**, 873–877.

Nomura, D., Koga, S., Kasamatsu, N., Shinagawa, H., Simizu, D., Wada, M. & Fukuchi, M. 2012. Direct measurements of DMS flux from Antarctic fast sea ice to the atmosphere by a chamber technique. *Journal of Geophysical Research: Oceans (1978–2012)* **117**, C04011.

Novakov, T. & Penner, J.E. 1993. Large contribution of organic aerosols to cloud-condensation-nuclei concentrations. *Nature* **365**, 823–826.

O'Dowd, C.D., Facchina, M.C., Cavalli, F., Ceburnis, D., Mircea, M., Decesari, S., Fuzzi, S.,Yoon, Y.J. & Putaud, J.P. 2004. Biogenically driven organic contribution to marine aerosols. *Nature* **431**, 676–680.

O'Dowd, C.D. & Smith, M.H. 1993. Physicochemical properties of aerosols over the northeast Atlantic: evidence for wind-speed-related submicron sea-salt aerosol production. *Journal of Geophysical Research* **98**, 1137–1149.

Pakulski, J.D. & Kiene, R.P. 1992. Foliar release of dimethylsulfoniopropionate from *Spartina alterniflora*. *Marine Ecology Progress Series* **81**, 277–287.

Peter, J.A., Blyth, A.M., Brooks, B., Lingard, J. & Smith, M.H. 2008. On the composition of Caribbean maritime aerosol particles measured during RICO. *Quarterly Journal of the Royal Meteorological Society* **134**, 1059–1063.

Polimene, L., Archer, S.D., Butenschön, M. & Allen, J.I. 2012. A mechanistic explanation of the Sargasso Sea DMS 'summer paradox'. *Biogeochemistry* **110**, 243–255.

Prospero, J.M., Savoie, D.L., Saltzman, E.S. & Larsen, R. 1991. Impact of oceanic sources of biogenic sulphur on sulphate aerosol concentrations at Mawson, Antarctica. *Nature* **350**, 221–223.

Pruppacher, H.R. & Klett, J.D. 1978. *Microphysics of Clouds*. Dordrecht, the Netherlands: Reidell.

Quinn, P.K. & Bates, T.S. 2011. The case against climate regulation via oceanic phytoplankton sulphur emissions. *Nature* **480**, 51–56.

Reed, R.H. 1983. Measurement and osmotic significance of beta-dimethylsulphoniopropionate in marine macroalgae. *Marine Biology Letters* **4**, 173–178.

Russell, L.M., Hawkins, L.N., Frossard, A.A., Quinn, P.K. & Bates, T.S. 2010. Carbohydrate-like composition of submicron atmospheric particles and their production from ocean bubble bursting. *Proceedings of the National Academy of Sciences of the United States of America*, **107**, 6652–6657.

Saigne, C. & Legrand, M. 1987. Measurements of methanesulphonic acid in Antarctic ice. *Nature* **330**, 240–242.

Saltzman, E.S., Dioumaeva, I. & Finley, B.D. 2006. Glacial/interglacial variations in methanesulfonate (MSA) in the Siple Dome ice core, West Antarctica. *Geophysical Research Letters* **33**, L11811.

Saltzman, E.S., Savoie, D.L., Zika, R.G. & Prospero, J.M. 1983. Methane sulfonic acid in the marine atmosphere. *Journal of Geophysical Research: Oceans (1978–2012)* **88**, 10897–10902.

Saltzman, E.S., Whung, P.Y. & Mayewski, P.A. 1997. Methanesulfonate in the Greenland Ice Sheet Project 2 ice core. *Journal of Geophysical Research: Oceans (1978–2012)* **102**, 26649–26657.

Saxena, P., Hildemann, L.M., McMurry, P.H. & Seinfeld, J.H. 1995. Organics alter hygroscopic behavior of atmospheric particles. *Journal of Geophysical Research* **100**, 18755–18770.

Simó, R., Grimalt, J.O., Pedrós-Alió, C. & Albaigés, J. 1995. Occurrence and transformation of dimethyl sulphur species in stratified seawater (western Mediterranean Sea). *Marine Ecology Progress Series* **127**, 291–299.

Simó, R. & Pedrós-Alió, C. 1999. Role of vertical mixing in controlling the oceanic production of dimethyl sulphide. *Nature* **402**, 396–399.

Smith, M.H. 2007. Sea-salt particles and the CLAW hypothesis. *Environmental Chemistry* **4**, 391–395.

Spiese, C.E., Kieber, D.J., Nomura, C.T. & Kiene, R.P. 2009. Reduction of dimethylsulfoxide to dimethylsulfide by marine phytoplankton. *Limnology and Oceanography* **54**, 560–570.

Stefels, J. 2000. Physiological aspects of the production and conversion of DMSP in marine algae and higher plants. *Journal of Sea Research* **43**, 183–197.

Stefels, J., Carnat, G., Dacey, J.W., Goossens, T., Elzenga, J.T..M. & Tison, J.L. 2012. The analysis of dimethylsulfide and dimethylsulfoniopropionate in sea ice: Dry-crushing and melting using stable isotope additions. *Marine Chemistry* **128**, 34–43.

Stefels, J., Gieskes, W.W. & Dijkhuizen, L. 1996. Intriguing functionality of the production and conversion of DMSP in *Phaeocystis* sp. In *Biological and Environmental Chemistry of DMSP and Related Sulfonium Compounds*, R.P. Kiene et al. (eds). New York: Plenum Press, 305–315.

Stefels, J., Steinke, M., Turner, S., Malin, G. & Belviso, S. 2007. Environmental constraints on the production and removal of the climatically active gas dimethylsulphide (DMS) and implications for ecosystem modelling. *Biogeochemistry* **83**, 245–275.

Stefels, J. & van Boekel, J.S.W. 1993. Production of DMS from dissolved DMSP in axenic cultures of the marine phytoplankton species *Phaeocystis* sp. *Marine Ecology Progress Series* **97**, 11–18.

Sunda, W.K..D.J., Kieber, D.J., Kiene, R.P. & Huntsman, S. 2002. An antioxidant function for DMSP and DMS in marine algae. *Nature* **418**, 317–320.

Tison, J.L., Brabant, F., Dumont, I. & Stefels, J. 2010. High-resolution dimethyl sulfide and dimethylsulfoniopropionate time series profiles in decaying summer first-year sea ice at Ice Station Polarstern, western Weddell Sea, Antarctica. *Journal of Geophysical Research: Biogeosciences (2005–2012)* **115**, G04044.

Todd, J.D., Curson, A.R.J., Dupont, C.L., Nicholson, P. & Johnston, A.W.B. 2009. The dddP gene, encoding a novel enzyme that converts dimethylsulfoniopropionate into dimethyl sulfide, is widespread in ocean metagenomes and marine bacteria and also occurs in some Ascomycete fungi. *Environmental Microbiology* **11**, 1376–1385.

Todd, J.D., Curson, A.R., Kirkwood, M., Sullivan, M.J., Green, R.T. & Johnston, A.W. 2011. DddQ, a novel, cupin-containing, dimethylsulfoniopropionate lyase in marine roseobacters and in uncultured marine bacteria. *Environmental Microbiology* **13**, 427–438.

Todd, J.D., Rogers, R., Li, Y.G., Wexler, M., Bond, P.L., Sun, L., Curson, A.R.J., Malin, G., Steinke, M. & Johnston, A.W.B. 2007. Structural and regulatory genes required to make the gas dimethyl sulfide in bacteria. *Science* **315**, 666–669.

Toole, D.A. & Siegel, D.A. 2004. Light-driven cycling of dimethylsulfide (DMS) in the Sargasso Sea: closing the loop. *Geophysical Research Letters* **31**, L09308.

Toole, D.A., Siegel, D.A. & Doney, S.C. 2008. A light-driven, one dimensional dimethylsulfide biogeochemical cycling model for the Sargasso Sea. *Journal of Geophysical Research* **113**, G02009.

Trevena, A.J. & Jones, G.B. 2006. Dimethylsulphide and dimethylsulphoniopropionate in Antarctic sea ice and their release during sea ice melting. *Marine Chemistry* **98**, 210–222.

Trevena, A. & Jones, G. 2012. DMS flux over the Antarctic sea ice zone. *Marine Chemistry* **134–135**, 47–58.

Trevena, A.J., Jones, G.B., Wright, S.W. & van den Enden, R.L. 2003. Profiles of dimethylsulphoniopropionate (DMSP), algal pigments, nutrients and salinity in the fast ice of Prydz Bay, Antarctica. *Journal of Geophysical Research* **108**, 3145–3156.

Turner, S.M., Malin, G., Liss, P.S., Harbour, D.S. & Holligan, P.M. 1998. The seasonal variation of dimethylsulfide and dimethylsulfoniopropionate concentrations in nearshore waters. *Limnology and Oceanography* **33**, 364–375.

Turner, S.M., Nightingale, P.D., Broadgate, W. & Liss, P.S. 1995. The distribution of dimethyl sulphide and dimethylsulphoniopropionate in Antarctic waters and sea ice. *Deep-Sea Research II* **42**, 1059–1080.

Twohy, C.H. & Anderson, J.R. 2008. Droplet nuclei in non-precipitating clouds: composition and size matter. *Environmental Research Letters* **3**, 045002.

Vairavamurthy, A., Andreae, M.O. & Iverson, R.L. 1985. Biosynthesis of dimethylsulfide and dimethylpropiothetin by *Hymenomonas carterae* in relation to sulfur source and salinity variations. *Limnology and Oceanography* **30**, 59–70.

Vallina, S.M. & Simó, R. 2007. Strong relationship between DMS and the solar radiation dose over the global surface ocean. *Science* **315**, 506–508

Vallina, S.M. & Simó, R. 2008. Re-visiting the CLAW hypothesis. *Environmental Chemistry* **4**, 384–387.

Vallina, S.M., Simó, R., Anderson, T.R., Gabric, A., Cropp, R. & Pacheco, J.M. 2008. A dynamic model of oceanic sulphur (DMSO) applied to the Sargasso Sea: simulating the dimethylsulfide (DMS) summer paradox. *Journal of Geophysical Research* **113**, G01009.

Van Bergeijk, S.A., Schonefeldt, K., Stal, L.J. & Huisman, J. 2002. Production and consumption of dimethylsulfide (DMS) and dimethylsulfoniopropionate (DMSP) in a diatom-dominated intertidal sediment. *Marine Ecology Progress Series* **231**, 37–46.

Van Bergeijk, S.A., van der Zee, C. & Stal, L.J. 2003. Uptake and excretion of dimethylsulphoniopropionate is driven by salinity changes in the marine benthic diatom *Cylindrotheca closterium*. *European Journal of Phycology* **38**, 341–349.

Van Diggelen, J., Rozema, J., Dickson, D.M.J. & Broekman, R. 1986. b-3-Dimethylsulphoniopropionate, proline and quaternary ammonium compounds in *Spartina anglica* in relation to sodium chloride, nitrogen and sulphur. *New Phytologist* **103**, 573–586.

Van Rijssel, M. & Gieskes, W.W. 2002. Temperature, light, and the dimethylsulfoniopropionate (DMSP) content of *Emiliania huxleyi* (Prymnesiophyceae). *Journal of Sea Research* **48**, 17–27.

Vila-Costa, M., Kiene, R.P. & Simó, R. 2008. Seasonal variability of the dynamics of dimethylated sulfur compounds in a coastal northwest Mediterranean site. *Limnology and Oceanography* **53**, 198–211.

Vila-Costa, M., Simó, R., Harada, H., Gasol, J.M., Slezak, D. & Kiene, R.P. 2006. Dimethylsulfoniopropionate uptake by marine phytoplankton. *Science* **314**, 652–654.

Watson, A.J. & Liss, P.S. 1998. Marine biological controls on climate via the carbon and sulphur geochemical cycles. *Philosophical Transactions of the Royal Society of London. Series B: Biological Sciences* **353**, 41–51.

Welsh, D.T. 2000. Ecological significance of compatible solute accumulation by microorganisms: from single cells to global climate. *FEMS Microbiology Reviews* **24**, 263–290.

Wolfe, G.V. & Kiene, R.P. 1993. Radioisotope and chemical inhibitor measurements of dimethyl sulfide consumption rates and kinetics in estuarine waters. *Marine Ecology Progress Series* **99**, 261–269.

Wolfe, G.V., Steinke, M. & Kirst, G.O. 1997 Grazing-activated chemical defense in a unicellular marine alga. *Nature* **387**, 894–897.

Woodhouse, M.T., Carslaw, K.S., Mann, G.W., Vallina, S.M., Vogt, M., Halloran, P.R. & Boucher, O. 2010. Low sensitivity of cloud condensation nuclei to changes in the sea-air flux of dimethyl-sulphide. *Atmospheric Chemistry and Physics* **10**, 7545–7559.

Yancey, P.H. 1994. Compatible and counteracting solutes. In *Cellular and Molecular Physiology of Cell Volume Regulation*, K. Strange (ed.). Boca Raton, Florida: CRC Press, 81–109.

Yang, Q., Mayewski, P.A., Twickler, M.S. & Whitlow, S. 1997. Major features of glaciochemistry over the last 110,000 years in the Greenland Ice Sheet Project 2 ice core. *Journal of Geophysical Research: Atmospheres (1984–2012)* **102**, 23289–23299.

Yoch, D.C. 2002. Dimethylsulfoniopropionate: its sources, role in the marine food web, and biological degradation to dimethylsulfide. *Applied and Environmental Microbiology* **68**, 5804–5815.

Yoon, Y.J. & Brimblecombe, P. 2002. Modelling the contribution of sea salt and dimethyl sulfide derived aerosol to marine CCN. *Atmospheric Chemistry and Physics* **2**, 17–30.

Yost, D.M.J., Ross, J. & Mitchelmore, C.L. 2010. Alterations in dimethylsulfoniopropionate (DMSP) levels in the coral *Montastraea franksi* in response to copper exposure. *Aquatic Toxicology* **98**, 367–373.

Yu, S. 2000. Role of organic acids (formic, acetic, pyruvic and oxalic) in the formation of cloud condensation nuclei (CCN): a review. *Atmospheric Research* **53**, 185–217.

Zemmelink, H.J., Dacey, J.W., Houghton, L., Hintsa, E.J. & Liss, P.S. 2008. Dimethylsulfide emissions over the multi-year ice of the western Weddell Sea. *Geophysical Research Letters* **35**, L06603.

Zhuang, G., Yang, G., Yu, J. & Gao, Y. 2011. Production of DMS and DMSP in different physiological stages and salinity conditions in two marine algae. *Chinese Journal of Oceanology and Limnology* **29**, 369–377.

Zubkov, M.V., Fuchs, B.M., Archer, S.D., Kiene, R.P., Amann, R. & Burkill, P.H. 2002. Rapid turnover of dissolved DMS and DMSP by defined bacterioplankton communities in the stratified euphotic zone of the North Sea. *Deep Sea Research II* **49**, 3017–3038.

AUTHOR INDEX

Page numbers referring to complete articles are given in **bold** type.

B

C

D

G

H

I

J

K

M

S

T

W

SYSTEMATIC INDEX

T

U

V

X

Z

SUBJECT INDEX

C

D

G

H

I

T

For Product Safety Concerns and Information please contact our EU
representative GPSR@taylorandfrancis.com
Taylor & Francis Verlag GmbH, Kaufingerstraße 24, 80331 München, Germany

www.ingramcontent.com/pod-product-compliance
Lightning Source LLC
LaVergne TN
LVHW080757270926
848950LV00046B/497
9781482220599